CAMBRIDGE LIBRARY COLLECTION

Books of enduring scholarly value

Earth Sciences

In the nineteenth century, geology emerged as a distinct academic discipline. It pointed the way towards the theory of evolution, as scientists including Gideon Mantell, Adam Sedgwick, Charles Lyell and Roderick Murchison began to use the evidence of minerals, rock formations and fossils to demonstrate that the earth was older by millions of years than the conventional, Bible-based wisdom had supposed. They argued convincingly that the climate, flora and fauna of the distant past could be deduced from geological evidence. Volcanic activity, the formation of mountains, and the action of glaciers and rivers, tides and ocean currents also became better understood. This series includes landmark publications by pioneers of the modern earth sciences, who advanced the scientific understanding of our planet and the processes by which it is constantly re-shaped.

Plant Life Through the Ages

Published in 1931 to complement Seward's magisterial four-volume textbook *Fossil Plants*, this book is a digest of his earlier detailed study, written for a non-specialist audience as an introduction to the field of palaeobotany. Seward begins by describing the basics of geology and palaeobotany in order to explain how the interpretation of fossilised plant remains found in rocks can shed light on the natural world of prehistoric times. He then covers geological periods in chronological sequence, from the Pre-Cambrian to the Quaternary. Throughout, he emphasises the fragmentary nature of the evidence and the difficulties in extrapolating from the surviving fossil record, but he also explains the great discoveries made in the field and how they came about. The accompanying drawings give an impression of the likely combinations of plants found in each period, allowing the reader to visualise the different landscapes evoked in Seward's engaging prose.

Plant Life Through the Ages

A Geological and Botanical Retrospect

A.C. SEWARD

CAMBRIDGE
UNIVERSITY PRESS

CAMBRIDGE UNIVERSITY PRESS

Cambridge, New York, Melbourne, Madrid, Cape Town, Singapore,
São Paolo, Delhi, Dubai, Tokyo, Mexico City

Published in the United States of America by Cambridge University Press, New York

www.cambridge.org
Information on this title: www.cambridge.org/9781108016001

This edition first published 1931
This digitally printed version 2011

ISBN 978-1-108-01600-1 Paperback

The original edition of this book contains a number of colour plates, which cannot be printed cost-effectively in the current state of technology. The colour scans will, however, be incorporated in the online version of this reissue, and in printed copies when this becomes feasible while maintaining affordable prices.

Additional resources for this publication at www.cambridge.org/9781108016001

PLANT LIFE
THROUGH THE AGES

Cambridge University Press
Fetter Lane, London

New York
Bombay, Calcutta, Madras
Toronto
Macmillan

Tokyo
Maruzen Company, Ltd

The Franz Josef Glacier, Westland, New Zealand, with Mount Roon in the background, a tree-fern (*Hemitelia Smithii*) in the foreground: photographed (1 mile from the terminal face of the ice) for the author by Dr E. Teichelmann at the request of Dr L. Cockayne, F.R.S.

PLANT LIFE THROUGH THE AGES

A Geological and Botanical Retrospect

BY

A. C. SEWARD, Sc.D., LL.D., F.R.S.

MASTER OF DOWNING COLLEGE
HONORARY FELLOW OF EMMANUEL COLLEGE
PROFESSOR OF BOTANY CAMBRIDGE

Including nine Reconstructions of
Ancient Landscapes drawn for the Author by

EDWARD VULLIAMY, M.A.
KING'S COLLEGE CAMBRIDGE

CAMBRIDGE
AT THE UNIVERSITY PRESS
1931

He is blessed over all mortals who loses no moment
of the passing life in remembering the past. *Thoreau*

PRINTED IN GREAT BRITAIN

To

MY WIFE, MARY

In grateful acknowledgment of
stimulating comradeship and
helpful criticism

CONTENTS

Chapter VIII

Chapter IX

Chapter X

Chapter XI

Chapter XII

Chapter XIII

PREFACE

The Preface to the fourth volume of my book *Fossil Plants*, published in 1919, concludes with the following words: "If it is possible to carry out my intention of supplementing the descriptive treatment of plants, which forms the basis of Volumes I–IV, by a general review of the Floras of the Past, the results will be published as an independent work more intelligible, I hope, to the general reader than the text-book which, with a certain sense of relief, is now brought to a conclusion". The attempt has now been made.

I have tried to keep before me the layman as well as the student of botany and geology, two different classes of readers who do not usually approach a subject from the same point of view. My aim has been to illustrate the nature of the documents from which geologists have compiled a history of the earth, or at least such scraps of history as can be written from the material that is available; to give some account of the methods employed in the interpretation of the documents; and to present in language that is not unnecessarily technical a summary of the more interesting results obtained from records of the rocks which throw light on the development of the plant-world.

Of the many friends from whom help has been received Mr Edward Vulliamy of King's College, Cambridge, has earned my warmest thanks. Knowing his skill as an artist and taking advantage of his generosity, I asked him to collaborate with me in drawing a series of landscapes which might serve to stimulate interest in the vegetation of former periods. The work, willingly undertaken, entailed much time and patience. In expressing my gratitude to him I take full responsibility for the form of the reconstructions: the artistic merit is Mr Vulliamy's; inconsistencies and inaccuracies either in the scenery or in the representation of individual plants must be attributed to me.

Comparatively few illustrations of actual fossils are given as these can readily be found in text-books and other sources referred to in the text and in the bibliography. In selecting subjects for illustration I have not hesitated to apply to friends as well as to colleagues personally unknown to me: my requests for photographs have invariably met with ready response. Formal acknowledgment of the source is made at the foot of most of the illustrations: I wish to supplement this with an expression of gratitude to all who have helped me to secure a representative set of photographs. The drawings of geological sections were made for me, through the kind offices of the late Dr Horne, by the senior draughtsman on the Edinburgh Staff of H.M. Geological Survey. By permission of the Director use has been made of the large collection of photographs in the possession of the Geological Survey of Great Britain. A word of special thanks is due to Dr Cockayne at whose request his friend Dr Teichelmann of Hokitika, New Zealand, took the excellent photographs reproduced respectively as the frontispiece and as fig. 138. My friend Prof. Collet, of Geneva, kindly obtained for me the photograph of the Alps (fig. 12) with permission for its reproduction from the Ad Astra-Aero of Zürich. For permission to reproduce fig. 50 I am indebted to the Council of the Geological Society of London and Prof. Gilligan of Leeds.

For other photographs and drawings or for various kinds of assistance I am indebted to the following friends and colleagues: Mr R. M. Adam of Edinburgh; Mr Denis Allen of Pembroke College, Cambridge, for calling my attention to the photograph published in the New Zealand *Free Lance* and reproduced (fig. 23) by permission of the proprietors of that paper who kindly supplied additional information; Prof. P. Bertrand (Lille); Prof. Bøggild (Copenhagen); Mrs Brindley (Cambridge); Prof. C. J. Chamberlain (Chicago); Dr J. M. Clarke (State Museum, Albany); Dr Cotton (The Royal Gardens, Kew); Dr R. Crookall; Prof. Sir T. W. Edgeworth David (Sydney); Mr D. Davies and the Editor of *Discovery* for permission to reproduce fig. 77; Mr T. N. Edwards of the British Museum (Natural History); Dr Fermor of the Indian Geological Survey; Dr Foxworthy of the Forest Research Office, Federated Malay States; Prof. Garwood (University College, London); Prof. J. W. Gregory; Dr Ogilvie Gordon; Prof. Gordon (King's College, London); Prof. Halle of Stockholm, who generously supplied several illustrations; Dr T. M. Harris (Cambridge); Mr Holttum, Director of the Botanic Gardens, Singapore; Prof. Howchin (Adelaide); Dr Howe of the New York Botanical Garden; Prof. Marr, Cambridge; Dr C. A. Matley (Edinburgh); Hofrat Prof. Hans Molisch (Vienna); Prof. F. W. Oliver; Mrs Clement Reid; Dr Sederholm

(Helsingfors); Prof. Reynolds (Bristol); Dr Skottsberg (Göteborg); Dr Tilley (Cambridge), on behalf of Sir Douglas Mawson; Prof. J. Walton (Glasgow); Dr Wanderer (Dresden); Prof. Watts (London); Prof. Went (Utrecht); Dr David White of the U.S. Geological Survey; Dr Wieland of Yale University; Dr Woodhead (Huddersfield), and Mr P. W. Wright. For the photographs of plane trees and bald cypresses (figs. 108, 115) in their natural surroundings I am indebted to the courtesy of the officers of the Forest Service of the United States Department of Agriculture, Washington. It was through the generosity of the late Dr Walcott that the photographs reproduced as figs. 38 and 40 were obtained. For the photograph by Mr Rickmers (fig. 126) I am indebted to Mr A. R. Hinks. The maps of various geological periods are based in part on those published in the encyclopaedic book by Dr Arldt, *Die Entwicklung der Kontinente und ihrer Lebewelt*, with modifications in boundaries suggested by Mr W. B. R. King, Fellow of Magdalene College, Cambridge, Sir T. W. Edgeworth David, and Mr Philip Lake.

Many friends have shared the labour of reading proofs and I thank them for much helpful criticism. I naturally applied to Prof. Lang for criticism of my account of Devonian vegetation: it is a pleasure to acknowledge valuable assistance also from younger colleagues, several of whom I like to think of as old pupils: Prof. Sahni of Lucknow; Dr Hamshaw Thomas, Dr T. M. Harris and Dr Godwin of Cambridge; Miss Chandler; and my son-in-law, Prof. J. Walton of Glasgow. For help in the geological chapters I am indebted to Mr T. C. Nicholas and Mr W. B. R. King of the Sedgwick Museum. A word of thanks is due to my secretary, Miss M. Gray, for the skill with which she deciphered my manuscript.

In a book written primarily for non-specialists references to scientific literature may seem out of place. I have not attempted to give a complete list of references: original sources are cited in order that readers desirous of pursuing any branch of the subject may know where to turn for fuller information. The numbers after authors' names in the footnotes refer to dates of publication: the full titles are given in the bibliography (p. 545).

For illustrations of the plants described reference should be made to sources mentioned in footnotes or to palaeobotanical text-books such as Scott (20), (23), chiefly, though not exclusively, concerned with Palaeozoic plants; Seward (98), (10), (17), (19), containing in addition to descriptions of fossil genera short accounts of classes of living plants; Potonié (21) which, unlike the other text-books, includes fossil flowering plants; and the *Handbuch der Paläobotanik*, Vol. I, by Hirmer (27), containing numerous photographic illustrations. References to earlier literature will be found in all these books. The attention of students is called to W. Jongmans' *Fossilium Catalogus, Plantae*, which is now being published in parts.

In trying to steer a middle course between a popular presentation of a subject and a presentation which may be acceptable to students disposed to become specialists, one is keenly conscious of the difficulty of deciding what to omit: I should be sorry to have to give reasons which have guided my action.

In conclusion I wish to express my appreciation of the generous treatment invariably received from the Syndics of the Cambridge University Press over a period of nearly forty years.

A. C. SEWARD

THE MASTER'S LODGE
DOWNING COLLEGE
January 1931

LIST OF ILLUSTRATIONS

The sources of illustrations are acknowledged in the fuller descriptions below the figures

NOTE: The frogs waving leaves of the maidenhair tree (*Ginkgo biloba*) reproduced on the cover are from a Japanese painting entitled "Turning the Tables: snakes crucified by frogs" by Kiosai (1831–83). (British Museum collection of Japanese Paintings, No. 1632.) I am indebted to the Director of the British Museum for the photograph from which the drawing was made. A. C. S.

CHAPTER I

INTRODUCTORY

The earth with no history to it—what it would be if it had all been made only last night and were not a worn ancient face, seamed, stained, and engraved with endless cross-hatching of documentary wrinkles, its mountains the ruins of more wondrous height now all but erased. *C. E. Montague*

In the following chapters an attempt is made to describe, with as few technical terms as possible, the salient features of the plant-world during successive periods of geological history. The great majority of men and women are content to enjoy plants as living members of the vegetable kingdom, to feel the stimulus and inspiration of their beauty, or to learn about their manner of life with a view to successful cultivation. If, on the other hand, we realize that the present plant-population is a legacy from the past—one term in a series stretching through the ages—the study of vegetation acquires an additional fascination. My object is to give to those who have little or no knowledge of botany or geology glimpses of the plant-world at the several stages of its development.

The documents at our disposal are preserved as fragmentary records in that outer and relatively thin layer of the earth's mass which is known as the crust of the earth, "the old jumble-box of history", which is in part accessible to investigation. Students of fossil plants who aim at something more than collecting, describing, and naming specimens are concerned with many and diverse problems: they search for light on the unsolved riddle of evolution; they desire information on the relation of plants that are extinct to plants which still exist, and on the relation of the vegetation of one period to that of another. As a French author[1] says: many times in the course of its history the earth has changed its green mantle and has left scraps of it scattered in the rocks as so many natural archives. We endeavour not only to reproduce the several phases in the development of the plant-kingdom, but also to restore the geographical features and the climatic conditions. If we wish to obtain a true conception of the men and women of a former age it is essential to supplement the study of individuals by an enquiry

[1] Depape (28).

into the conditions under which they lived; to picture as accurately as possible the environment which influenced their thoughts and actions. Similarly in order to gain a true picture of ancient floras we must visualize the geographical setting, the background to the great drama of life.

The face of the earth, as we see it now, presented a slightly different appearance when our remote ancestors emerged from a lower plane of existence as the pioneers of the human race. In still earlier days the geographical features bore less resemblance to those of the present. The farther we penetrate into the past the more difficult it is to reconstruct the boundaries of land and sea. By piecing together the facts contributed by the rocks themselves and the fossils they contain it is possible in some measure to bring to life the things that are dead, to see the earth as it was and to reconstruct the protean landscape. "It is the organic remains, no doubt, which afford us our first and most important aid in the elucidation of the past. But the goal of investigation must still remain the recognition of those great physical changes, in comparison with which the changes in the organic world only appear as phenomena of the second order, as simple consequences."[1] As General Smuts says: "Matter like life is intensely active, indeed is Action in the technical sense; the difference is not between deadness and activity, but between two different kinds of activity. Through their common activities the fields of matter and life thus overlap and intermingle, and absolute separation disappears".[2]

The historian may fall into serious error unless he appreciates the nature and the limitations of his sources: the student of the records of the rocks should not begin his investigations until he is able to estimate the value of the documents which he handles. He cannot hope to see in true perspective successive steps in evolution unless he is able to follow the procession of floras across the stage that was repeatedly reset. It is one thing to study the fossils by themselves; it is another and a more attractive thing to think of them as living plants in a real world.

A comparatively slight acquaintance with geological phenomena is sufficient to inspire sympathy with Darwin's insistence on the imperfection of the geological record, and to place the student on his guard against the danger of drawing far-reaching conclusions from evidence that can only be partial and lamentably incomplete.

[1] Suess (04–09). [2] Smuts (26).

Another consideration, not only of interest for its own sake but important in relation to problems of evolution, is the time-factor; the relative duration of each chapter of geological history and, so far as it can be ascertained, the intervals, expressed in years, which separate the different periods from our own time.

In the first few chapters some of these questions are considered; features of the inorganic world relevant to the main thesis are briefly described. Geological history includes both the history of the organic world and that of the earth as a whole. Questions of climate, the distribution of land and water, the vagaries of the mobile crust of the earth, the chances of preservation of plants as fossils, the nature of fossils, and other problems come within the scope of an enquiry into the development of successive plant-dynasties, their relation to one another and to the world in which they flourished. The end in view is to recover from the records all that can be recovered, "to realize the past story as if it were now passing before us".

In order to trace the wanderings of plants at successive stages in the history of the world we must be able to picture the consecutive aspects of the earth's surface; we must know the form and extent of the land-masses, and the position of the oceans and seas which formed barriers to the migration of terrestrial plants. By following to their sources series of sedimentary rocks, tracing the passage of old shingle beaches into the sands and muds of gradually deepening water; and by noting the substitution of calcareous accumulations on the floor of a clear sea for material derived from the waste of land, we learn something of the relative position of sea and land. The ancient sediments and the fossils which they contain, whether marine, freshwater, or terrestrial in origin, furnish the data on which geologists construct maps of the world as it was. While it is obvious that maps of geological periods are to a large extent based on assumption, the form and size of continents plotted over existing oceans being necessarily hypothetical, careful reconstructions of former geographies are helpful if they are accepted as guesses at the truth based on such evidence as is available.[1]

It is easy to suggest lines of investigation; but it may fairly be asked, Can we by a critical survey of the data hope to arrive at any reasonably accurate results commensurate with the labour of the enquiry? Hooker in a letter to Darwin, written after reading the

[1] For maps of the world at various stages of geological history see Arldt (07).

chapters on geological evidence in *The Origin of Species*, expressed this opinion: "I would say that you still in your secret soul underrate the imperfection of the geological record, though no language can be stronger or arguments fairer and sounder against it. Of course I am influenced by Botany, and the conviction that we have not in a fossilized condition a fraction of the plants that have existed, and that not a fraction of those we have are recognizable specifically". Granting the correctness of this depressing estimate, we derive encouragement from the fact that it has been possible from the interpretation of fragmentary documents and palimpsests to throw a flood of light upon the earlier ages of human history. However meagre the material may be unbiased research reveals something of value. By eliminating all that is untrustworthy, we can at least place the student in a position to form his own conclusions and supply him with a broken outline which can be modified or improved as fresh discoveries are made.

CHAPTER II

THE CRUST OF THE EARTH

Nature's history-book, which she hath torn as ashamed of. *Robert Bridges*

It was formerly believed that the rocks accessible to us on continents and islands form a stony film covering a molten globe: a crust was formed and solidification gradually extended to greater depths. As years passed in their millions cooling continued, the primeval stores of potential energy were dissipated, and the environment of the organic world at each successive stage was conditioned by forces diminishing in intensity. The coarsely crystalline rocks, such as granites and the banded granitic rocks, or gneisses, were regarded as the foundation stones of the primitive crust. The great Ice Age, an event antedating the present era by a comparatively insignificant fraction of geological time, was accepted as evidence of a moribund earth approaching a condition of "cold senility",[1] a world that had lost its youth.

The conception of a store of potential energy slowly running down is in itself wholly inadequate to serve as a basis of geological history. The material of our earth was once merged in a completely gaseous sun. The earth was born, partly gaseous, partly liquid: "the world was once a fluid haze of light". We are not concerned with the origin of the earth: we accept the infant world from the astronomers at a stage in its growth when the surface was cool enough to be partially, or it may be completely covered with water. The discovery of radioactive minerals, "the last surviving vestiges of more vigorous primeval matter", disclosed an unsuspected and almost limitless source of energy to which the earth owes its prolonged youth.

Rocks are the source-books of geological history: the geologist classifies them according to their relative age, using the fossils and the rocks as data from which to reproduce as far as he can the life and physical conditions of each stage of earth-history. The imperfection of the records is in part responsible for the universal practice of dividing the history of the earth into eras, periods, and stages of varying duration; on the other hand the structure of the crust

[1] Chamberlin (16).

clearly reveals a succession of cycles,[1] periods of stress alternating with periods of relative quiescence, and these are expressed in the current geological time-scale and table of contents (fig. 2). Much of the earth's crust consists of sedimentary rocks, strata of coarse or fine detritus; conglomerates that are old pebble beaches, sandstones, shales or muds—results of the action of rain, frost, and other instruments of destruction which operate with fluctuating though ceaseless activity on hill and dale. With these materials are associated limestones and chalk, upraised coral reefs, banks of coralline seaweeds, or masses of marine shells accumulated during long ages on an ocean bed. Other rocks are formed of the almost indestructible skeletons of diatoms, a group of ubiquitous, microscopic algae; while certain calcareous and siliceous strata owe their origin to the precipitation of carbonate of lime or silica by the action of living plants, chiefly bacteria and simple algae. Beds of peat, lignite, and coal recall swamp-covered areas or lagoons and lakes on the floors of which generations of aquatic plants have left their carbonized tissues and spores. Sandstones composed of well-rounded grains of quartz and piles of wind-blown sand are legacies from desert regions. Beds of salt, gypsum, and other minerals mark the sites of inland seas exposed to long-continued desiccation; suncracks, footprints, and rain-pits on sandstones and hardened muds speak of surfaces exposed to the air.

It is obvious that if in any region rock-building had continued without a break from the dawn of geological time to the present day, a vertical section of the crust to a depth of fifty miles or more would furnish a fairly complete record without any gaps in the sequence. The history of the earth is not so recorded. The crust is often a mosaic, masses of older rocks inlaid with patches of newer formations, folded and crumpled strata altered by intrusions of molten material, and the records of life obliterated.

Sedimentary rocks form only a part, though an important part, of the surface. There have been recurrent manifestations of igneous activity, the pouring out over vast areas of volcanic material hundreds or even thousands of feet in depth as illustrated by the Deccan traps of India, a succession of basaltic lava-flows of late Cretaceous or early Tertiary age which originally covered an area of about half a million square miles;[2] the great sea of black lava on the Snake River desert of Idaho; the Tertiary sheets of basalt

[1] Barrell (17); Schuchert (23); Joly (25).

[2] Washington (22).

and beds of volcanic ash which once extended as a continuous lava-field from the north of Ireland, the west of Scotland and the Faroe Islands to far within the Arctic Circle.

Geological researches afford little evidence of progressive decline in the intensity of forces which might fairly be expected to decrease *pari passu* with a gradually cooling globe. Igneous activity has continued spasmodically from the earliest ages to the present time; its history "in no way bears out any theory of a gradual decline of energy with the lapse of ages".[1] "We are dwellers upon a world", writes Prof. Joly, "in the surface materials of which there exists an all but inexhaustible source of heat." The geologist is often able to recognize that rocks superficially resembling products of vulcanicity were originally formed as sediments, which by exposure to intense and prolonged heat generated in the course of heavings and straining of the crust have been metamorphosed, or reconstituted, their fossil contents being destroyed in the process. The stresses to which the earth's surface has repeatedly been exposed have caused folding of the strata, "coiled plungings of the crest hither and thither", stretching and compression so intense that the rocks have snapped and slipped over one another, thus reversing the original order of the beds. Intensity of terrestrial convulsions is not the monopoly or distinguishing feature of the earlier ages; it is equally revealed in the jumble of rocks uplifted, during one of the later phases of geological history, into the Himalayas, the Andes, and the Alps, mountain chains which "challenge credulity by the evidence of their extreme youth". Much of the crust is inaccessible, much has been destroyed, and much of it that is accessible has been subjected to repeated foldings and fractures which have destroyed the documents essential to the historian of the past. In some regions an overstrained crust found relief in vertical movement accompanied by the formation of parallel lines of fracture and the consequent subsidence of blocks of broken ground. A remarkable example of this class of phenomenon is furnished by the Great Rift Valley,[2] a trough-like gorge extending from the Jordan and the Dead Sea through some of the African lakes and still farther south. History is written not only in the rocks and fossils but in the breaks which produce irregularities in the sequence of geological formations. A regular series of parallel sedimentary rocks built up during a prolonged period of gradual subsidence of the floor of a sea may be

[1] Harker (09).

[2] Gregory (96), (20); Parsons (28).

overlain by a series of much younger strata, the two series being in juxtaposition but separated by an uneven dividing line. Sediments that were originally horizontal have often been upraised and folded by pressure causing contraction: the rocky waves are planed down by denudation and the land sinks below the sea in which a fresh series of sedimentary beds is gradually deposited. When once more

FIG. 1. The Danish Arctic station overlooking Disko Bay off the west coast of Greenland (about 200 miles north of the Arctic Circle). The rocks in the foreground and immediately behind the station are Pre-Cambrian gneiss: the mountain consists of flows of Tertiary basaltic lava alternating with thinner and lighter bands of volcanic ash, with talus slopes below, resting unconformably on the ancient land-surface. (Phot. A. C. S. from *A Summer in Greenland*, Seward (22[2]), Cambridge Univ. Press.)

the crust is uplifted the two sets of strata are seen to be separated by an uneven dividing line. This type of discordance, known as an unconformity, is evidence of an interval—it may be an interval of some hundred million years—the history of which is recorded in the rocks of other regions. The lower part of the cliff of Disko Island off the west coast of Greenland seen in fig. 1 consists of igneous, Pre-Cambrian rocks, which represent the eroded surface of one of the oldest land-masses in the world: above are some of the Tertiary basaltic lava-flows and beds of ash to which reference has already been made. The lapse of time represented by this unconformity, between the Pre-Cambrian and Tertiary rocks, is inconceivably great and comprises almost the whole of geological history since the dawn of life.

The simplest and most direct method of illustrating the nature of geological evidence, and of demonstrating the changing face of the earth, is by means of brief descriptions of a few concrete examples. Incidentally the examples help us to realize that the development of the organic world has been accomplished on a stage with an ever-changing background, and that revolutions in the inorganic world have been disastrous to the preservation of a continuous series of records from which to compile a history of successive floras. Before describing a few scenes from the past it will be helpful to glance at the list of chapters into which the history of the earth has been divided, and in very general terms to consider the question of geological age, both the relative age of successive periods and, as far as facts are available, their absolute age in years.

The larger divisions adopted by geologists are arranged in the order of their relative ages in fig. 2: this table is based on data

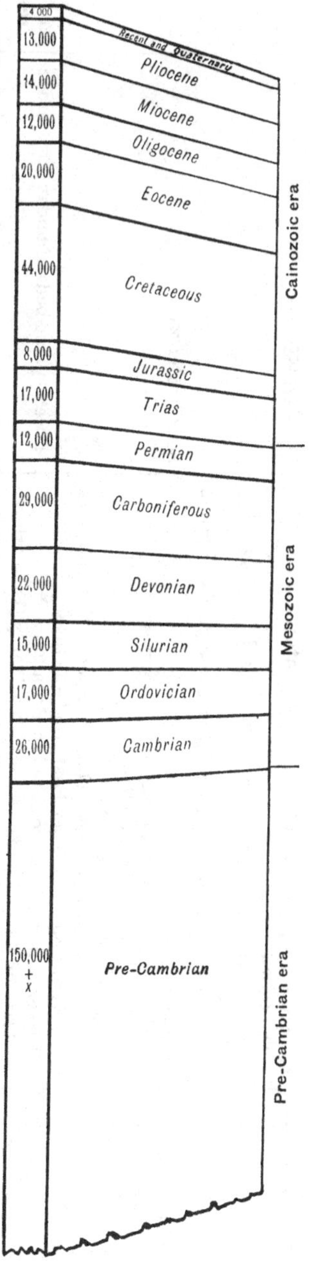

FIG. 2. Diagram, in the form of a slab seen in perspective, showing the order of the geological eras and periods, and the relative thickness in feet of the rocks of each period. (Based on estimates by Prof. Sollas.) The thickness of the Pre-Cambrian system is much greater than the diagram indicates.

GEOLOGICAL TABLE

Era	Period	Deposits	Plants	Events
Cainozoic Era	Quaternary	The present stage passing imperceptibly into: The Post-Glacial stage River-gravels, cave-deposits, submerged forests, peat The Glacial stage Boulder clay and other glacial and interglacial deposits	See Chapter XVII	
	Tertiary	Pliocene stage Cromer forest bed (Cromerian) Teglian plant-beds (Teglian) Reuver plant-beds (Reuverian) Shelly sands and clays, pebble-beds, etc. (Norwich Crag, Red Crag, and other East Anglian crags)	See Chapter XVI	
		Miocene stage. Unrepresented in Britain	Miocene plant-beds abundant in Europe	The Alpine revolution
		Oligocene stage. Plant-beds of Bovey Tracey (Devonshire), Bembridge (Isle of Wight), etc. Eocene stage. London Clay, Plant-beds of Alum Bay (Isle of Wight), Bournemouth (Hampshire); Plant-beds of Mull (western Scotland) and Antrim (northern Ireland)		
Mesozoic Era	Cretaceous	Upper Cretaceous: Chalk (Senonian, Turonian, Cenomanian); Upper Greensand, Gault (Albian)	See Chapter XV	Laramide revolution
		Lower Cretaceous: Lower Greensand (Aptian); Wealden (Neocomian)	The oldest examples of flowering plants of *modern type*	
	Jurassic	Upper Jurassic: Purbeckian, Portlandian, Kimmeridgian, etc. Middle Jurassic: Inferior Oolite and Estuarine series Lower Jurassic: Lias	See Chapter XIV	
	Rhaetic	In Britain a comparatively thin series of beds between the Jurassic and Triassic systems	Rich plant-beds in Greenland, southern Sweden, Tongking, and many other parts of the world	
	Triassic	Keuper: Marls, gypsum, rock-salt, etc. Muschelkalk (unrepresented in Britain): Dolomites and limestones; salt deposits Bunter: Red sandstones, pebble-beds, etc.	See Chapter XIII	

Palaeozoic Era	Permian	Upper Permian	Marls and gypsum; Magnesian Limestone and red sandstones; Marl Slate (Kupferschiefer)	Thuringian Zechstein	Permian plants few and fragmentary in England. The best material is from North America, France, Germany, Asia and the southern hemisphere	
		Lower Permian	Rothliegendes	Saxonian Autunian	See Chapters x and xii	
	Carboniferous	Upper Carboniferous (Pennsylvanian)	Coal Measures[1]: Radstockian: Stephanian: Upper Coal Measures; Staffordian: Transition series; Yorkian series: Middle Coal Measures; Lanarkian: Lower Coal Measures			Hercynian and Appalachian revolutions
			Millstone Grit (in part)			
		Lower Carboniferous (Mississippian)	Millstone Grit (in part); Carboniferous Limestone and Calciferous Sandstone		Some of the best preserved remains are in the volcanic series of southern Scotland	
	Devonian	Upper Devonian and Upper Old Red Sandstone			See Chapter ix	Caledonian revolution
		Middle Devonian and Middle Old Red Sandstone			Rhynie plant-beds of Scotland: Elberfeld beds of Germany, etc.	
		Lower Devonian and Lower Old Red Sandstone			The oldest known undoubted land-plants	
	Silurian	Downtonian; Salopian; Valentian			Traces of terrestrial plants	
	Ordovician	Bala series (Coniston limestone); Llandeilo series (Borrowdale volcanic series); Arenig series (Skiddaw slates)			See Chapter viii; Algae (mainly calcareous)	
	Cambrian	Upper Cambrian; Middle Cambrian; Lower Cambrian			Algae, very few of which are determinable	
Pre-Cambrian Era	Pre-Cambrian	Algonkian: Torridonian; Archaean: Lewisian			A few traces of obscure, simple plants; Sec Chapter vii	Laurentian revolution

[1] For classification of the Coal Measures see also pp. 161, 261.

from various sources.[1] The numbers on the edge of the column convey a general idea of the maximum thickness of the piles of rock comprised within each period. It is impossible to form even a rough estimate of the thickness of the strata included in the earliest geological era, but it is safe to assume that the number of feet, 150,000, assigned to the Pre-Cambrian era is much too small. The term era is used for the longest division of time bounded by critical periods in the earth's history. The table shows the order, thickness, and duration of each period: it is a vertical boring through the whole series of sedimentary and other strata nearly seventy miles in depth. This vast pile is made up of material derived from the wear and tear of rocks, and is the result of the "more or less rapid wearing away almost to sea-level, one after another, of more than twenty ranges of mountains like the present European Alps or the American Rockies".[2]

The fuller table on pages 10 and 11 shows the geological periods into which the three eras are divided, also some of the subdivisions of the periods. The relation in time of the major revolutions to the geological periods is indicated in the right-hand column.

The Age of the Earth

Estimates of the age of the earth cover a wide range and are based on very different kinds of evidence. Prof. Eddington[3] states that "the geologist may claim anything up to 10,000 million years without provoking a murmur from astronomers". Sir Ernest Rutherford says that the age of the earth cannot exceed 3400 million years and is probably much less. Several years ago Lord Kelvin, from calculations based on the earth's initial store of heat, estimated that the consolidation of the crust occurred about 100 million years ago: this estimate he afterwards reduced to 40 or even to 20 million years. The discovery of stores of potential energy in the rocks, which being liberated as heat would counterbalance the loss by radiation, rendered previous calculations as valueless as those of Archbishop Ussher. The poet Cowper took a more modern view:

"Some drill and bore
The solid earth, and from the strata there
Extract a register, by which we learn
That He who made it, and revealed its date
To Moses, was mistaken in its age."

[1] Sollas (05), (09); Hobbs (21).
[2] Barrell, Schucert and others (18).
[3] Eddington (23).

It has since been shown that the register quoted by Cowper needs revision. Recent researches into the nature of radioactivity have furnished a new and more exact method of estimating the age of the earth: radioactivity implies a spontaneous breaking up of the nuclei

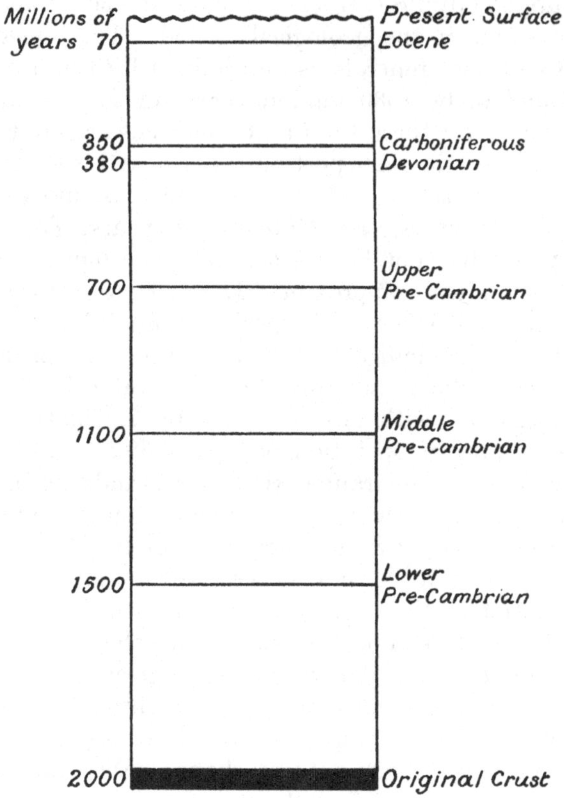

FIG. 3. The age of a few geological periods estimated in millions of years. Scale: 1 cm. = 200,000,000 years. (Based on data from Prof. Holmes.)

of atoms. The complex atoms of uranium are constantly changing into the simple atoms of lead and helium and setting free radiation; the rate at which uranium is now changing into lead is known. By measuring the amount of uranium in a sample of the crust and the amount of derived lead associated with it, it has been possible to calculate the age of the uranium-containing rocks. As Prof. Holmes says,[1] "the discovery of radioactivity not only revealed with dramatic suddenness the unjustified restrictions which had been placed on geologists, but it led directly to the elaboration of the most elegant and refined method of measuring geological time that has yet been devised". The accompanying diagram (fig. 3) based upon data supplied by Prof. Holmes shows the relative and absolute antiquity of some of the geological periods. The age of certain Lower Pre-Cambrian minerals as determined by the uranium-lead ratio was found to be 1580 million years: in the diagram this is reduced to 1500. But these Pre-Cambrian rocks are known to have been intruded into still older portions of the earth's crust, and it is therefore clear that the age of the primeval crust must be greater than 1500 million years, say 2000 million years. The age of all geological epochs has not been estimated; the figures on the left of the diagram indicate approximately the time measurements in millions of years for a few of the periods. Much has been written on the length of time indicated by the present rate of denudation and by changes in the organic world demonstrated by fossils. Time cannot be measured with any accuracy by calculations founded on the present rate of operation of agents of erosion or of rock-building: there is no guarantee that our standards are correct. Neither can we measure the progress of evolution in definite periods of years. New forms of flowering plants are produced by the breeder; others appear from time to time as the result of natural causes, but so far as evolution in the wider sense is concerned, the plant-world appears to have been almost static during the span covered by man's direct observation. We can follow the growth of an individual from cell to cell; we can picture evolution through simple to less simple and more complex stages in a family or group, but we have no absolute measure of the rate of change. We know that "the period of organic evolution is but a moment when compared with the tremendous duration of inorganic evolution". The physical conditions under which life is possible "form only a tiny fraction of the range of physical conditions which prevail in the universe as a whole"

The Grand Canyon of Arizona

From the cliffs of the Grand Canyon of Arizona[2] (fig. 4), over a mile in depth, it is possible to trace a series of events extending

[1] Harker (14); Holmes (15), (20), (27); Holmes and Lawson (27). See also Eddington (23); Jeans (29).
[2] Frech (95), (02); Noble (22).

through a few geological periods and representing some hundreds of millions of years. At the base (*a–b*) are the weather-worn roots of a mountain range consisting of folded granitic rocks penetrated by intrusions (*s*) from a substratum of molten or semi-molten

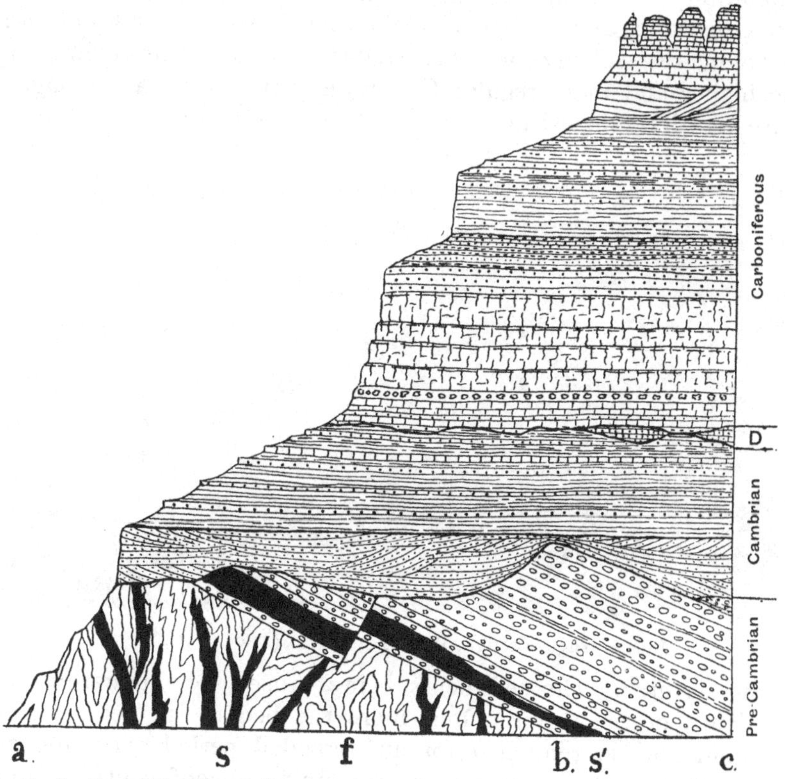

FIG. 4. Section of the Grand Canyon, Arizona. *a–b*, folded Lower Pre-Cambrian rocks penetrated by dykes, *s*; *b–c*, Upper Pre-Cambrian strata and a sheet, *s′*, of igneous rock; *f*, fault; *D*, patches of Devonian beds. (After Frech.)

material. Eventually the hills were submerged, and on their eroded flanks were deposited in a slowly subsiding area a series of sediments, several thousand feet in thickness, derived from the destruction of the land and spread out over the flood-plains of rivers or in a fresh-water lake. At an early stage in the building-up of this sedimentary series a sheet of igneous rock, a subaqueous lava-flow, was intruded along the plane of bedding as seen at *s′*. The whole Pre-Cambrian mass, including the basal granites and the overlying sediments, was upraised: the originally horizontal strata (*c*) were tilted, and the oblique line (*f*) along which the sheet *s′* is displaced, is one of many fractures caused by the upward movement of the crust. Later, after long exposure to denudation and the carving of the land into undulating scenic features, the whole Pre-Cambrian complex was submerged and, as the varying inclination of the Lower Cambrian beds indicates, shifting currents sorted the shallow-water sediments. The higher and more regular Cambrian strata mark a prolonged phase of orderly deposition of different kinds of material, differences

FIG. 5. Ingleborough, West Yorkshire. (Phot. A. C. S.)

in the nature of the sediments being caused by fluctuations in the depth of the water. Once more the bed of the sea became dry land and so remained for an extended period. The uneven surface of the Cambrian rocks represents an unconformity, a time interval of long duration. With the exception of the two small patches of Devonian rocks (*D*) there are no records in the section of strata belonging to the periods between the Cambrian and the Carboniferous. The uneven boundary line which indicates missing chapters of geological history affords evidence of land conditions and a consequent deficiency of documents relating to the unrecorded periods. Eventually the land subsided to form the floor of the Carboniferous sea on which layer after layer of material accumulated, limestones in clear water, sandstones and shales in the shallows until the slowly subsiding floor emerged, and the scene shifted from water to land. This section of the canyon tells us nothing of post-Carboniferous history; but it is known from other evidence that the rocks were upraised to a height of more than 10,000 ft. after the close of the Tertiary period, and that still more recently a river carved out the gorge to a depth of 6500 ft.

Ingleborough, West Yorkshire

The section shown in fig. 6 tells another story: it represents the structure of a portion of the earth's crust in a district of West Yorkshire (fig. 5) from Simon Fell across Ingleborough (2373 ft.) to the small coalfield of Ingleton, a distance of about five miles. Fig. 5 is a view of flat-topped Ingleborough and the low-lying district near Ingleton. In the upper part of the section (fig. 6) the regular disposition of the approximately horizontal beds suggests an elevated plateau from which denudation sculptured the present hills and the sloping contours of their flanks. The truncated hills rest on a broad plinth of limestone, and this is abruptly underlain by a basal platform of steeply inclined rocks exposed in the dales. The older and lower portion of the horizontal series consists of about 600 ft. of limestone rich in calcareous shells, corals, and the hard parts of other lime-secreting marine creatures. Above the limestones are sandstones and other products of erosion, with a slab of grit which forms the cap of Ingleborough. This series of beds furnishes a continuous record of deposition on the floor of the Carboniferous sea. For a long time the sea was relatively deep and in it accumulated masses of calcareous material. At a later stage the depth of water decreased and the calcareous ooze and shell-banks were covered with sand and other detritus from the neighbouring land. The grit on the summit of Ingleborough, the Ingleborough grit, is a small relic of a series of beds of coarse sand deposited over an area of some thousands of square miles in the delta of a great river, which

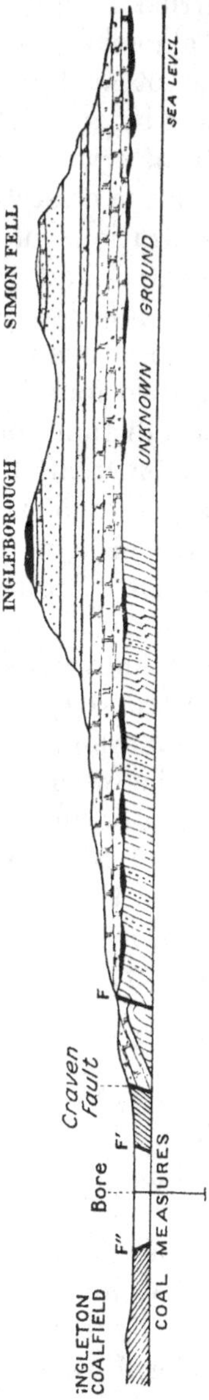

Fig. 6. Section through Ingleborough and the Ingleton coalfield. (After Strahan.)

carried innumerable tons of grit and sand from the granitic highlands of a northern continent. The basal platform of Silurian, Ordovician, and other Pre-Carboniferous rocks, separated from the overlying mass by an unconformity, is the worn-down surface of a land which was afterwards destined to form the floor of the Carboniferous sea.[1]

In the dissected plateau of West Yorkshire we have one of many illustrations of the effect of widespread earth-movements which elevated and fractured the earth's crust after the close of the Carboniferous period, forming among other mountain ranges the Pennine chain of England. The stresses to which the crust was exposed found relief in fractures, or faults, which have been traced for a distance of eighty miles: the rocks on the opposite sides of the Craven fault (fig. 6) were displaced through a vertical distance of more than 5000 ft. The evidence is as follows: the grit on the top of Ingleborough is part of the Millstone Grit of the Carboniferous system, a formation which in many other localities is overlain by a regular succession of sedimentary rocks and associated seams of coal known as the Coal Measures. The Coal Measures formerly existed over the whole of the district and were swept away by denudation. In the Ingleton coalfield to the west, below a superficial covering of drift, we find representatives of the Coal Measures; and a boring in the coalfield has shown that the Ingleborough grit cannot be less than 2800 ft. below the present surface of the ground. Strata that were some thousands of feet above sea-level have been thrown down along a line of fault a vertical distance of at least 5000 ft.

The North West Highlands of Scotland

The removal by denudation of a series of beds may restore to the light of day an ancient land-surface which has been hidden for millions of years. An impressive illustration of this is seen in the photograph reproduced in fig. 7. The mountain Slioch in the North West Highlands of Scotland is built of strata assigned to the Upper Pre-Cambrian, or Torridian (from Loch Torridon) stage: the lower and lighter bosses on the left, Meall Each, are made of the still older Lower Pre-Cambrian, or Archaean, gneiss which has been revealed by the removal of the original covering of Torridonian strata.[2] This unconformity between the Lower and Upper Pre-

[1] For a fuller account of West Yorkshire see Strahan (10); Kendall and Wroot (24).

[2] Peach and Horne (07).

FIG. 7. Slioch and Meall, Loch Maree, North West Highlands of Scotland: the highest mountains consist of Pre-Cambrian Torridonian beds: the lower humps on the left are part of an older (Archaean) land-surface exposed by denudation. (Geological Survey Collection of photographs.)

Cambrian series forms "one of the most marked stratigraphical breaks in the geological structure of the British Isles". The older gneiss was folded into lofty mountains intersected by valleys: in the course of ages the mountains were brought low until on their uneven surface were accumulated in an arid climate thousands of feet of coarse detritus from the waste of land far to the north. Prolonged erosion removed much of the Torridonian material and uncovered the old foundations, while elsewhere in the district, from Loch Maree to Loch Torridon, masses of the Torridonian grit remained as hills from two to three thousand feet in height. The older Archaean gneiss deeply buried under a load of younger deposits "has been once more exposed and has revealed some of the topography of the oldest land-surface known to exist in western Europe, while at the same time the successive protrusions, dislocations, and displacements of the remotest Archaean ages have been laid bare to our eyes".

CHAPTER III

GEOLOGICAL CYCLES

Ruin upon ruin, revolution upon revolution. *Robert Bridges*

"THE whole course of nature is dominated by the existence of periodic events." Geological history contributes its full share of evidence in support of the truth of this statement. Illustrations have already been given of the instability of the everlasting hills, of the recurrent interchange of sea and land over wide regions, and of the operation of forces which fashion the changing aspect of the earth's surface. It is now proposed to describe, in as general terms as possible, a few additional examples of mountain-building and other phenomena associated with the rhythmic convulsions, or revolutions, of the earth's crust which are a dominant feature of the world's development: the problem is to discover whether it is possible to correlate revolutions in the organic world with transformations of the physical background.

Geological Revolutions (see table, pp. 10 and 11)

Geologists are generally agreed on the approximate dates, as measured by the succession of periods, assigned to the major revolutions recorded in the rocks. Despite the difficulty of accurately comparing the ages of the oldest rocks in different parts of the world and of interpreting the tattered documents they provide—documents which have passed through the furnace and have again and again been the sport of eroding agents—it is clear that during the enormous space of time represented by the Pre-Cambrian era there were several cycles of rock-folding and mountain-building on a large scale. The oldest of these is known as the Laurentian revolution. Towards the close of the Silurian period was initiated the Caledonian revolution which created chains of mountains on a grander scale than the Alps, stretching from Ireland and Scotland to Scandinavia, the northern edge of Greenland, and to the borders of the Pacific in the Far East. The next crustal disturbance on a grand scale, the Appalachian and Hercynian revolutions, occurred at the end of the Carboniferous period: this gave birth to the Appalachian chain in eastern North America, the Pennine chain of England, and other ranges on the continents of Europe and Asia.[1] Towards the close of the Cretaceous period the Laramide revolution produced ridges of hills from Cape Horn to Panama, from south Mexico to Alaska.[2] Finally during the latter half of the Tertiary period were uplifted the greatest mountain chains in the present world, the Himalayas, the Andes, the Alps and other ranges.

A discussion of the causes of revolutions, the periodic manifestations of activity in the framework of the crust, is beyond my power, and fortunately it is outside the scope of this book. The assertion previously made, that the old conception of the earth as a mass which has been slowly parting with its original store of energy is inadequate, finds justification in the discovery of radioactive minerals and their importance as sources of heat. It is not intended to convey the impression that crustal dislocations have no connexion with the theory of a gradually cooling earth. Dr Jeffreys[3], in his book on *The Earth* and in subsequently published papers, maintains the adequacy of this theory. He disagrees with the opinion of Prof. Holmes that the continuous cooling theory has failed to provide a satisfactory explanation of revolutions: he also definitely declines to accept a theory attractively developed by Prof. Joly[4] in his stimulating book on *The Surface History of the Earth*, a theory involving the assumption that after long intervals the pent-up heat generated by radioactive substances fused portions of the crust's substratum; the consequent decrease in density caused a sinking of the land and transgression of the sea; after the dissipation of the store of heat the substratum again became solid, its density was increased and mountain chains were upraised.

The Laurentian revolution is so named from the St Lawrence River: Pre-Cambrian rocks are the foundations of the Labrador peninsula; they extend southwards to the Great Lakes, and north-west to the Arctic sea, covering an area of about 2,000,000 square miles. The Lower Pre-Cambrian rocks of this Canadian shield and rocks of the same type in other parts of the world were formerly thought to be the foundation-stones of the earth's crust; but it is now known that the granites have been intruded in great dome-like masses into the sedimentary beds above them. "Geology", as Sir Archibald Geikie said, "has not yet revealed and is not likely to reveal a portion of the first solid crust of our globe." There is still

[1] Bailey, E. B. (28).
[2] Pirsson and Schuchert (20).
[3] Jeffreys (24), (26), (26^2).
[4] Joly (25).

no answer to the question, "Whereupon are the foundations thereof fastened?"

An illustration of the effects of the Caledonian revolution is afforded by Pre-Cambrian and later rocks in the North West Highlands of Scotland.[1] Reference has already been made to the exposure of a Pre-Cambrian land-surface in northern Scotland by the removal of the overlying Torridonian strata (fig. 7). The scene reproduced in fig. 8, considered in the light of the section shown in fig. 9, exhibits a later phase of geological history. The mountain Meall A' Ghiubhais consists mainly of Torridonian beds (fig. 9, *B*), but the left shoulder includes some older Pre-Cambrian gneiss and marine strata of Lower Cambrian age (fig. 9, *Ca*, *Cb*, *Cc*). The foreground (fig. 8) is made of Lower Cambrian rocks; the greater part of the mountain is built of rocks which once formed part of the basal Pre-Cambrian complex, a small piece of which is seen at the bottom of the section. After the deposition of many thousand feet of Torridonian sediments (fig. 9, *B*) the land was raised with its burden of detritus, exposed to denudation and eventually worn down by the action of the sea to a featureless plain. This plain became the floor of the Cambrian sea and on it were deposited sands and mud (fig. 9, *Ca*, *Cb*, *Cc*) derived from adjacent hills. Much of the Cambrian material, in places 2000 ft. thick, indicates clear and relatively deep water and includes relics of marine life. The subdivision labelled *Cc* is known as the *Olenellus* zone from the occurrence of the trilobite *Olenellus* which, with the exception of some North American types, is the oldest representative of this extinct class of crustacea. The rocks of the *Olenellus* zone form the foreground in fig. 8.

At a later date the earth's crust was convulsed and fractured by gigantic thrusts which folded the rocks, crumpling and breaking them into enormous blocks, portions of which were moved bodily for miles along the lines of fracture (fig. 9, *T.P.*), the so-called thrust-planes. As shown in the section, masses of gneiss and Torridonian strata came to rest as lofty hills on a foundation of the younger Cambrian beds, the whole resting on a foundation of Torridonian rocks, *B*, seen at the base of the section. Subsequent to this violent reversal of the normal order of succession, the whole series, including the main mass of the mountain—the Cambrian beds with the Torridonian and the older gneiss—was rent by steeply inclined

[1] Peach and Horne (07).

FIG. 8. Meall A' Ghiubhais, Loch Maree, North West Highlands of Scotland, composed mainly of Pre-Cambrian strata; Lower Cambrian (*Olenellus*) beds in the foreground. (From the Geological Survey Collection of photographs.)

fractures. The dislocations responsible for these overthrusts and attendant phenomena occurred in the latter part of the Silurian period and in the earlier stages of the Devonian period, when the Scottish Highlands were an outlying portion of a vast northern continent with mountain chains higher than the modern Alps.

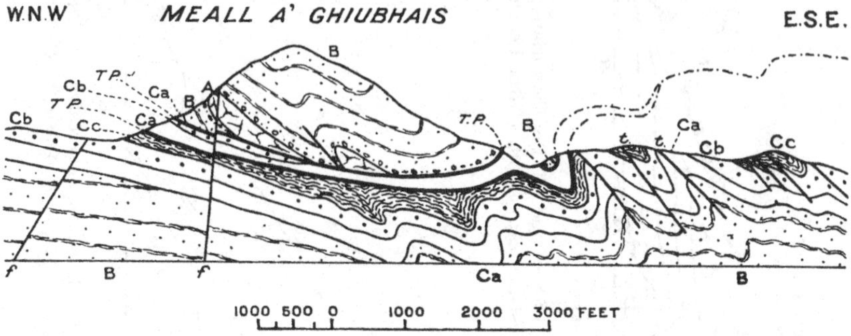

FIG. 9. Section through the mountain shown in fig. 8. *A*, Lower Pre-Cambrian gneiss; *B*, Upper Pre-Cambrian Torridonian rocks; *C*, Cambrian strata; *Ca*, basal quartzite; *Cb*, Pipe rock; *Cc*, *Olenellus* zone; *T.P.* thrust-planes; *t*, minor thrusts; *f*, faults. (After Peach and Horne, with the omission of certain details.)

The section seen in fig. 10 affords a further demonstration of the almost incredible extent of the displacement of pieces of the crust in the North West Highlands. At the western end of the section the mountains consist for the most part of folded Pre-Cambrian gneiss: the junction of the overlying patches (*B*, *B*) with the folded rocks below shows very clearly the uneven nature of the eroded surface on which the Torridonian sediments were deposited. On the western slope of Mullach Coire Mhic Fhearchair the sandstones and grits (*B*), 2000 ft. thick, accumulated round and above the hill of gneiss (*A*) 1000 ft. in height, which they completely buried. On the eastern side of the mountain Cambrian beds, *C*, rest unconformably on the almost horizontal Torridonian series, *B*. The line, *TT*, marks the trend of the thrust-plane along which masses of Pre-Cambrian gneiss, traversed at *D* by dykes of igneous rock, have been carried upwards from the parent rock on the east of that seen at *AA* in the middle and on the left of the section. On the western side of Loch an Nid is a separate peak formed of an outlier of gneiss isolated by denudation from the mass farther to the west on the slopes of Mullach Coire Mhic Fhearchair. To the east of Loch an Nid the Cambrian beds, *C*, are truncated by the thrust-planes *T*; and above this is a long block of gneiss, *A*, truncated by a second

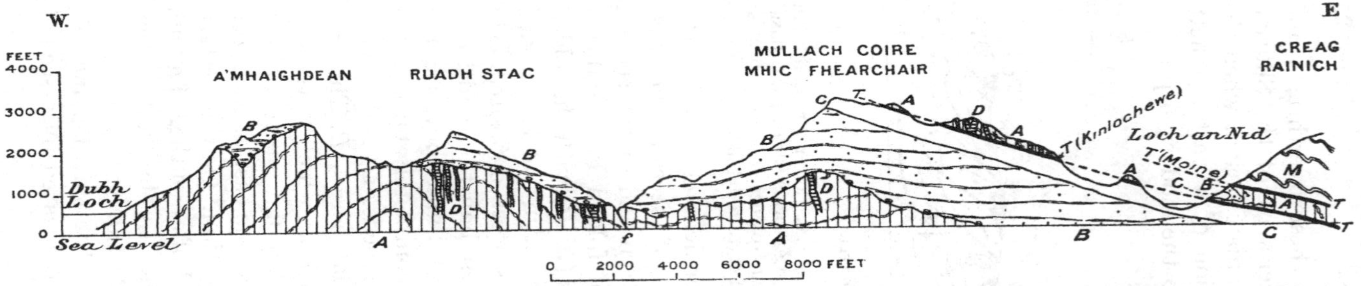

FIG. 10. Section of Pre-Cambrian and Cambrian rocks in the North West Highlands of Scotland. *A*, Pre-Cambrian gneiss; *B*, Torridonian beds; *C*, Cambrian strata; *D*, dykes in the gneiss; *M*, Moine schists; *TT'*, thrust-planes; *f*, fault. (After Peach and Horne, with the omission of certain details.)

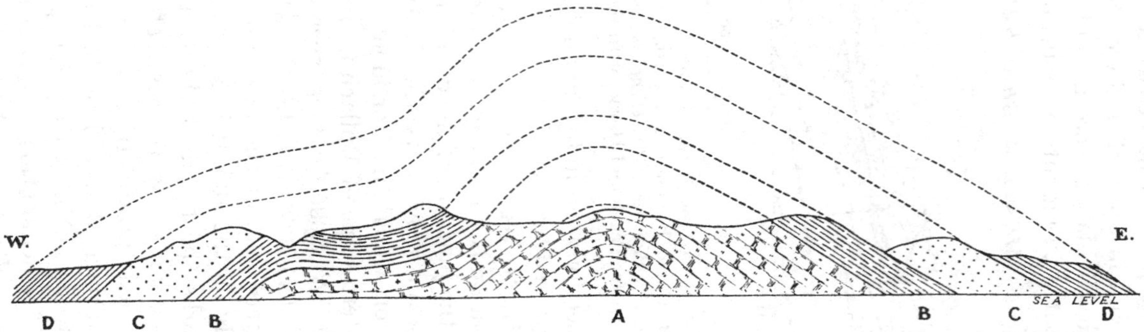

FIG. 11. Section across Derbyshire showing folded Carboniferous strata and illustrating erosion. *A*, Carboniferous Limestone; *B*, Pendleside series (grits and limestone); *C*, Millstone Grit; *D*, Coal Measures. (After Lake and Rastall.)

thrust-plane, T', along which the mass M, composed of a series of metamorphosed sediments (the Moine schists), was transported *en bloc*.

Another of the major revolutions, the Appalachian[1] and Hercynian, which was initiated in America rather earlier than in Europe, calls for a passing reference because of its importance as a factor in the formation and preservation of our coalfields. At the close of the Carboniferous period a large area of the earth's surface came under the influence of stresses which caused the elevation of mountain ranges across the middle of Europe; in the Mendip Hills; in South Wales, and in the south of Ireland; in the Pennine chain, in Belgium, Germany, the Donetz district of southern Russia and in many regions in Siberia and the Far East.[2] Over the British Isles a double set of folds was produced, one running east and west and the other north and south. The truncated arches of the Derbyshire hills[3] serve as a simple example of one phase of these crustal disturbances: the completion of the lost crowns (fig. 11) enables us to appreciate the results achieved by the ordinary instruments of erosion—rain and frost, streams and rivers. We, who measure time by the insignificant span of human history, find it difficult to grasp the cumulative effect of ceaseless and protracted operations which to our limited vision seem wholly unfitted for the task of laying bare to their roots ranges of alpine heights. The Mesozoic era was ushered in by the Hercynian and Appalachian revolutions; its close was marked by the Laramide revolution. Throughout the Mesozoic era Europe was not implicated in any crustal disturbances comparable in size and intensity with those so far described, but in the course of the Tertiary age the long period of immunity from major revolutions was succeeded by a display of mountain-building which rivalled the revolutions of earlier days.

During periods of intense folding and mountain-building, molten material from deep-seated reservoirs was intruded into the upper crust of sedimentary rocks. Some of these intrusions occur as dykes which, by their greater resistance to weathering, stand out as jagged walls or projecting ridges; some form sheets of basalt or other crystalline rock interbedded with strata entirely different in origin. One of the most famous of the late Palaeozoic sheets is the great

[1] Schuchert (23); Daly (26); Evans (26[2]); Bailey, E. B. (28).
[2] Gregory (25).
[3] Lake and Rastall (13).

Whin Sill,[1] a mass of basaltic material which can be traced over a distance of seventy-five miles, from the Farne Islands, off the coast of Northumberland to the Eden valley in Westmorland. At one locality it forms a bold scarp on the crest of which the Romans built part of their great wall; farther to the south-west the Sill is exposed in the valley of the Tees where it has caused the waterfalls of Cauldron Snout and High Force. The Whin Sill was forced into the upper layers of the earth's crust when the Carboniferous period was near its close.

The Building of the Alps

It has long been known that the material of which the Alpine complex consists includes old crystalline rocks and a tangle of sedimentary strata of different ages, folded, overturned, and uprooted during the throes of the Alpine revolution when the superb and disorderly ranks of peaks were called into being. The Alps are made in part of land-derived detritus and calcareous sediments piled up in thousands of feet on the slowly subsiding floor of the Tethys Sea, a great central sea, of which the Roman Mediterranean is the diminutive descendant: this material was deposited in an enormous trough on the earth's surface between an ancient northern continent and a still greater southern continent, the Gondwanaland of geologists. The sediments of the world-encircling Tethys with intruded masses of the still more ancient igneous rocks are the stuff of which the Alps are made. In the heart of Switzerland there is a region known as the Pre-Alps (fig. 12), a well-defined zone stretching from the Lake of Thun to the Lake of Geneva and the River Arve. The majority of Swiss geologists believe that in the Pre-Alps we have a block composed of sedimentary material "entirely foreign to the district", a travelled mass derived from the southern shallows of the Tethys Sea on the sloping flanks of the African edge of Gondwanaland.[2] The northern border of the southern continent was pushed forward by the irresistible forces of crustal contraction and drove before it the yielding sedimentary rocks which lay in its path in the Tethys Sea: as the forward movement continued, fractures formed in the crust provided an outlet for tongues of semi-molten basalts from the substratum (sima). Still advancing, the edge of Africa was lifted up as a stupendous arch and some of the far-

[1] Holmes, A. and H. F. Harwood (28).

[2] Collet (26), (27); see also Heritsch (29).

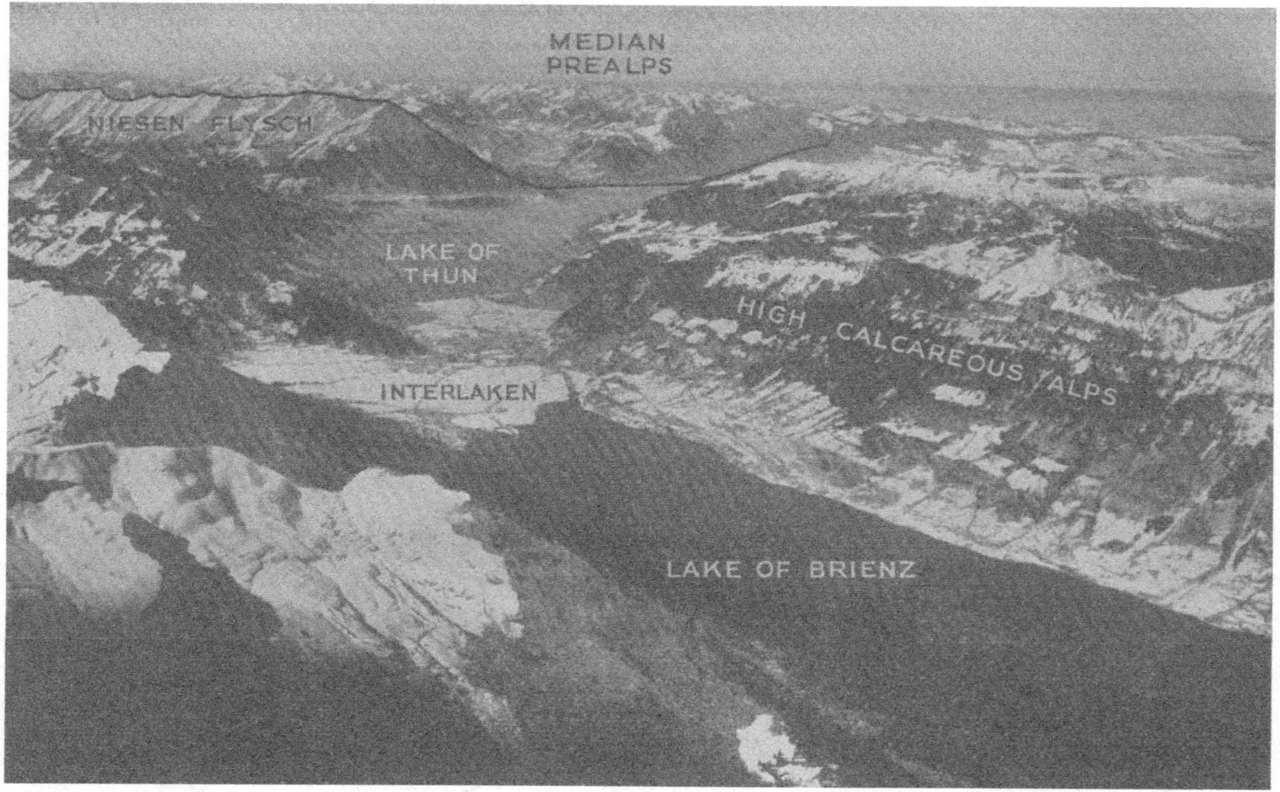

FIG. 12. A view from the air of the Swiss Alps showing the Median Pre-Alps beyond the Lake of Thun with the Jura Mountains in the distance. (Photograph by Ad Astra-Aero, Zürich.)

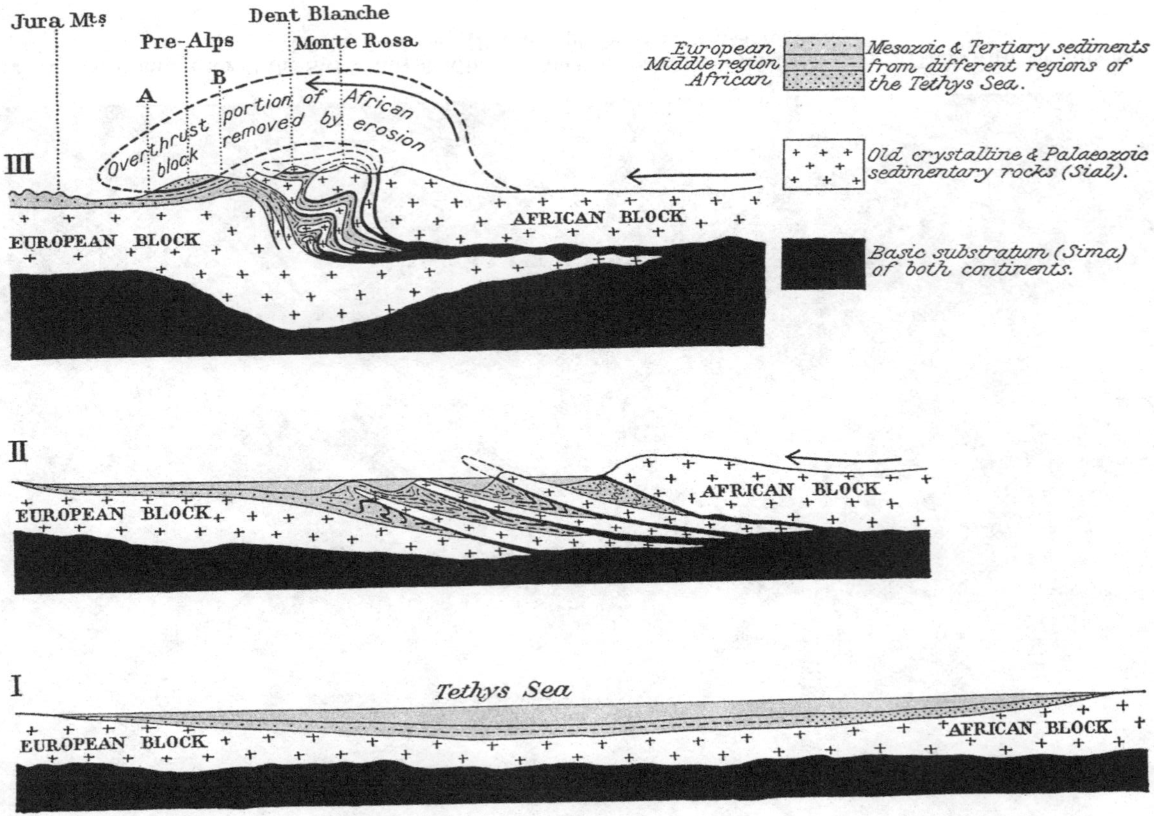

FIG. 13. Diagrammatic sections illustrating stages in the building of the Swiss Alps. These sections, suggested to me on reading a description by Prof. Collet, of Geneva, were drawn from sketches made for me by Mr T. C. Nicholas of Cambridge. For explanation, see the text.

travelled sedimentary material on its front at last found a resting-place in that part of Europe that we call the Pre-Alps. The connecting bridge of rock was eventually destroyed by denudation (fig. 13, III).

The three diagrammatic sections are intended to illustrate this amazing story. Section I (fig. 13) shows the Tethys Sea several hundred miles in breadth sundering the two continents. A single layer (coloured blue) represents the pile of sediments on the sea-floor: below this is a foundation of sedimentary beds of an earlier age and still older crystalline rocks; underneath is the basaltic substratum. The revolution was initiated in the Carboniferous period: slowly, it may have been almost imperceptibly, the edge of the African continent advanced, crumpling and overturning the opposing rocks; from the sea-floor and from the lower face of the moving continent were detached long rib-like shafts of the older rocks which rose above the surface of the dwindling sea as festoons of islands forming the advanced guard of the main mass (section II). As the jaws of the vice—the African land and the resisting block of Europe—came nearer together, the shortening crust was crushed into overfolds, fractured, and kneaded like dough beneath the mighty dome of the overriding and invading continent (section III). The sea was obliterated: in its place was an arched and heterogeneous jumble of rock. Later dislocation produced the present trough of the Mediterranean. The overthrust mass slowly yielded to the action of denudating agents, and from the folded and riven complex were hewn the Alps as we know them.

Far to the east the same Alpine revolution gave birth to the Himalayas. In the folds of the Indian ranges and in the Tibetan plateau there is a continuous series of Palaeozoic and Mesozoic strata which once lay beneath the waters of the Tethys. Sir Francis Younghusband[1] thus describes a view in the Himalayas from an altitude of 20,000 ft.: "Over the great sea of cloud which filled up the valleys all the most famous peaks could be seen rising like glistening pearly islands from a fleecy ocean". Substituting sea for cloud, we see the giant hills, "types and symbols of eternity", as they first rose above the waters of the Tethys.

These summaries of events at different periods of earth-history serve to emphasize not only the kaleidoscopic changes which have

[1] Younghusband (26).

taken place over the face of the world; they also make clear the meaning of the imperfection of the geological record. There are, however, portions of the earth's crust which have almost completely escaped the devastating effects of crustal disturbance; blocks upraised or left as more resistant pedestals when neighbouring areas sank along rifts and faults. It is in these relatively undisturbed refuges, or asylums, that we find the nearest approach to an unbroken sequence of records of plant-life: in the Indian peninsula, in many parts of Africa, in the heart of Siberia, and elsewhere there are thick series of freshwater deposits stored with samples of the "green mantle" which, inconstant in pattern, has for long ages covered wide regions of the earth's surface.

CHAPTER IV

OTHER ASPECTS OF GEOLOGICAL HISTORY

Foreshortened in the tract of time. *Tennyson*

THE scenes from geological history described in Chapter III furnish convincing evidence of a certain periodicity in the forces which have moulded the salient features of the earth's surface. It is possible that the revolutions which affected the crust during the early stages of the Pre-Cambrian era were more world-wide and greater in intensity than those of subsequent eras: be this as it may, it is a fair statement that the study of the framework of the world does not demonstrate any appreciable or regular decline in the potency of the forces of nature in the course of geological history.

Ice Ages

We now turn to another question which is relevant to the main purpose of this book: is there any reason for regarding the great Quaternary Ice Age as an exceptional event, an indication of a permanent lowering of temperature, or even a foreshadowing of conditions which would render continuance of life impossible? The recognition of geological deposits belonging to many different periods and in widely separated parts of the world which demonstrate ice-action on a large scale is one of the most interesting results of recent research. In the glaciers of the Swiss Alps, in the Rocky Mountains, and in many regions of perpetual snow, on the edge of the ice-sheet of unknown depth covering the greater part of Greenland, on the ice-bound coast of Antarctica it is possible to gauge the power of ice as an instrument of rock erosion, and to acquire such knowledge as enables us to recognize signs of former glaciation in places where ice is now unknown. The grooved and polished rock faces, the masses of clay full of ice-scratched boulders varying in size from pebbles to huge blocks many tons in weight, layers of sand and mud in which the alternation of relatively coarse and fine sheets (figs. 37, *G*, *H*; 53) of almost papery thinness registers seasonal changes in the velocity of glacial streams—these are some of the familiar marks of glaciers and ice-sheets. The Quaternary glacial deposits are among the most recent accumulations on the earth's crust. The scored and smoothed surface of grit shown in fig. 14 is a common though none the less an impressive illustration of the rasp-like effect of a mass of ice studded with etching and eroding stones moving over solid rock.

A few examples will serve as justification for the statement that the prevalence of glacial conditions over extensive areas is a "normal feature of the earth's history". To the plant palaeographer records of Glacial epochs are of special importance as evidence of "periods of fierce testing of organisms".[1] Evidence has been obtained of the occurrence of glacial conditions in several parts of the world during certain stages of the Pre-Cambrian era. In the provinces of Ontario and Quebec a sheet of boulder clay or tillite (the term tillite is usually applied to the older glacial deposits agreeing in origin and structure with the boulder clays or tills of the present day) has been traced for a distance of at least 500 miles. A typical ice-scratched boulder from one of the oldest Pre-Cambrian tillites is seen in fig. 15.[2] Traces of ice-action in the Pre-Cambrian era have been discovered in many other parts of the world. One of the oldest known examples of what appears to be an Archaean (older Pre-Cambrian) glacial deposit has been described from Finland:[3] a phyllite, that is an altered and hardened clay, which consists of alternate layers of coarser and finer sediment (fig. 37, G) presents in its seasonal banding a striking resemblance to the material deposited by streams issuing from the snout of a modern glacier. The enlarged photograph (fig. 37 H) shows the juxtaposition of larger particles, characteristic of the sediment when melting ice in the summer increases the volume and transporting power of the water, and the much finer detritus carried by the more sluggish streams in the winter.

In recent years much attention has been paid to these finely laminated sediments: about twenty years ago the Swedish geologist Baron de Geer[4] gave an exceptionally interesting account of the employment of them as a measure of time and as guides to the rate of recession of glaciers when the amelioration of climate in the latter part of the Quaternary period caused the ice to retreat from large areas in the northern hemisphere. The examination of a section

[1] Coleman (26).

[2] Bain (25) calls in question the Pre-Cambrian age of the boulders and tillites described by Coleman.

[3] Sederholm (26).

[4] de Geer (12). See also Chapter XVII.

FIG. 14. Glaciated pavement in Cumberland: the striae indicate the direction—NNE to SSW—in which the Quaternary ice moved. (Geological Survey Collection of photographs.)

through a mass of banded glacial sediments reveals a regular alternation of coarser and finer layers: each pair of layers is called a varve (*varve* or *hvarf*, a periodic repetition). A varve varies from about 0·2 to 3 cm. in thickness and represents one year: the coarser basal layer being the thin film of sediment deposited in a lake, on the floor of the sea, or in some other place from a glacier-stream swollen by the summer melting; the upper layer, the film of finer

FIG. 15. Ice-scratched boulder from a Pre-Cambrian tillite at Cobalt, Ontario. (From a photograph supplied by Prof. Coleman.)

particles deposited in the cold season from water with less velocity and less carrying power. Each year fans of clay and sand are formed as a glacier retreats: the varves, when traced over a strip of country which had been covered by ice, may be compared with broad and thin, overlapping tiles. The uppermost extends so much over the underlying varve as the ice-border had receded. It is therefore possible by accurate measurement and laborious correlation of varves in different places to estimate the rate of ice-movement. De Geer, for example, found that Stockholm was finally thawed from the last Quaternary ice about 9000 years ago. Varves are now being formed in glaciated countries: they are a characteristic feature of Quaternary glacial deposits[1] and—this is the point of special interest—precisely similar, seasonally banded layers (fig. 53) occur in many parts of that enormous region in the southern hemisphere which was overridden by ice in the later stages of the Palaeozoic era (fig. 49): also, as fig. 37 (G, H) shows, they afford strong evidence of the occurrence of Pre-Cambrian glaciers.

It has been suggested by Prof. Coleman,[2] a leading authority on glacial phenomena, that the Pre-Cambrian era may have been colder than the later eras. Cambrian tillites occur in southern Norway; on a large scale in China;[3] in South Australia;[4] and elsewhere. Ordovician and Silurian tillites are recorded from Scandinavia and Alaska respectively.

The greatest of all Ice Ages is that which is generally spoken of as the Permo-Carboniferous Glacial epoch. It probably coincided with the latter part of the Carboniferous period and the early part of the Permian period: as shown in a later chapter its precise age is still under discussion. In the Indian peninsula, well within the tropics in regions where snow and ice are now unknown, there are thick and widespread beds of tillite containing boulders up to 15 ft. in diameter, and glaciated pavements no less convincing than that shown in fig. 14. The mud-covered limestone on the edge of the Penganga River seen in fig. 16 owes its gentle curvature to the action of an ice-sheet which passed over a large area in India and over regions beyond the present Indo-Gangetic plain: the bank on the right of the photograph is formed of the ancient boulder clay. In Afghanistan, in Kashmir, in the Salt range to the north-west of the Himalayan chain there is similar evidence of the northern extension of glaciers and ice-fields. In South Africa tillites corresponding in age with those of India occur in Cape Colony, the Transvaal, South-West Africa, Tongoland, and Madagascar. Fig. 17 illustrates a glaciated platform in Griqualand West, covered on the right by a typical tillite; the glacial origin is "as unmistakeable as is that of the drift-sheets of North East America and North West Europe"[5] left by the Quaternary ice. The greater part of British

[1] Good photographs of Quaternary varves have been published by the Geological Commission of Finland. See M. Sauramo (23).

[2] See Coleman (26) for references to descriptions of glacial deposits of various ages.

[3] Willis, B., and others (07).

[4] Howchin (08), (12).

[5] Davis (06). See also Mellor (05); Maufe (22).

FIG. 16. The east bank of the Penganga River, near the Irai valley, Hyderabad State. The slightly curved surface of glaciated Penganga Limestone is covered with a thin layer of mud left by the river in the monsoon floods; the bank on the right consists of the Upper Carboniferous (or Permo-Carboniferous) Talchir tillite. (From a photograph by Dr L. L. Fermor of the Indian Geological Survey.)

Fig. 17. Late Palaeozoic glaciated pavement covered, on the right, by the Dwyka tillite (old boulder clay): Griqualand West, Cape Province. (From a photograph by Mr R. B. Young supplied by Dr Rogers.)

South Africa was under ice, probably about the end of the Carboniferous period, and in some places the glacial deposits are more

FIG. 18. Upper Carboniferous glaciated pavement in the Inman valley, eight miles north of Victor Harbour, South Australia: surface exposed 9 ft. by 3 ft. (From a photograph by Prof. Howchin, Adelaide.)

than twice as thick as those in the northern hemisphere formed during the latest Glacial epoch. Equally impressive boulder beds cover a vast area in Australia; in Tasmania, in Queensland, in the Northern Territory, in Western and South Australia. Prof. Howchin, who kindly supplied the photograph reproduced in fig. 18, writes as follows: "the Inman valley (South Australia) is a most extraordinary combination of glacial features revealing a Palaeozoic topography that is but little altered from what it was when the ice-sheet withdrew. The valley is closed on three sides, open to the south from which direction the ice-flood came. The ice-sheet filled the valley and overflowed the enclosing heights, to accomplish which it could not have been less than 2000 ft. in thickness. Thousands of large erratics have weathered out from the tillite and lie on the surface, and wherever the floor is exposed it is highly glaciated". The rock shown in the photograph, Prof. Howchin adds, "had not seen the light since Permo-Carboniferous times until I removed the covering of boulder clay". In Australia[1] the Carboniferous ice-sheet moved from south to north; in India and South Africa it travelled south. Tillites of the same epoch have been found in Brazil,[2] the Argentine, Uruguay, Bolivia, and the Falkland Islands; scourings from the rocks left by sheets of ice which had their source farther to the south. In a later chapter further reference is made to this Permo-Carboniferous Ice Age which is attested by the clearest evidence over thousands of square miles in India, Australia, Africa, and South America, all of which regions are the dismembered blocks of the ancient continent of Gondwanaland. The approximate areas under glacial conditions in the closing stage of the Palaeozoic era are shown on the map reproduced on p. 156 (fig. 49).

At intervals during the ages which separated the Carboniferous and Quaternary periods there were local glaciers in regions which are now ice-free, but there are no grounds for assuming the occurrence of glacial conditions comparable in extent or intensity to those revealed by Palaeozoic and Pre-Cambrian rocks. A general survey of glacial phenomena recorded in the successive stages of the world's history leaves the impression of a remarkable harmony in the varying climatic conditions and gives no support to the idea of a progressive lowering of temperature.

The Wegener Hypothesis

In later chapters devoted to the description of successive floras an attempt is made to recreate the earth's surface, to substitute

[1] Howchin (10), (12), (24), (29).
[2] Woodworth (12).

for the present oceans and continents the lands and seas of former ages. The self-evident fact that these restorations may be far from the truth has already been stated: moreover maps of former periods, as usually drawn, may be based on a false assumption, that the continents as we know them are in the main fixed portions of the earth's crust. A contrary view has been expressed; that the continents have drifted into their present positions, and that the sundering oceans are a measure of the distance to which the disrupted pieces of former, much larger, land-masses have floated over a viscous floor farther and farther apart. This conception of the earth's surface has been much discussed: its relevance to problems associated with the study of ancient floras, particularly the wanderings of plants from one region of the world to another, is a cogent reason for a brief consideration of the theory which in its present form we owe to Prof. Wegener.[1] The Viennese geologist Suess described the crust as consisting of two zones, an upper and lighter shell, including sedimentary rocks of all ages and crystalline rocks such as gneiss and granite which belong to the type known as acidic, that is, containing 60–75 per cent. of silica; and a substratum composed of heavier basic rocks such as basalt, with a lower silica content: the upper zone he called the sial; the lower the sima. Wegener, who adopts the view of a dual crust, visualizes the continents as a rocky scum of sial (or sal) floating on the denser sima which, beyond the limits of the submerged edges of the continents, forms the floor of the ocean. Originally, so Wegener believes, the sial was more or less equally spread over the whole surface of the earth and formed the floor of a Panthalassa or universal ocean with an average depth of nearly 3000 metres. At a later stage the granitic crust was crumpled up into a Pangaea, a world-continent occupying perhaps half of the entire surface. It may be that the masses of granites, gneiss, and associated crystalline rocks which form wide regions in Canada, Finland, Scandinavia, South America, Africa and in the Far East are the remains of some of the nuclei of the upraised and folded crust, nuclei which have either remained as uncovered and stable blocks or have been hidden and buried deep under thousands of feet of later rocks.

The basaltic sima probably rests on some still heavier material, iron or an iron-nickel alloy: the earth is believed to be solid throughout.[2]

[1] Wegener (24); Köppen and Wegener (24).
[2] Daly (26); Clarke and Washington (24).

Under the stress of forces set up by contraction of the crust, and in part the result of a rise in temperature produced by the disintegration of radioactive substances, the fused basic sima is from time to time forced along lines of weakness in the fractured sal or poured out over the surface as lava. Wegener's contention is that the continents are comparable to drifting icebergs detached from a once continuous mass, the sea corresponding to the sima. He draws attention to a general resemblance in outline and in geological structure between continents which are now separated by broad oceans, e.g. the eastern edge of South America and the western border of Africa. The Atlantic is described as a great rift in the sal, gradually broadened in the course of ages. Similarly, he regards South America, Africa, India and Australia as the detached portions of the old continent of Gondwanaland, as it were pieces of a jig-saw puzzle that have drifted apart. There was almost certainly a vast southern continent which persisted through several geological ages: the question is, whether the present continents are pieces of much greater land-masses saved from the wreck when the rest foundered; or whether, as Wegener believes, practically the whole of Gondwanaland is represented by the land-areas that remain, their present positions being the result of drifting apart from a disrupted whole. There is much that is attractive in Wegener's view, but it rests on assumption unsupported by definite proof. The theory has been severely criticized: one geologist[1] asserts that Wegener has suggested much, but has proved nothing. Whether or not the regions of the world, which are usually held to be relatively immovable in a lateral direction, are on the contrary gigantic floating islands, is a problem which cannot be solved until we are in possession of a series of geodetic observations sufficiently accurate to eliminate the possibility of experimental error.

If we rule out the possibility of the evolution, more or less simultaneously, of the same species independently at different places, that is, the doctrine of multiple centres of creation, and accept the orthodox view of single centres of creation, it is difficult, indeed impossible, to account for the natural occurrence of the same species of animals and plants living in countries separated by barriers of water. On the other hand, regions now far apart may have been connected by land-bridges or chains of islands which served as routes of migration or stepping-stones. Recourse to Wegener's view

[1] Lake (22), (23).

makes it easier to explain anomalies in plant-geography. One thing at least must be conceded: the present distribution of plants over the earth's surface demands former lines of communication between regions now separated by thousands of miles of ocean. We must for the present be content to leave the choice of solutions until more trustworthy evidence is available.[1]

Wegener favours not only a drifting of the sal, but a shifting within certain limits of the poles, apart from the movement of the land relative to the poles. It has long been the practice of geologists, puzzled by the discovery within polar regions of fossil plants and animals closely allied to species now living in subtropical or even tropical countries, to assume instability of the earth's axis. By altering the position of the poles it is possible to bring strata, which are now well within the Arctic Circle, into situations more in agreement with the climatic requirements of the plants preserved as fossils. The interesting questions raised by extinct Arctic floras are more fully considered in later chapters. By taking liberties with the axis of the earth we can no doubt find partial answers to some of the questions, but not by any means complete answers. There is a further difficulty: in the opinion of astronomers it is very unlikely that the position of the poles has materially changed in the course of geological time. Short of accepting Wegener's hypothesis, we are left with two further possibilities: a redistribution of land and water with a consequential rearrangement of oceanic circulation, the substitution of low-lying for mountainous areas, and other changes in physical geography which would produce slightly more favourable climatic conditions. Or, alternatively, we may adopt the view that too high a value has been placed on plants as "thermometers of the ages". A fuller treatment of the difficult questions raised by the occurrence of fossil floras in unexpected places is reserved for another chapter.

[1] For an interesting discussion on Wegener's hypothesis the reader is referred to the *Theory of Continental Drift, a Symposium on the origin and movement of land-masses both inter-continental and intra-continental, as proposed by Alfred Wegener*, published in 1928 by the American Association of Petroleum Geologists. See Willis, B. (28); also Krige (29) who gives references to other sources, and Rastall (29).

CHAPTER V

PRESERVATION OF PLANTS AS FOSSILS

In Biography you have your little handful of facts, little bits of a puzzle, and you sit and think, and fit 'em together this way and that, and get up and throw 'em down, and say damn, and go out for a walk. ***R. L. Stevenson***

THE great majority of fossil plants occur as leaves or pieces of leaves, foliage-shoots, cones, seeds; or as broken stems without appendages—samples taken at random by the wind and "rain furies"—flung into rivers and carried with washings from the land to be deposited in a delta, on the floor of a lake or shallow sea. The lighter and more buoyant fragments come to rest in one place, while over another area are scattered the heavier contributions from forests and the fern-covered banks of rivers. The waifs are strewn light-heartedly by nature; it is for us to sort them as best we can.

Many leaves are preserved as thin carbonized films on the surface of the hardened mud: by different methods of treatment with chemical reagents it is often possible to examine microscopically the more resistant cells which have retained their shape and, to some extent, their structure. If the black or dark brown leaf or other part of a fossil shows a tendency to peel off the surface of the rock on which it lies it is easy to detach the mummified pieces; after bleaching with chlorate of potash and nitric acid, and then making them more transparent by washing in ammonia, the fragments can be mounted in Canada balsam for microscopical examination. Another method, first employed by Prof. Walton, of dealing with carbonized plants when the fossil cannot be detached from the rock, is to make a transfer: a piece of the matrix with the specimen to be examined is trimmed to a convenient shape, covered with a solution of cellulose acetate in amylacetate and allowed to dry: it is then transferred to a slide in hot Canada balsam and, after covering with melted paraffin wax, the whole is placed in a bath of hydrofluoric acid. The acid disintegrates and removes the matrix, leaving the fossil isolated on the slide. After removing the wax and washing, the preparation can be examined under the microscope. For a fuller account of this and other methods of treatment reference should be made to the sources mentioned in the footnote.[1] In other instances the original plant-substance has been destroyed, leaving only an imprint of form and venation. Less frequently, stems, twigs, leaves and other odds and ends have been petrified, not merely encrusted by a covering of lime deposited on the evaporation of the water of so-called petrifying springs or from the spray of a waterfall, but turned into stone by the infiltration of calcareous or siliceous solutions.

Among the most fruitful sources available to the botanical historian are the calcareous nodules, or coal-balls,[2] occasionally found in beds of coal, especially in the Lower Coal Measures of England, and in Carboniferous plant-bearing beds in several other parts of the world. These balls (fig. 19) are petrified patches of vegetable débris from the marshes and swamps of the closing stages of the Carboniferous period. The calcareous nodules, composed mainly of carbonate of lime, vary from an inch to two feet in diameter and are scattered irregularly through a seam, the lines of bedding of the enclosing coal arching over them as the result of the greater compression of the yielding matrix, in contrast to the greater resistance to pressure offered by the hard concretions. The peaty layers were covered by incursions of sea water as the tree-studded swamps and the low-lying banks of the estuaries subsided. As water charged with salts percolated through the submerged ground samples were petrified and left as permanent memorials. The section of a large coal-ball reproduced in fig. 19 shows many well-preserved pieces of a great variety of plants, a few of which are indicated by the lettered lines.

Microscopical examination of thin, transparent sections cut from the coal-balls enables us to see the most delicate tissues, occasionally even traces of cell-contents. We can reconstruct the framework; and from the analogy of existing plants, it is often possible from a study of the anatomical details to obtain glimpses of the original environment. As we look into the intricate network of cells the fossil becomes a living thing: in certain tissues we recognize a mechanism adjusted to the reception of sunlight; other structural features suggest a plant living in water or on swampy ground. In the arrangement of the thicker-walled and stronger cells, which serve as mechanical

[1] Bather (07), (08); Walton (28), (28²), (28³), (30); Lang (26²); Kräusel (29) with full references; Radley (29); Barnes and Duerden (30); Hoskins (30).

[2] Stopes and Watson (08) with references. See also Koopmans (28); Hirmer (28); Absalom (29).

FIG. 19. Section of a large coal-ball from the Coal Measures of Shore, Lancashire, prepared by Mr J. Lomax. *Bot.* petiole of fern (*Botryopteris*); *Cal.* stems of *Calamites*; *Cd.* root of *Cordaites*; *Cr.* crystalline material; *Ld. Lepidostrobus*; *Lyg.* petiole of *Lyginopteris*; *Sa.* rootlet of *Stigmaria*; *Sd.* seed; *Sg.* leaf-base of *Sigillaria*; *Sp.* stem of *Sphenophyllum* ; *X.* piece of wood. Nat. size.

supports to the feebler parts, we see in the architecture of stem and leaf the same accuracy of correlation between structure and function as in plants of the present day; the same principle of construction which we, "the afterthoughts of creation", have adopted in buildings and bridges—maximum strength with the minimum expenditure of material. The usual method of preparing transparent sections of coal-balls for microscopical examination has been to cut by machinery thin slices of the rock and grind them down to the necessary thinness. Recently a more convenient method has been discovered by Prof. John Walton of Glasgow in collaboration with Dr Koopmans of Utrecht. This consists in cutting or grinding a coal-ball to a flat, smooth surface and then immersing it in an etching solution of hydrofluoric acid: the acid dissolves a film of the carbonate of lime of which a coal ball is mainly composed and the actual plant-substance is left in relief. After washing and carefully drying, a substance which hardens on drying into a tough film, such as cellulose esters, e.g. the trade preparation "Durofix", or gelatine, is poured over the surface and left to set. The plant-substance, separated by the etching process from the matrix, becomes embedded in the film of gelatine or cellulose, which is easily peeled off and can be mounted in Canada balsam under a cover-slip. By the employment of this method, which is much simpler and cheaper than cutting sections, it has been found possible to obtain fifty films from a thickness of 2 mm. of coal-ball.

From the investigation of the anatomy of extinct plants we discover many architectural types that are now unknown; but even in the oldest representatives of terrestrial vegetation we find the same types of cell, the same grouping of cells into tissues, and the same adaptation—if one may venture to use the word—of the mechanism of the plant-body to the activities of the living organism as those with which we are familiar in the plants of the present day. It is hardly necessary to emphasize the importance of searching for specimens in which the tissues of the plant are preserved: external form alone is often insufficient as a guide to affinity and may be seriously misleading. If petrified material is not available, much useful information on the structure of the superficial cell-layers, and not infrequently of the tissues below the surface, can be obtained by the microscopical examination of fossils, which at first sight appear to be unpromising, by the employment of some of the methods referred to above. Many instances might be given of the danger of trusting to external form: one will suffice. Dr Florin[1] of Stockholm, who has long been engaged in the comparative examination of the structure of the surface-layers of fossil conifers, was able to prove that certain Tertiary leaves assumed, on the

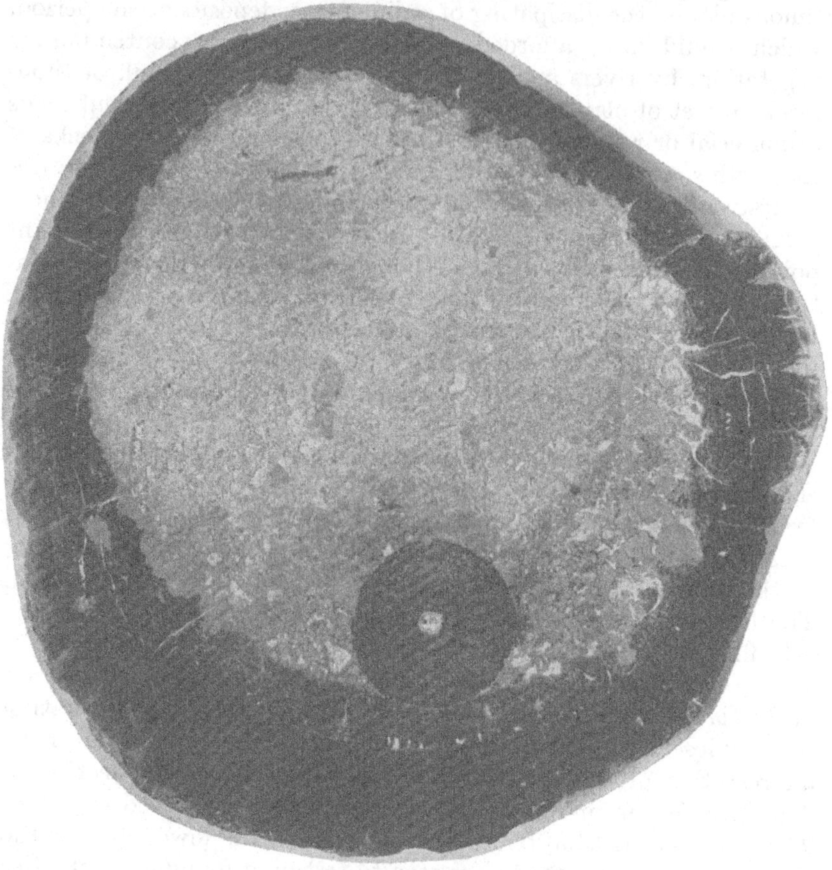

FIG. 20. Transverse section of a stem of a *Lepidodendron* preserved in volcanic ash (Lower Carboniferous) at Dalmeny, Scotland: the cylindrical axis has fallen against the bark owing to the destruction of the inner cortical tissue. $\frac{1}{3}$ nat. size. (From Seward and Hill (00).)

evidence of form, to belong to a conifer (*Podocarpus*), are leaves of a flowering plant.

It needs but a slight acquaintance with the circumstances in which preservation of samples of the existing vegetation is now

[1] Florin (26).

being effected by natural agencies to make us grasp the reality and degree of the imperfection of ancient botanical records. We think of the disastrous operations of erosion, the chiselling and planing of mountain ranges until highlands are reduced to monotonous plains; the dissipating of sedimentary deposits of one period, which would have afforded valuable evidence of contemporary vegetation, by rivers and floods of a later age. We think of thousands of feet of plant-bearing rocks buried under more recent loads of material or rendered inaccessible by the foundering of blocks of the earth's crust. On the other hand, there are not a few instances of destructive forces which have left as legacies some of the most precious records we possess. Showers of volcanic ash with attendant poison gases and flows of lava, though causing wholesale devastation, have saved many fragments from complete destruction. Exceptionally well-preserved plants occur in beds of volcanic ash, which served as a source of lime and silica for the mineralizing solutions which caused the petrification of entombed stems or foliage-shoots.

The fossil shown in fig. 20 is part of the stem of an extinct Palaeozoic tree (*Lepidodendron*) which was found in a bed of Carboniferous volcanic ash at Dalmeny, near Edinburgh:[1] the outer bark is preserved in wonderful perfection, also the woody cylinder which, as the more delicate tissue decayed, fell from its central position. The lighter coloured material, consisting of fragments of volcanic ash, fills the space formerly occupied by the softer portion of the stem. An impressive example of the partial destruction of a series of Tertiary forests during periods of volcanic activity, alternating with quiescent intervals long enough to permit the natural re-afforestation of the devastated country, is seen in the section of Amethyst Mountain in the Yellowstone Park[2] reproduced in fig. 21. The mountain is 9400 ft. above sea-level: in the lower part of the section the volcanic strata are seen to rest on a foundation of older rocks *A*, and these in turn were laid down on the eroded surface of a still more ancient foundation, *B*. The numerous instances of trees and other plants of various geological periods overwhelmed by volcanic ash are merely additional illustrations of "Nature's unchanging harmony"; they bear witness to the recurrence in the past of events which occur spasmodically at the present day. In

[1] Seward and Hill (00).
[2] Knowlton (99).

fig. 22, reproduced from a photograph taken by Dr C. A. Matley on a visit to the crater of Papandajan in Java, we see the remains of a forest partially buried to a depth of several feet in volcanic ash. "The trees are dead and stripped entirely of their leaves and small twigs by the volcanic bombardment; but are not burnt." A pre-

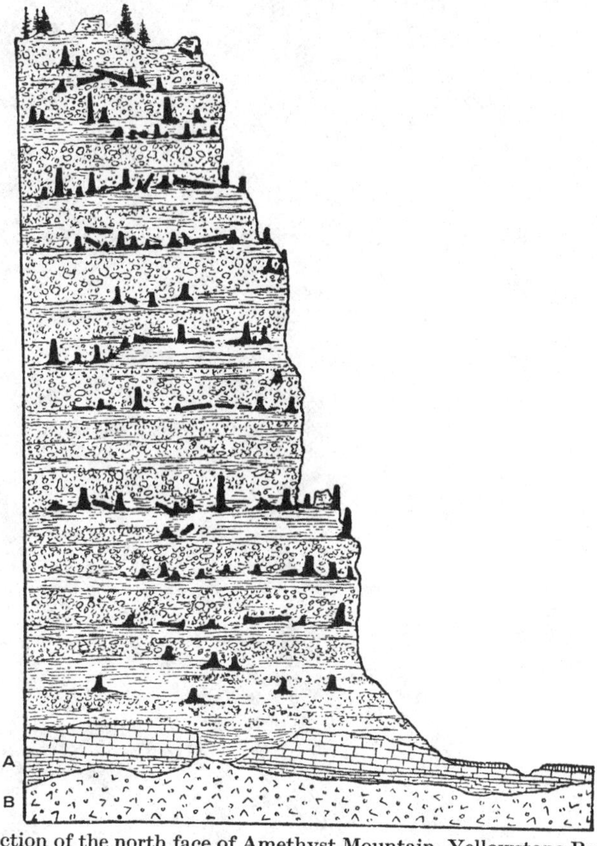

FIG. 21. Section of the north face of Amethyst Mountain, Yellowstone Park, showing a series of forests buried in volcanic ash reaching a thickness of over 2000 ft. The steepness of the slope is exaggerated. (After W. H. Holmes.)

cisely similar scene is shown in fig. 51 which represents an early Carboniferous landscape in southern Scotland.

The streets of Pompeii with rows of shops and houses enable us to picture the life and amenities of a Roman town: similarly, the impressions of leaves and twigs, with butterflies and other winged insects, in the volcanic sediments which mark the site of the vanished

FIG. 22. The crater of Papandajan, Java; in the foreground a forest partially buried in volcanic ash. (From a photograph by Dr C. A. Matley.)

FIG. 23. An old forest in the bed of the Waikato River, New Zealand, recently brought to light by the erosion of at least 40 ft. of sediment by the overflow water from a dam. (Photograph reproduced by permission from the New Zealand *Free Lance* (1929).)

lake of Florissant in an open valley in Colorado,[1] tell us of the trees and humbler plants which flourished on its shores in the Miocene period.

Occasional references are made in the subsequent descriptions of fossil plants to instances of preservation by petrifaction or other means: for a fuller account of the various ways in which plants may be preserved in rocks the reader should consult palaeobotanical or geological text-books. A different method of preservation has recently been described in New Zealand: the rapid wearing away by water of the old bed of the Waikato River has laid bare the ruins

FIG. 24. Petrified wood, mainly stems of an extinct dicotyledon, *Nicolia*, scattered over the Libyan desert, Wadi Tih. (From a photograph taken for me by Prof. F. W. Oliver.)

of a forest buried long ago under a thick mass of water-borne sediment where presumably it would have remained hidden had it not been for floods caused by human agency. Just such another example of the uncovering of a forest site may be seen near Glasgow where in quarrying Carboniferous rock men have brought to light the stony stumps of trees rooted in soil where they grew perhaps 200 million years ago.

The trees recently exposed in the bed of the New Zealand river were probably alive a few thousand years ago: the amount of sediment removed might have been deposited in the old river-bed in the course of three or four thousand years. There appears to be no difference except in age between the trees of the buried forest and those still living in the district. The photograph, taken for me by Prof. F. W. Oliver, reproduced in fig. 24, shows innumerable pieces

[1] Cockerell (08).

of petrified stems scattered over the Libyan desert in Wadi Tih as far as the eye can see: they are a few of an enormous number of trees, embedded in sand millions of years ago, which in the course of ages have been brought to light through the weathering away of the less resistant, encasing rock. The fossils take us back to a time when over the land that is now a wind-swept waste was an abundance of water with drifting forest trees that have long been extinct, their stems and branches buried in the sandy sediment, turned into stone and after millions of years littered as broken remnants over the desert of a later age.

The huge trunk of an extinct tree (*Pitys Withami*[1]) erected in the grounds of the British Museum (Natural History) at South Kensington was found many years ago in a quarry of early Carboniferous sandstone near Edinburgh. Passing over an interval measured in many millions of years we see an old yet vigorous tree drawing from the soil and air the elements of its food and substance. Near its base the flooded waters of a rapid river undermine the banks: in the lower reaches of the river a broken stem, bare of branches and leaves, is borne on the turbid stream to a resting-place in the sands of a delta. Stripped of the bark the wood slowly decayed, but patches were rendered permanent by the petrifying action of percolating water: in the course of time sand and trees became stone, and eventually by the uplifting of the unstable crust of the earth were brought within man's reach.

The Interpretation of Records of Ancient Plants.

For our present purpose it is more pertinent to discuss methods of interpretation. The guiding principles for the student of ancient vegetations must, as far as possible, be the same as those employed in the study of living plants. In attempting to decipher the records of the rocks imagination must be kept within reasonable bounds. One of the functions of the historian concerned with human societies and institutions has been described as "the imaginative or speculative, when he plays with the facts that he has gathered, selects and classifies them, and makes his guesses and generalisations". In the words of another writer, "the greatest instrument at the disposal of the investigator is the imagination. The imagination creates the hypotheses which experiment must test". Palaeo-

[1] For a brief account of the genus *Pitys* see Chapter XI, p. 201: for a photograph of the British tree see Scott, D. H. (24).

botanical research tries to the utmost the self-control of the student: the passion for the past is strong; evidence is meagre, and the temptation to substitute fancy for fact is hard to resist. One of the aims of the interpreter of botanical records is to discover guides to the conditions under which the plants lived. We are as a rule unable to determine whether a plant grew on low ground or on a mountain slope: habitats can be inferred only in exceptional instances.

It may be possible in certain circumstances to ascertain the relative representation of different families or individuals in the vegetation of a former period by counting and classifying the fossils; but there are many considerations to be taken into account. The fact that pieces of certain kinds of plants are more numerous than samples of other kinds does not necessarily mean that dominance among the fossils is an index of dominance in the living forests. Plants growing near a river which transported both sediment and scraps of vegetation would be in an exceptionally favourable position for preservation: it is also clear that the distance to which twigs, leaves, and seeds were carried would be mainly determined by relative buoyancy and weight or power of resistance to decay. The miscellaneous contents of plant-bearing sands and muds in the estuary of a river do not as a rule give a picture of a natural assemblage growing in one environment: they are mere random samples of associations which occupied different and it may be not contiguous situations. Leaves and the branches or stems which bore them seldom retain their organic unity in the fossil state; it is rare to find specimens such as those seen in figs. 56 and 70 (pp. 194, 237). Reference is made in a subsequent chapter to an interesting piece of research on the composition of a Coal-period forest based on the statistical method.[1]

The long distances to which branches and trunks are carried by rivers and currents warn us against the assumption that a piece of wood in a bed of sandstone was once part of a tree growing near the place where it occurs as a fossil. On the west coast of northern Greenland stems of firs and other trees are often stranded on the beach of what is now a treeless land where the tallest willows rarely exceed 3 ft. in height. Most of the Greenland drift-wood comes from Siberian rivers,[2] which carry logs into the sea where they are swept

[1] See p. 262. Davies, D. (21), (29); see also Chaney (24), (27) for a similar method in the interpretation of Tertiary records.

[2] Ingvarson (03).

southward by the polar current and, after rounding Cape Farewell, travel some hundreds of miles along the west coast. Before the importation of wood from Denmark the Eskimo was dependent upon the transport of logs by currents. In the Copenhagen Museum is a trunk of a fir (*Picea*) 18 ft. long and more than a foot in diameter which was washed up on the east coast of Greenland. Drift-wood is abundant on parts of the coast of the island of Jan Mayen.[1] In January, 1912, in the course of a geological journey by members of Capt. Scott's second Antarctic expedition, a piece of petrified wood was discovered in sandstone near latitude 74° S.:[2] where was the original home of the tree? Did it grow near the place of its burial, or was it drifted from some far distant forest? It is often difficult to decide whether the fossil plants at a particular locality may be accepted as the remains of vegetation which grew there, or whether they were transported from a foreign source: we have to be guided by the state of preservation of the material, which may or may not show signs of much wear and tear, by evidence of growth on the spot afforded by underground stems or roots preserved where they grew, or by the close association of fragments which are clearly the *disjuncta membra* of one individual.

In order to visualize the vegetation of the world at any one phase of geological history we must obviously be able to correlate the plants of one region with those of another. This is no easy task. If identical species are found at two or more localities, whether widely separated geographically or not, the strata containing them are referred to the same position in the geological time-scale. In Huxley's words, "palaeontology[3] has established two laws of inestimable importance: the first that one and the same area of the earth's surface has been successively occupied by many different kinds of living beings; the second, that the order of succession established in one locality holds good, approximately, in all". Palaeobotanical research has demonstrated a very wide geographical range for certain floras, or parts of floras. Plants from Jurassic rocks in Europe agree closely with those found in Asia, in Graham Land in the far South, in western North America, and elsewhere. The vegetation of that period is often described as remarkably uniform, as indeed it was in some respects; and it has been assumed without justification that in the Jurassic period climatic zones were very

[1] For another example see Kindle (21). [2] Seward (14).
[3] Huxley (96).

feebly marked. There is undoubtedly a striking resemblance in general facies among Jurassic floras throughout the world, though fuller knowledge of the members of the floras may reveal differences consistent with the axiom that in all ages there has been a zonal distribution of temperature over the world. We know that the bracken fern (*Pteridium aquilinum*) has now an almost world-wide distribution; and the same is true of some other ferns and a few flowering plants. But there is another and a more important consideration: each species of plant began its career at one place, the place where it was evolved, whence it spread over an expanding area.

Rocks in widely separated localities containing the same or closely allied species are not necessarily contemporaneous. Huxley,[1] following Herbert Spencer, laid stress on the danger of regarding strata in different countries yielding identical species as contemporaneous in the strict sense, and suggested the term homotaxis (similarity of order). "Similarity of organic contents", he said, "affords no proof of synchrony, but is compatible with the lapse of a most prodigious interval of time." Sir Joseph Hooker[2] expressed the same opinion: "finding similar fossil plants at places widely different in latitude and therefore in climate is in the present state of our knowledge rather an argument against than for their having existed contemporaneously".[3] Huxley's warning led to an over-suspicious attitude towards the value of evidence furnished by fossils: the fact that very similar floras occur in regions far apart does not necessarily mean that the floras flourished in those places at the same time. Plants migrate and colonize fresh ground. Granting that migration was slow, there is no necessity to assume that by the time each species had reached a point far from its birthplace it had ceased to exist in its original home. We speak of Jurassic, Cretaceous, and other floras, and in so doing imply that the floras occur in rocks that are homotaxial: we think of them in relation to a certain geological period, a period it may be which represents some millions of years; and we picture the floras as waves of vegetation spreading over new territories. We cannot estimate in time the duration of the progressive march of the floras. The plant-world is not in any sense stationary; it is a moving panorama over

[1] Huxley (96). [2] Hooker (91).

[3] Asa Gray (78), went so far as to say that identity of plant-contents affords good evidence against synchrony.

the earth's surface. By designating any stage in the movement by the name of a geological period we do not commit ourselves to the idea of contemporaneous existence; we recognize a particular assemblage of plants as a characteristic feature of a long period of geological history extending over thousands or millions of years.

Finally there is the ever-present problem of specific identity. Fossil species from widely separated regions may appear to be identical; but the appearance of identity may be the result of insufficient knowledge of the plant's structure or of its reproductive organs. We may readily refer to one species specimens found within a restricted area, while we hesitate to draw the same conclusion when they come from localities far apart. Making due allowance for geographical separation, the most rational course would seem to be to employ different specific names whenever there is any reason to doubt actual identity, and to adopt group-names for the purpose of giving expression to the view that certain forms, which may not be specifically identical in the strict sense, agree so closely as to warrant inclusion under a common group-name. The investigation of such data as can be gleaned from an examination of ancient floras is of vital importance in all questions connected with the present geographical distribution of plants. The plant-world, as we see it, is but a phase of an age-long and ever-changing procession of floras the progress and modification of which has been affected in different degrees by a number of influences, changes in climate, retarding or accelerating changes in the physical background. As an American author[1] says, we must regard "the present distributional phenomena of the earth as merely a moment in the prolonged and incessantly active procession of change due to secular or sudden changes of climate, or to destructive or constructive events in surface geology, and discontinuously marked by evolutionary activity". To understand the present it is therefore necessary to turn to the records of other days.

It is clear that we cannot hope to do more than trace in broken outlines the history of the rise and fall of plant-dynasties; but if we keep in view the limitations imposed by the nature of the documents, it may be of some interest, and possibly of some value to students of evolution, to gather together into a connected story such scraps of information as are available.

[1] Shreve (16).

CHAPTER VI

A CLASSIFICATION OF PLANTS

We do but learn to-day what our better advanced judgments will unteach to-morrow.
Sir Thomas Browne

At this stage it may be useful to give a list of the larger groups and some of the families of the plant-kingdom. Many of the subdivisions include both living and fossil representatives. The resemblance of some fossils to existing plants is close enough to justify the inclusion of both in the same family or genus: a common and convenient practice is to add the termination *-ites* to a generic name instituted for recent (living) plants when it is applied to a fossil plant, or by some other slight modification of the name to show that the plant in question is not represented in the vegetation of the present day. The generic name *Equisetites* is substituted for *Equisetum*; *Magnoliaephyllum* is used instead of *Magnolia* for a leaf which probably belonged to an extinct species of that genus. A large number of fossil plants do not conform in their more important features—that is features which are believed to be criteria of relationship—to any of the subdivisions designed primarily for existing members of the plant-kingdom. For these extinct and less familiar types, the most interesting to students of evolution, special family- or group-names have been adopted. In the following list families and genera which include only extinct plants are underlined. The classification is deliberately incomplete. The brief descriptions should be amplified by reference to text-books. The table will enable readers to allocate to their respective positions in the plant-kingdom many of the genera mentioned in the description of the fossil floras.

THALLOPHYTA — The simplest and more primitive plants. The plant-body is a single cell, a filament, a surface of cells, or a cell-mass. With few exceptions, there is no differentiation into stem, root, and leaf; the body in many instances is a *thallus*.

ALGAE — Chlorophyll present: in many algae the green of the chlorophyll is masked by the presence of red or brown pigments.

Cyanophyceae (Myxophyceae or Schizophyceae): Blue-Green Algae See Chapter VII. The colour is by no means always blue-green. Many of these algae are enclosed in a firm and almost leathery sheath which is more resistant to decay than the cells within it.

Chlorophyceae: Green Algae Certain calcareous genera are mentioned and briefly described in Chapter VIII. Most of the calcareous genera belong to the family Dasycladaceae, a subdivision of the section of Green Algae known as the Siphoneae, characterized by a plant-body (thallus) which is not divided into cells; it is unseptate and tubular, usually branched, and the protoplasm contains numerous nuclei, whereas in cellular plants each cell has usually one nucleus. (Figs. 83, 113.)

Rhodophyceae: Red Algae The Red Algae include calcareous seaweeds, such as *Lithothamnium* and *Goniolithon* (fig. 25), in which the plant-body is coral-like and hard and the walls of the minute cells form a stony framework. Extinct algae of similar form include Solenopora and allied genera. (See Chapter VIII.)

Phaeophyceae: Brown Algae Some resemble higher plants in the differentiation of the body into stem, "leaf", and holdfasts; also in their complex internal structure. The seaweed *Macrocystis* exceeds in length the height of the tallest trees.

Diatomaceae: Diatoms The membrane is impregnated with silica. Diatoms occur in the sea from Arctic to Antarctic regions and are abundant in freshwater lakes and pools, on peaty ground, etc.

FUNGI

BACTERIA

Chlorophyll absent: with the exception of some Bacteria, which can build up their bodies from inorganic sources, all Fungi are parasitic or saprophytic.

CHAROPHYTA This very ancient class, the best known example of which is *Chara* (stonewort), occupies an isolated position. The plants live in fresh water or brackish water: the slender stem reaches a height of 50–60 cm. and bears whorls, or circles, of delicate branchlets.

BRYOPHYTA In several genera the plant-body resembles in its simple form that of the typical Thallophyta: in most there is a slender leaf-bearing stem. Some mosses and a few liverworts have an axial strand of elongated conducting cells, but in none is there any true vascular tissue consisting of typical xylem (wood) and phloem.

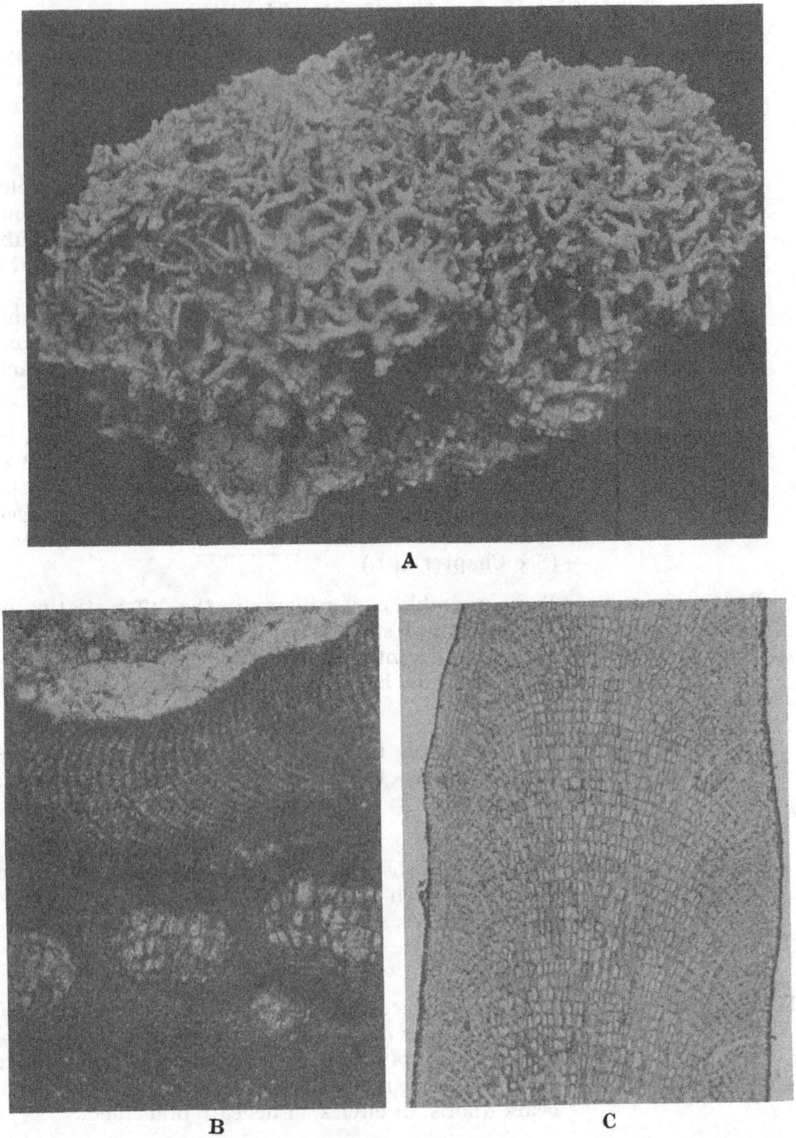

FIG. 25. Calcareous Algae superficially resembling coral. A. *Goniolithon*—from Tucker Island, Bermuda. Slightly reduced. B. A section through part of a fossil *Lithothamnium* from Tertiary beds (Oligocene) in the Panama zone. × 75. The lighter areas show the positions occupied by reproductive organs. (After Howe.) C. A section through a piece of *Goniolithon*, after the removal of the lime, showing the cells. × 50. (These photographs were obtained by courtesy of the New York Botanical Garden.)

BRYOPHYTA *cont.*	
HEPATICAE: Liverworts MUSCINEAE: Mosses	The fossil records are very meagre; it has, however, recently been shown that representatives of the Bryophyta existed in the latter part of the Palaeozoic era. (Walton (25), (28[4]).)
PTERIDOPHYTA[1]	With a very few exceptions Pteridophytes have true roots in place of the simpler absorbing cells (rhizoids) of the lower plants. True vascular, or conducting, strands are present. Some produce spores of one size only; others bear two kinds, microspores and megaspores, as in the still higher groups, and are spoken of as heterosporous.
FILICALES	
EUFILICINEAE:[2] True ferns	Characterized by the large and usually branched leaves (fronds). The main divisions of the frond are known as pinnae; the smaller (leaflets) as pinnules.
Ophioglossaceae	E.g. *Ophioglossum*, the adder's tongue fern.
Marattiaceae	A few, mainly tropical, genera.
Psaronieae	Palaeozoic ferns agreeing generally in structure with the Marattiaceae, but distinguished by well-marked peculiarities. Psaronius, a Carboniferous and Permian genus.
Osmundaceae	The widely distributed *Osmunda* (e.g. *Osmunda regalis*, the royal fern) and the South African and Australasian *Todea* are the best known genera.
Schizaeaceae	No representatives in Europe: some in North America and in the southern hemisphere.
Gleicheniaceae	Very common in the southern hemisphere. The forked fronds are a characteristic feature of *Gleichenia*. (Fig. 26.)
Matoniaceae	One living genus, *Matonia* (fig. 94), confined to the Malayan region.
Hymenophyllaceae: Filmy ferns	E.g. the widely distributed *Hymenophyllum tunbridgense*.
Dicksoniaceae	Including *Thyrsopteris* (fig. 97), a genus confined to the island of Juan Fernandez.
Cyatheaceae Subdivided into smaller families [Bower (26)]	None in Europe. Several tree-ferns in tropical and south temperate regions.

[1] For a recent classification see Fritsch, K. (29).

[2] A few families are omitted. For a full account of ferns see Bower (23–28); also Christ (10), especially for distribution; and Seward (22).

FIG. 26. *Gleichenia glauca*. Western Hill, Penang. 2500 ft. (Photo. Mr R. E. Holttum.)

PTERIDOPHYTA *cont.*

Dipteridaceae	A single Indo-Malayan genus, *Dipteris*. (Fig. 95.)
Polypodiaceae	The most abundant family in the northern hemisphere: some, e.g. the bracken fern (*Pteridium aquilinum*), are cosmopolitan.
COENOPTERIDEAE[1]	Carboniferous and Permian, and one Devonian representative. In some of these extinct ferns the stems grew in thickness by means of a cambium, a zone of permanently juvenile cells which added to the conducting tissue of the plant.
Botryopterideae Zygopterideae	For an account of these families see Scott (20); Seward (10); also Hirmer (27); Bower (23–28).
HYDROPTERIDEAE: Water ferns	Aquatic plants, probably derived from the true ferns.
Marsiliaceae	*Pilularia*, the pillwort, is a British genus; *Marsilia*, a tropical and temperate genus.
Salviniaceae	*Salvinia*, not a member of the British flora; occurs in some parts of Europe. *Azolla*, a tropical and subtropical genus, has become naturalized in Britain.
EQUISETALES (ARTICULATAE)	Articulated (jointed) stem and branches. Leaves and branches are borne at regular intervals, at the nodes, in whorls; each whorl is separated from the next by a bare portion of stem, the internode.
EQUISETACEAE: Horsetails	*Equisetum* (fig. 27): characterized by inconspicuous leaf-sheaths clasping the stem and divided into teeth on the upper margin. Some species almost cosmopolitan: all living species, one of which attains a height of 20–40 ft., are herbaceous.
CALAMARIEAE	Arborescent; much taller and thicker than *Equisetum* and differing in the possession of a cambium—a cylinder of permanently young cells which produced new layers of conducting tissue each growing season. Asterocalamites, Calamites, Asterophyllites, Annularia, Lobatannularia, etc. See Chapters XI and XII.
SPHENOPHYLLALES (ARTICULATAE)	Articulate. Differ from Equisetales in the anatomy of the stem and in the reproductive organs: whorls of wedge-shaped or deeply divided leaves. See Chapter XI.
Sphenophylleae	The leaf-bearing stems of Sphenophyllum superficially resemble those of *Galium aparine* (goose grass).

[1] Seward (10), p. 433: from κοινός = common or general, indicating the combination in these extinct ferns of characters now found in different families. The name Primofilices had been proposed by Dr Arber (06).

FIG. 27. Shoot of an *Equisetum* (*E. maximum*, a common horsetail) with whorled branches attached to the underground rhizome. Much reduced.

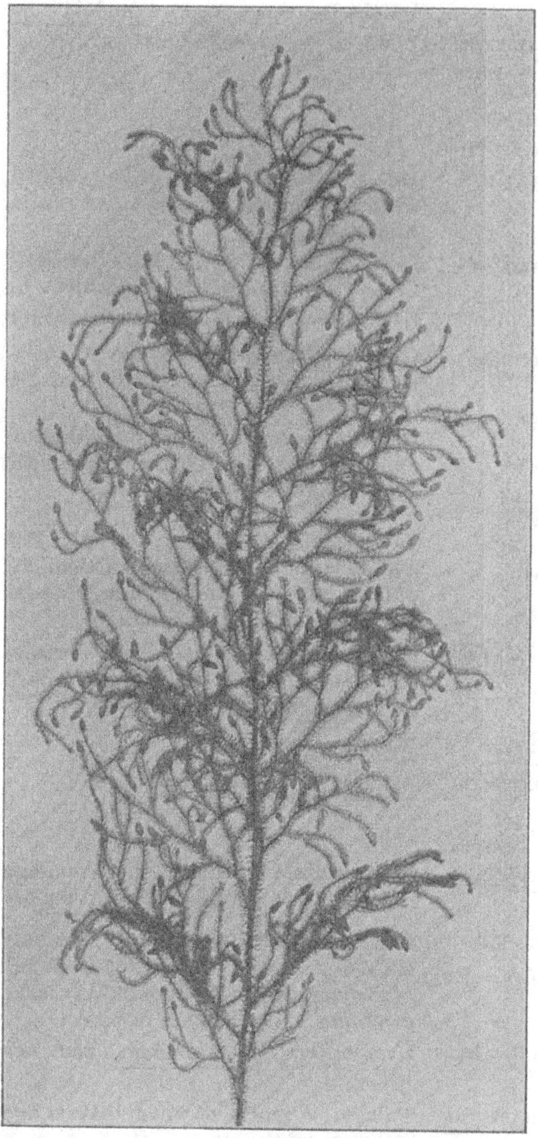

FIG. 28. *Lycopodium cernuum*, a common tropical club "moss". About ½ nat. size.

PTERIDOPHYTA *cont.*

ARTICULATAE INCERTAE SEDIS. Genera for which the group-name Protoarticulatae has been instituted	E.g. the Devonian genera Pseudobornia, Hyenia, and Calamophyton. See Chapter IX.
LYCOPODIALES	Differ from Filicales in having crowded, small leaves which are spirally disposed and not in whorls as in Equisetales.
LYCOPODIACEAE	*Lycopodium* (fig. 28). Club "mosses", stag's horn "moss"; *Selaginella, Phylloglossum.*
Lepidodendreae	Arborescent lycopods; most species were trees.
Lepidospermeae	A group founded on seed-like organs borne on cones belonging to extinct lycopodiaceous plants.
ISOETALES	
ISOETACEAE: (Quillworts)	*Isoetes* (quillwort) grows in freshwater lakes; characterized by its short, tuberous stem with a cambium[1] and the tuft of quill-like leaves.
PSILOTALES	
PSILOTACEAE	Two genera, *Psilotum* (fig. 29) and *Tmesipteris*, both rootless; they usually occur as epiphytes. *Psilotum* widely spread in the Tropics; *Tmesipteris* native of Australia and New Zealand.
PSILOPHYTALES	Devonian plants comparable in certain respects to the Psilotaceae. E.g. Psilophyton, Pseudosporochnus, and several others. See Chapter IX.
Rhyniaceae	Rhynia, Hornea, etc.
Asteroxylaceae	Asteroxylon, Schizopodium.
PLANTS OF UNCERTAIN AFFINITY	
Cladoxyleae	Characterized by a complex stem-anatomy. For a description of the genus Cladoxylon see Chapter IX.
PTERIDOSPERMOPHYTA	Many members of this group have the habit of ferns, but differ in the possession of seeds and in the power of forming secondary conducting tissue by means of a cambium. See Chapters IX–XIII.
Lyginodendreae	Lyginopteris, Heterangium, and other Palaeozoic genera.
Calamopityeae	Palaeozoic genera of which little is known except the structure of the stems.
Medulloseae	Large plants with woody stems and fern-like foliage bearing seeds.

[1] See above, under *Calamarieae.*

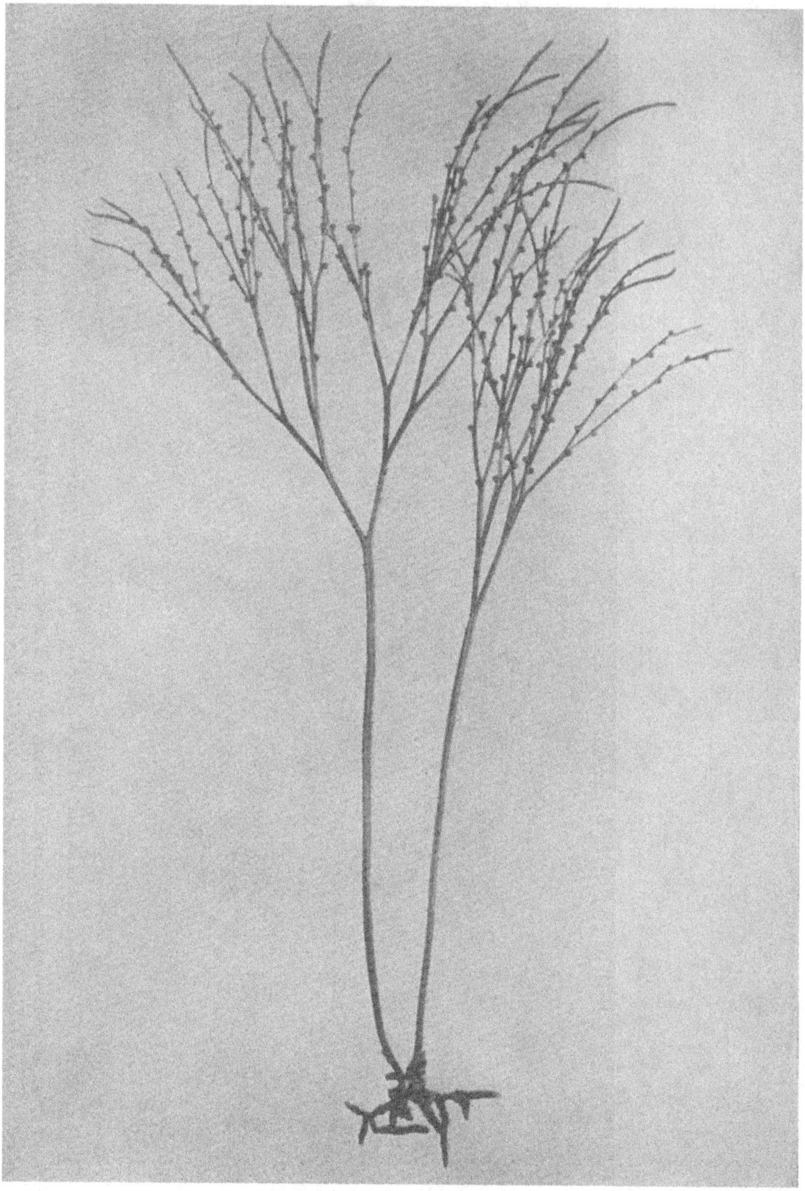

FIG. 29. *Psilotum triquetrum*. Slightly reduced.

Fig. 30. *Microcycas calocoma*; a Cuban cycad. The stem is about 10 ft. high. (Photograph supplied by Prof. C. J. Chamberlain.)

GYMNOSPERMAE

CORDAITALES	Arborescent Palaeozoic plants.
Poroxyleae Cordaiteae Pityeae	For an account of these and other fossil gymnosperms see Scott (23); Seward (17), (19); also Chapter XII. See also Walton (23[2]) for a description of the remarkable Triassic genus Rhexoxylon.
CYCADOPHYTA	This designation is convenient for certain fossil stems and leaves which cannot be referred with certainty to a more precise position in the class; it includes also the living cycads.
CYCADALES	The few living genera are mostly tropical; in habit many resemble palms or tree ferns. See Chamberlain (19).
Cycadaceae: Cycads	There are nine genera, the most widely distributed of which is *Cycas*; *Microcycas* (fig. 30) is confined to Cuba. The other genera are *Encephalartos* and *Stangeria* (Africa); *Dioon* and *Ceratozamia* (Mexico); *Zamia* (Florida, South and Central America); *Macrozamia* and *Bowenia* (Australia). (Seward (17).)
BENNETTITALES	In habit and in stem-structure these plants agree closely with cycads, but differ very widely from the Cycadales in the reproductive organs.
Bennettiteae Williamsonieae	These families reached their maximum development in the Jurassic and the older Cretaceous floras.
GINKGOALES	*Ginkgo biloba*, the maidenhair tree, is the sole survivor of an ancient and widely distributed group (figs. 31, 135). Extinct genera—Saportaea, Baiera, Phoenicopsis, etc.
Ginkgoaceae	*Ginkgo* (fig. 31). Fossil leaves referred to Ginkgoites are many of them no doubt generically identical with the existing *Ginkgo*.
CONIFERALES	The majority bear cones, e.g. pines, cedars, larches; several genera bear the seeds singly or not in cones, e.g. *Taxus*, the yew. See Seward (19), p. 106.
Araucarineae	*Araucaria* (monkey puzzle, Norfolk Island "pine", etc.) and *Agathis* (e.g. kauri pine of New Zealand). The family is now restricted to South America, Australia, Malaya, etc.
Cupressineae	Leaves small and whorled. Cypresses, junipers, *Libocedrus*, etc.
Sciadopitineae	A single genus: *Sciadopitys*, the umbrella "pine" of Japan.

FIG. 31. One of the oldest maidenhair trees (*Ginkgo biloba*) in Europe: the Botanic Garden, Utrecht. (Photograph supplied by Prof. Went.)

GYMNOSPERMAE *cont.*	
Sequoiineae	Represented by the genus *Sequoia*; the mammoth tree and redwood of the Pacific coast of California.
Callitrineae	*Callitris* (Australia, etc.), *Tetraclinis* (North Africa), *Widdringtonia* (Equatorial and South Africa), *Actinostrobus* (Western Australia). Leaves small and whorled.
Abietineae	The largest and most widely spread family in the northern hemisphere: pines, cedars, firs, larches, etc.
Podocarpineae	*Podocarpus*, a common southern genus; *Dacrydium* and some other southern conifers.
Taxineae	*Taxus* (yews); *Torreya*; *Cephalotaxus*.
Taxodineae	*Cryptomeria*; *Taxodium* (North America); *Glyptostrobus* (China); *Athrotaxis* (Tasmania); etc.
GNETALES	A group represented by three living genera, *Gnetum*, *Ephedra* and *Welwitschia*, which in certain features shows the closest resemblance to the Angiosperms. The fossil record is very meagre. Pearson (29).
CAYTONIALES	This group[1] includes two Jurassic genera, Caytonia and Gristhorpia, which alone among Pre-Cretaceous plants have afforded evidence of the possession of the angiosperm type of fruit, i.e. seeds enclosed in a seed-vessel provided with a stigma.
ANGIOSPERMAE: Flowering plants	The ovules (which become the seeds) enclosed in a seed-case or carpel. This is now the dominant race. For accounts of the families see Wettstein (23–24); Rendle (04), (25); Hutchinson (26); Parkin (27).
MONOCOTYLEDONEAE	Grasses, palms, and many other plants: leaves usually parallel-veined. Seedlings with one seed-leaf (cotyledon). The stems and roots except in a very few instances, e.g. the venerable dragon trees (*Dracaena*) of Teneriffe, do not grow in thickness year by year; there is no cambium in the conducting tissue.
DICOTYLEDONEAE	Trees, shrubs, and herbs: leaves usually net-veined. Seedlings with two cotyledons. The stems of dicotyledonous trees like those of conifers grow in girth by annual increments of wood and phloem (phloem, the outer conducting tissue in which much of the food is transported, is included with cork in the term bark as used in a wide sense) produced by a cambium.

[1] Thomas, H. H. (25).

CHAPTER VII

THE EARLIEST RECORDS OF PLANT-LIFE: THE PRE-CAMBRIAN ERA

The mightiest moments pass uncalendared. *Thomas Hardy*

SEVERAL different names have been employed for the earliest chapters of geological history. For present purposes the term Pre-Cambrian is chosen for the long succession of ages preceding the Cambrian period, ages which, it is believed, exceeded in duration the whole of the time represented by the rest of geological history. The term Azoic (without life), a designation which stirs the imagination, though inappropriate because it implies more than can be proved, has been used for the older part of the Pre-Cambrian era. Similarly the term Eozoic is sometimes applied to an age with which it is attractive to associate the "dawn of life"—the first beating of a feeble pulse. More recently it has been proposed to call the rocks of the primeval world Pampalaeozoic.[1] Another term that is often adopted for the latter part of the Pre-Cambrian era is Proterozoic: this connotes the existence of primitive organisms in contrast to the supposed absence of life in the still older Archaean stage. A classification based partly on negative evidence and in part on the insecure foundation of what is assumed to have been is unsatisfactory. We know that in the Cambrian seas life was not only abundant, but we are told that the animals were "intensely modern" and belonged to "the same order of nature as that which prevails at the present day". It is almost true to say that as we descend from Cambrian to Pre-Cambrian the thread of life is snapped; records are lacking, or such as occur cannot be interpreted with any degree of certainty. This is not wholly true: the later Pre-Cambrian rocks have yielded a few fossil animals, and, as we shall see later, there are good reasons for assuming the existence of plants in the seas. We have to admit that definite dividing lines are figments of the imagination; a world that was lifeless is not sharply delimited from a world inhabited by the humbler forms of life, nor from an epoch in which were evolved less simple organisms foreshadowing the numerous, parallel lines of progress stretching more or less continuously through the ages.

[1] Gregory and Barrett (27).

The Origin of Life

Geology tells us nothing of the origin of life. Cambrian rocks reveal seas teeming with living creatures: Pre-Cambrian rocks give us glimpses of a few precursors, but they are very few; and we are left with the conviction that the roots of the tree of life penetrate deep down into an apparently barren crust, how deep we cannot say. The almost complete absence of life in the enormous thickness of Pre-Cambrian rocks has often been attributed to the obliteration of any fossils there may have been by the destructive action of heat; but this cannot be the true explanation in regions where Pre-Cambrian sediments are practically unaltered. In the British Isles, in America, Africa, China, India, and in many other parts of the world the strata are equally barren. It is certain that if the Pre-Cambrian beds had been formed as ordinary marine deposits, like those of the succeeding Cambrian period, they would contain records of life unless the inhabitants of the sea, both animal and plant, were too fragile or unprotected for preservation. It has been suggested that the primeval oceans were practically limeless and that it was not until a later age, when river-borne minerals provided the material, that certain kinds of animals and plants acquired the habit of wearing calcareous armour. There are reasons for supposing that some of the Pre-Cambrian rocks were formed on land, not under water, and in an arid climate. The remarkable freshness of the felspars, minerals abundant in the Torridonian grits of the north-west Highlands of Scotland, points to very rapid, mechanical disintegration rather than by decomposition due to chemical action: "it would almost seem to indicate cold climatic conditions accompanied by periodic floods of great intensity". The occurrence of similar undecomposed pieces of felspar crystals is recorded from the Antarctic continent at the present day. Many geologists have expressed the opinion that during a part at least of the Pre-Cambrian era the climate was dry: pebbles from the Torridonian grits of Scotland and from strata of similar age in Sweden[1] often show smooth sloping faces meeting in prominent ridges, a feature characteristic of rocks exposed to sand-blasts. Similar pebbles may be seen in places other than deserts: in an old shingle bed exposed on the beach a short distance north of Corrie in the Isle of Arran (western Scotland) many of the pebbles, which project above the surface

[1] Högbom (10). See also Walther (09).

of the cementing matrix, have acquired the desert form through exposure to wind-driven sand. The frequent occurrence of red sediments[1] in different regions is adduced as additional evidence of drought.

We may picture a land bare of vegetation, mountain sides exposed to the destructive influence of sharply contrasted day and night temperatures, avalanches of rock, restlessly drifting sand dunes, floods following torrential rains spreading the disintegrated rock and wind-blown sand as fans and sheets in the valleys. If the conditions were even approximately such as these, it is hardly surprising that we search in vain for the remains of a land-vegetation. Deserts probably covered wide tracts of ground; and, as we have seen, some regions lay for long periods under an ice-field: but these conditions were by no means universal on Pre-Cambrian continents. The unfossiliferous state of vast piles of sedimentary strata is significant, and it would seem improbable that if there had been a vegetation, some relics of it should not have been preserved.

"A time there was when Life had never been;
A time will be, it will have passed away."

Employing the name Pre-Cambrian for the whole of the period antecedent to the Cambrian, it is convenient to give expression to the well-established conclusions of geologists by the recognition of two major groups within the Pre-Cambrian era. There is an older group in which granitic rocks predominate: with them are associated finer grained, crystalline rocks of volcanic origin and some conglomerates and other sedimentary material: for this complex the term Archaean is used. The overlying series consists of many thousands of feet of sedimentary rocks and a small proportion of crystalline igneous rocks: this is sometimes called the Algonkian series (the name of a linguistic aboriginal group widely spread in North America). Another term for this upper series is Proterozoic. The Archaean series cannot be regarded as the foundation-stones of the earth. In the main the two-fold division holds good. It is important to note that the Algonkian system does not consist of a regular uninterrupted set of beds; there are many gaps in the sequence, unconformities indicating a succession of earth-movements on a grand scale, and a lapse of time beyond our power of comprehension. A comparative study of the Archaean and Algon-

[1] Barrell (08); Walther (12); Tomlinson (16); Case (19); Twenhofel, W. H., and other authors (26).

kian rocks throughout the world has demonstrated that they have a wider distribution over the earth's crust than those of any subsequent age. It would seem that at least one-fifth of the whole land-surface consists of Pre-Cambrian rocks; and no doubt were it possible to bore through the masses of younger strata we should find deep down a basal pedestal of coarsely crystalline igneous foundations. Not only is the Pre-Cambrian era distinguished from all succeeding eras by the greater geographical range of its rocks, but there is reason to believe that at no subsequent stage in geological history were the revolutions of the crust on so grand a scale.

What was the appearance of the face of the earth at the beginning of geological history when the crust became cool enough to permit of the condensation of aqueous vapour? Was there a world-sea encompassing the whole earth, or were there, as now, continental blocks and sundering oceans? It has been suggested that the first scene of all was a waste of deep waters, too deep for rays of light to reach the submerged and wrinkled crust below. In this world-ocean,[1] assuming that it existed, protoplasm had its birth: in the course of time, as the infinitely small acquired the power of growth, ultra-microscopic particles of living matter rose to the dignity of free-floating single cells, the pioneers of the plant and animal worlds. As time passed, foldings of the earth's crust brought within reach of sunlight arched portions of the sea-floor; free-floating cells attached themselves to the rocks and progress in development proceeded under new conditions. Single units or colonies of cells became intimately united into filaments, cell-surfaces, and cell-masses: the principle of division of labour was established, and the evolution of multicellular plant-bodies was inaugurated. At a still later period portions of the crust emerged from the universal sea as islands, the forerunners of continents: plants born and nurtured in water, and which had reached a comparatively advanced stage of evolution, were lifted into a strange environment. As the land slowly rose the plants anchored to the rocks were for regular intervals, as the tides receded, exposed to the air; and as the uplift continued they were permanently deserted by the sea. It is conceivable that some of the translated plants successfully grappled with the new conditions and adjusted their mechanism to meet the changed circumstances. Thus, it may be, the advanced guard of terrestrial vegetation took

[1] A much less probable suggestion is that terrestrial life began on mountain tops (*Nature*, January 26, 1922, p. 107).

possession of its kingdom. In barest outline this is a conception of the origin of land plants elaborated with much ability and ingenuity by Dr Church[1] of Oxford.

The occasional occurrence of typical seaweeds in places which are rarely submerged at high tide enables us to realize the possibility of a gradual transformation of an aquatic into a terrestrial plant.

FIG. 32. *Fucus vesiculosus* var. *muscoides* with *Armeria maritima* (thrift) on Clare Island, Co. Mayo, Ireland. (Photo. Dr A. D. Cotton.)

Fig. 32 shows a colony of small seaweeds, described as a variety of one of the brown bladderwracks of our coasts (*Fucus vesiculosus* var. *muscoides*), established in a situation slightly above the mean high-tide level where the plants are flooded only at spring tides.[2] The flowering plant in the mat of *Fucus* is the thrift, *Armeria maritima*. It is interesting to find that Hugh Miller,[3] nearly eighty

[1] Church (19). For a criticism of Dr Church's views, see Fritsch, F. E. (21).
[2] Cotton (15).
[3] Miller, H. (49).

years ago, quoted a view propounded by the author of the *Vestiges of Creation* that marine algae had been transformed into the higher terrestrial plants; Miller mentioned a *Fucus* which he saw on a rocky beach where it remained uncovered by water for days together.

It is highly probable that life had its origin in the primeval sea: how and when we do not know. As the author of one of the more recent discussions on the mystery of life says: "Science, truly understood, is not the death, but the birth of mystery, awe, and reverence".[1] We have no adequate reason for assuming that the waters ever covered the earth. It is more likely that the first oceans filled hollows and deeps among scattered islands and continents. Water would appear to be a more favourable setting for the creative act than the bare surface of the world's first crust: a bare, treeless expanse of rock lacking the harmony of colour with which lichens now decorate the wildest landscape; bare of herbs and mosses; lifeless. To our restricted vision a world without human beings is difficult to picture; a much greater demand is made upon the imagination by a world in which inorganic forces operated alone without the interaction of matter and life.

Problematical and Organic Pre-Cambrian Remains

Leaving the realm of sheer speculation and the hopeless endeavour to trace the plant and animal worlds to their common starting-point, we pass to a general consideration of the evidence of plant-life, for the most part indirect evidence, which the Pre-Cambrian rocks have afforded.

The term Agnostozoic, though probably not intended to be taken seriously, has also been suggested[2] for the earliest chapter of geological history: it is a fitting comment on the terms Azoic, Eozoic, and Proterozoic which imply more than is known, and aptly describes the true state of our ignorance. We do not know when or how life began; we cannot measure the rate of the earlier stages of evolution, nor can we accept as proof of the existence of plants much of the evidence that has been adduced, and not infrequently presented with a confidence worthy of a better cause. During some part at least of the enormously long stretch of time comprised within the limits of the Pre-Cambrian era, life undoubtedly

[1] Donnan (28).

[2] Walther (19–22).

existed; but a critical examination of the records usually fails to give precision to this conclusion. The publication of a paper by the late Dr Walcott[1] on the discovery of some supposed algae in Algonkian rocks, in the words of another author, "gave the problem of life in the ancient formations a new aspect"[2]; though it may fairly be said that the correctness of his conclusions is open to question. Dr Walcott described several forms of rock-structure, which he believed to be of algal origin, from a series of freshwater limestones in Montana, in which they form reefs and banks through a thickness of several thousand feet of strata. For these structures he proposed generic names and adopted an admittedly artificial classification, because he found no clue to the various kinds of algae supposed to be responsible for their production. He did not describe the genera as so many different types comparable to the lime-secreting algae of later geological periods, several of which still exist in the seas of the present day: he attributed the structures to the activity of some primitive algae, and correlated the different patterns assumed by the calcareous masses with hypothetical differences in the unknown plants believed to be concerned in their formation.

Algae as Rock-builders

There is no doubt that certain kinds of algae have a share in the production of grains and pebbles of carbonate of lime: submerged plants obtain carbon from the carbon dioxide present in the water in which they live: when bicarbonate of calcium occurs in solution the removal of carbon dioxide causes the precipitation of the insoluble carbonate of lime. This brings us to a much-discussed topic which has produced considerable divergence of opinion, namely the part played by algae in forming the small calcareous bodies known as oolitic grains, because of their resemblance to the roe of fishes. These grains and pebbles are being formed in lakes to-day, and rocks of different ages are wholly or in part made up of such units. The grains have a concentric and often a superposed radial structure, a series of hollow shells usually of carbonate of lime encircling some central nucleus, a grain of sand or other foreign body. Sometimes the calcareous material occurs in cake-like masses, the so-called "water-biscuits"[3] of certain authors. Microscopic bodies of

[1] Walcott (14); Grout and Broderick (19).
[2] Moore (25).
[3] Clarke, J. M. (00).

algae are often found on the surface of the grains, and by dissolving the lime algal filaments may be released from the stony matrix. Some striking examples of calcareous biscuits have recently been described from South Australia by Sir Douglas Mawson:[1] they occur in abundance on low ground subject to inundation in the winter months and vary in size from tiny particles to discs 20 cm. in

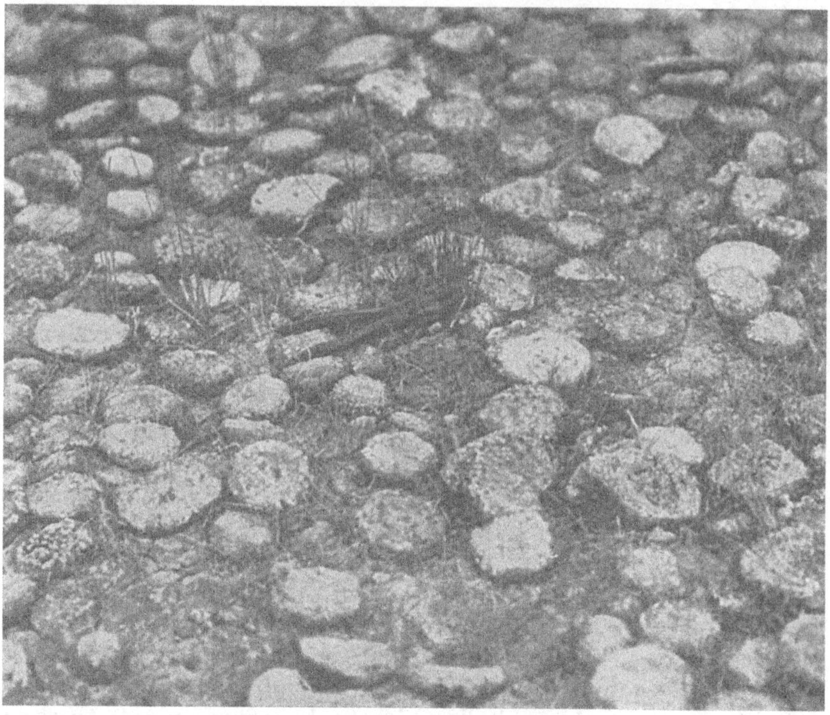

FIG. 33. Calcareous "biscuits" on the Biscuit Flat, near Robe, South Australia. The size is indicated by the pliers in the centre of the photograph. (Photo. Sir Douglas Mawson.)

diameter; some are thicker and resemble buns. Dried cells of Blue-Green Algae were found on the upper surface of the biscuits which was exposed to the light: on dissolving the lime an organic residue was left containing recognizable remains of well-known members of that family, *Gloeocapsa* and *Schizothrix*. If a biscuit is cut across, the calcareous material is seen to be porous and arranged in con-

[1] Mawson, D. (29).

centric layers; but it has no cellular framework. Fig. 33 shows the ground covered with these curious bun-like stones: similar structures have been described from localities in North America and from an Irish lake.[1]

The knowledge that algae are often intimately associated with calcareous grains now in process of deposition led to the examination of sections of oolitic grains in rocks, and to the discovery in oolites of minute and tortuous tubes (fig. 41, C.) which were identified as remains of algae. Before discussing the significance of these facts a brief reference is desirable to the nature of the plants now associated with calcareous deposits. The plants are for the most part members of a group of algae known as the Cyanophyceae or Blue-Green Algae. They occur almost all over the world and are represented by numerous forms, consisting of minute single cells, groups of cells or colonies, or of more closely connected cell-rows often enclosed in a tough, resistant sheath. I recall a striking scene between northern Australia and Java: broad lines of cinnamon-brown, stretching as far as the eye could see on the blue surface of the Pacific, consisting of floating millions of a Blue-Green Alga. The Cyanophyceae live under very diverse conditions; on moist rocks, as mat-like sheets on soil, in semi-arid localities, in hot springs, and within the Antarctic Circle. They are among the lowest and most primitive of plants. It may be that the Blue-Green Algae represent one of the earliest and most conservative branches of the plant-kingdom, a group tracing back its ancestry through almost the whole of geological time. On the other hand, there is much to be said in favour of the view that certain kinds of bacteria, of a type still represented in the bacterial flora of the present day, were among the first members of the plant-kingdom. These bacteria, though without the green colouring matter, chlorophyll, which enables green plants to seize the energy of sunlight, are able in certain circumstances to build up their bodies, utilizing as the driving force energy liberated in the course of chemical reactions.

It has been shown that members of the Cyanophyceae play a part in building up the calcareous travertine and the siliceous sinter in the hot springs of the Yellowstone region: they are factors in the precipitation of lime in many places and under many different forms. But it is equally certain that concentrically constructed grains and

[1] Barclay (86); see also Baumann (12); Lemoine (11); Rothpletz (13); Brown, T. C. (14); Roddy (15); Tuyl, van (16); Bucher (18).

larger masses of carbonate of lime are very often formed without the intervention of plants: the stones in animal bladders, the deposit of lime in a boiler fed with hard water are some of many instances. The small tubes found in some oolitic grains to which reference has already been made are known as species of the genus *Girvanella*[1] (fig. 41, C.): it may be that they are only sheaths of a Blue-Green Alga, the actual filaments of which have not been preserved. The more important question is, does the occasional presence of tubules or other plant-remains in oolitic grains afford satisfactory evidence of any causal connexion? It has been pointed out[2] that the tubular structures, which may be parts of plants, are found either on the surface, as in the calcareous biscuits of South Australia and in the peripheral portion of the calcareous pebbles, or they occur as a central nucleus; they are not always present throughout the whole mass. One cannot decide whether an alga in a calcareous grain had a share in the precipitation of the lime, or whether it was passively imprisoned in the mineral material deposited through inorganic forces. The presence of the plant may be accidental, not causal. My own view is that the occurrence of oolitic grains or larger pebbles built up in successive layers does not necessarily afford evidence of plant agency; algae, no doubt, often had a share in the formation of calcareous pebbles, but it would be going too far to say that their presence is essential.

The Interpretation of certain Rock-Structures: Organic or Inorganic?

In 1883 the name *Cryptozoon proliferum* was given by James Hall[3] to some large concentric structures, comparable to groups of compact cabbages 1 to 2 ft. in diameter, forming reef-like masses in a Cambrian limestone at Saratoga, New York. As seen in fig. 34, which is a photograph of a smoothed limestone, *Cryptozoon* is a series of concentric layers or coats forming single or intersecting groups. A specimen isolated from the matrix is shown in fig. 35: as seen in side-view (A) it resembles a large concretionary nodule; the concentric layers are clearly seen in surface-view (B). Hall's original specimen was subsequently re-figured by Walcott.[4]

[1] See Garwood (13) and Pia in Hirmer (27).

[2] Cayeux (09).

[3] Hall (47), (83).

[4] Walcott (14). See also Rothpletz (16); Rothpletz and Giesenhagen (22).

FIG. 34. Glaciated surface of a "*Cryptozoon* reef" in the Cambrian limestone, near Saratoga, New York. About $\frac{1}{10}$ nat. size. (From a photograph supplied by Dr J. M. Clarke.)

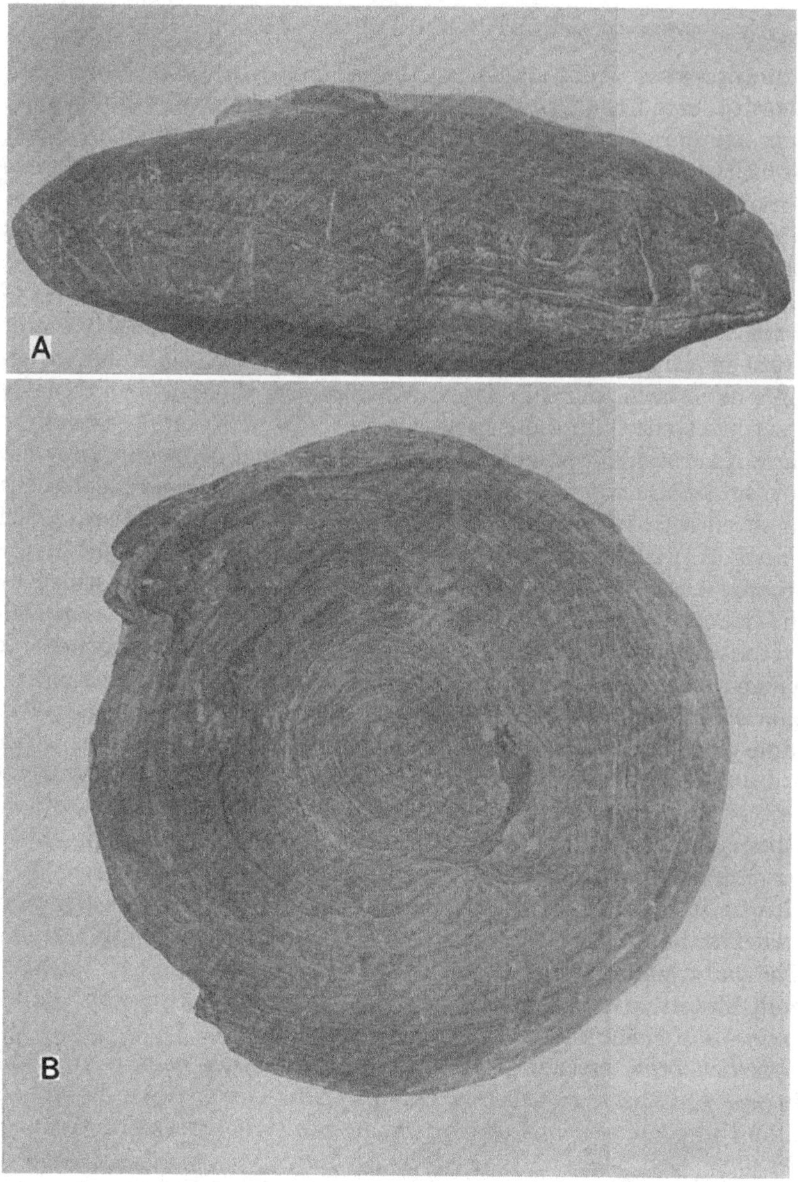

FIG. 35. *Cryptozoon*. Side-view (A) and surface-view (B) of a specimen from Saratoga, in the British Museum. $\frac{1}{6}$ nat. size.

Structures essentially similar to the original *Cryptozoon* have been recorded from Pre-Cambrian rocks in Ontario, from the Gobi Desert, from northern Greenland, from Cambrian Rocks in Brittany,[1] and from other parts of the world. A specimen of the same type has been found in South Australia. Dr Wieland, who described examples from Cambrian strata in Pennsylvania, suggested *Cryptophycus*[2] as a more appropriate name since it definitely implies an algal nature. On the other hand, Prof. Holtedahl,[3] who records the occurrence of *Cryptozoon* in Cambrian rocks in Norway, in Lower Palaeozoic strata in Ellesmere Land and elsewhere, states that similar structures have been found in Triassic strata. Dr Wieland has recently described some Cretaceous rock-structures from Wyoming under the name *Phormidioidea*[4] which he regards as algal in origin: they are not unlike the much older *Cryptozoon*. This author advocates the employment of Kalkowsky's term *Stromatolith*.[5] The general belief among American geologists and several European authors in the organic origin of *Cryptozoon* is, I venture to think, not justified by the facts. The *Cryptozoon* structures differ essentially from Calcareous Algae, such as *Lithothamnium* (fig. 25) and other genera, to which reference is made in a later chapter, in the absence of any characters suggesting a cellular framework. They are precisely the same in their series of concentric shells as many concretions which are universally assigned to purely inorganic agencies. They agree closely with the Australian calcareous biscuits described by Sir Douglas Mawson (fig. 33) which he believes owe their origin to the precipitation of lime by the activity of Blue-Green Algae. It is clearly impossible to maintain that all such concentrically constructed bodies are even in part attributable to algal activity. The only connecting link with plants is the possibility that algae contributed to the deposition of the carbonate of lime by the extraction of carbon dioxide from the water.[6]

It has been asserted that cells and chains of cells comparable in size and shape to those of existing Blue-Green Algae have been found in some sections of *Cryptozoon*-like structures. The term cell

[1] Bigot (25). [2] Wieland (14). [3] Holtedahl (19), (21).
[4] Wieland (30). [5] Kalkowsky (08).
[6] After this chapter was written a paper was received from Dr Høeg (29) of exceptional interest and importance on the origin of structures in limestone closely resembling *Cryptozoon*. The author describes a Stromatolith of post-glacial (Quaternary) age from south-eastern Norway which he regards as in all probability marine in origin and formed of carbonate of lime precipitated by Blue-Green Algae or Bacteria.

may be correctly used, but one would like to have evidence more convincing than the photographs and drawings which have so far been published. Microscopic spherical particles are not necessarily organic in origin. If *Cryptozoon* is not an alga, and if there is no adequate reason for assuming that algae had anything to do with its production, how was it formed? In practice it is not always possible to distinguish organic forms from artifacts.

Experiments made by Dr Liesegang[1] demonstrate the important bearing of diffusion phenomena upon the general question of rock-structure. He found that if a coagulated colloidal material, a gel, contains a substance in solution, and a second solvent capable of reacting with the former is allowed to diffuse into it, reaction takes place, but not continuously; with the result that the product is deposited in layers separated by apparently clear intervals. If a solution of sodium carbonate is added to a test-tube partly filled with 1 per cent. agar-gelatine containing calcium chloride, calcium carbonate is deposited in a succession of rings. The Liesegang phenomena can be demonstrated in many other ways: if a large drop of 20–30 per cent. solution of silver nitrate is placed in the centre of a plate of gelatine containing a small amount of potassium dichromate, circles of silver chromate alternating with circles free from precipitate are gradually produced as the silver nitrate diffuses through the jelly (fig. 36). Though the Liesegang figures are most easily demonstrated in a medium such as gelatine, a jelly is not essential.[2] A deposit of a colloidal calcareous mud on the floor of a sea might provide conditions favourable to the formation of concentric shells of carbonate of lime and the ultimate development of masses constructed on the plan of *Cryptozoon*. The meaning of the "reaction pulses" expressed in the Liesegang phenomena has not been fully elucidated. It is in the highest degree probable that Liesegang does not exaggerate the geological importance of the phenomena bearing his name. Cyanophyceae or similar primitive algae may have flourished in Pre-Cambrian seas and inland lakes; but to regard these hypothetical plants as the creators of reefs of *Cryptozoon* and allied structures is to make a demand upon the imagination inconsistent with Wordsworth's definition of that quality as "reason in its most exalted mood".

[1] Liesegang (13), (15); Hatschek (19), (25); Hedges and Myers (26); Lloyd and Moravek (28).

[2] See *Nature*, February 19, 1927, p. 267.

It is hardly necessary to consider in detail the various types of rock-structure from Pre-Cambrian beds to which Walcott[1] gave generic names. The genus *Newlandia* is characterized by concentric rings rather wider apart than those of *Cryptozoon*. Some of the specimens illustrated in his paper, e.g. *Greysonia* and *Copperia* have a more tubular structure. The genus *Collenia* is represented by

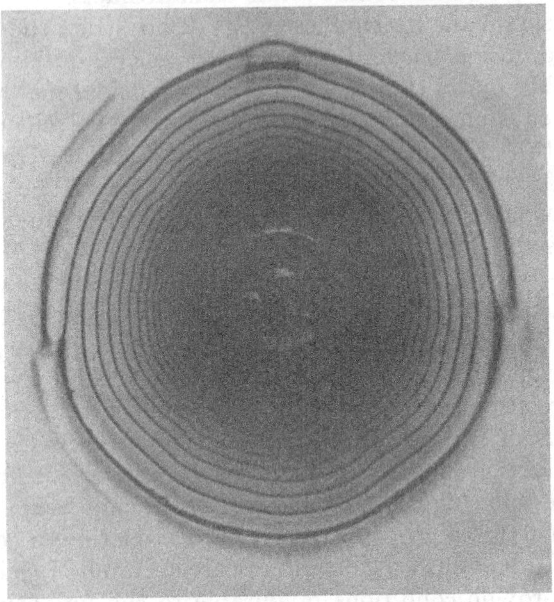

FIG. 36. Liesegang's rings: silver chromate. (Photo. Mr G. E. Briggs.)

irregular dome-shaped, lamellar bodies similar to *Cryptozoon*. In a very few examples the residue left after treating the rock with acid revealed the presence of a small number of minute cell-like structures, the organic nature of which cannot be said to have been established. If they are cells there is no proof that they had any share in the production of the enclosing matrix. The occurrence of a great variety of concretions has long been recognized as a characteristic feature of the yellow, magnesian limestones of Permian age which form cliffs on the coast of Durham: as Adam Sedgwick said many years ago, there is no end to the forms assumed by the

[1] Walcott (14).

dolomitic structures. My attention was drawn by Mr W. N. Edwards to the striking resemblance of many of the magnesian limestone concretions to Walcott's genera. Prof. Holtedahl,[1] who was independently struck by this resemblance, published some photographs of the Permian concretions indistinguishable in any essential features from illustrations in Walcott's paper. There would seem to be no reason to suspect the co-operation of algae in the formation of the Permian concretions.

From Pre-Cambrian dolomites in Finland Dr Metzger[2] described some branched, tubular stem-like forms, 0·5 mm. in diameter, which he named *Carelozoon jatulicum*: an examination of specimens at Helsingfors leads me to think that the author of the species is correct in regarding it as an animal fossil. Some small phosphatic nodules recorded from Torridonian rocks in north-west Scotland[3] are probably organic; but more cannot be said. The genus *Atikokania*,[4] founded on concentrically and in part radially constructed specimens from Pre-Cambrian rocks of Ontario, is described as probably spongioid: it may be included in the category of inorganic structures to which *Cryptozoon* belongs.

There are many records of supposed Pre-Cambrian plants, but none of them, in my opinion, afford absolutely convincing evidence of actual plant-structure. Dr Kräusel[5] recently described some carbonaceous material obtained from a boring made in search of coal near Prague which shows vegetable tissue: the weak point is that the Pre-Cambrian age of the fossil, *Archaeoxylon Krasseri*, has not been proved. The tissue elements are compared by Kräusel with the tracheids of a conifer; but the structure is too imperfect to admit of any satisfactory determination.

Geological literature contains many allusions to indirect evidence of plant-life during the latter part of the Pre-Cambrian era. Graphite, the most important constituent of black lead pencils, in its chemically pure form consists wholly of carbon.[6] Graphite occurs in nature in igneous and sedimentary rocks: some is unquestionably inorganic in origin, for example the rich stores in crystalline rocks in Ceylon. Its definition as "the last term of the coal series"[7] predisposes us to regard graphite as a product of plants; but such interpretation can be applied only very partially. Layers of graphite, 3–13 ft. thick,

[1] Holtedahl (21). [2] Metzger (24). [3] Peach and Horne (07).
[4] Walcott (12). [5] Kräusel (24). [6] Frauenfelder (24).
[7] Walcott (99).

occur in the district of Lake George, New York, interbedded with Pre-Cambrian sediments which have been called "a fossil coal bed".[1] Prof. Sederholm[2] considers that a bed of almost pure carbon of Pre-Cambrian age in Finland is certainly organic in origin: whether plant or animal we cannot say. Other examples might be quoted, but from none do we learn what we most wish to know.

Reference may be made here to certain problematical bodies figured by Sederholm[3] as *Corycium enigmaticum* from finely laminated, Pre-Cambrian rocks in Finland. The fossils occur as slightly compressed sacks (fig. 37, A–D) with clearly defined boundary walls which, on magnification, show the presence of much carbonaceous matter (fig. 37, F). The specimen reproduced in fig. E is a *Corycium*-like body in which the carbonaceous material is more abundant. Reference has previously been made to the glacial origin of the rock in which these fossils occur: it is probable, as Dr Sederholm suggests, that the nature of the matrix may be responsible for the preservation of the carbonized remains of what are probably plants, but whether complete and very simple plants or parts of plants it is impossible to say.

The presence of iron-ore in Pre-Cambrian rocks has been repeatedly mentioned as indicative of plant-life. It is well known that some bacteria and other micro-organisms cause the deposition of ferric hydroxide.[4] Ferruginous deposits may be the result of chemical or organic agents. We cannot with certainty determine whether or not the iron deposits in Pre-Cambrian rocks in North America, Brazil, India, and elsewhere owe their origin to bacterial action. It has been suggested that the iron formation in the Hudson's Bay[5] district is organic in origin: it may be; but that is as far as we can go. Rocks may owe their richness in iron or carbon to the action of bacteria despite the fact that no actual cells are revealed. Similarly a rock composed mainly of carbonate of lime, limestones, or dolomites (formed of the double carbonate of calcium and magnesium) without any vestige of plant-fossils may be built up of minute calcareous grains deposited by the action of bacteria. Calcareous deposits now being formed on the Great Bahama Bank[6] afford an interesting example of rock-building which helps us to

[1] Walcott (99). [2] Sederholm (99).
[3] Sederholm (25) with references. See also Sederholm (10).
[4] Ellis (19).
[5] Moore (18). See also Harder (19) and Pia (28).
[6] Black, M. (30); Drew (14).

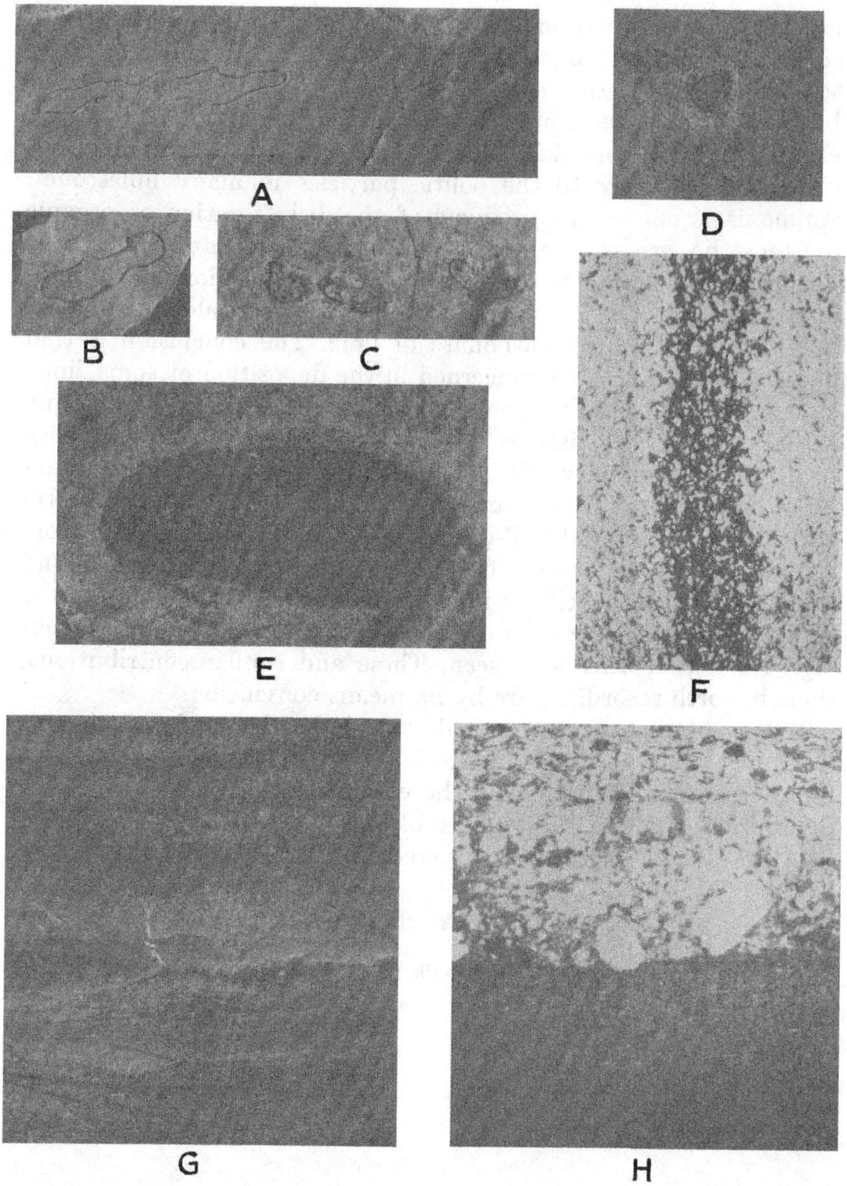

FIG. 37. *Corycium enigmaticum*. A, B, D *ca.* $\frac{1}{6}$ nat. size. C. *ca.* $\frac{1}{4}$ nat. size. E. *Corycium*-like fossil. F. Magnified section of the wall of *Corycium*: Archaean, Finland. G. Finely laminated Archaean clay (Phyllite) showing varves. Finland. H. Magnified section showing winter and summer layers of the clay. (Photographs supplied by Dr Sederholm.)

understand the formation of some of the limestones of past ages from sheets of chalky ooze on the floor of shallow seas. A calcareous mud stretching over a wide area to the west of the Bahama Bank is believed to have been precipitated through the action of bacteria, especially *Bacterium calcis*.[1] The mud consists in part of small grains similar in structure to the oolitic particles in many limestones. Ammonia is one of the products of the disintegration of organic matter—the process we speak of as decay—effected by certain bacteria; the ammonia combines with carbon dioxide to form ammonium carbonate, and the action of this on calcium sulphate in sea water produces carbonate of lime. The conclusion is that bacteria may have been concerned in the deposition of some limestones. We can hardly expect to find in Pre-Cambrian rocks any actual proof of the existence of bacteria though it would be foolish to deny the possibility. It is claimed that sections of a Pre-Cambrian limestone from Montana show minute bodies similar in form and size to cells and cell-chains of existing *Micrococci*.[2] More recently it has been stated that micro-organisms have been found in sections of chert pebbles in a Pre-Cambrian conglomerate:[3] the supposed organic structures are described as worm-like and branching. No actual cells were seen. These and similar contributions, though worth recording, are by no means convincing.

It may be that a too-sceptical attitude has been adopted in this incomplete summary. My desire is to lay stress on the need of a more critical examination of the evidence which has led to the description of the earliest phase of geological history as an "Age of Algae"—algae with doubtful credentials:

> "Creatures borrowed and again conveyed
> From book to book—the shadows of a shade."

[1] Blackwalder (13). See also Pia (28).
[2] Walcott (15).
[3] Gruner (25).

CHAPTER VIII

THE EARLIER PALAEOZOIC SEAS THE CAMBRIAN, ORDOVICIAN AND SILURIAN PERIODS

Before the world had passed her time of youth. *Wordsworth*

IT is convenient, and indeed logical, to deal with the first three sections of the Palaeozoic era in one chapter. They mark a phase of geological history recorded in many thousand feet of sedimentary rocks interrupted in some regions by unconformities indicative of changing conditions, but not separated from one another by widespread discordance such as we associate with the greater revolutions in the earth's crust. In various parts of the world a well-defined unconformity marks the boundary between the Pre-Cambrian and Cambrian rocks; the last of the Pre-Cambrian crustal movements led to an extension of the continental areas far beyond their present limits. By slow degrees the sea encroached over portions of the continental margins; the marine invasion was punctuated by slow advances and recessions throughout the prolonged interval represented by the first three Palaeozoic periods. The boundary between the Pre-Cambrian and Cambrian ages finds expression also in the striking contrast between the almost complete barrenness of Pre-Cambrian rocks and the wealth of marine animals entombed in Cambrian sediments. The three periods under consideration form a connecting link between the latest stage of the preceding era, represented in some regions by strata believed to be continental in origin—that is strata formed either on land or in land-locked waters, and not in the ocean—and the Devonian period when, over wide areas, inland lakes replaced the open sea.

The Cambrian Period

Cambrian strata, as the name implies, are well developed in Wales: in Merionethshire they rise to the surface as the Harlech dome where the thickness of the beds is estimated at about 11,000 ft. Rocks containing fossils agreeing closely with those from North Wales occur in South Wales, in Shropshire, in the Malvern Hills, and elsewhere. In some places conglomerates indicate proximity to an old Pre-Cambrian shore: other sediments consist of sandstones and shales often altered by pressure and heat into quartzites and slates. The slates of Llanberis and other localities in North Wales are examples of the finer sediments of Cambrian seas in which the mineral particles acquired a parallel arrangement causing the rocks to split into clean-cut plates no thicker than a fairly stout piece of paper.

In North America the Pacific Ocean filled the Cordilleran trough, a broad depression passing through California to British Columbia, in which Cambrian and later sediments accumulated, to be upraised after long ages as the Rocky Mountains. On the other side of the continent a similar stretch of sea occupied the Appalachian trough where foundations of another mountain range were laid. Mount Robson (13,068 ft.) "the most majestic peak of the Canadian Rockies" is an impressive relic of a huge block of marine Cambrian and Ordovician strata (fig. 38) elevated from the Cordilleran sea.[1] At the beginning of the Cambrian period the Asiatic continent is believed to have been a featureless land. Cambrian rocks occur in north-west India and in the northern Himalayas, in Australia, in the Argentine, and in China; they demonstrate by their fossils that certain genera of marine creatures ranged from one end of the world to the other.

The Ordovician Period

The term Ordovician (from the Ordovici, an ancient British tribe) was applied by Prof. Lapworth to a series of marine sediments and volcanic rocks which had previously been included by some authors in the Silurian system, and by others in the Cambrian. In the British Isles Ordovician rocks cover a wider area than that occupied by the Cambrian. The Ordovician sea transgressed over part of the Cambrian land and on the sea-floor were deposited sandstones and shales; while in the deeper and clearer water beds of limestone, extending as far north as Arctic Alaska, were built up in part at least of the calcareous skeletons of animals and seaweeds. Lava was poured out from volcanic islands and interbedded with products of land erosion. In the English Lake District Ordovician strata attain a thickness of at least 30,000 ft., and it is to these beds, formed on a sea-floor and subsequently upraised as mountains, that much of the

[1] Walcott (13). See also Burling (23).

FIG. 38. Mount Robson, British Columbia; an upraised and eroded mass of Cambrian and Ordovician marine rocks. (From a photograph supplied by the late Dr Walcott.)

most attractive scenery of Lakeland[1] is due (fig. 39). Ordovician lavas and ashes form "the highest and most picturesque mountains of North Wales, culminating in the noble cone of Snowdon".[2] Rocks of this period occur in the Southern Uplands of Scotland; also in the Arctic regions, Scandinavia, Russia, in many other parts of Europe, in North and South America, China, the Himalayas, and Australia.

FIG. 39. Windermere in winter. The hills in the background consist mainly of Ordovician volcanic rocks. (After Marr: Cambridge Univ. Press.)

Much of the North Atlantic was, probably, land; the coast-line extended from the south-west to the north-east through part of Ireland and across Scotland to the North Sea. A considerable proportion of the British Isles was submerged: the greater part of Asia was dry land; and in northern Siberia beds of gypsum and salt point to the prevalence of a dry climate.

The Silurian Period

The Silurian period derives its name from the Silures, an ancient tribe inhabiting part of Wales and the English border where strata of this age are well represented. Geologists who still employ the name Silurian for all the strata between the Cambrian and Devonian periods speak of the Ordovician as Lower Silurian, and what is here

[1] Marr (16). [2] Geikie (97).

called Silurian they regard as Upper Silurian. In some regions Silurian beds rest conformably on Ordovician; in others there is a break in the sequence. But there was no far-reaching change in the earth's surface during the whole of the three-fold phase (Cambrian, Ordovician and Silurian). Silurian rocks are exposed in Shropshire, Herefordshire, and in parts of Wales where they consist of conglomerates and other shallow-water deposits formed near the coasts of Ordovician islands which were gradually invaded by the Silurian sea. Beds of limestone, built up of reef-forming corals and other organisms, occur in the English midlands. Marine strata with similar fossils occur in the south of Sweden, in Bornholm, the island of Gothland, the Baltic provinces, and on the Russian plain. Silurian rocks are well represented in many parts of central and southern Europe, in the State of New York, and over a wide area extending from the mouth of the Saint Lawrence River to the south-west and northwards well into the Arctic Circle. The waters of Niagara pour over the receding edge of a precipice of Silurian limestone. Beds of the same age occur in China, Australia, New Zealand, Brazil, and in North Africa. The occurrence of the same species of marine fossils in New Zealand, Europe, and North America is evidence of free intercommunication in Silurian seas.

From North America, through Greenland to the north of Scotland and Norway across the North Atlantic, there stretched a large continent, though there seems to have been a polar connexion between the seas of Europe and the interior of America. The ancient Tethys Sea spread over a large part of Europe and North America, and to the south lay a vast continent comprising a portion of Asia, Australia, and most of Africa. In contrast to the extensive outpourings of lava which characterized the Ordovician period in some parts of the world, the Silurian strata were deposited during a quiescent phase. In the words of Sir Archibald Geikie, "up to the close of the Silurian period the long history embodied in the rocks presents a constant succession of slowly sinking sea-floors. Wide tracts of ocean stretched over most of Europe, and across the shifting bottom, sand and mud washed from lands that have vanished, spread in an ever-accumulating pile. Now and then, some terrestrial movement of more than usual potency upraised this monotonous sea-bed, but the old conditions of ceaseless waste continued and fresh sheets of detritus were thrown down upon the broken-up heaps of older sediment. All through the vast cycles of time demonstrated by these accumulations of strata, generations of sea creatures came and went in long procession, leaving their relics amidst the ooze of the bottom".[1]

The long succession of fluctuating transgressions and recessions of the sea was at length brought to a close. Before the end of the Silurian period there were premonitory signs of an approaching revolution, a revolution which found its most intense expression in the upheaval of the Caledonian range to which reference was made in a former chapter. Far-reaching changes in the distribution of land and water were a consequence of the heavings of the crust: mountains rose from the sea, oceans contracted and continental conditions, practically unrecorded in the rocks since the latter part of the Pre-Cambrian era, were again established. Upper Silurian beds pass gradually into the basal strata of the Devonian period: typical marine sediments merge into the deposits of shallower seas, land-locked bays, and lagoons which contain some of the oldest known samples of terrestrial vegetation.

The Earliest Traces of Terrestrial Plants

The almost exclusively marine strata of the Cambrian, Ordovician and Silurian periods, and the absence of any satisfactory remains of undoubted terrestrial plants give an unnatural emphasis to what appears to be a sudden creation of a land flora revealed by the fossil plants in the continental Devonian sediments. Where light is most needed, where records would be of absorbing interest we find blank pages, only blurred lines here and there; we know practically nothing of the immediate precursors of the land vegetation. On the other hand, the Cambrian and other sediments of the Lower Palaeozoic Age have furnished an amazing variety of marine animals: almost all the groups of the lower animals known to us to-day are represented, and by forms that had reached a complexity and a degree of specialization far beyond our idea of primitive organisms.[2] From one block of Cambrian shale not more than 6 ft. by 40 ft. in area and 7 ft. thick in the Rocky Mountains Walcott obtained fifty-six genera. The crustacean *Marrella* (fig. 40), (allied to the Trilobites), named by Dr Walcott[3] after my old friend and teacher Prof. Marr of Cambridge, illustrates the excellent state of preservation of some of these Cambrian fossils. In examples of soft-bodied animals described by Walcott it is possible to detect the

[1] Geikie (97). [2] Macgregor (27). [3] Walcott (12).

form of the alimentary canal, the water-vascular system, and other organs. The Cambrian bivalve, *Lingula*, is a representative of a genus "still living in the sand-bars and mud-flats of Chesapeake Bay under conditions which have not effected any essential change in its structure since the time of the Lower Cambrian".[1] Before the close of the Silurian period vertebrates, represented by certain fishes, had become associated with the then dominant class of invertebrates. Faith in the fundamental principles of evolution necessitates the assumption that the Cambrian species were derived from Pre-Cambrian ancestors antedating by many millions of years their

FIG. 40. *Marrella*, a Cambrian crustacean from the Rocky Mountains. × 4. (Photograph supplied by the late Dr Walcott.)

Palaeozoic descendants. The more primitive forms were no doubt ill adapted for preservation: as Daly[2] and others imagine, the conditions may have been such as precluded the secretion of lime to form protective armour or resistant skeletal systems. It has also been suggested that a protective skeleton would be superfluous in the absence of predatory animals. The earliest marine organisms probably lived in the surface waters: at a later stage the discovery

[1] Brooks, W. K. (94).

[2] Daly (07), (12); Lane (08).

of the sea bottom led to a much more rapid development of higher types; competition became keener and new forms were evolved.

We have already seen that there are grounds for assigning to the Pre-Cambrian era a duration at least equal to that of the sum of all subsequent periods, and probably even longer. The threads of life are suddenly lost as we follow them back to the dawn of the Cambrian period: that they were deeply rooted in antiquity there can be no shadow of doubt. A few relics have been discovered; others will be found. Some of the Pre-Cambrian sediments were deposited in shallow water not far from the coastal margins, and in them one would expect to find waifs and strays of a land vegetation, if such existed.

The world-wide distribution of certain Palaeozoic animals and the occurrence of coral reefs in limestones within the Arctic Circle have induced authors to speak of a remarkably uniform climate. But we have seen that glacial conditions prevailed either in the form of local glaciers or as widespread ice-sheets during a portion of the Cambrian period. Moreover the occurrence of salt deposits in North America, Siberia, India and Persia demonstrates climatic conditions different from those which generally prevailed. We do not know enough of the relation of the earliest organisms to their physical environment to use them as trustworthy tests of climate. There is no valid reason to suppose the climate to have been either much more uniform than at present or, in the main, essentially different from that of later ages.

The plants from Cambrian, Ordovician and Silurian sediments throw very little light on the nature of the terrestrial vegetation. We cannot say whether the dearth of records means that the land was bare or supported only simple ancestors of future trees and herbs, or whether the conditions of deposition of the rocks were such as precluded the preservation of transported vegetable débris. It is possible that in Pre-Devonian ages the land had not been colonized by migrants from the ocean or by descendants of the primitive inhabitants of the pools and rocks: on the other hand, it is difficult to believe that the Devonian floras were entirely derived from highly organized marine algae which had solved the complex problems presented to them as the sea-floor was upraised during the Caledonian revolution. The probability is that before the end of the Silurian period, and perhaps at a much earlier date, the continents were sparsely clothed with relatively simple pioneers of a land flora. The nature of the sediments of the three periods we are considering is, we may assume, one of the chief explanations of the lack of data. Reference is made on a later page to a few fragmentary relics of plants which lived on land. It is rash to prophesy what the future will unveil, though one may venture to predict the discovery of comparatively advanced precursors of the Devonian vegetation.[1]

Algae: Living and Extinct

For the most part the plants obtained from the rocks of the three older Palaeozoic periods are members of the class Algae. Before attempting to summarize the information afforded by the fossils it may be helpful to give a brief account of the more striking features of some groups of algae to which reference is made in the sequel. The records of the rocks throw little light on the evolution of the algae as a whole. Remains of certain types are abundant throughout the greater part of the geological series; other types, exceedingly common to-day, are represented by very few fossils which can be identified with any degree of certainty. The predominance in the strata of certain groups must not be interpreted as evidence of their greater antiquity; it is most probably due to the fact that the majority of algae are delicate organisms, lacking the stronger and more efficiently protected tissues of terrestrial plants. Calcareous Algae and not the soft-bodied forms bulk largely as fossils. The term Calcareous alga is applied to those in which the cellular framework is impregnated with carbonate of lime and thus adapted to preservation. The groups of algae represented at the present day are: the Blue-Green Algae (Cyanophyceae) to which reference has already been made; the Green Algae (Chlorophyceae); the Red Algae (Rhodophyceae); the Brown Algae (Phaeophyceae). The deposition of carbonate of lime in small crystals on the surface of an alga may be seen in some members of all groups; but it is only those in which the plant-body is completely, or almost completely, encrusted with lime that are spoken of as Calcareous Algae. This designation is employed for several living genera of Green Algae, particularly members of the families Dasycladaceae[2] (or Siphoneae Verticillatae) and Codiaceae included in the larger division Siphonales, a group

[1] Scott (22).

[2] For descriptions and illustrations, see Pia (20), also the section on Algae by this author in Hirmer (27). For records of Cambrian Siphoneae see Lorenz (04).

in which the plant-body is not built up of small, separate cells as in most plants, but consists either of one large "cell" (coenocyte), often several centimetres in length, which forms a central axis more or less thickly encased in lime and bearing series of delicate branches. Examples of different types of this group of Calcareous Algae, characteristic of the warmer seas, are the genus *Penicillus*, the merman's shaving-brush, in which a short stem tapering upwards bears a tuft of slender forked branches at the summit; *Dasycladus* (fig. 83) and many similar forms with a central tubular stem, a few centimetres high, giving off at regular intervals circles of lateral appendages; *Acetabularia* (fig. 113, E), like a greenish white, lime-encrusted toadstool; *Halimeda*, a larger alga in which the spreading branches are divided into oval or wedge-shaped segments. The calcareous Chlorophyceae furnish a remarkable example of the persistence of a certain architectural plan, varying in detail from age to age, which has persisted from the earlier stages of the Palaeozoic era to the present day (see fig. 136).

Another and less simple group of Calcareous Algae is represented by the Corallinaceae, a family of the Red seaweeds (Rhodophyceae). The most widely distributed and the best known genus is *Lithothamnium*[1] (fig. 25): some species resemble clumps of coral, others form nodular masses. It occurs in all parts of the world, from Arctic Ellesmere Land to South Victoria Land in Antarctica. In company with large Brown Algae it flourishes in Arctic seas, and off the coasts of Spitsbergen it forms calcareous banks many miles in extent where the temperature does not usually rise above 0° C. *Lithothamnium* and other genera are often the dominant partners with corals in coral reefs. A section through a *Lithothamnium* (fig. 25, B) shows that the basal portion of the plant consists of cell-filaments more or less parallel to the substratum on which it grows succeeded in the upper part by regular, vertical or radiating rows of small cells forming a compact mass superficially resembling the wood of a conifer. The specific determination of specimens is by no means easy: an authority on the group says that in a single oblique section one may recognize several species of one or more genera. This fact is significant in relation to the still more difficult task of distinguishing fossil forms. In many living species the reproductive organs become embedded in the calcified mass, but in some they are borne superficially and unprotected. It is noteworthy that in

[1] Kjellman (83); Lemoine (17), (29); Howe (12), (18), (19), (22).

many fossil algae agreeing generally with *Lithothamnium* and its allies, no trace of fertile cells has been found: this may indicate that in the older types the reproductive organs were not covered by a subsequent development of calcified tissue.

In other genera of the Red Algae the calcareous body consists of an articulated stem and branches, e.g. in *Corallina*, in which bundles of filaments are encased in cylindrical calcareous sheaths.

The number of early Palaeozoic plants represented by satisfactory specimens is considerably less than one would infer from the published lists of species. Names have been freely given on very slender grounds to impressions on shale or other rocks which may or may not have been made by plants. Form alone is often an inadequate and misleading criterion: plants and animals cannot always be distinguished from one another. Some of the numerous patterns exhibited by Graptolites,[1] a cosmopolitan group of primitive animals in Ordovician and Silurian seas, cannot be distinguished from impressions of branched algae when the actual structure is not preserved.

Granting the existence of algae in the older Palaeozoic oceans, the more important questions to be considered are: What kinds of algae flourished? What do we know of their relationship to living genera? The facts at our disposal do not provide more than very incomplete answers. Dr Walcott[2] described and illustrated several impressions on pieces of Cambrian shale from British Columbia to which he gave various generic and specific names. Some of the fossils show a close resemblance in form and in the symmetry of branching to familiar examples of recent seaweeds; but his distribution of the specimens among the different families rests on wholly insufficient foundations. In the absence of structure or other convincing evidence of relationship to existing families or genera it is preferable, if a family name must be used, to adopt a non-committal designation such as Protophyceae, suggested by Dr Lindenbein,[3] a name implying that the fossils are most probably algae but not referable to a definite systematic position. Sections of examples of Walcott's algae revealed the presence of minute bodies which he compared with cells and cell-filaments of Blue-Green Algae: they may be plant-cells; but an inorganic origin is not

[1] Ruedemann (04), (08), who gives a map of the Ordovician world illustrating the distribution of land and water; also illustrations of Graptolites.

[2] Walcott (19).

[3] Lindenbein (21), (21[3]).

excluded. A specimen from British Columbia, presented to the Sedgwick Museum, Cambridge, was examined by Dr Walton by a method which he had formerly described: he was able to detect features suggesting affinity to some Blue-Green Algae.[1] An examination of other specimens might furnish data on which to base a revised classification of Walcott's material.

The name *Buthotrephis* has been given to some fossils of Upper Silurian age from Indiana[2] which serve as illustrations of a not uncommon form of impression resembling an algal body but not necessarily vegetable in origin. Equally unconvincing fossils have been described as species of *Sphenophycus* and *Delesserites* from Ordovician rocks in the State of New York. These and similar specimens, though possibly correctly included in the plant-kingdom, throw no light on the problems we wish to solve.

Geological literature contains many allusions to examples of a certain type of impression met with in marine sediments of various ages, from the Cambrian to the Tertiary period, which, according to the preference of authors, are described under such names as *Spirophyton*, *Taonurus*, *Cancellophycus*,[3] etc. The old name *Fucoides cauda galli* implies an algal fossil, and expresses a resemblance to the outspread feathers of a cock's tail. *Spirophyton* is mentioned in some text-books as a persistent type of marine alga: by some authors it is believed to be inorganic, and with good reason. The only specimen known to me in which carbonaceous matter is abundantly associated with the surface markings is one in the Kidston Collection (now in the Geological Survey Museum) from Lower Carboniferous rocks in Scotland. Various suggestions have been made on the method of formation of *Spirophyton*;[4] but the important point is that this and similar impressions have little or no value as botanical documents.

The nature of the large concentrically layered structures usually named *Cryptozoon* has already been discussed. *Cryptozoon* occurs in Cambrian rocks in North America (figs. 34, 35), and in several other countries: similar though smaller examples of the same structure are common in limestones of many geological periods. They have no claim to be called algae. Algae may in some instances have contributed to the deposition of the carbonate of lime, but an in-

[1] Walton (23).
[2] White, D. (01), (02[2]).
[3] Sarle (06). See also Kryshtofovich (11) who discusses similar fossils of Jurassic age.
[4] Seward (03).

organic origin is at least equally probable. Reference has already been made to the genus *Girvanella*.[1] The name was proposed for the tangled mass of vermiform tubules, about $\frac{1}{600}$ of an inch in diameter, discovered in sections of nodules of Ordovician limestone from the Girvan district in Scotland (fig. 41, C). *Girvanella* is undoubtedly organic: it may be an alga related to the family Codiaceae of the group Siphoneae, or, as previously suggested, the tubes may be sheaths which originally contained cell-filaments of a Blue-Green Alga. The genus is recorded from Cambrian, Ordovician and Silurian rocks in many parts of the world: it is abundant in Carboniferous and Jurassic Oölites; whatever its nature, it is not characteristic of any one chapter of geological history.

Another example of what is in all probability a member of the Blue-Green Algae is illustrated in fig. 41, D. An argillaceous rock of Ordovician age, known as Kuckersite and well developed in Esthonia, contains in addition to marine animals innumerable yellow bodies, 0·01–0·08 mm. in length, made up of several cell-groups embedded in a mucilaginous matrix. To these cell-aggregates Zalessky[2] gave the name *Gloeocapsomorpha* and compared them with the living Blue-Green Alga *Gloeocapsa*. Dr Lindenbein,[3] who made an exhaustive study of Kuckersite, assigns the alga to the Protophyceae. *Gloeocapsomorpha* must have existed in countless millions in the Ordovician ocean, either attached to the sea-bed or possibly as free-floating colonies on the surface.

Among the more satisfactory examples of Cambrian fossils which are usually believed to be algae are certain specimens included in the genus *Epiphyton* (fig. 41, A). The genus is recorded from a moraine on the Beardmore glacier in Antarctica[4] and from Cambrian rocks in Siberia and Sardinia. The best example is a species, *Epiphyton grande*, described by Prof. Gordon from a boulder of Cambrian rock dredged up from the Weddell Sea (lat. 62° 10′ S.). It consists of clustered groups of small radiating and branched tubes circular in section (fig. 41, A). As Gordon says, it would seem that somewhere in Antarctica there must have been a development during Cambrian times of limestone containing marine algae. *Epiphyton* may be related to some later Calcareous Algae such as the

[1] Rothpletz (08); Yabe (12); Chapman (07).

[2] Zalessky (18²), (20), (26), (28³). I am indebted to Prof. Zalessky for specimens of *Gloeocapsomorpha*.

[3] Lindenbein (21²).

[4] Gordon, W. T. (20).

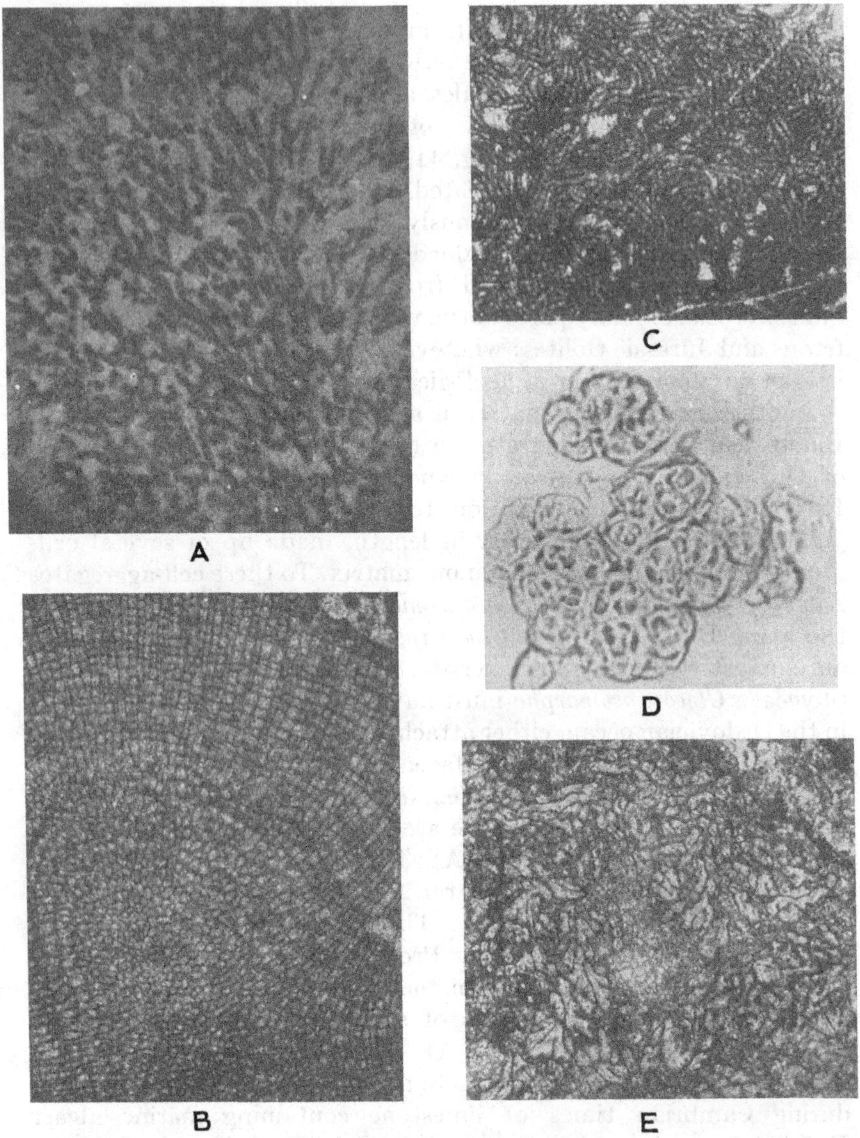

FIG. 41. A. *Epiphyton grande*. Cambrian. B. *Solenopora Garwoodi*. C. *Girvanella problematica*. Ordovician. D. *Gloeocapsomorpha*. E. *Sphaerocodium gotlandicum*. A × 20; B × 30; C × 90; D × 500; E × 40. (A, Photo. Prof. Gordon; B, C, E, Photo. Prof. Garwood; D, Photo. Prof. J. Walton.)

genus *Hedströmia* described by Rothpletz[1] from Silurian beds in the island of Gothland or to Garwood's Carboniferous genus *Ortonella*.[2] Ordovician limestones in the Baltic States and in Norway appear to be largely composed of the small calcareous bodies of algae agreeing with recent genera of the Siphoneae (Dasycladaceae). One of the Ordovician forms, the genus *Palaeoporella*, is represented by hollow cylinders or funnel-shaped bodies, 2–14 mm. long, which give off slender forked branches from a stouter axial cell. Another type is illustrated by *Cyclocrinus*:[3] this alga, 7 cm. long, resembles a small ball seated on the top of a stalk: from the swollen upper part of the stalk are given off radially spreading branches with hexagonal ends which together form a faceted surface. It bears a close resemblance to the living siphoneous genus *Bornetella*. From several localities in Norway evidence has been obtained of the abundance of algae similar to those from Ordovician limestones. Dr Høeg[4] has recently described some small cylindrical pieces of a calcareous alga from Middle Ordovician rocks in Norway as a species of a new genus, *Dimorphosiphon rectangulare*. The fragments, approximately 10 mm. long, are built up of branched tubular cells without cross-walls and embedded in a calcareous matrix. *Dimorphosiphon* is believed to be the oldest member of the family Codiaceae and related to the genus *Halimeda*, a calcareous alga which plays a prominent part in tropical seas in the formation of "coral" reefs.[5] The discovery of an Ordovician alga allied to *Halimeda* is of special interest because hitherto no undoubted representatives of the Codiaceae have been found in strata below the Cretaceous system. Calcareous species of the same tubular form are recorded from Cambrian rocks in northern China. Impressions of small plants, discovered in Silurian rocks in Norway and described under the generic name *Chaetocladus*,[6] appear to be uncalcified Siphoneous algae. *Chaetocladus capillata* has a cylindrical stalk, reaching 7 cm. in length, with whorls of hair-like branches. Examples of similar articulated plants are recorded under several names from Ordovician beds in Wisconsin and elsewhere in North America which suggest relationship with the recent Dasycladaceae. Reference may be made to some Ordovician algae from the Trenton Limestone in the State of New York described under the generic names *Primicorallina* and *Callithamnopsis*[7]

[1] Rothpletz (13).
[2] Garwood (14).
[3] Pia in Hirmer (27).
[4] Høeg (27).
[5] Rutten (20).
[6] Høeg and Kiaer (26).
[7] Ruedemann (09).

which resemble in form and, so far as can be seen, in structure, some existing members of the Dasycladaceae. In each the plant-body consists of a tubular axis a few centimetres long bearing whorls of much more slender branches. It is probable that algae such as these had a considerable share in the formation of some of the calcareous deposits in Palaeozoic seas which were subsequently uplifted as beds of limestone.[1]

Another type, *Sphaerocodium*, forms nodules (fig. 41, E), 3 cm. or more in diameter, in the Silurian limestones of Gothland: as the name implies, the structure resembles that of the living genus *Codium*.

Under the name *Solenopora* are included numerous species, probably not all generically identical, agreeing closely in their compact cellular structure with *Lithothamnium*. *Solenopora* was first described as an animal from Ordovician rocks in Esthonia, but a more thorough examination of its structure led Dr A. Brown[2] to refer it to the Calcareous Algae. It forms nodules varying in size from a pea to several centimetres in diameter, and the body is composed of contiguous rows of cells in vertical or radiating series. The structure of *Solenopora* is well shown in the section of a Lower Carboniferous species reproduced in fig. 41, B. Some authors have described what they believe to be traces of reproductive organs, but the evidence is unconvincing. Prof. Garwood,[3] whose papers on Palaeozoic calcareous organisms should be consulted for fuller information, speaks of the Woolhope (Silurian) Limestones in the Old Radnor district of Wales as the most remarkable example in Britain of an algal rock. *Solenopora* is recorded from Ordovician rocks in southern Scotland, in the Baltic States, in North America and other parts of the world: it played a prominent part also in the construction of Jurassic limestones. In later geological periods *Lithothamnium* and allied genera carried on the tradition established in earlier times by *Solenopora*.

This brief summary, deliberately incomplete as a description of genera, may suffice to emphasize one point: the comparatively slight difference in their broader features between some of the older Palaeozoic Calcareous Algae and algae which are still living. It is true that no proof has been given of the relationship of *Solenopora*

[1] For an account of algae as rock-builders see Pia (26).
[2] Brown, A. (94); Hinde (13).
[3] Garwood (13), (14); Garwood and Goodyear (19).

to *Lithothamnium*; the all-important test of affinity supplied by reproductive organs is lacking; but, despite the fact that in most species the cells of *Solenopora* are rather larger than those of *Lithothamnium*, the close similarity in the plant-body predisposes us in favour of a direct relationship.

Let us imagine ourselves voyagers on the Ordovician sea a few hundred million years ago over the site of what is now Wales and the West of England: we should look over an expanse of water; clouds, blue sky, and sunshine as we see them to-day. Not far from the edge of the mainland we can picture a scene differing in no essential feature from that off the coast of northern Queensland at the present time; curved lines or rings of calcareous rock, consisting in part of coral-like masses of algae, thrown up by the waves to form the margins of lagoons, the light vivid green of the enclosed shallow water forming a striking contrast to the darker blue of the deeper water of the surrounding sea. In the older Palaeozoic oceans, as in the seas we know, Calcareous Algae formed reefs of limestone. Differences in structure between the ancient and the modern plants would not be apparent on a superficial inspection. Closer examination would disclose differences in detail; but there is no reason to suppose that the plan of organization was fundamentally distinct from that which has persisted along separate lines of development through the long succession of geological ages. On the rocks at low tide we should probably see Green, Brown and Red seaweeds: on the cliffs, perhaps, scattered clumps of small green plants—but we know nothing of Ordovician terrestrial vegetation, not even whether it existed. To a distant observer the country would seem to be lifeless; a land of desolation, bearing on its surface or receiving from the waters the germs of a vegetation which was soon to clothe the continents with a new mantle. There is a fascination in looking backward into the ages antecedent to the unfolding of the plant-world in its more complex and grosser manifestations: we search in vain for the potential sources of future floras, but we are conscious of a world "dreaming on things to come".

Remains of Higher Plants

One of the few fossils from the early Palaeozoic periods, which may perhaps be a fragment of a land plant, was found in Silurian beds in the island of Gothland. Prof. Halle[1] describes a slender axis, 50 mm.

[1] Halle (20).

long, bearing a lateral branch and covered with short linear appendages and names the plant ***Psilophyton*** **(?)** ***Hedei*** because of its resemblance, which may be only superficial, to the genus *Psilophyton*, a plant characteristic of early Devonian floras. This Gothland fragment is very similar to specimens from Upper Silurian beds in England.[1] The actual nature of the appendages has not been determined and we do not know whether the fragments belong to a land plant or to a marine alga. Several specimens of plants have been described from beds of Upper Silurian age in Victoria, Australia;[2] some consist of pieces of stem bearing small, crowded leaves; others are branched fragments without leaves; the former are compared with a Devonian type known as *Thursophyton*, and the latter resemble another Devonian genus *Hostimella*. In the same rock were found larger specimens of stems or branches with slender leaves 2 cm. long and 1 mm. broad, differing in appearance from any previously recorded plant. These Silurian fossils are the oldest examples of what appear to be terrestrial plants with the exception of the smaller fragments already mentioned from Gothland and England. They do not tell us very much; but they afford evidence of the existence of two Silurian types, probably terrestrial, which agree closely with forms characteristic of the earlier Devonian floras and of a third type that appears to be peculiar to this meagre Pre-Devonian flora.

There are some obscure specimens in the Sedgwick Museum, Cambridge, from the Skiddaw Slates (Ordovician) of England, which were originally placed in the genus *Buthotrephis* and subsequently named *Protoannularia*[3] because of what may well be a misleading resemblance to the whorled shoots of the Carboniferous genus *Annularia*; an examination of them convinced me that their affinity cannot be determined. The best preserved specimens of Silurian plants are pieces of petrified stems of *Nematophyton*, *Nematophyton Hicksi* from Wenlock beds in North Wales and *N. Storriei* from the same series in South Wales. This puzzling genus is more fully described in the next chapter.

[1] Specimens may be seen in the Fossil Plant Gallery in the British (Natural History) Museum.

[2] Lang and Cookson (27).

[3] Nicholson (69). See Steinmann and Elberskirch (29).

CHAPTER IX

A LAND VEGETATION: THE DEVONIAN PERIOD

And God said, Let the earth put forth grass, herb yielding seed, and fruit tree bearing fruit after its kind, wherein is the seed thereof, upon the earth: and it was so.
Genesis i, 11

In the Devonian system are included rocks of many kinds: some are limestones rich in corals and other creatures of the sea; others are red sandstones and grits, conglomerates and shales, sediments in shallow water containing occasional fragments of terrestrial plants in company with strange armoured fishes and gigantic crustaceans, the profusion of which in some places led Hugh Miller to speak of the rocks as "platforms of sudden death". The Devonian strata form two groups, two sets of documents: one set from which we read of the persistence of the Silurian ocean over a comparatively stable region of the earth's crust carrying on the traditions established in former ages. The other set of documents reveals a changed world, large areas of the crust deserted by the sea, land reclaimed as the waters retreated; enormous piles of sediment spread over the floors of inland lakes and arms of the sea by torrential streams carrying, with scourings from the rocks, broken pieces of plants from the mountain slopes and river banks.

Marine Devonian and Continental Old Red Sandstone Rocks

These two groups of rocks constitute respectively the marine, or typical Devonian facies, and the continental or Old Red Sandstone facies, two sharply contrasted sources of history. The latter furnishes the data with which we are more immediately concerned. In some parts of the world there is no sharply defined boundary between Silurian and Devonian rocks; similarly it is not always easy to determine whether certain rocks and their meagre fossil content should be assigned to the later stage of the Devonian or to the early stage of the Carboniferous period. The Devonian system as a whole is usually divided into three series, Lower, Middle, and Upper: here too the imperfection of the records forms an obstacle to precision. In describing events chronicled in a succession of strata, many thousand feet thick and representing a prolonged epoch, we seem to envisage as contemporary happenings stages that were separated by thousands or even millions of years. In Forfarshire[1] the Lower Devonian beds alone are said to exceed 14,000 ft. in depth, and the Middle series of Caithness has a thickness of 16,000–18,000 ft.[2] The map shown in fig. 42 is an attempt to reconstruct the main features of the world at about the middle of the Devonian period when the seas transgressed over wide areas which in the earlier part of the period were dry land. Localities where most of the plants have been found are indicated by crosses: some of these are far from the continental edge and appear unfavourably situated for the preservation of samples of a terrestrial vegetation. The distribution of land and water fluctuated in the course of a single period within fairly wide limits: it is, therefore, clear that an attempt to give a reconstruction of geographical conditions which prevailed at any one part of the period can hardly fail to be inconsistent with fact.

A stupendous and almost world-wide movement of the earth's crust, which is most clearly demonstrated by the rocks in north-western Europe, began before the close of the Silurian period and continued well into the Devonian. This crustal convulsion, the Caledonian revolution,[3] created a new set of conditions which must have had a far-reaching influence on the plant-world. The elevation of the Caledonian chain may have been a dominating factor in bringing about the invasion and subsequent colonization of the land by translated members of an ocean flora. As the mountains rose, climatic equilibrium would be disturbed; currents in the atmosphere and in the water would be diverted. In recent years it has become increasingly probable that land plants existed before the Devonian period, but the facts at present available do not justify the inference that the land had been widely colonized by a Silurian vegetation. It may confidently be predicted that the scraps of evidence in favour of a Pre-Devonian land vegetation which have so far been discovered will be substantially supplemented. The probability is that some of the older Devonian plants are examples of transformed algae: others can hardly be so regarded and are more likely to be descendants of simpler ancestors which had long been accustomed to live above the sea. As yet we know very little of Pre-Devonian vegetation; but it may safely be asserted that future search will add a few more links to the broken ends of the chains of life. For the present it is consistent with fact to say that in the course

[1] Hickling (08), (12). [2] Crampton and Carruthers (14). [3] See p. 21.

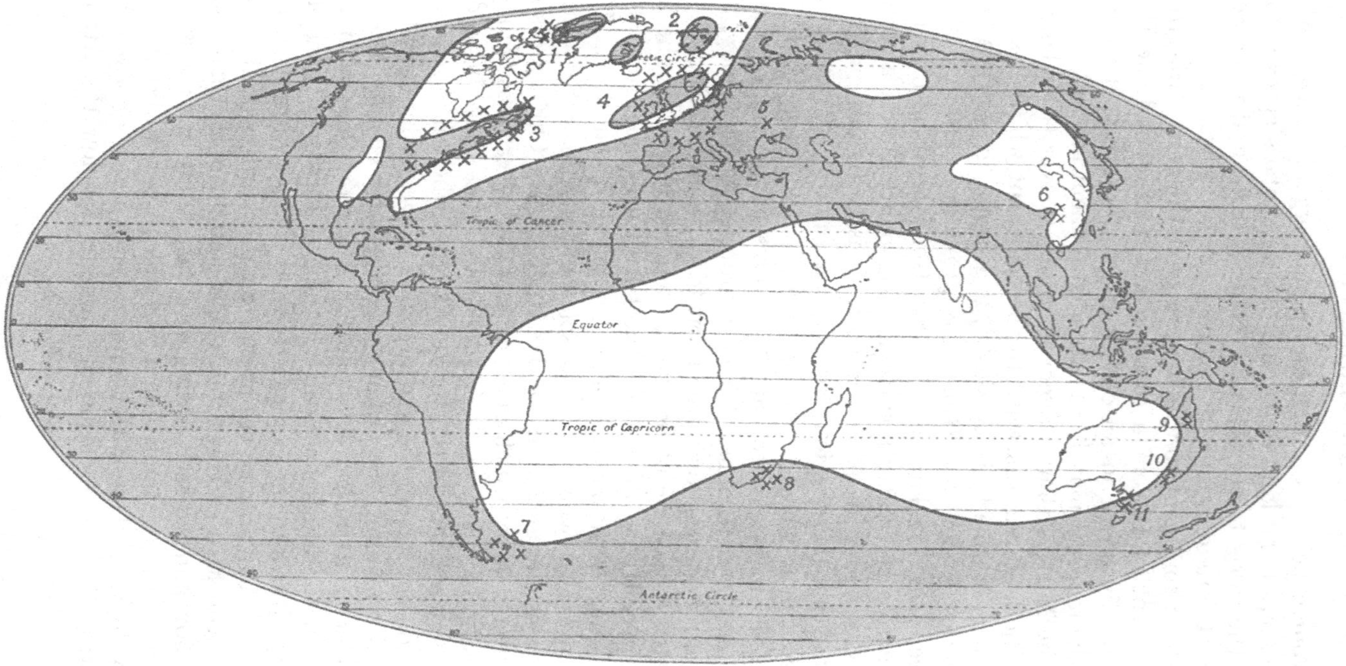

FIG. 42. Map of the Devonian world. Crosses mark localities where plants have been found. The shaded part represents water. (With acknowledgment to Mr P. Lake.)

of the Devonian period a varied and highly organized vegetation took possession of the land and some of its members spread from one end of the world to the other. Whence it came and what were its antecedents are questions that cannot yet be answered.

The northern continent to which Suess gave the name Eria, from the development of Devonian rocks in the district of Lake Erie, was undoubtedly larger in an earlier stage of the period than is shown on the map. It may have extended without any substantial gap from the North Atlantic region of America over Greenland to the old Baltic land and eastward to the continent which has been called Angara. Marine beds in Alaska, in Europe, in North Africa, South America, and in the Himalayas prove the existence of relatively deep seas. The large continent of Gondwanaland may have stretched across the water now separating South America, India, and Australia; or, if Wegener's hypothesis is to be our guide, we must assume that these regions were formerly huddled together as a compact continent which, after long ages, was disrupted and the dismembered blocks floated into their present positions. Land-encircled lakes, comparable to those in tropical Africa to-day, and narrow arms of the sea were prominent features on the Erian land. The Old Red Sandstone of Scotland, rendered classic by the life-long labours of Hugh Miller, is the most fruitful source from which to obtain data for the reconstruction of physical conditions. That Scottish naturalist who combined scientific insight with poetical imagination is immortalized in the Hugh Miller Cliff in eastern Canada "in a country which he never saw, but over whose foundation stones he laboured well". The Old Red phase may be said to begin on the north side of the Bristol Channel where it is separated by a short distance from the marine phase in Devonshire. But it is in Scotland, including the Orkney and Shetland Islands, that the Old Red reaches its fullest development. The mountains of Morven (2313 ft.) in Caithness and other detached blocks of horizontal strata bear witness to the former extension of the sediments beyond the limits of their present occurrence on the borders of the Highlands. The Old Man of Hoy[1] (fig. 43), a column more than 600 ft. high of yellow and red rock resting on a base of lava on the western edge of the Orkneys, stands like a partially preserved table of contents of volumes of Upper Devonian records, many of which have been destroyed by denudation or submerged below the Atlantic.

[1] Geikie, A. (79).

Fig. 43. The Old Man of Hoy and cliffs of Upper Devonian (Old Red Sandstone) rocks. (Photo. Messrs York and Son, London.)

Igneous rocks interbedded with layers of sandstone in the Ochill Hills, the Cheviots, and in other places afford evidence of contemporary volcanic activity.

The prevailing red colour of the Old Red Sandstone has often been regarded as evidence of an arid or semi-arid climate.[1] The nature of the sediments, their structure and manner of occurrence, and the frequent lack of fossils lend support to the view that the climatic conditions, both on the northern continent and in some parts of the southern continent, were similar to those in fairly dry regions and subject to occasional and violent rain-storms. It has been suggested[2] that the desert conditions may have been due not so much to lack of rain as to the barrenness of rocks not yet clothed with a protecting and moisture-giving covering of vegetation. Some authors speak of an exceptionally uniform climate throughout the world; but on inadequate grounds. The great majority of Devonian plants have been obtained from rocks in the far north, in the north temperate zone, from the southern borders of South America and Africa, and from the southern half of Australia. We know practically nothing of the floras which flourished within the tropics. Moreover, the occasional occurrence of gypsum and the presence of glacial deposits in the Table Mountain series of Cape Colony give us glimpses of conditions like those at the present day on the shores of the Caspian Sea and of ice-filled valleys.

A few early Devonian plant fragments are recorded from the Falkland Islands; but it is the close agreement of the much more abundant marine fossils with those of South Africa which arrests attention. Du Toit[3] says: "the wonderful lithological, palaeontological, and structural parallelism between South Africa and South America during this epoch cannot be sufficiently emphasized. Taken in conjunction with the evidence of a similar kind during the Permian and Triassic, it proves the intimate connexion of these two continents over an enormous period of time". This agreement is a point in support of the Wegener hypothesis. Table Mountain, which towers as a gigantic bastion above Cape Town, is built of rocks that are probably Lower Devonian in age. The Table Mountain series is overlain by the marine Bokkeveld series, and above this is the Witteberg series. From these three sets of strata only a few

[1] Barrell (13); Crampton and Carruthers (14), who describe a Mid-Devonian landscape; Tomlinson (16).

[2] Evans, J. W. (26).

[3] du Toit (27).

fragmentary plants have been collected. The non-marine beds, about 4000 ft. in thickness, represent material transported by torrential floods and spread over a vast delta much larger than that of the Ganges. It is not surprising that under conditions such as these records of contemporary vegetation are scanty.

A. The Earlier Devonian Floras

The space allotted in this chapter to the older Devonian plants may seem to be out of proportion to that given to the richer floras of later periods, but in view of the exceptional importance to students of evolution of the earliest decipherable records of a terrestrial vegetation this inconsistency is not unreasonable. Readers unfamiliar with the various groups of plants may find it helpful to refer to Chapter vi and to keep before them certain genera of existing plants as standards of comparison. The genus *Lycopodium* (fig. 28), including the so-called club "mosses" and the stag's horn "moss", and many species from Arctic to tropical lands, illustrates one of the characteristic features of the Lycopodiales, namely the small size of the crowded, spirally disposed leaves. *Psilotum* (fig. 29) of the group Psilotales, a southern, tropical genus, is distinguished by the lack of roots, its slender, forked green shoots with sparsely scattered and hardly visible scale-leaves. *Equisetum* (fig. 27), ranging from high Arctic regions to the southern tropics, is the living type of a long-established articulate plan of construction. In the great majority of ferns, e.g. in the cosmopolitan bracken fern (*Pteridium aquilinum*), the most striking feature is the large size of the repeatedly branched fronds. It must, however, be remembered that similarity in plan is not in itself proof of relationship between extinct and recent plants: reproductive organs and anatomical characters have greater value as criteria of affinity.

In 1889 an American writer[1] described in the following words the impression made upon him by a study of Devonian plants: "While they have given us fascinating glimpses of the head of the column of terrestrial vegetation that has marched across the earth's stage during the different geological ages, they have given us little insight into the spirit of the movement". Since this was written more has been discovered; we have learnt much, but the veil which hides the secret of creation has been only partially lifted. We are "ever learning, and never able to come to the knowledge of the truth".

[1] Newberry (90).

In order to reconstruct the changing landscapes and to follow the rise and fall of plant-dynasties it is necessary to have a general acquaintance with the components of the several floras. The difficulty is to avoid excess of detail without omitting facts that are essential. Most people who wish to know something of the vegetation of the past naturally ask: In what respects and to what extent were the plants of former ages different from those of the present? In order to answer this question, so far at least as the limitations of the subject permit, an attempt has been made to reproduce successive phases of geological history by means of restorations both of the plants and of the conditions under which they lived. Reconstructions necessarily rest on insecure and shifting foundations: our conception of the habit of an extinct plant to-day may be proved entirely wrong by a new discovery to-morrow. The imperfection of the documents at our disposal cannot be too strongly emphasized. The statement that certain groups, families, or genera did not exist in former periods means that no traces of them have been found in the rocks. We believe that the most highly organized plants in the Devonian floras were gymnosperms: the Devonian age antedated by some hundred millions of years the strata which have afforded the first recognizable remains of members of the present dominant class, the angiosperms (flowering plants). Primitive ancestral forms of this class may have existed at a much earlier stage of geological history than is generally admitted. We know that before the close of the Devonian period a vegetation in which most of the larger groups of the plant kingdom were represented had colonized wide areas of the earth's surface.

Adopting the three-fold division of the Devonian system into Lower, Middle, and Upper, we find that while there are certain well-marked differences between the Lower and Middle floras, these are insignificant in comparison with the much greater and more fundamental contrasts between them and the later Devonian floras. As one would expect, the botanical dividing lines between one period and another, and between the successive phases of a single period, are not as a rule sharply marked: there is some overlapping. At the outset of our attempt to follow the progress of plant-life through the ages it is worth while to draw attention to the possibility of misinterpreting contrasts between collections of plants from beds of the same, or approximately the same geological age: at first sight contrasts may appear to denote differences in age or to mark stages in progressive evolution, whereas they may be merely expressions of dissimilar environments. The plants of a swamp differ from those of a wood, but these and other communities are units which together make up one contemporary mosaic. Before describing the best preserved Devonian plants—those made known to us during the last few years through the masterly investigations of the Middle Devonian peat-beds of Aberdeenshire by Dr Kidston and Prof. Lang—followed by the work of Dr Kräusel in Germany—a brief account will be given of genera that are characteristic of the oldest Devonian vegetation.

The genus *Nematophyton*, already mentioned as one of the few Silurian plants, was first described from Canada by Sir William Dawson under the name *Prototaxites* and regarded by him as a conifer allied to *Taxus* (the yews). Subsequently it was generally accepted as an alga and renamed *Nematophycus* or, more appropriately, *Nematophyton*. In a recent text-book[1] the older term *Prototaxites* has been revived on grounds that are hardly adequate. The largest specimens, in the Montreal Museum, are pieces of stems up to 2 ft. in diameter. Relatively thick tubes, varying in size and reaching a diameter of $70\,\mu$,[2] without cross-partitions and with smooth walls lacking the pits or spiral or annular bands characteristic of the conducting cells of vascular plants, make up the bulk of the stem. Between these more or less loosely arranged tubes, which follow a vertical and rather sinuous course, are much narrower tubes less regular in their arrangement and similar in structure to the filaments of some fungi. The lack of cohesion between the larger tubes reminds one of the stem of a conifer that has been partially destroyed by the ravages of a fungus. This comparison is not intended to be more than an analogy. In some species the mass of large tubes is interrupted by what are known as medullary spots, small areas occupied exclusively by the smaller tubes. The only evidence of the existence of a superficial tissue different from that of the main mass of the stem is furnished by a fragment from the Middle Devonian chert-bed of Rhynie in which the vertical tubes are enclosed by a peripheral zone of tubes with the long axes at right-angles to the surface. Nothing is definitely known of the reproductive organs or of the appendages borne on the stem. Although Dawson acknowledged that his original comparison with a conifer was incorrect, he adhered to the view that *Nematophyton* grew on land. The

[1] Hirmer (27).

[2] μ (micron) = $\frac{1}{1000}$ of a millimetre.

occurrence of the genus in a peat deposit is a point in favour of Dawson's contention. This "most remarkable and puzzling"[1] extinct plant was abundant in Scotland in the Lower Devonian period: it flourished in Canada and, as shown in the table on p. 153, in other regions. Hugh Miller[2] recorded the occurrence of pieces of wrinkled, carbonized stems, 3–4 in. in diameter in the Arbroath flagstones of Scotland without giving them a name: a recent examination of these stems by modern methods has enabled Prof. Lang to recognize the characters of *Nematophyton*. The genus is recorded from Middle and Lower Devonian rocks in Scotland and Canada, from Upper Silurian rocks in England, and from Upper Devonian beds in America. A restoration of *Nematophyton* was published by Dawson,[3] but we have not ventured to include one in the reconstruction of the older Devonian vegetation shown in fig. 45. Anatomically the genus conforms more closely to certain algae and fungi than to any known vascular plant; it probably grew on swampy ground and must have reached the dimensions of a tree. It is one of the oldest types, a type far removed from anything living to-day and possibly a derivative of some marine alga which was able to accommodate itself to a terrestrial life when portions of the sea-floor were raised above the reach of the tides.

The next genus to be described has been assigned by authors to both the animal and the plant kingdom. *Parka decipiens*,[4] so-called by Dr Fleming from Parkhill on the Firth of Tay, has not been found in rocks higher than the Lower Devonian; it is recorded from a few localities in England, in beds usually regarded as Upper Silurian (the Downtonian series of the Ludlow district), which it has recently been suggested are probably of Lower Devonian age or, as Dr Evans[5] prefers to call them, passage beds between the Silurian and Devonian systems. The genus is founded on carbonized specimens preserved as flat impressions more or less circular in form though occasionally lobed, varying in size from 5 mm. to about 7 cm. in length. Scattered over the surface of the impressions are numerous circular discs giving the fossils a resemblance to crushed raspberry fruits. The discs are groups of spores embedded in a thin carbonaceous layer partially preserved as a mummified framework of small cells. It is noteworthy that the spores, unlike those of most

[1] Kidston and Lang (24[2]); Lang (24), (26). See also Penhallow (97); Hörich (15).
[2] Miller, H. (57).
[3] Dawson (88).
[4] Don and Hickling (17).
[5] Evans, J. W. (26).

existing plants, do not occur in groups of four but as a common mass: the resistant nature of the spore-wall would seem to be an indication that the reproductive cells were fitted for dispersal in air rather than in water. The statement that *Parka* is not a complete plant, but was borne on a stalk,[1] is not supported by convincing evidence. In the simplicity of its construction *Parka* is comparable with some algae; it also resembles some of the liverworts with which it agrees in the nature of the spore-membrane. The plant probably lived on mud or soil exposed for long periods to the air: its spores may have been among the first of countless millions of wind-borne particles which have strewn the earth's surface since the advent of terrestrial vegetation.

Another genus, smaller than *Parka* and more definitely algal in affinity, is *Pachytheca*: this is recorded from Silurian (Wenlock) rocks and from the Downtonian passage beds in England, also from Lower Devonian beds in Canada and Scotland. It has been found in the Middle Devonian Rhynie peat.[2] The plant consists of small, spherical bodies, 0·5 cm. in diameter; there is a central, or medullary, region occupied by a plexus of small branching tubular cells, and a cortical zone of radially disposed tubes, each of which in some species encloses a varying number of still narrower algal filaments. *Pacytheca* may be an algal colony which rolled about the floor of a lake or sea, comparable to some existing Blue-Green Algae or to the balls of green filaments of a *Cladophora*[3] (Chlorophyceae).

The Psilophytales

Psilophyton has been a favourite name to apply to more or less fragmentary remains of axes bearing small spinous, or occasionally smooth, appendages, which occur in the older Devonian rocks and afford no definite evidence of their systematic position.[4] Two of the more satisfactory species may be included in this summary: *Psilophyton princeps* (fig. 45, *Pn.*), originally described by Dawson from Lower and Middle Devonian rocks in Canada, was a gregarious plant on swampy ground, a plant with a creeping stem (rhizome) provided with hairs in place of roots and bearing slender, erect, forked branches clothed in the lower parts with thickly set spinous appen-

[1] Edwards, T. N. (21).
[2] Kidston and Lang (24[2]). See also Barber (89), (90).
[3] West and Fritsch (27), p. 170.
[4] Solms-Laubach (95).

dages: from the tips of the smooth and more delicate apical forks hung oval spore-capsules. Dawson's restoration[1] of this North American and European species has in the main been confirmed by more recent discoveries. The name *Psilophyton* was adopted because of a resemblance to the recent genus *Psilotum* (fig. 29), a genus that is also reminiscent of *Rhynia* and other types in the Rhynie peat. It is an astonishing fact that *Psilotum* is not represented by any satisfactory fossil forms among the thousands of plants preserved in the enormous thickness of sedimentary strata deposited since the Devonian period. If, as seems possible, *Psilotum* is distantly related to certain Devonian genera,[2] it affords an impressive demonstration of the imperfection of the geological record. The alternative explanation is hardly tenable, namely that a type, which had become extinct, reappeared after the lapse of several hundred millions of years.

Another species, *Psilophyton Goldschmidti*, was discovered by Halle[3] in Norwegian rocks which, from their fossils—and these are exclusively plants—are believed to be of Lower Devonian age. In this form the spinous stem bore repeatedly forked lateral branches without spines, comparable except in the absence of leaflets to the fronds of ferns. It may be, as Halle suggests, that in these bare branch-systems we have an early stage in the development of the fern type of leaf which is more clearly foreshadowed in another member of the older floras, *Aneurophyton* (fig. 45, *An.*). In the later Devonian floras the typical fern-plan of foliage became definitely established (e.g. in *Archaeopteris*; fig. 48, A). The smaller *Lycopodium*-type of leaf may have arisen as a simple emergence from a stem or branch.

Pieces of stem have been described from Lower Devonian rocks in Scotland, Canada, Germany, Norway, and elsewhere under the generic name *Arthrostigma* (fig. 45, *Ar.*): these differ from *Psilophyton* in the greater diameter of the stems, which may be 2 cm. wide. Prof. Halle, whose illustrations convey the best idea of the genus, points out the difficulty of drawing a clear line between *Arthrostigma* and *Psilophyton*: fossils described under these two names may be parts of one and the same genus. *Arthrostigma* is characterized by short and broad spines, not unlike the prickles of a rose stem, and on other parts of the stem by simple slightly curved leaves. As in *Psilophyton* the tips of the shoots were spirally coiled like the young fronds of ferns, and in both genera a narrow

[1] Dawson (59), p. 479. [2] Zimmermann (26). [3] Halle (16).

strand of conducting tissue occupied the centre of the stem. Reference may be made at this point to some casts of stems from Devonian rocks in South Africa which cannot be precisely correlated with European horizons but are probably homotaxial with Middle or Lower beds in the northern hemisphere. These stems have been provisionally described as lepidodendroid, but their affinities have not been definitely determined. A characteristic feature of the Lycopodiales is the presence of a connecting strand between each leaf and the axial conducting tissue of the stem or branch: on the scars of stems from which the leaves have fallen a median dot shows that a vascular strand or vein entered each leaf. In the South African stems and in similar stems from the Falkland Islands the leaf-scars, though superficially similar to those of a *Bothrodendron*, afford little or no indication of a definite mark denoting the presence of a vein.[1] It is significant that in *Arthrostigma* we have no evidence of a vascular supply to the leaves, and this is equally true of other early Devonian Psilophytales. The probability is that some at least of the stems from localities in the southern hemisphere are allied to *Arthrostigma* more closely than to any of the true Lycopodiales, though we cannot in these older types recognize any very clear dividing line between the Psilophytales and the primitive lycopods. Certain Lower Devonian fossils from Belgium and Canada originally described as *Lepidodendron gaspianum* should be included in *Arthrostigma.*

One of the oldest Devonian plants assigned to the Psilophytales is the genus *Zosterophyllum* (fig. 45, Z.).[2] This name was given by Prof. Penhallow to some fossils, from the Lower Devonian rocks of Scotland, which he described as *Zosterophyllum myretonianum*: specimens had previously been compared by Hugh Miller with leaves of *Zostera*, a common marine flowering plant. *Zosterophyllum* may be described as a tufted plant with crowded, leafless branches 2 mm. broad, reaching a length of about 15 cm., radiating from a confused basal region and bearing near their tips spirally disposed short and broad kidney-shaped appendages attached by a short stalk. The appendages may be spore-capsules, but no spores have been found.[3] Prof. Lang has considerably advanced our knowledge of the genus: he found evidence of the existence of an axial strand

[1] For stems of this type the name *Cyclodendron* has been used: Kräusel and Range (28).

[2] Lang (27).

[3] The presence of sporangia and spores has now been demonstrated: see Lang and Cookson (30).

of annular woody tubes (tracheids). Fig. 44 A, shows a slightly enlarged vertical section of a piece of a *Zosterophyllum* branch, and in fig. 44, B, is reproduced on a much larger scale one of the regularly pitted conducting elements. The latter photograph is an illustration of one of the oldest known examples, in Britain at least, of well-preserved tracheids,[1] tubular cells agreeing essentially with those of water-transporting tissue in vascular plants at the present day. A peculiar feature of *Zosterophyllum*, which can often be readily recognized, is the apparent anastomosis of branches. A figure given by Miller shows parallel branches connected by narrow cross-pieces

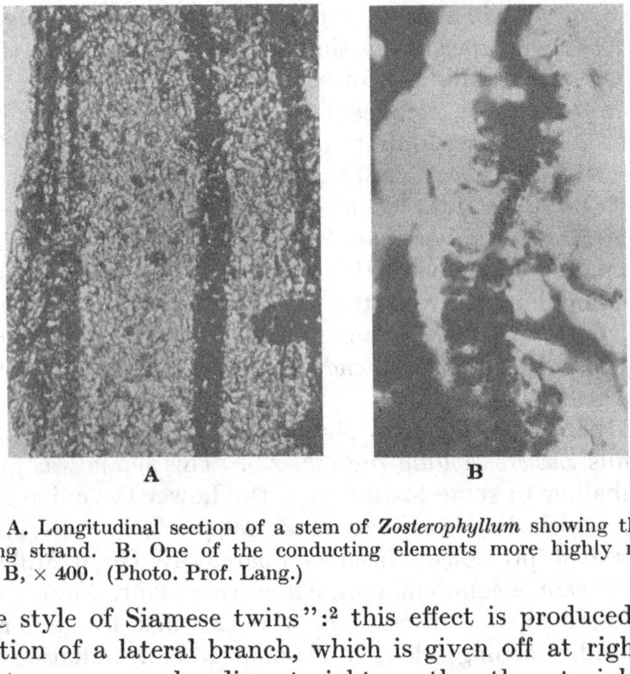

A B

FIG. 44. A. Longitudinal section of a stem of *Zosterophyllum* showing the central conducting strand. B. One of the conducting elements more highly magnified. A, × 20; B, × 400. (Photo. Prof. Lang.)

"in the style of Siamese twins":[2] this effect is produced by the bifurcation of a lateral branch, which is given off at right-angles to the stem, one arm bending straight up, the other straight down: the two arms of the bifurcation together with the parent stem giving the impression of twin branches. An incomplete fossil figured by White as *Psilophyton alcicorne*[3] from the Perry Basin, North America, may well be a piece of a *Zosterophyllum*, but it occurs in a collection of specimens which are mainly Upper Devonian. It is

[1] Tracheids were found by Halle in the Lower Devonian *Arthrostigma* [Halle (16)].
[2] Miller (57). [3] Smith and White (05), Pl. v.

probable that the so-called *Psilophyton* and some other fossils figured by White came from a lower horizon. *Zosterophyllum* may be compared with the Middle Devonian plant *Hicklingia*, and with the Lower Devonian *Gosslingia* with which it may be generically identical. Our knowledge of the genus *Hicklingia*[1] (fig. 45, *Hk.*) is based on a fossil from the middle Old Red Sandstone of Caithness in the Victoria University Museum, Manchester, which reminds one of a complete herbarium specimen carefully pressed on paper: it consists of a tuft, 17 cm. long, of slender, ribbon-like branches occasionally forked and arising from an obscurely preserved basal region. At the tips of some of the branches are small oval sporangia.

A plant very similar to *Zosterophyllum* has been described from Lower Devonian rocks in South Wales under the generic name *Gosslingia*,[2] from Mr Gossling who discovered the plant-bed. The internal structure is fortunately preserved: a comparatively large strand of conducting tissue runs up the centre of the stem and is separated from the surface-layer (epidermis) by a broad cortex. The occurrence of stomata on the epidermis shows that the plant was not submerged. It was rootless and leafless. Associated with the stems and branches were some stalked reniform bodies very like those of *Zosterophyllum*: traces of spores were detected. *Gosslingia* and *Zosterophyllum* may be identical: they are certainly very closely allied early Devonian genera.

Of the plants so far mentioned *Parka*, *Arthrostigma*, and *Zosterophyllum* are the safest guides to the Lower Devonian age of plant-bearing beds. *Psilophyton* occurs also in Middle Devonian rocks, and this is true of *Nematophyton* which extends into the Upper Devonian.[3]

Higher Plants

Prof. Lang[4] has recently described some fossil wood discovered in south Cornwall by Miss Hendriks at two localities near the Lizard mapped respectively by the Geological Survey as Lower Devonian and Ordovician. From the nature of the material it would seem most probable that the beds belong to the middle part of the Devonian system. It was possible after treating the wood with hydrofluoric acid and other reagents to embed fragments in paraffin and cut numerous sections from a specimen 1 mm. in thickness.

[1] Kidston and Lang (23^2).
[2] Heard (27).
[3] Penhallow (97).
[4] Lang (29).

The wood consists of secondary xylem, that is conducting elements arranged in regular radial series as in the wood of a pine tree, with small medullary rays, one cell broad, extending in a slightly oblique direction through the wood. The elements of the wood are pitted on all the walls, not on the radial walls only as in living conifers. This wood resembles in some of its characters that of the Middle Devonian *Palaeopitys* which is described below; but it differs in some details. With the wood were found slender branches bearing spirally arranged appendages resembling *Thursophyton* and *Asteroxylon* (fig. 45, *Ast.*). The Cornwall plant differs from all known Lower Devonian genera in having conducting tissue made up of secondary wood, not merely of a primary strand of irregularly arranged elements: it agrees generally in the structure of the wood with stems from later geological horizons included in the genus *Dadoxylon*, which in general terms may be described as a comprehensive type of stem-structure resembling that of certain existing conifers though differing more or less widely in the finer anatomical features. The occurrence of a species of *Dadoxylon* (*D. Hendriksi*)—using the term as it is employed by Lang in a rather wider sense than usual—suggests an age not older than Middle Devonian. Wood of similar type is unknown in rocks as old as the Lower Devonian.

This discovery affords a good example of the employment of fossil plants as indices of geological age: on botanical grounds the Cornish rocks are referred to a Middle Devonian horizon; on purely geological evidence they were assigned to Lower Devonian and some of them to the Ordovician period. We cannot, however, confidently assert that no trees with stems of the *Dadoxylon* type existed before the middle of the Devonian period.

Plants from a Devonian Marsh

By far the most satisfactory records of a Middle Devonian flora are those furnished by petrified material preserved in a flinty rock (chert) discovered in 1913 by Dr Mackie near Rhynie in Aberdeenshire, Scotland. In the siliceous matrix are embedded stems, branches, and spore-bearing organs of vascular plants, which in company with numerous fungi and other microscopic organisms built up a considerable thickness of peat. In some places the chert is full of the remains of a single species; in other layers the creeping stems of one plant are seen penetrating the tissues of another. Here and there in the petrified mass are thin partings of sand indicating flooding from neighbouring streams or the overflow of a lake over swampy ground; or possibly sand was occasionally carried by wind from adjacent sand dunes. The peaty soil may have been permeated by water rich in silica derived from hot springs or fumaroles. There must have been some such source whence the petrifying solution was derived; a fortunate accident which caused the sealing up of samples of a bed of peat formed about 300 million years ago. In the reconstruction (fig. 45) an attempt is made to reproduce the Rhynie landscape. In the background is a range of Pre-Cambrian mountains: at their base sand and gravel in mounds, ridges, and talus-slopes—products of erosion in a semi-arid climate sorted by torrential streams. On the left sand dunes line the beach: here and there smoking fumaroles, manifestations of volcanic activity, and possibly the source of the flinty preservative which permeated successive layers of vegetation on the marshy flats. Most of the plants are represented as growing on a swamp studded with pools, and some in the foreground are shown on a drier soil at a slightly higher level.

A Devonian swamp on casual inspection would seem familiar: the ground carpeted with slender, rush-like plants 6–20 in. high, with occasional patches of taller clumps. Neither in the peat nor in the pools should we see any flowers: a sheet of uniform green relieved by the brown tints of dead or decaying herbage: no cries of birds; no sounds other than those made by fitful breezes or violent storms.

The higher or vascular plants will be described first: *Rhynia*, *Hornea* and *Asteroxylon*. None of them can be accommodated in existing groups without violence to definitions based on living types: they are assigned to the extinct group Psilophytales. *Rhynia*[1] (fig. 45, *R.* and *H.*) occurs in two forms, *Rhynia Gwynne-Vaughani* and *R. major*, the former about 7 in. high and the latter three times as large. The plants were rootless and leafless, the erect shoots rising from creeping underground stems superficially resembling the leaves of a Pillwort (*Pilularia*). Some of the cylindrical branches bore apical spore-capsules similar in form, though not in structure, to those of some liverworts. The wonderful preservation of the *Rhynia* stems is illustrated in fig. 46: the continuity of the surface-layer of cells is interrupted by an occasional stoma, the small pore by which plants regulate gaseous exchange with the outer air. Below the

[1] Kidston and Lang (17), (20), (21).

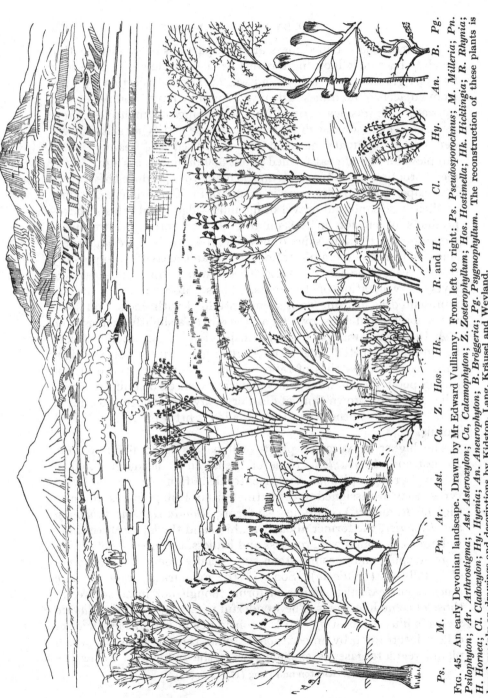

FIG. 45. An early Devonian landscape. Drawn by Mr Edward Vulliamy. From left to right: *Ps. Pseudosporochnus*; *M. Milleria*; *Pn. Psilophyton*; *Ar. Arthrostigma*; *Ast. Asteroxylon*; *Ca, Calamophyton*; *Z. Zosterophyllum*; *Hos. Hostimella*; *Hk. Hicklingia*; *R. Rhynia*; *H. Hornea*; *Cl. Cladoxylon*; *Hy. Hyenia*; *An. Aneurophyton*; *B. Bröggeria*; *Pg. Psygmophyllum*. The reconstruction of these plants is based mainly on drawings and descriptions by Kidston, Lang, Kräusel and Weyland.

epidermis a few layers of relatively large and apparently empty cells are succeeded by a broad cortical region in which the petrified contents probably represent the bearers of the chlorophyll and grains of starch. The axial region is occupied by a cylindrical strand of woody tubes surrounded by a zone of more delicate tubular cells

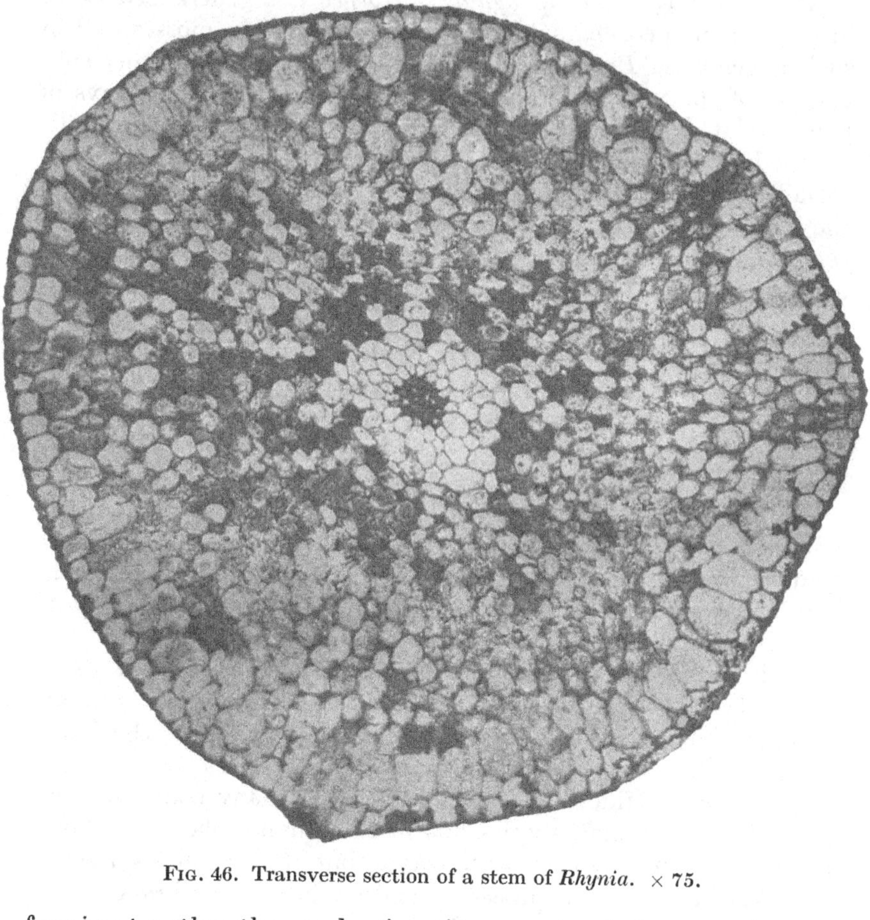

FIG. 46. Transverse section of a stem of *Rhynia*. × 75.

forming together the conducting tissue, a simple form of the conducting apparatus which in varying form, though with little change in the structure of the essential elements, has been retained by vascular plants through the ages. *Hornea* (fig. 45, *R.* and *H.*), similar to *Rhynia* in habit and in size agreeing with *Rhynia Gwynne-Vaughani*, is distinguished by tuber-like swellings on the creeping stem. The most distinctive feature is in the structure of the terminal sporangia: these do not consist merely of a wall enclosing a mass of spores, as in *Rhynia*, but the central part of a sporangium is formed of a column of sterile cells surrounded and overarched by a spore-forming region, an arrangement closely resembling that in the spore-capsule of a bog moss (*Sphagnum*). *Hornea* is a generalized type: in it are combined characters now found in certain mosses and in such a genus as *Psilotum*. Evidence, though by no means conclusive, of the wide geographical range in early Devonian days of plants related to *Hornea* in the structure of the spore-capsule, is afforded by a specimen described under another name by Prof. Halle[1] from the Falkland Islands. A well-preserved sporangium of the same type as that of *Hornea* was described by Halle as *Sporogonites exuberans*[2] from Norwegian beds, which may be of Lower or possibly of Middle Devonian age. In the characters of the organs so far known *Rhynia* and *Hornea* closely resemble the Bryophyta (mosses and liverworts): "*Hornea*", Dr Church[3] says, "is in fact little more than a slightly ramified and free-growing *Anthoceros*" (a liverwort); but modern bryophytes have no conducting tissue of the relatively advanced type characteristic of the Devonian genera.

The third genus, *Asteroxylon* (fig. 45, *Ast.*), rather larger than *Rhynia major*, is characterized by an underground stem bearing two kinds of branches; slender forked branches which burrowed in the peat or grew in water, and erect, branching shoots clothed with small, overlapping leaves like those of a *Lycopodium* (fig. 28). In their description of the Rhynie *Asteroxylon Mackiei* Dr Kidston and Prof. Lang[4] regarded certain leafless and repeatedly bifurcate axes associated with the leafy shoots as most probably the fertile portions of the plant. The discovery of another species, *Asteroxylon elberfeldense*, in Middle Devonian beds in Germany confirmed this view. In the German species,[5] as in the Rhynie plant, the lower part of the stem is covered with leaves, but at a higher level leaves are replaced by spinous processes, and farther up the branches are smooth and bear sporangia at the tips.

Evidence of the occurrence of a similar though not identical type of plant in the southern hemisphere is furnished by the photograph

1 Halle (11).
2 Halle (16).
3 Church (26).
4 Kidston and Lang (20).
5 Kräusel and Weyland (26), (29).

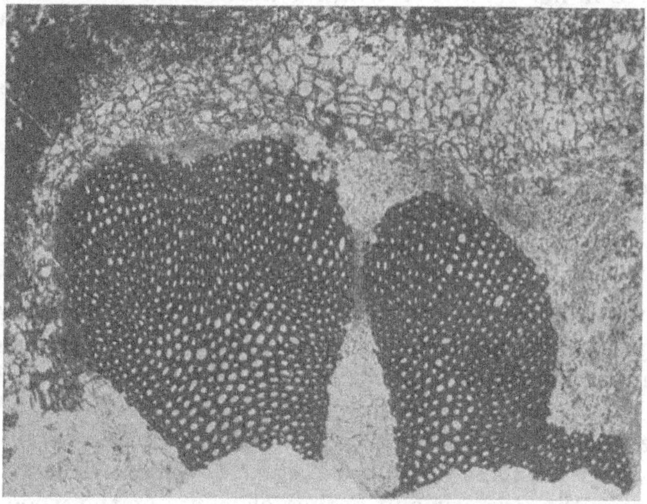

A

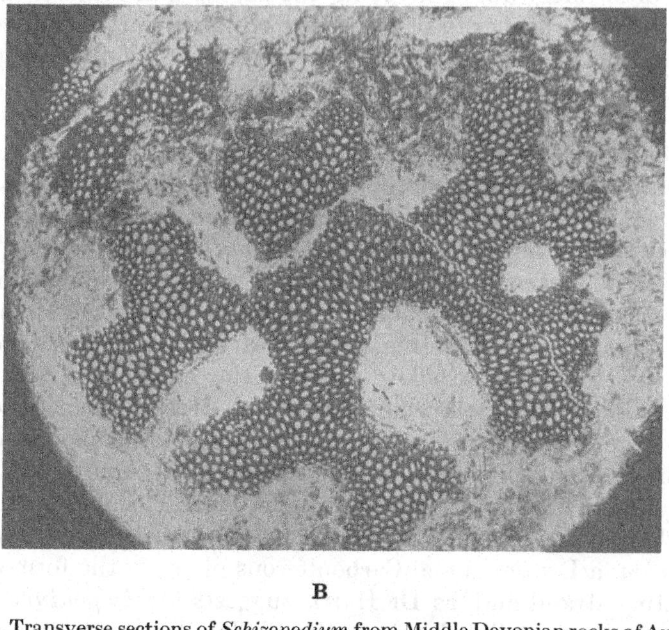

B

FIG. 47. Transverse sections of *Schizopodium* from Middle Devonian rocks of Australia. The conducting tissue (wood) of the lower stem, B, consists entirely of primary elements; the wood of the upper stem, A, shows additions of regular rows of secondary elements. × 30. (Photographs taken in Prof. Lang's laboratory.)

reproduced in fig. 47. The section, for which I am indebted to Prof. Sir Edgeworth David, was cut from a fragment found in Middle Devonian rocks in New South Wales and photographed for me in Prof. Lang's laboratory. The irregularly stellate group of woody conducting elements agrees closely with the vascular strand of the more complete Scottish specimens, but the Australian specimen reveals certain distinguishing characters of special interest. A full account of the Australian plant based on material supplied by Prof. Sir T. W. Edgeworth David has recently been published by Dr T. M. Harris.[1] In a block of chert from Middle Devonian rocks in Queensland were found twenty-five silicified axes closely packed together, probably the lower parts of upright stems. The stems are cylindrical, 3–15 mm. in diameter, bearing forked, leafless branches. In the centre is a strand of conducting tissue of lobed outline, the lobes separated by sinuses (figs. 47, 64, C): the pattern assumed by the strand varies in stems of different ages. The more delicate tissue surrounding the stronger and darker elements of the wood is described as phloem, that is the tissue which in living plants is made up mainly of slender tubes conveying the sugars and other food-substances from the leaves to places where building material is needed. As seen in fig. 47, the woody part (xylem) of the conducting strand differs from that of *Asteroxylon* in having elements disposed in regular radial rows, an arrangement characteristic of secondary wood that has been formed by a cambium layer; but in the Australian strand there is no cambium. A comparison with *Asteroxylon* (cf. fig. 64, B, C) reveals certain differences, and a distinctive name, *Schizopodium* (from the Greek σχιζόπους, which means "with divided toes", after the form of the wood in transverse section) has been given to the southern type. In *Schizopodium* the stems are leafless; the reason for this may be that the parts preserved are stem-bases which were below ground: the conducting strand is stellate, the phloem following the outline of the xylem into the sinuses between the arms. In *Asteroxylon* the whole stele is circular, only the wood (xylem) being stellate. *Schizopodium* agrees with *Palaeopitys*[2] in having radially arranged wood-elements with pits on all the walls, a character already noted in *Gosslingia*; it resembles *Cladoxylon*, a Devonian and Carboniferous plant, in the form of the conducting strand and, as Dr Harris suggests, *Schizopodium* seems to be a connecting link between the Cladoxyleae and *Asteroxylon*

[1] Harris, T. M. (29). [2] See p. 136.

which is one of several Devonian plants included in the group Psilophytales.

The association in a single individual, as described by Kräusel in *Asteroxylon elberfeldense*, of leaf-bearing and smooth axes illustrates a common difficulty in palaeobotanical work. Various forms of leafless and smooth branch-systems have been figured from Middle Devonian rocks in Scotland, the Orkney Islands, Bohemia and elsewhere under the generic name *Hostimella*:[1] some of them are dichotomously branched and show a bud-like prominence near each fork; others have a different habit. Specimens resembling a laterally branched axis of a bare fern frond were assigned by Nathorst to a new genus *Aphyllopteris*.[2] As Prof. Lang says, it is not always possible to recognize a clear distinction between *Aphyllopteris* and *Hostimella*. It is convenient to adopt provisionally the term *Hostimella* for smooth, branched axes from the older Devonian rocks pending a fuller knowledge which will enable us to allocate them to definite systematic positions and to refer to them under less artificial names. An example of a fertile species is afforded by *Hostimella racemosa* founded by Lang[3] on material from the Middle Devonian flora of Orkney consisting of smooth axes, some of which are characterized by widely divergent forks, bearing sporangia at the tips of short lateral branches. It was specimens of this type that Salter[4] described as rootlets provided with tubercles, and it may well be that a fossil figured by Halle[5] as an unknown plant from the Falkland Islands is a fragment of a closely allied form. A restoration of *Hostimella racemosa*, based on data supplied by Prof. Lang, is included in the older Devonian landscape (fig. 45, *Hos.*).

Another fairly common type of stem in Middle Devonian beds is that known as *Thursophyton*, from Thurso in Scotland, and formerly described as *Lycopodites*. We know that some at least of the *Thursophyton* fossils are the leafy stems of *Asteroxylon*, but in the absence of anatomical evidence it is safer to retain the less committal name *Thursophyton*. Before describing other vascular plants reference may be made to the simpler members of the Rhynie flora.

Though most of the Palaeozoic fossils assigned by authors to the

Typical examples may be seen in the Royal Scottish Museum, Edinburgh, and in other collections.

[2] Nathorst (15).
[3] Lang (25).
[4] Salter (57).
[5] Halle (11).

group Charophyta[1] are unconvincing, there is little doubt of the existence in the Devonian period of plants which agreed generally with the existing genera *Chara* and *Nitella*. Specimens described by Kidston and Lang as *Algites* (*Palaeonitella*) *Cranii*[2] from Rhynie are almost certainly fragments of a species allied to surviving members of the Charophyta. In the Devonian swamps, as in similar circumstances to-day, lower forms of plant-life found a favourable nidus and played an important part as members of a peat-forming association. The tissues of *Rhynia* and of the other vascular plants contain in their cells innumerable tubular threads, which often occur as tangled clusters swelling into vesicles that developed into well-protected resting spores. The thick-walled spores are often invaded by still more delicate threads of other fungi, and these in turn produced spherical reproductive cells within the cavities of the larger spores. Some of the fossils exhibit a strong family likeness to such common fungi as *Saprolegnia*, *Pythium*, and *Peronospora*, genera familiar to all botanical students and, under other names, well known as garden pests. Several forms are described from Rhynie under the comprehensive term *Palaeomyces*. Bacteria, in the form of densely packed colonies of minute single cells, have also been detected. There were in addition slender, unbranched and curved filaments of small, discoid cells hardly distinguishable from examples of the Blue-Green Alga *Oscillatoria*, which is widely spread at the present day.

We have passed in review the plant-population of a Devonian peat-bog: the vascular plants have certain primitive traits, but in their organization and in the detailed structure of their mechanism they are surprisingly modern.

Other Devonian Plants

Pseudosporochnus (fig. 45, *Ps.*) is one of several genera described from Middle Devonian beds in Bohemia:[3] it reached a height of 10 ft. and more, far taller and more robust than any of the Rhynie plants. From a bulbous base, attached by slender outgrowths to the soil, a columnar stem rose to a height of several feet before it divided into a crown of digitate branches clothed with filiform, forked appendages. On the older parts of the stem there is a well-marked

1 See Groves and Bullock-Webster (24) for an account of the Charophyta.
2 Kidston and Lang (21²).
3 Potonié and Bernard (04).

vertical ribbing. Some of the branches of *Pseudosporochnus* may, as Kräusel suggests, be represented by fossils described as *Hostimella*: others, similar to the finer ramifications of *Asteroxylon*, bore small terminal sporangia. The genus has been recognized by Lang[1] in Orkney. A different type, known only as imperfectly preserved impressions, was described by Nathorst from Norway as *Bröggeria*[2] (fig. 45, B). The stem and branches were leafless: some branches bore loose cylindrical cones in place of single sporangia. It is possible that the superficial resemblance of the fertile shoots to the spore-bearing arms of *Archaeopteris* (fig. 48, A) may be evidence of affinity. From the Middle Devonian beds of Scotland and some other regions specimens have been referred to *Protolepidodendron*,[3] a name indicative of relationship to *Lepidodendron* which was one of the noblest trees in the Coal-period forests. The branches are covered with oval or hexagonal, cushion-like areas to which were attached narrow leaves. This genus is recorded also from Upper Devonian rocks in New South Wales[4] and North America.

A plant from the Middle Devonian flora of Germany—*Calamophyton germanicum* (fig. 45, *Ca.*)[5]—in some respects foreshadows the living *Equisetum* (fig. 27) and the calamites of the Coal Measures: the leaves disposed in whorls are slender and linear, simple or repeatedly bifurcate, and on younger shoots forming more crowded tufts. Some branches bore cylindrical catkins (strobili) with bifurcate appendages having small sporangia at the ends of the recurved ultimate branchlets. *Calamophyton* resembled *Pseudosporochnus* in habit. The genus *Hyenia* (fig. 45, *Hy.*),[6] represented by two species, one from Norway and the other from Germany, is also characterized by its whorled thread-like and forked leaves and repeatedly forked branches. The sporangia are pendulous from the slender stalks of loose catkins. *Hyenia* and *Calamophyton* are the oldest examples of articulate plants, i.e. plants of the *Equisetum* type but with a less regular and a less completely whorled construction: they are referred to a group, Protoarticulatae.

Among the more puzzling fossils from Middle Devonian rocks are some which were described by Miller as fucoids (seaweeds): they were subsequently included by different authors in *Psilophyton*, *Ptilophyton*, *Spiropteris*, and other genera. The most recent account

[1] Lang (27).
[2] Nathorst (15).
[3] Lang (26).
[4] Walkom (28).
[5] Kräusel and Weyland (26), Pl. xv.
[6] *Ibid.* Pl. xvi.

of these fossils with many names, but probably all belonging to one species, is by Prof. Lang, who rechristens the plant *Milleria Thomsoni* (fig. 45, *M.*).[1] It is known only in the form of comparatively slender axes over 2 ft. long with alternate lateral members and, nearer the apex, incurved branches bearing clusters of elongated sporangia. There is no evidence amounting to proof of the nature of the stem supporting the fertile branches; it is, however, possible, as Kidston believed, that the spirally arranged lateral branches of the specimens long known as *Caulopteris Peachii* may be the bases of the leafless branch-systems assigned, after many christenings, to *Milleria.* We do not know whether *Milleria* was a fern or a pteridosperm.[2] Possibly the association with it of *Hostimella pinnata* affords a clue to the habit of the lateral members of the fertile axes: this form of *Hostimella,* which may be part of *Milleria,* bears small, divided ultimate branchlets of the type familiar as substitutes for leaflets in many of the older Devonian plants. The genus *Aneurophyton* (fig. 45, *An.*)[3] (so called because of the lack of veins—the slender conducting strands—in the leaflets), from the Middle Devonian of Germany, illustrates a further stage in the development of true fronds. It was an arborescent plant with two kinds of leaves, or possibly one frond bore two kinds of lateral branches; some with short forked appendages, incomplete leaflets; others with smaller appendages provided with sporangia. The conducting tissue of the stem consisted not only of a simple rod of tubes, as in *Rhynia,* but of a central core surrounded by a cylinder of secondary wood, in this respect agreeing to some extent with the Australian *Schizopodium.* In living ferns the conducting tissue of the leaf-stalks forms a pattern different from that in the stem: in *Aneurophyton* stem and leaf are, structurally, almost identical. We have in this feature an indication that division of labour among the members of the plant-body was less pronounced than it afterwards became.

In *The Footprints of the Creator* Miller described a petrified fossil from Cromarty as a piece of a gymnosperm stem: it was named by M'Nab *Palaeopitys Milleri.* Kidston and Lang[4] made a thorough examination of this exceptionally interesting specimen, which consists of a fragment 2 in. in diameter. There is a narrow axial region of primary wood surrounded by a cylinder of secondary elements

[1] Lang (26). [2] For description of pteridosperms see pp. 147, 189.
[3] Kräusel and Weyland (26), (29). [4] Kidston and Lang (23).

penetrated by rays of smaller cells. As seen in transverse section, the secondary wood agrees closely with the wood of a conifer; but when viewed in longitudinal section it reveals a peculiarity, namely the occurrence of pits on all the walls, and not only on the radial walls. Another feature is the absence of a pith. Nothing is known of the habit of the plant, nor of its reproductive organs: in the structure of the conducting tissue it resembles *Aneurophyton*. *Palaeopitys* is a striking illustration of the high structural level attained by a member of the Middle Devonian flora. A similar type is illustrated by *Dadoxylon Hendriksi* described on a previous page (p. 126).

An exceptionally interesting member of a contemporary flora has been described by Kräusel and Weyland as *Cladoxylon scoparium* (fig. 45, *Cl.*).[1] This is a species of a genus previously believed to be confined to Upper Devonian and Lower Carboniferous floras: the plant was a shrub or small tree with a columnar stem, or possibly a short stem bearing a cluster of divergent branches. The leaves were wedge-shaped, simply forked or like fans divided into many segments and bearing spore-capsules on the margin; some leaves were stalked, in others the flat lamina was attached directly to the supporting branch. The genus *Cladoxylon*[2] was originally established on petrified stems. A characteristic feature is the occurrence in the stem of separate strands of wood instead of the concentration of the wood, as in most plants, into an axial cylinder. The conducting tissue as seen in a transverse section has the form of bands, horse-shoe shaped or U-like strands, which unite with one another as they pass through the stem: the general plan is similar to that in *Schizopodium*. In the stouter axes of *Cladoxylon scoparium* the conducting tissue forms a tangle of irregularly placed groups of woody tissue. An analogous arrangement occurs, with modifications, in some existing tropical flowering plants with climbing stems: in Palaeozoic gymnosperms it was much commoner. An interesting suggestion has been made that the leaves of the Devonian *Cladoxylon* may be branch-systems which had not become completely transformed into organs having the distinctive features of foliage-leaves: if this is a correct interpretation it affords support to the view that the larger fern-like forms of leaves were evolved from repeatedly

[1] Prof. Sahni in a recent paper on the genus *Asterochlaenopsis* (Sahni (30)) gives reasons for transferring this plant to a new genus and not including it in *Cladoxylon*.

[2] For a general account, with figures, of this genus see Scott (23) and Seward (17).

branched lateral members of stems such as we see in an earlier state in some examples of *Psilophyton.* Another genus similar in the structure and arrangement of the conducting tissue to *Cladoxylon* is *Duisbergia*,[1] described from Middle Devonian rocks in Germany. The stem bore crowded lateral organs, 5 cm. long, flat and frayed at the edge into narrow segments. These leaves were probably spirally attached, their stalks being decurrent, that is running down the main axis and giving the appearance of a stem split down its length into narrow pieces each of which produced flat, leafy appendages. Between the sterile leaves, which may be compared with the larger leaves of *Psygmophyllum*,[2] occur smaller appendages which appear to be fertile.

Petrified stems of an even more highly organized type have been described from North America as *Callixylon* and *Dadoxylon*. The name *Callixylon* was applied by Zalessky[3] to an Upper Devonian stem from Russia which agrees in many respects with stems of living araucarian trees but differs in details: a species is recorded from beds, which may be Middle Devonian in age, in the State of New York as *Callixylon Marshii*.[4] Stems of the *Dadoxylon* type, that is with wood resembling in various features that of existing conifers of the *Araucaria* family, have been described by Penhallow[5] from several North American localities. The differences in detail between the two genera *Callixylon* and *Dadoxylon* need not be discussed. These Middle Devonian trees furnish an impressive proof of the advanced state of differentiation of the plant-body; there is little to choose in anatomical complexity between them and living araucarias. The araucarian character, in the structure of the wood, was well represented in the Carboniferous vegetation: it was on stems with wood of this type that the strap-shaped leaves known as *Cordaites*[6] were borne. The genus *Cordaites* has been recorded by authors from Devonian rocks, but the leaves so named have no value as evidence of the occurrence of the genus before the Carboniferous period, when it became a conspicuous forest tree. Wood of the *Callixylon* type—characterized by the occurrence of groups of pits like those on the tracheids (elongated conducting cells) of *Araucaria* separated by smooth unpitted portions of the tracheid walls—is

[1] Kräusel and Weyland (29). [2] See p. 139. [3] Zalessky (11).
[4] Hylander (22). See also Arnold (29) and Arnold (30) published after this chapter was written. [5] Penhallow (00). See also Lang (29) and p. 126 of this chapter.
[6] See p. 225.

fairly abundant in Devonian rocks of North America, both Middle and Upper.

Most of the plants enumerated above were leafless or provided with small, scale-like foliage. The largest leaves so far discovered among the older floras are those known as *Psygmophyllum* (fig. 45, *Pg.*). In a species described from Middle Devonian rocks in Belgium[1] the leaves are 10–14 cm. long and 10–15 cm. broad and have a long stalk: smaller leaves from Norway[2] reach 3 cm. in length. The fan-shaped leaves of *Psygmophyllum* are very similar to those of the maidenhair tree (*Ginkgo biloba*) and in shape are not unlike those of *Duisbergia*. There is, however, no adequate reason for assuming relationship between *Psygmophyllum* and *Ginkgo*. A larger leaf, similar in shape to the Norwegian species, was found in the Middle Devonian of Caithness. The discovery by Lang[3] in Middle Devonian rocks of Caithness of more than eight different types of spore, varying in size and in form, which cannot be assigned to their respective parents, is a striking confirmation of the belief that the samples of the older floras so far obtained afford a very imperfect picture of the vegetation.

B. The Later Devonian Floras

Most of the material on which our knowledge of the later Devonian floras is based was gathered from the sedimentary rocks of Bear Island, the southernmost island of Svalbard Archipelago, a projecting piece of the continental shelf which extends from Norway to Spitzbergen (Svalbard) (map, fig. 42, 2); from rocks of the same age in Ellesmere Land (fig. 42, 1), still nearer the pole and not far from the most northerly land in the world; from the rich plant-beds of southern Ireland; from North America (fig. 42, 3); from Belgium, Germany, France, and southern Russia (fig. 42, 4 and 5). Devonian plants from Australia (fig. 42, 9–11) demonstrate the world-wide distribution of *Archaeopteris* (fig. 48, *A*, *A′*) and some other genera. The existence as far north as lat. 77° N. of a luxuriant vegetation, including plants in no way inferior in development to the closely allied or identical species in the Donetz basin of Russia, raises the difficult question of climatic contrasts between the past and the present, a problem which confronts us through the whole of geological history and still awaits a satisfactory solution.

[1] Leclercq (28[2]). [2] Nathorst (20[2]). [3] Lang (25).

Upper Devonian floras differ in several respects from those we have already considered: the transition from the earlier to the later vegetation cannot be described as gradual: a few of the older genera remain; others appear under a slightly different guise; but on the whole the difference is great, and whether or not it may be mainly the result of the substitution of more normal conditions for the swamps and marshes of the earlier days, we are conscious of a marked change in the fashion of the plant-world, the prevalence of types with modern in place of archaic characters. The leafless or almost leafless, spinous or smooth, branch-systems which are the distinguishing features of the Psilophytales are barely represented: in their stead we find *Cyclostigma* (sometimes called *Bothrodendron*)[1] (fig. 48, *Cy.*), *Lepidodendron*, and *Leptophloeum* (fig. 48, *L.*) bearing witness to progressive evolution within the great group Lycopodiales. The generalized character of some of the Psilophytales suggests a close relationship between plants which we may compare with *Psilotum* and types more akin to *Lycopodium*. Plants with fern-like fronds like those of *Aneurophyton* and *Sphenopteridium* (fig. 48, *Sm.*) were still fairly common in the later floras, but it was in Upper Devonian floras that fronds of the modern type—that is, fronds characterized by well-developed leaflets supplied with veins —came into prominence and set a fashion in construction which has persisted ever since. The genus *Psygmophyllum* (fig. 48, *Pg.*), first recorded from the Middle Devonian, is represented in the later floras by species for the most part with larger and more varied leaves, though the Belgian leaves referred to on a previous page are as large as those of any of the younger examples.

In the American *Asteropteris* we have the earliest member of the Filicales, the first step in the gradual development of a group destined for long periods to remain subordinate to the pteridosperms. In *Pseudobornia* and *Sphenophyllum* (fig. 48, *Pb.*, *Sp.*) we see a fuller realization of the articulate type. In a word, the Upper Devonian vegetation brings us nearer to what is familiar: many, though not all, of the older plants give the impression of types new to the land and, perhaps, not yet completely adjusted to a strange environment; whereas in the later floras the correlation of plant-body

[1] The slight difference between these two genera hardly concerns us: for fuller definitions of genera of fossil plants reference should be made to text-books. The true *Bothrodendron* is a characteristic member of the late Carboniferous vegetation. For *Cyclostigma* see Haughton (60).

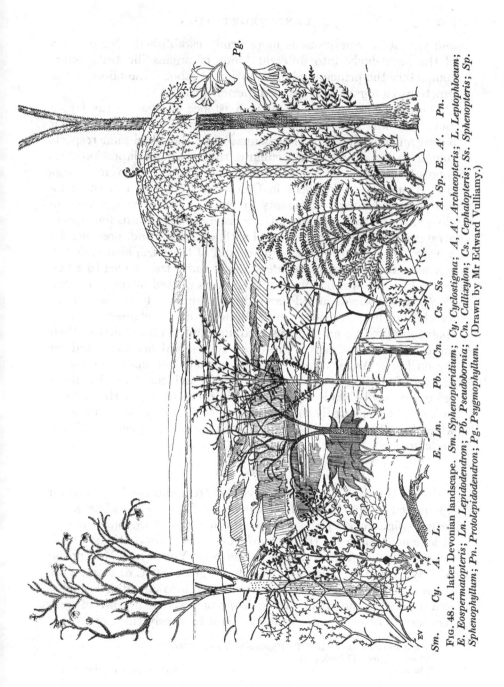

FIG. 48. A later Devonian landscape. *Sm. Sphenopteridium*; *Cy. Cyclostigma*; *A, A′. Archaeopteris*; *L. Leptophloeum*; *E. Eospermatopteris*; *Ln. Lepidodendron*; *Pb. Pseudobornia*; *Cn. Callixylon*; *Cs. Cephalopteris*; *Ss. Sphenopteris*; *Sp. Sphenophyllum*; *Pn. Protolepidodendron*; *Pg. Psygmophyllum*. (Drawn by Mr Edward Vulliamy.)

and terrestrial conditions is more firmly established. The division of the plant-body into different kinds of organs illustrates more completely the principle of distribution of labour, the allocation of functions to organs best fitted to perform them.

Though we are convinced of the existence of algae in the Upper Devonian seas and lakes, we know practically nothing of their nature and affinities. The older genus *Parka*, though in some respects similar to certain algae, is peculiar in the apparent adaptation of its spores to dispersal on land. The same peculiarity is seen in an alga from Upper Devonian beds in Ohio: the specimens consist of incomplete carbonized fragments in which nests of spores are enclosed in a delicate cellular matrix. Each group contains four spores arranged as a tetrad and provided with a resistant and, presumably, a waterproof membrane. A full account of these fragments, originally named *Sporocarpon furcatum* and recently transferred to a new genus *Foerstia*, will be found in the papers quoted in the footnote.[1] *Parka* and *Foerstia* afford examples of imperfectly known plants which may be transitional forms between algae, characterized by spores adapted to an aqueous medium, and plants which in their more resistant and better protected spores had become fitted for life on land. As an example of another and not uncommon form that is often assigned, though on slender evidence, to the algae, reference may be made to some large impressions, more than 30 cm. in length, of repeatedly forked ribbon-like fossils, good examples of which have been described from North America as species of *Thamnocladus*.[2]

Forest Trees

One of the most impressive Devonian trees and the largest so far discovered in Pre-Carboniferous rocks is *Protolepidodendron primaevum*[3] (fig. 48, *Pn.*) preserved in the muds of a Devonian lagoon or lake near Naples in the State of New York. Restorations[4] of this remarkable tree are exhibited in the New York State Museum at Albany. The stem probably reached a height of 20 ft. or more, a tall column, possibly bifurcating near the summit into spreading branches and covered in the upper part with simple leaves like those of some lycopods. The occurrence of leaves on a relatively thick stem sug-

[1] White and Stadnichenko (23); Kidston and Lang (24).
[2] White, D. (02); Hollick (10).
[3] White, D. (07).
[4] Berry (20); Goldring (27).

gests a comparison with the recent *Araucaria imbricata* (monkey puzzle) which retains its leaves for several years. Special attention is drawn to the association on one trunk of surface-features that subsequently became the distinguishing marks of two separate genera, *Lepidodendron* and *Sigillaria* which flourished in the Carboniferous forests. *Lepidodendron* and *Sigillaria* share a character in common, namely the possession of what are called leaf-cushions. On the fall of a leaf a clean-cut scar is left, and in some of the earlier Devonian plants, such as *Arthrostigma* and the so-called Lepidodendroid stems from South Africa and the Falkland Islands, there is no prominence or cushion associated with the leaf-scar. An analogous difference is illustrated by a comparison of the twigs of the spruce fir (*Picea*) and the silver fir (*Abies*): in *Picea* the leaf-scar is at the summit of a small peg-like prominence which remains on the leafless twig; in *Abies* the scar is merely a circle on the smooth surface of the twig. In *Protolepidodendron*,[1] as represented by Middle Devonian forms, the branches are covered with well-defined cushions similar to the more elaborate cushions of *Lepidodendron* which became more abundant in the Carboniferous period (fig. 48, *Ln.*). On the thinner stems of the Naples tree the leaf-cushions are arranged spirally as in *Lepidodendron*; lower down they occur in regular vertical series as in *Sigillaria*. At a still lower level near the base the exposed surface shows a reticulate pattern, no doubt the result of exfoliation of the surface-tissues and the laying bare of a network of strengthening fibres in the deeper cortex. The trunk ended in a bulbous base from which innumerable rootlets radiated through the soil. In the swollen base we have a feature, comparable on a large scale to the small tuberous stem of the Devonian genus *Hornea*, and perhaps representing an early stage in the development of the forked and spreading root-like organs (*Stigmaria*) of *Sigillaria* and *Lepidodendron*. It is possible that some large fossils from Ellesmere Land called by Nathorst[2] *Lyginodendron Sverdrupi*, characterized by a reticulate pattern on the stem, may be the older portions of a tree allied to *Protolepidodendron*. We know nothing of the reproductive organs nor of the anatomy of the Naples tree. A *Protolepidodendron* closely resembling the American species has recently been recorded from Devonian rocks in New South Wales.[3]

A tree of which we have a fuller knowledge is represented by the

[1] Lang (26). See *ante*, p. 135.
[2] Nathorst (04).
[3] Walkom (28).

genus *Cyclostigma* (fig. 48, *Cy.*), or, as it is sometimes called, *Bothrodendron*, allied to *Protolepidodendron.* The best examples are from Bear Island and southern Ireland. In its forked stem and branches and in the compact cones *Cyclostigma* may be compared with the common tropical species of *Lycopodium* shown in fig. 28. The main stem rose from a large bifurcate rhizome, similar to the Carboniferous stigmarias, like a tall pine; but in place of formal tiers of branches it divided into a spreading canopy of forked shoots thickly clothed with long and narrow leaves disposed in spirals or occasionally in whorls. On the thicker branches clean-cut scars mark the position of fallen leaves, and at a lower level the smooth surface is replaced by a wrinkled bark. Cones of fertile leaves bearing sporangia containing both large and small spores hung from the ends of slender twigs.

The genus *Lepidodendron* differs from *Bothrodendron* in the association of leaf-cushions with the leaf-scars. A petrified specimen, described by Zalessky[1] from Russia, demonstrates the occurrence of a Devonian species with a conducting strand composed of a solid rod of tracheids, a primitive type that is found also in some Carboniferous *Lepidodendra.* Another member of the lycopod group is recorded from North America and Spitsbergen, also from both Upper Devonian and Lower Carboniferous rocks in Australia: the northern form is known as *Leptophloeum rhombicum*[2] and the southern as *Lepidodendron australe*[3] (fig. 48, *L.*). The stems—the only part that is known—are characterized by a pattern resembling a tesselated pavement formed of four-sided pieces set edgewise; the pieces are simple cushions and each bore a narrow leaf. Stems of similar type are known from the Coal Measures of Europe, and for some of them a new generic name *Phialophloios*[4] has been proposed. We know very little of the common Australian species, which is widely spread in rocks ranging from Middle Devonian to Lower Carboniferous, beyond its surface features, and it is doubtful whether the name *Lepidodendron* is appropriate.

Pteridosperms and Ferns

We may now pass in review some plants with the habit of ferns. More than fifty years ago a disastrous flood near the village of Gilboa in the Catskill Hills exposed a number of erect trees pre-

1 Zalessky, M. and G. (21).
2 Smith and White (05).
3 McCoy (74); Walton (26); Seward (07).
4 Hörich (15).

served where they grew. More recently additional specimens have been disinterred, and it is now possible to picture with unusual clearness the appearance of a grove on the shore of the Devonian sea. Dawson referred the first Gilboa trees to *Psaronius*, a genus of ferns that flourished in the later Carboniferous and in the Permian period; but it has been shown by Miss Goldring[1] that the fern-like appearance is misleading and that, in all probability, the fronds bore two kinds of reproductive organs, sporangia attached to special, saucer-like leaflets, and small seeds, 6–8 mm. long, at the tips of divergent branches of another type of leaflet. The internal structure of the plant is unknown. A cylindrical stem, 30–40 ft. high and 3 ft. or more in diameter bore a crown of tripinnate fronds 6–9 ft. in length (restoration, fig. 48, *E.*): but the canopy was less dense than that of a modern tree fern because of the sparsely scattered, bilobed appendages which had little or no web of flat green tissue. Near the base of the stem, below the region clothed with stumps of leaf-bases and marked by spirally set scars of fallen fronds, a network of strengthening fibres was exposed on the destruction of the superficial tissues, as in an old *Protolepidodendron.* This Gilboa tree affords the best evidence of the pteridosperm nature of a Devonian genus and it suggests a similar systematic position for other fern-like fronds of the same geological age: it is named *Eospermatopteris textilis.* Other fern-like fronds, borne on stems which have not been identified—represented almost entirely by sterile specimens—are known as *Sphenopteridium condrusorum,*[2] a widely spread species with very slender and slightly webbed leaflets, and a closely allied form *Sphenopteridium Keilhaui*[3] from Bear Island (fig. 48, *Sm.*). Accurate specific discrimination between these large fronds which were scattered over the world is hardly possible. A further advance towards the modern type of leaflet is illustrated by a few specimens included in the genus *Sphenopteris*, a general term applied to fern-like fronds differing, though in some instances very slightly, from *Sphenopteridium* in the greater development of a green web on the wedge-shaped leaflets. *Sphenopteris Bailyi*[4] from Ireland and *S. flaccida*[5] (fig. 48, *Ss.*) from Belgium are examples of the oldest known representatives of the genus.

In *Cephalopteris*[6] (fig. 48, *Cs.*), another genus founded on frond-

[1] Goldring (24), (26), (27); Clarke, J. M. (21). [2] Gilkinet (22).
[3] Nathorst (02). [4] Baily (61). [5] Gilkinet (22).
[6] Formerly called *Cephalotheca*: Nathorst (10), p. 278.

like branch-systems, the distinctive feature is the occurrence of opposite pairs of fertile branches, each branch bearing near its attachment to the main axis of the frond a dense mop of sporangia. The alternately arranged sterile branches of this Bear Island plant had a few very small bifurcate appendages similar to those of *Sphenopteridium* and other genera. It seems to have been a common feature in Devonian fronds to be differentiated into sterile and fertile regions as in a minority of modern ferns.

The realization of the typical form of a large fern frond is seen in *Archaeopteris* (fig. 48, *A.*), one of the most characteristic and widespread members of Upper Devonian floras. The largest examples are from the Kiltorcan beds of Ireland in which *Archaeopteris hibernica* is abundant: the fronds, 50–80 cm. long, bear two series of long and tapering branches (pinnae) covered with rows of closely set wedge-shaped leaflets as in some species of *Adiantum*. Specimens with leaflets agreeing more or less closely with those of the Irish species are recorded from several other European localities, from North America and Australia: another form is represented by *Archaeopteris fimbriata* and *A. fissilis*, from Ellesmere Land and Russia, in which the leaflets are cut into slender divergent segments (fig. 48, *A.'*).[1] A few leaflets occur on the frond-axis between the lateral branches, and the presence of two wing-like expansions at the base of each leaf-stalk is another characteristic feature which this extinct genus shares with existing members of the Marattiaceae, a family of tropical ferns. A more important character is the nature of the fertile region of the frond: on the lower branches of most species the flat leaflets are replaced by slender, branched stalks bearing sporangia. It is probable that in some forms the fertile pinnae were on the apical part of a frond. A species from Ellesmere Land and Russia, *Archaeopteris archetypus*,[2] is distinguished by the unusual dimensions of the leaflets, which are hardly smaller than the leaves of a *Ginkgo*. In this species there is a broad, collar-like expansion near the base of the leaf-stalk. It is impossible clearly to distinguish between the larger leaflets of this type and some of the leaves described as *Psygmophyllum*: the resemblance may be an expression of affinity. We know very little of the stem of *Archaeopteris*: an Irish specimen[3] affords evidence of the attachment of fronds to a relatively slender rhizome. It may be that in some species the stem

[1] For references see the table on p. 153. [2] Schmalhausen (94); Nathorst (04).
[3] I am indebted to Prof. Johnson of Dublin for a photograph of this specimen.

was creeping and in others erect. Fronds with leaflets practically identical with those of some forms of *Archaeopteris* have been found in Carboniferous beds, but none are known with fertile pinnae like those of the Devonian species. The marked difference between the sterile and fertile pinnae in *Archaeopteris*, in *Dimeripteris*,[1] a North American and Russian type, and in some other Upper Devonian genera reminds one of the similar separation of the two regions in the fronds of the royal fern (*Osmunda regalis*).

The question is, Was this very fern-like foliage borne on a plant that in its reproductive methods and in its anatomical structure differed from true ferns? It may well be asked, Why should we not accept *Archaeopteris* as a fern, especially as no seeds have been found attached to the fronds? For many years the beautifully preserved impressions on the hardened Carboniferous muds associated with seams of coal were naturally regarded as pieces of fern fronds, and it was customary to describe many of them under the generic names of living ferns. About forty years ago evidence began to accumulate which eventually proved conclusively that the great majority of the Carboniferous "ferns" were seed-bearing plants —pteridosperms.[2] Having regard to the fact that the number of Carboniferous fronds in English rocks alone is greatly in excess of the Devonian fronds from all over the world, it is not surprising that we have still in most instances to be content with circumstantial evidence. It is noteworthy that some stems from Bear Island and Ellesmere Land agree very closely with those of *Heterangium*[3] and other undoubted pteridosperms of Carboniferous age. Reference may be made to the discovery in Devonshire, in beds which may be Upper Devonian, of some slender forked axes bearing at the tips deep cup-like expansions with a toothed margin: these fossils, named *Xenotheca devonica*[4], are hardly distinguishable from Carboniferous specimens which are known to have enclosed seeds and are comparable to the husks of a hazel nut. The nearest approach to proof of the pteridosperm nature of the fern-like Devonian plants is furnished by *Eospermatopteris* described by Miss Goldring and by Kräusel's description of *Aneurophyton*. We may confidently predict that additional evidence will be forthcoming.

So far as I am aware, only one example of a fern in the strict sense has been recorded from Devonian rocks. This is a small piece

[1] Schmalhausen (94); Smith and White (05).
[2] See also Chapter XI, p. 189.
[3] Nathorst (02).
[4] Arber and Goode (15).

of a petrified stem from the State of New York which was named *Asteropteris* because of the stellate form, as seen in transverse section, of the woody conducting tissue. A similar form of conducting strand occurs in the later and more elaborate fern, *Asterochlaena* (fig. 64, L). From the ends of the narrow arms of the star were detached small strands for the supply of the leaves. Prof. P. Bertrand,[1] who added considerably to Dawson's original account of the fragment, describes *Asteropteris* as the stem of an herbaceous fern with leaves, the form of which is not known, borne in superposed whorls and not in the usual spiral fashion. In its stellate form the conducting tissue of *Cladoxylon* resembles that of *Asteropteris*, but in the latter genus the strands supplying the leaves are much simpler.

It has been stated that another fern, *Clepsydropsis* (fig. 59), occurs in Upper Devonian rocks in Australia; but the age of the beds is almost certainly Carboniferous.

Other Late Devonian Plants

Upper Devonian strata have yielded a few examples of articulate plants. By far the largest of these is *Pseudobornia ursina* (fig. 48, *Pb.*) described by Nathorst from Bear Island. The main stem grew horizontally, as in *Equisetum*, and at regular intervals gave off erect branches with tiers of leaves divided into long and narrow, fringed and spreading wedges. Fertile shoots, reaching 30 cm. in length, bore sporangia on the lower faces of reduced leaves. In habit *Pseudobornia* may be compared with a large *Equisetum*, but it was more robust; its leaves were much larger, and it differed in the form of the fertile branches. A few Devonian examples are known of the genus *Sphenophyllum* (fig. 48, *Sp.*) which, however, reached its maximum development in the Carboniferous period: whorls of wedge-shaped leaves, barely 1 cm. long, cut into forked filaments were borne on slender, jointed stems. In habit, though not in other respects, *Sphenophyllum* resembled *Galium aparine* (goose grass) of our hedgerows, and like *Galium* it probably sprawled over the stouter stems of other plants.[2] The genus, which is represented in Carboniferous floras by well-preserved material, is assigned to a special group of vascular cryptogams, the Sphenophyllales.

[1] Bertrand, P. (14).

[2] By some palaeobotanists *Sphenophyllum* is believed to be an aquatic plant. See White, D. (29).

Attention has already been called to the genus *Psygmophyllum* (fig. 48, *Pg.*) Two species are recorded from Upper Devonian beds in Spitsbergen: one of them in the shape, size, and venation of the leaves—no other parts of the plant have been recognized—bears a striking resemblance to the maidenhair tree (*Ginkgo biloba*), the "living fossil" as Darwin called it. Another form of *Psygmophyllum*, described by Johnson[1] from Ireland as *Ginkgophyllum hibernicum*, is characterized by its greater length and more deeply divided lamina. We are not entitled to assume a real relationship between these fossils and *Ginkgo*, since nothing is known of their structure or reproductive organs; but it is tempting to think of the *Gingko* stock, which has been traced to the early days of the Mesozoic era, having its roots in the Devonian period. (See fig. 136.)

There is no need to insist on the fact that Upper Devonian floras included gymnosperm trees of a type already mentioned in the account of the earlier floras. Species of *Callixylon* (fig. 64, Q) occur in Russia and North America. The discovery by Nathorst of pieces of large and thick stems in Ellesmere Land illustrates one of the contrasts between Devonian vegetation of the north polar regions and the "scrubbed shoots" and dwarfed prostrate stems of the present Arctic flora.

Among the plant remains of unknown affinity which are omitted from the above summary of the floras there are some large specimens of vertically ribbed stems, some reaching a diameter of 12 cm., from Middle and Upper Devonian rocks in Scotland and the Shetland Islands. These are often spoken of as the "corduroy plant"[2] and have been compared to casts of calamitean stems:[3] they are certainly not *Calamites* and are probably pieces of some tree in which a ribbed fibrous tissue was exposed after the loss through partial decay, or as the result of natural exfoliation, of the superficial bark. Partially decorticated lepidodendroid stems figured by Nathorst[4] from Lower Carboniferous beds in Spitsbergen resemble the Devonian fossils; specimens with similar surface features are recorded from Upper Devonian beds of France and compared to *Archaeocalamites*.[5] The "corduroy plant" may be the stem of a tree such as *Pseudosporochnus* or *Protolepidodendron* (fig. 45, *Ps.*; fig. 48, *Pn.*).

[1] Johnson (14[2]).

[2] Specimens may be seen in the Edinburgh Museum and in the Museum of the Geological Survey, London.

[3] Hooker (53); Miller, H. (57). [4] Nathorst (14), Pl. VIII. [5] Bureau (14).

There is no real evidence of the existence of *Calamites* in Devonian floras. The fossils described by Gilkinet[1] from Upper Devonian beds in Belgium as *Asterocalamites* (*Archaeocalamites*) do not appear to show any well-defined calamitean features.

In several Upper Devonian plants we recognize the advanced guard of hosts destined to wander over the northern hemisphere and to furnish the forests of the Coal Age with a vegetation far more varied and luxuriant than the world had previously known. A comparison of Upper Devonian floras shows no definite indication of well-marked provinces or of differences in the vegetation of widely separated regions equal in degree to those between similarly scattered floras at the present time. In the Distribution Table (p. 152) an attempt is made to indicate the geographical and geological range of most of the Devonian genera. It is not as a rule possible to say with confidence whether or not fronds or other parts of plants agreeing closely in external form, though discovered in widely separated localities, are specifically identical. The appended table summarizes the results of a comparison of the floras; it shows that certain genera had established themselves in lands far apart, and that by far the richer floras are those in the northern hemisphere; it reveals local peculiarities as well as a general uniformity in the facies of the vegetation as a whole. Most of the genera are allocated to groups: the temptation to classify is strong, however little we know of the characters on which a classification should be based: "when I cannot satisfy my reason I love to humour my fancy". But so long as the classification is regarded as tentative, it may serve a useful purpose. In the last column a few selected references are given to sources from which fuller information both on the plants and on the earlier literature may be found. References are given also in the notes preceding the table.

[1] Gilkinet (22).

NOTES ON TABLE A.

For map of the Devonian world, see fig. 42, p. 113.

In addition to the sources mentioned in the last column reference may be made to Dr Arber's book on Devonian Floras [Arber (21)] which contains illustrations and brief descriptions of several genera. It should, however, be remembered that the book had not been finally revised at the time of the author's death; that it was written before the publication of all but the first of the series of memoirs by Kidston and Lang; that the age of the Caithness beds is Middle Devonian and not, as Arber states, Upper Devonian. Short reviews of Devonian floras will be found in Scott (22), Seward (23).

I. Ellesmere Land. (Lat. 77° N.) Upper Devonian. The plant-bearing beds, associated with thin coal seams, were probably deposited in an estuary, and some of the specimens were undoubtedly preserved near the places where they grew. Nathorst (04), (11).

II. Spitsbergen. Lower Devonian. A few branched, leafless axes associated with fossil animals. Nathorst (94). Upper Devonian. A few plants with fish remains. Nathorst (04).

III. Bear Island. (Slightly south of lat. 75° N.) Upper Devonian. Numerous and well preserved impressions; coal seams and fish remains. Nathorst (99), (02), (10). For an account of the geology see Horn and Orvin (28).

IV. Canada and the United States of America. Lower to Upper Devonian. The most important regions are:

(1) The Gaspé peninsula and New Brunswick: plants often preserved *in situ* and in strata ranging from Lower to Upper Devonian. Some of the New Brunswick beds, formerly believed to be Devonian, were recognized by Dr White and by Dr Kidston as Upper Carboniferous [Geol. Surv. Canada; Summary Report for 1899. Ottawa 1900; Stopes (14)]. Dawson (71), (81), (82); White, D. (13); Coleman and Parks (22).

(2) South-eastern Maine. Upper Devonian. Estuarine beds with plants and interbedded lavas. Smith and White (05). Many of the figured specimens are undoubtedly from Upper Devonian rocks, but some suggest a Middle or even a Lower Devonian horizon.

(3) New York State; Pennsylvania, etc. Upper Devonian. Penhallow (89), (93), (00); Prosser (94). A few plants are recorded from Upper and Middle Devonian rocks in Ohio and other States: Newberry (90); Penhallow (96), (00); Elkins and Wieland (14); Scott and Jeffrey (14). For a general account of the conditions of deposition of the sediments in North America see Barrell (13).

V. Ireland. Upper Devonian. Freshwater strata (Kiltorcan beds) in Co. Kerry and adjacent regions have yielded many large examples of *Archaeopteris*, *Cyclostigma*, and other plants. Johnson (11–14).

VI. England. Lower to Upper Devonian. A few fragmentary fossils have been found in Cornwall, Devonshire, and Somersetshire. Arber and Goode (15); Edwards in Rogers (26).

VII. Scotland. Lower to Upper Devonian. The voluminous literature, including the classical descriptions by Hugh Miller in *The Testimony of the Rocks*

TABLE A. *Geological and Geographical Distribution of Devonian Plants.*

	Silurian	Devonian Lower	Devonian Middle	Devonian Upper	Carboniferous		Ellesmere Land 1*	Bear Island : Spitsbergen 2	North American region 3	Europe (Western) 4	Russia 5	China 6	Falkland Islands 7	South Africa 8	Australia 9–11	References and Notes
Thallophyta																
ALGAE																
Foerstia				—			.	.	×	.	.	.	.	.	.	White and Stadnichenko (23); Kidston and Lang (24)
Pachytheca	—	—	—				.	.	×	×	.	.	.	.	.	Barber (89), (90); Lang (24)
Parka	—	—					.	.	.	×	.	.	.	.	.	Don and Hickling (17)
Thamnocladus				—			.	.	×	.	.	.	.	.	.	White (02); Hollick (10)
ALGAE or BACTERIA																
Archaeothrix			—				.	.	.	×	.	.	.	.	.	Kidston and Lang (21²)
FUNGI																
Palaeomyces			—	—	—	→	.	.	.	×	.	.	.	.	.	Kidston and Lang (21²)
Charophyta																
Algites (*Palaeonitella*) *Cranii*		—	—				.	.	×	×	.	.	.	.	.	Kidston and Lang (21); Groves and Bullock-Webster (24); Karpinsky (06); Knowlton (89)
Psilophytales																
Psilophyton	—	—	—				.	.	×	×	.	.	.	.	.	Dawson (59); Bertrand, P. (13); Halle (16); Kidston and Lang (17); Halle (20); Carpentier (20); Edwards (24)
Arthrostigma, etc.		—	—				.	.	×	×	.	×	.	×	×?	Halle (16); Schwarz (06²); Seward (09); Seward and Walton (23); Cookson (26)
Rhynia			—				.	.	.	×	.	.	.	.	.	Kidston and Lang (17), (20), (21)
Hornea		—	—				.	.	.	×	.	.	×	.	.	Kidston and Lang (20), (21); Halle (16)
Hostimella	—	—	—				.	.	×?	×	.	.	×	.	×	Lang (25), (26); Lang and Cookson (27); Smith and White (05); Steinmann and Elberskirch (29)
Asteroxylon			—				.	.	.	×	.	.	.	.	.	Kidston and Lang (20), (21); Kräusel and Weyland (26), (29)
Schizopodium			—				.	.	.	.	.	.	.	.	×	Harris (29)
Hicklingia			—				.	.	.	×	.	.	.	.	.	Kidston and Lang (23²); Kräusel and Weyland (29)
Pseudosporochnus			—				.	.	.	×	.	.	.	.	.	Potonié and Bernard (04); Lang (27)
Zosterophyllum		—					.	.	×	×	.	.	.	.	×	Penhallow (93); Lang (27); Heard (27)

Lycopodiales																
Lepidodendron caracubense				—			.	.	.	×	×	.	.	.	.	Schmalhausen (94); Zalessky, M. and G. (21)
Cyclostigma				—			.	×	.	×	.	.	.	×?	.	Nathorst (02); Johnson (14); Schwarz (06²)
Protolepidodendron			—	—			.	×	×	×	×?	.	.	.	×	White, D. (07); Lang (26); Kräusel and Weyland (29); Schmalhausen (94)
Leptophloeum rhombicum: *Lepidodendron australe*			—	—	—		.	×	×	.	×	.	.	.	×	Smith and White (05); Seward (07); Walton (26)
Protoarticulatae																
Calamophyton			—				.	.	.	×	.	.	.	.	.	Kräusel and Weyland (26), (29)
Hyenia			—				.	.	.	×	.	.	.	.	.	Kräusel and Weyland (26), (29); Nathorst (15)
Articulatae																
Pseudobornia				—			.	×	.	.	.	.	.	.	.	Nathorst (02)
Sphenophyllum				—	—	—	.	×	.	×	.	.	.	.	.	Nathorst (02); Bureau (14)
Filicales																
Asteropteris				—			.	.	×	.	.	.	.	.	.	Bertrand, P. (14); Scott (20)
Pteridospermae and other gymnosperms																
Archaeopteris				—	—		×	×	×	×	×	.	.	.	×	Lesquereux (79); Prosser (94); Dun (97); Nathorst (02), (04); Smith and White (05); Johnson (11), (11²)
Eospermatopteris				—			.	.	×	.	.	.	.	.	.	Clarke (21); Goldring (24), (26), (27)
Cephalopteris				—			.	×	.	×	.	.	.	.	.	Nathorst (02); Bureau (14)
Dimeripteris				—			.	.	×	.	×	.	.	.	.	Schmalhausen (94); Smith and White (05); Gilkinet (22)
Aneurophyton			—				.	.	.	×	.	.	.	.	.	Kräusel and Weyland (26), (29)
Sphenopteridium				—	—		×	×	.	×	×	.	.	.	.	Nathorst (02), (04); Gilkinet (22); Arber and Goode (15); Schmalhausen (94)
Sphenopteris				—	—	—→	.	.	.	×	×	.	.	.	×	Baily (61); McCoy (74); Schmalhausen (94); Dun (97); Gilkinet (22)
Milleria			—	—			.	.	×	×	.	.	.	.	.	Dawson (71); Potonié and Bernard (04); Lang (26); Miller, H. (49), fig. 54
Callixylon			—	—			.	.	×	.	×	.	.	.	.	Zalessky (11); Elkins and Wieland (14); Seward (17); Hylander (22)
Dadoxylon			—	—	—	—→	.	.	×	×	.	.	.	.	.	Penhallow (00); Lang (29)
Cladoxylon			—	—	—		.	.	.	×	.	.	.	.	.	Kräusel and Weyland (26), (29). See Gothan (27⁵) for description of an Upper Devonian plant under a new generic name, *Pitzschia*
Calamopitys				—	—		.	.	×	.	.	.	.	.	.	Scott and Jeffrey (14)
PLANTAE INCERTAE SEDIS																
Palaeopitys			—				.	.	.	×	.	.	.	.	.	Miller, H. (49); Kidston and Lang (23)
Psygmophyllum			—	—	—		.	×	×	×	×	.	.	.	.	Nathorst (94), (15), (20²); Smith and White (05); Johnson (14²); Leclercq (28²)
Bröggeria			—	—			.	.	×	×	.	.	.	.	.	Nathorst (15); Smith and White (05)
Nematophyton	—	—	—	—			.	.	×	×	.	.	.	.	.	Kidston and Lang (21²), (24²); Solms-Laubach (95); Penhallow (96), (97); Hörich (15) Church (19)

* Numbers on map, fig. 42:

and other books, is cited in the various memoirs by Kidston and Lang (17–24), and by Lang (25), (26). See also Hickling (08); Crampton and Carruthers (14).

VIII. Belgium. Lower Devonian. A few plants from the Poudingue de Burnot. Gilkinet (75), (75²); Crépin (74), (75). Upper Devonian. A much richer flora from the Psammites du Condroz, a lagoon formation with some marine beds. Gilkinet (22).

IX. Germany. Lower Devonian. The Hartz district: Weiss, E. (85); Hörich (15); Steinmann and Elberskirch (29). Middle Devonian. Elberfeld and Gräfrath: Kräusel and Weyland (26), (29); Scott (26). Upper Devonian. For references see Frech (97–02); also Gothan (27⁵). The Thuringian beds, sometimes referred to as Devonian, are probably Lower Carboniferous. Solms-Laubach (96).

X. France. Lower Devonian. Pas de Calais: Bertrand, P. (13). Upper Devonian. Loire Basin and Maine et Loire: Bureau (11), (14); Carpentier (20). Only a few fragmentary remains are known from France.

XI. Bohemia. Middle Devonian. Drifted plants: Potonié and Bernard (04).

XII. Norway. Lower Devonian. Lake Röragen, E. Norway (some of the plants *in situ*): Halle (16). Middle Devonian. West Norway: Nathorst (15); Nathorst and Goldschmidt (13); Nathorst and Kolderup (15).

XIII. Russia. Upper Devonian. Lagoon sediments in the Donetz basin: Schmalhausen (94); Zalessky (11). Karpinsky (06); Kryshtofovich (27³).

XIV. China. Halle (16), (27²) records the occurrence of *Arthrostigma*. See also Colani (19).

XV. Falkland Islands. A few Middle or Lower Devonian plants: Halle (11); Seward and Walton (23).

XVI. South Africa. The determination of the precise position in the Devonian system of the barren rocks is not possible: Schwarz (06), (06²); Seward (03), (09); du Toit (26).

XVII. Australia. The distinction between Upper Devonian and Lower Carboniferous rocks is not always well marked: Feistmantel (90); Jack and Etheridge (92); David and Pitman (93); Dun (97); Benson, with good bibliography (27); Süssmilch (22); Walton (26); Cookson (26).

CHAPTER X

THE CARBONIFEROUS AND PERMIAN PERIODS

And the earth shall yield her increase. Ezekiel xxxiv, 27

We have now reached a stage in the history of the world exceptionally rich in records which have raised many problems relating to the history both of the inorganic and the organic realms. In order to obtain a general idea of the physical background and of the surprising vagaries of climate demonstrated by the rocks included in the Carboniferous and Permian systems, we will consider some of the geological facts before attempting a review of the several floras.

Carboniferous and Permian Rocks: the Geological Background

The accompanying map is intended to convey a general idea of the distribution of land and water during the latest phase of the Carboniferous and in the earlier part of the Permian period. As in other geographical restorations the outlines are necessarily in large measure hypothetical. A well-established fact is the existence of a central Tethys Sea dividing the northern and southern continents. The geographical distribution of certain Permian plants leads us to assume either a land-bridge across the Tethys or a group of islands which served as stepping-stones. The shaded areas in India, South America, Africa, Madagascar, and Australia show the probable extent of glacial conditions over the southern continent. The discovery of fragments of *Glossopteris* 300 miles from the South Pole by members of Captain Scott's expedition affords evidence of a former land connexion between Antarctica and the continent to the north.[1] *Glossopteris* is the name given to fossil leaves, which were almost certainly borne by pteridosperms, preserved in great abundance in the rocks of Gondwanaland.

As shown in the Geological Table (fig. 2, p. 9) the Carboniferous system embraces a long period: within it are included strata of very different origin which enable us to follow with comparative ease a succession of transformations in the aspect of the world. The gradual transition from Devonian to Carboniferous beds illustrates the

[1] Seward (14). See also Edwards (28).

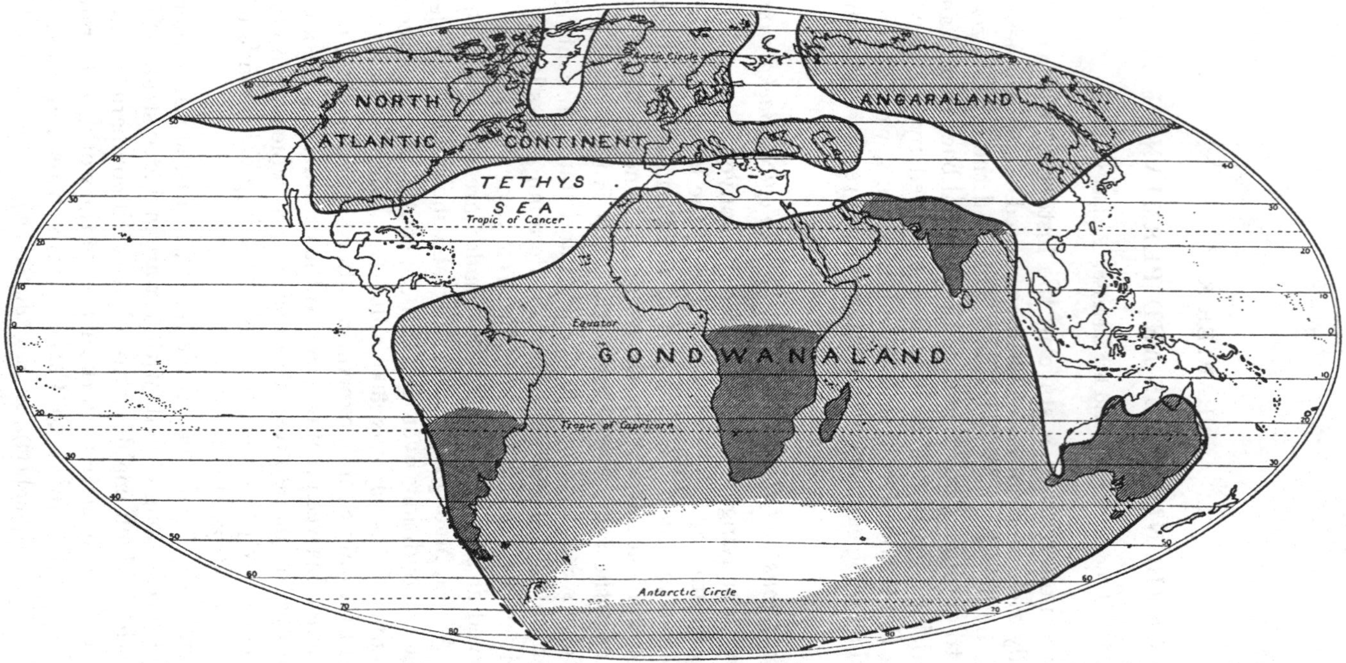

FIG. 49. Map of the world at the end of the Carboniferous period showing areas (darker shading) within which evidence of extensive glaciation has been found. The shaded areas are land. (Based on a restoration given by Arldt.)

arbitrary nature of the divisions of geological history into separate chapters. One period, or chapter, is not necessarily separated from the next by any well-defined interval represented by a discordance in the succession of strata or even by any clearly marked changes in the records of life. In some parts of the world the oldest Carboniferous rocks are shallow-water deposits—sands and pebble beds—which indicate a persistence of the conditions prevailing at the close of the Devonian period. These basal beds are overlain by some thousands of feet of bluish grey limestones composed of calcareous skeletons of marine animals and, in places, of reefs of lime-secreting algae. It is these deeper sea deposits which form the limestone cliffs of Derbyshire and West Yorkshire, the Mendip Hills in Somersetshire, and the cliffs at Dinant in Belgium. Similar rocks dominate the landscape in the Mississippi basin of North America, in central Europe, and elsewhere. In the early days of the period the older Devonian land was invaded by a sea which spread over most of northern and central Europe and far to the east: on the other side of the Atlantic it extended to Alaska and the Arctic regions. The land that is now England was represented by a few scattered islands, and by a narrow ridge across the Midlands (fig. 50), relics of which are seen in the eroded granitic crags of Leicestershire. From the limestones formed by the rain of calcareous shells on to the floor of the Carboniferous sea and by the activity of coral-forming polyps, we learn nothing of the vegetation of the land; but they demonstrate the existence of calcareous algae similar in their general plan of construction to those in the warmer seas at the present day. It is in the sands and muds of shallow bays and in the sediments along the shelving coast-lines that we discover samples of the contemporary vegetation. As the Carboniferous Limestone is followed towards the north of England and the south of Scotland its character changes; the calcareous strata are replaced by sandstone and shale transported by rivers from the northern continent. In southern Scotland there were long alternating periods when the waters retreated and advanced: low-lying swamps and marshes covered with vegetation were from time to time hidden under the sea. Similarly in North America, in Silesia, on the flanks of the Ural Mountains, in Spitsbergen, and in many other regions the nature of the rocks demonstrates proximity to the coast. Drifted stems and leaves, as well as plants preserved where they grew, enable us to follow the development of the land floras.

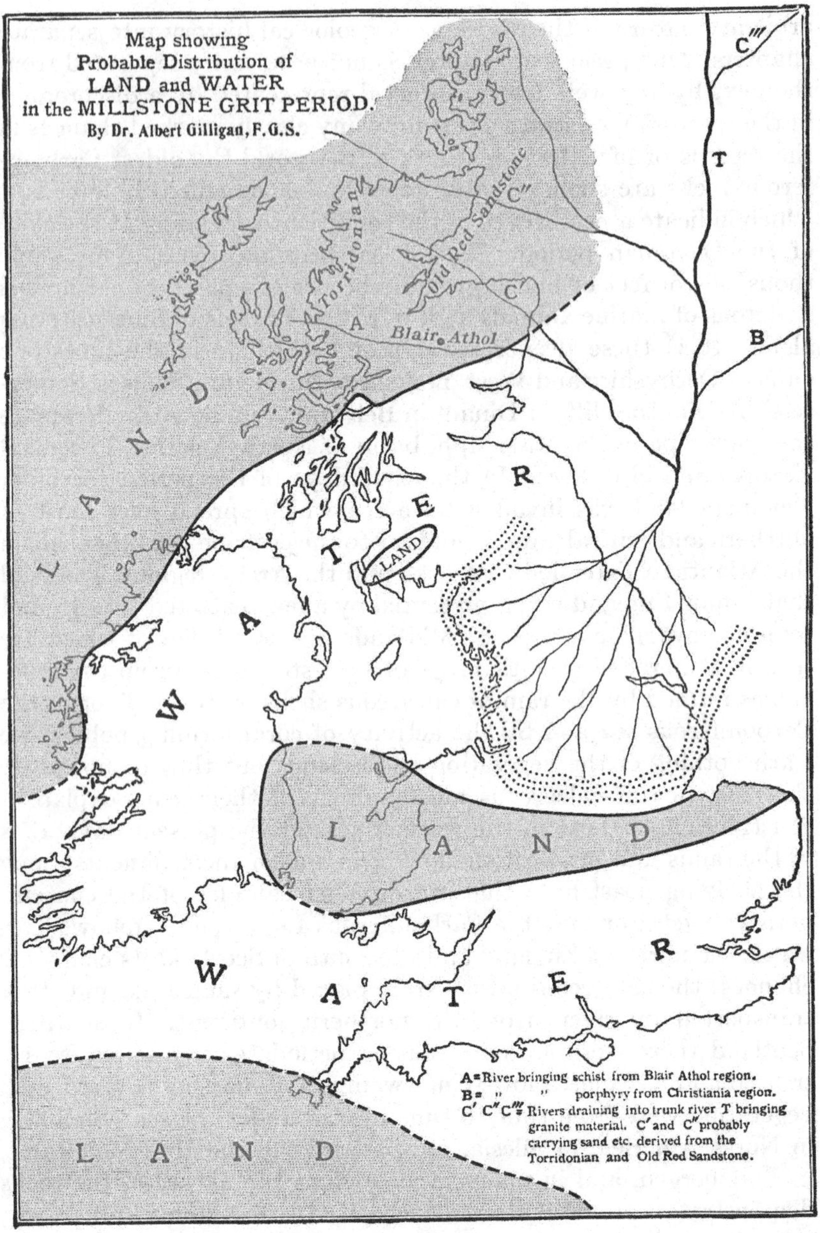

FIG. 50. Map of the British Isles in the Millstone Grit stage of the Carboniferous period. (After Prof. Gilligan.) See page 11.

There is a feature of the Lower Carboniferous series in southern Scotland, and to a less extent of the strata in northern England, which is of special interest to students of fossil plants; namely the evidence of widespread volcanic activity. The central valley of Scotland to the north of the Carboniferous sea was studded with volcanoes: North Berwick Law, the Bass Rock, Arthur's Seat at Edinburgh[1] mark a few of many centres of eruption. Beds of volcanic ash on the north shore of the Firth of Forth at Burntisland, in the Isle of Arran, and in other places have furnished exceptionally good botanical records. Temporarily inactive volcanic cones, like Vesuvius in its quiet phases, were clothed with vegetation, which was overwhelmed by showers of ash, and the charred remains were petrified by the precipitation within the tissues of mineral material dissolved by percolating water from the ejected ash. Fig. 51 illustrates a scene in the volcanic area almost identical with that in Java reproduced in fig. 22: a thick deposit of volcanic ash bordering a lake; charred and broken stems of *Lepidodendron* protruding from the cliffs and lying on the beach; farther to the right a single tree which escaped destruction. The large *Lepidodendron* stem, described in Chapter v (fig. 20), found in a bed of volcanic ash not far from Edinburgh, affords a good example of the wonderful preservation of delicate tissues. Portions of exceptionally well-preserved pieces of ferns from the volcanic ash of Burntisland are shown in fig. 58.

The geology of the Ingleborough district described in a previous chapter[2] illustrates the sequence of events in western Europe and in many parts of North America in the course of the Carboniferous period. The succession of strata is an epitome of the shifting scenes over a wide area on both sides of the Atlantic Ocean: clear seas after a long interval becoming shallower; limestones gradually replaced by masses of grit (Millstone Grit) and sand swept by rivers from a northern continent. The map[3] reproduced in fig. 50 shows the position of the delta in which the Millstone Grit was deposited, and the course of the rivers. In the latter part of the period the deltas became clothed with forests, which endured for a time and were replaced at intervals by basins of shallow water; an alternating series of seams of coal, layers of sand and mud registering gentle

[1] It is customary to speak of localities under the place-names now in use: but it must be remembered that the geographical features of this and most of the other geological periods bore little or no resemblance to those with which we are now familiar.

[2] See p. 17.

[3] Gilligan (20).

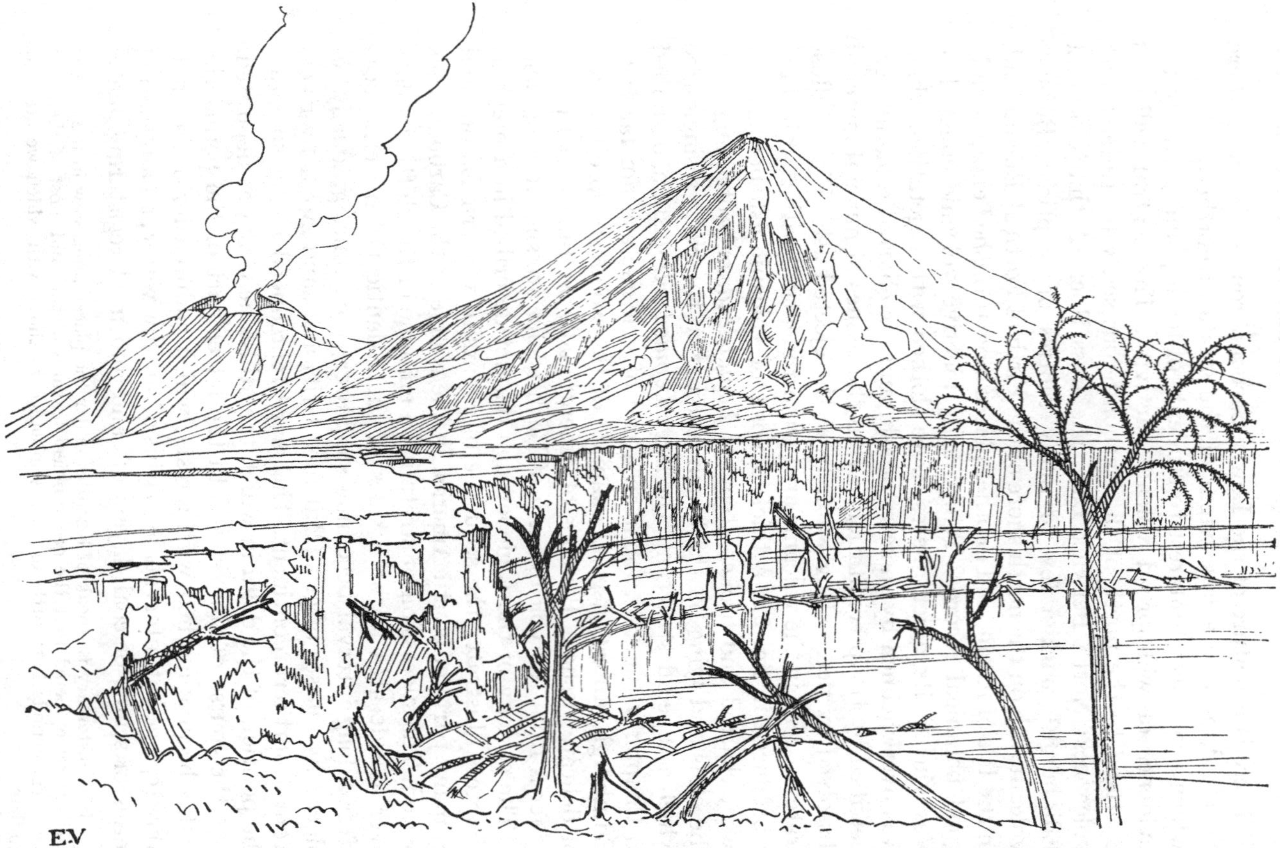

Fig. 51. A *Lepidodendron* forest overwhelmed by volcanic ash: reconstruction of a scene in southern Scotland in the earlier part of the Carboniferous period. (Drawn by Mr Edward Vulliamy.)

and frequently recurring oscillations of the crust which characterized the final stage of the Carboniferous period. Remains of the older Carboniferous plants are for the most part found in sediments deposited along the shores of the mainland and adjacent islands. By degrees the boundaries of the land were extended; the shallow seas and estuaries were converted into lagoons and forest-covered swamps, the material of which supplied the essential constituents of coal. The twofold division of the system into Lower and Upper Carboniferous serves to express the two phases: the lower series is mostly marine while the upper, though in part marine, includes a considerable thickness of beds of freshwater and of terrestrial origin. In Britain the following classification[1] is adopted:

UPPER CARBONIFEROUS	Coal Measures	Upper Coal Measures or Radstockian series	Stephanian
		Transition series or Staffordian	Westphalian
		Middle Coal Measures or Yorkian	
		Lower Coal Measures	
	Millstone Grit (upper part)		
LOWER CARBONIFEROUS	Millstone Grit (lower part)		
	Carboniferous Limestone		
	Calciferous Sandstone		

The Calciferous Sandstone series represents the basal sediments formed at the beginning of the marine transgression recorded in thousands of feet of limestone. The subdivisions of the Upper Carboniferous are based in the main on the floras. Considered as a whole, the later Carboniferous vegetation over the northern continents is fairly uniform, but by a critical comparison of the species and genera it has been possible to employ the plants as indices of time relations. American geologists[2] employ the term Mississippian for the Lower Carboniferous because of the development of marine strata in the region of the Mississippi basin. The Upper Carboniferous beds, which are termed Pennsylvanian, consist of seams of coal, sandstone and other shallow-water sediments which are either marine or freshwater. The contrast between comparatively deep seas, shallow basins and far-reaching swamps, though typical of many regions in the northern hemisphere during the latter part of the period, is by no means universal. In Russia, for example, there are limestones containing marine shells which prove the existence

[1] Kidston (23–25). See also page 261.

[2] For an account of American Carboniferous strata, see Chamberlin and Salisbury (04–06); Pirsson and Schuchert (20); Willis, B (22).

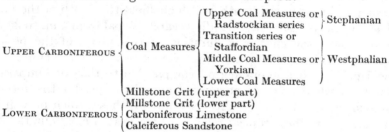

of an open sea contemporary with the lagoons and forests of western Europe and North America.

The Coal Measures

Most of the seams of coal are believed to have been formed by the accumulation of vegetable débris either on the site of a forest, or in rare instances from material deposited as fine organic sediment on the floor of a lake or estuary.[1] The conversion of vegetable matter into coal was accompanied by a great reduction in bulk. It has been estimated that a bed of coal 1 ft. thick represents a mass of vegetable débris 15–20 ft. thick. When we remember that in the coalfield of Saarbrücken in western Germany[2] there are about 230 different seams; and that in the South Welsh coalfield the depth of the Coal Measures is over 8000 ft., including seams of coal from 1 in. to 30 ft. in thickness, the enormous duration of this phase of the period becomes obvious. The question is often asked, Was the climate of the Upper Carboniferous period comparable to that of temperate countries where peat is now most abundant, or did the forests flourish under climatic conditions nearer to those which prevail in Sumatra and other tropical regions where peat is also found? We shall recur later to this difficult problem. It is probable that during the latter part of the period the landscape over wide areas in eastern North America, western Europe, and in some parts of the Far East resembled the present conditions seen in the Dismal Swamp of Virginia and Carolina (figs. 60, 115), and the swamps of Florida, or in the wide stretches of forest on the deltas of the Amazon and the Irrawaddy Rivers. The land was low and in part submerged; the surface-features were such as characterize the last phase of the planing down and erosion of an elevated portion of the earth's crust. Through the "wide and melancholy waste of putrid marshes" meandered sluggish streams carrying natural rafts of wood and tangled masses of vegetable débris: lakes and lagoons were conspicuous features, and there is evidence that at intervals the low-lying ground was invaded by the sea. Almost without exception each seam of coal in North America and Europe rests on a bed of clay, known as the underclay, penetrated by the spreading and forked "roots" (*Stigmaria*) of *Lepidodendron* and *Sigillaria*. The under-

[1] For a general, semi-popular, description of coal and the Coal Measures, see North (26) who gives many references to literature; also Stevenson (11–13).

[2] von Bubnoff, (26) describes the coalfields of Germany.

clay is the surface-soil of the forests which in the course of ages became seams of coal. In some seams, especially in Yorkshire and Lancashire, as also in Westphalia, in Spain, in central Europe, and in some North American coalfields, the coal contains calcareous nodules, known as coal-balls,[1] crowded with petrified fragments of the forest plants preserved in wonderful perfection. Reference has already been made to the preservation of a miscellaneous assortment of plant fragments—pieces of stems, slender twigs, spore-capsules and spores, roots and seeds—in a large coal-ball found in a Lancashire coalfield (see fig. 19).

The strata of the Coal Measures usually occur in basins: these, with few exceptions, were not depressions in which sedimentary material was deposited, but were formed by the folding of a series of horizontal beds into troughs and arches, subsequently separated into detached basins by the removal through denudation of the connecting rocks. The coalfield of Commentry[2] in Central France is one of the best examples of an original basin in which the seams of coal do not, as in most coalfields, represent the compressed and altered peaty material accumulated through long ages on the site of a forest, but were deposited as layers of carbonaceous sediment free from sand and mud on the floor of an enclosed body of water. The Upper Carboniferous rocks of southern England are continued beneath the English Channel into France and to the Ardennes; they are a part of the earth's crust involved in the process of mountain-building which succeeded a long quiescent period culminating in a far-reaching crustal revolution.[3]

This brings us to a critical stage of geological history. The Caledonian revolution, which began in the Silurian period and continued into the Devonian period, was probably a potent factor in relation to the development of the earliest known terrestrial vegetation. The next great revolution (Hercynian) began in the latter part of the Carboniferous period: the coal forests succeeded one another on the worn-down and almost featureless plains which were the final expression of the geological cycle initiated by the uplift of the Caledonian ranges. Eventually the spread of the forests was checked by the Hercynian revolution, a revolution which folded the crust

[1] See Chapter v, p. 46. [2] Grand'Eury (12), (13).

[3] The extent of the Kent coalfield hidden under the chalk and other rocks of the surface is shown on a map published in *Nature*, vol. CXXIII, p. 608, 1929.

into ranges of Alps represented to-day by diminished relics in the south of Ireland, in the Mendip Hills, in Belgium, and the Hartz Mountains (see map, fig. 78). Similarly in North America the Palaeozoic era was brought to a close by the Appalachian revolution. The effect of these convulsions can hardly be overestimated. The luxuriant forest-covered swamps were converted into a barren land: in the English midlands and in the central part of North America the Upper Coal Measures are overlain by red sandy beds in which fossils are rare, beds of sediment that were laid down in lakes and seas encircled by desert or semi-desert lands.

The Permian Period

The name Permian was adopted by Sir Roderick Murchison from the Government of Perm in Russia where he worked out the succession of strata intermediate in age between the Carboniferous and Triassic systems. In many parts of the world the transition from the Carboniferous to the Permian period is marked by a series of beds uninterrupted by any discordance: in other regions, as in England, the boundary-line is defined by an unconformity; the folded Coal Measures are overlain unconformably by Permian rocks. This lack of parallelism between the two systems indicates a relatively long interval. When the Pennine Hills were elevated at the close of the Carboniferous period, the seas retreated and the emergent land was exposed to prolonged erosion before another movement of the crust set the scene for the deposition of the oldest Permian sediments. There were contemporary changes in the climate: the denuded Pennine range traversed an arid country. In North America there is a greater continuity in the passage beds between the two systems, and this shows that the Appalachian revolution was not precisely synchronous with the Hercynian folding in western Europe. The important point is that except in the Tibetan zone of the Himalayas and in southern Europe, where normal marine conditions persisted without a break, and in Gondwanaland where a similar uniformity is seen in the continuous succession of sedimentary beds, the Permian period furnishes proof of a gradually changing world. Over a wide tract in North America and in Europe the land extended its boundaries; folding of the earth's crust produced chains of mountains; land-locked seas replaced the wider stretches of open water, and climatic conditions favoured the deposition of layers of salt on the floors of lake basins. The accumulation of thick sheets of red sand was another result of the change from a humid to a relatively dry climate. The forest-covered swamps were gradually transformed into an arid land where talus fans were spread by torrential rains and by avalanches of rock over the flanks of the hills. The transformation of the geological background left its impress on the vegetation: there was no sudden interference with the order of nature, no abrupt collapse of one dynasty and a swift rise to power of another. The new and more rigorous conditions caused the extinction of many of the older types, while a comparatively small number were able to persist as dominant forms in an impoverished flora. This was the final act in the Palaeozoic era, the first of a series of transformations which in the course of the succeeding Triassic period led to the inauguration of a new and very different plant-world.

In North America, where Permian rocks follow the Carboniferous without a break, the highest beds of the latter period are overlain in Texas, Kansas, Nova Scotia, New Brunswick, and elsewhere by the Upper Barren Coal Measures, strata in which the altered circumstances are demonstrated by layers of gypsum and salt. In England the Permian system is represented on the east of the Pennine Chain from Northumberland to Durham by a set of beds beginning with the Marl Slate, a stage equivalent to the copper-bearing bituminous series which extends as a comparatively thin band from the Hartz Mountains to the Thuringian forest. This basal Permian deposit was formed on the floor of an inland sea. Some Lower Permian sands in Durham, which are said to have the characters of a dune formation, afford additional evidence of a tendency towards more arid conditions. Resting on the Marl Slate is the Magnesian Limestone[1] displayed in the yellow cavernous cliffs on the Durham coast and formed in the clearer waters of a sea richer in salt than the open ocean which stretched across the world farther to the south. The uplifting of the Hercynian range not only cut off arms of the sea; it intercepted the rain-bearing winds from the southern ocean—as the Himalayan chain is now responsible for the deserts of Tibet. In the thick series of salt beds of various kinds at Stassfurt in Germany,[2] which it is estimated must have required 10,000 years for their deposition, we have an impressive sign of the

[1] See Geological table, p. 11.
[2] Lück (13).

times. On the west of the Pennines, along the Eden valley into Scotland, the Permian system is represented in the main by red sandstones associated here and there with volcanic rocks, which afford evidence of accumulation in a dry climate. The red Permian sandstones of Penrith in Cumberland consist of "desert sands with grains beautifully rounded" and reach a thickness of 1000 ft.

In the course of the period there were oscillations of the land, fluctuations in the boundaries of seas and of relatively dry land areas. In southern Europe and stretching eastwards to the Himalayas the Tethys Sea remained unchanged. It is the rocks of North America,[1] and of north and central Europe which furnish us with a picture of seas like the Caspian, and of desert regions with oases and occasional swamps occupied by descendants of the later Carboniferous floras. Permian rocks occur over large areas in Siberia where they form part of a succession of beds known as the Angara series (from the Angara River), ranging in age from Carboniferous to Tertiary. Some of the Asiatic Permian sediments contain plants closely related to European species but with them are other types peculiar to the Far East: the lower beds, in which the plants occur, were probably deposited under humid climatic conditions; whereas strata higher in the Permian series afford evidence of a semi-desert environment. In the Far East, as in Europe and North America, the close of the Palaeozoic era was a time of mountain-building; a chain of hills was laid across Mongolia in an east and west direction between lat. 45° and 50° N. On the continent of Gondwanaland a different history is recorded: the Glossopteris flora was gradually spreading farther north; southern types invaded the domain of the northern floras.

The division of the Permian system in Germany into a lower series—the Rothliegendes, in which conglomerates and sandstones predominate, and an upper series, the Zechstein, the equivalent of the Magnesian Limestone of England—is widely adopted. It expresses the clearly marked distinction in northern and western Europe between the earlier and later phases of the period, a contrast which is lacking in more southern countries.

On the whole, the vegetation of the northern hemisphere during the Permian period was comparatively uniform, as it was in the preceding Carboniferous period; but the discovery thirty years ago

[1] For recent descriptions of American Permian rocks see White, D. (09), (29); also Schuchert (28), who gives many references to literature.

in the north of Russia[1] of Upper Permian plant-beds containing *Glossopteris* and some other members of the southern flora afforded proof of a great northward migration and a commingling of two previously separated types of vegetation. We are driven to the conclusion that there must have been some connecting bridge by which the wanderers were able to cross the Tethys Sea.

This brings us to the end of the Palaeozoic era, to the threshold of a new world where we find few survivors from the past. We see an apparently sudden rise to prominence of families and genera foreshadowing the steady progress towards the evolution of the modern plant-world which is gradually revealed in Mesozoic and Tertiary rocks.

India and the Southern Hemisphere: Gondwanaland

Attention has so far been confined to the Carboniferous and Permian systems as they are developed across the northern hemisphere excluding India. Over large areas on the great northern continents the physical conditions and the vegetation were surprisingly uniform: an expanse of low swampy ground studded with scattered lakes and sluggish waters and here and there patches of higher ground favourable to the growth of plants requiring drier situations. Calcareous nodules in shales, which form the roof of some English coal seams, have supplied well-preserved specimens of plants that grew above the swamps.

As Suess said, after the events chronicled in the Coal Measures of the northern hemisphere, in the south "the outlines of a great continent become disclosed to us, and from the closing days of the Carboniferous this remains for a long period one of the most prominent features of the earth, Gondwanaland". This ancient continent stretched across the world to the south of the Tethys Sea (map, fig. 49): as envisaged by Wegener it was originally a compact mass subsequently broken into blocks that drifted apart and are now known as South America, South Africa, India, Australia and Antarctica; or, it may be, a continent which occupied as a continuous area a region which now includes both the sundered lands and the intervening oceans. The most valuable records of the early Carboniferous floras in the southern hemisphere are those from marine sediments in Australia deposited near the edge of the land. The Australian plants agree in the main with Lower Carboniferous

[1] Amalitzky (01).

genera in the northern hemisphere. This marine phase, characteristic of the earlier part of the period in Australia, was brought to an end by an upheaval of the crust. Sea was replaced by land; and this change in the physical world was an important contributory cause of the later Palaeozoic glaciation. Evidence has also been obtained of the existence in Argentina of representative genera of the Lower Carboniferous vegetation of the northern hemisphere. In South Africa no specimens of Lower Carboniferous plants have been discovered, and only a few fragments at one locality in India. Reference has already been made to the general uniformity in the character of the early Carboniferous vegetation over both hemispheres: differences in facies correlated with differences in latitude there undoubtedly were; but, so far as the scanty records from southern countries enable us to form an opinion, there is no definite indication of any fundamental distinction between a northern and a southern vegetation.

We pass now to the consideration of the difficult problem presented by the later Palaeozoic boulder beds (tillites) in India and the southern hemisphere which have been briefly described in Chapter IV. The first question to be considered is the relation in age of the great Ice Age, which in extent and duration surpassed the Quaternary glacial period in the northern hemisphere, to the later Palaeozoic plant-bearing rocks in North America and Europe. There are indications that the reign of ice began in certain parts of Australia earlier than in India and South Africa. It is probable that ice-caps were formed and reached their greatest development in different regions at different times. Geologists have been accustomed to speak of the southern Glacial period as Permo-Carboniferous; but this vague term is unsatisfactory as it implies inability to apply to events in the southern hemisphere the more precise terminology used in the northern hemisphere. Some authors refer the Ice Age to the Permian period; others believe that it began in the latter part of the Carboniferous period.

It would be out of place in a general review of the world's vegetation fully to consider the various arguments advanced in favour of divergent views on the geological age of the Gondwanaland Ice Age. The glacial deposits of India, Australia, South America, and South Africa are overlain by strata containing fossils, a few of which are specifically identical with plants widely spread in the late Carboniferous or early Permian floras of the northern hemisphere, but the majority are essentially different and are characteristic of the Glossopteris flora. Among the latter the following are the most numerous and distinctive: *Gangamopteris*, *Glossopteris*, *Schizoneura*.[1] The important point is this: What relation does this southern vegetation (Glossopteris flora) bear to the Permo-Carboniferous (using this term to include the latest stage of the Carboniferous and the earlier stages of the Permian period) of the north? Were the two phases of the plant-world contemporaneous, or was the Glossopteris flora evolved after the close of the Carboniferous period? Some geologists believe that the Glossopteris flora came into being in the Permian period, certain authors would say in the earlier days of the period; others assign its first appearance to a later stage. The most recent protagonist of the view that the Ice Age and the Glossopteris flora are not older than Middle Permian is Prof. Schuchert[2] of Yale University: it is certain, he says, that the widely spread glacial deposits of Gondwanaland are of Permian time and "in all probability of late Middle Permian age". Whether one agrees with this opinion or not, the value of the service he has rendered by his full statement of the case is freely admitted. Other authors maintain, and with them I am in agreement, that the Ice Age had begun and the earlier members of the Glossopteris flora had colonized wide areas before the beginning of the Permian period. This view is briefly stated in the following pages and a general discussion of the situation will be found in an Address to the Botanical Section of the British Association given at the South African Meeting.[3] An important contribution to the palaeobotanical evidence bearing on the correlation of northern and southern floras has recently been made by Dr John Walton,[4] and Dr Dighton Thomas[5] has dealt with other aspects of the problem. Dr Walton records the occurrence of *Glossopteris* in the coalfield of Wankie in Rhodesia in association with several well-known species of plants that are widely distributed in Europe or North America in the latest Carboniferous or in the earliest Permian floras. Dr Thomas states that among British geologists there is a general belief in the Upper Carboniferous age of the late Palaeozoic glacial deposits, and he advances arguments in support of this view.

[1] For descriptions see Chapter XI.

[2] Schuchert (28).

[3] Seward (29). See also de Cotter (29).

[4] Walton (29), (29[2]). For an account of the geology of Southern Rhodesia see Maufe (24).

[5] Thomas, D. (29).

Australia

The most complete records are probably those displayed in sections of the earth's crust in New South Wales. I am greatly indebted to my friend Prof. Sir Edgeworth David for supplying information which enables us to follow the geological history in Australia through the Carboniferous and Permian periods.[1] A simplified diagram of the succession of beds in the Hunter River area of New South Wales is shown in fig. 52. At the base we have sediments from comparatively deep and from shallow seas, known collectively as the Burindi series. In some of the beds pieces of drifted *Lepidodendron* stems occur in association with marine shells. Resting on the uppermost strata of the Burindi series is the Wallarobba conglomerate, a mass of pebble-beds 1500 ft. thick, which marks the initial stage of a great upheaval and the beginning of a new geological cycle recorded in the rocks of the Kuttung series. In the volcanic material immediately above the conglomerate were found petrified specimens of a Lower Carboniferous fern *Clepsydropsis* (fig. 59),[2] and of the gymnosperm *Pitys*, also a petrified *Lepidodendron* indistinguishable from a Lower Carboniferous species from Scotland. The Kuttung series, approximately 10,000 ft. in depth, furnishes an impressive record of a prolonged period of volcanic activity, with recurrent quiescent intervals when the valleys were occupied by glaciers which have left traces in beds of boulder clay and erratics. A block of folded glacial mud showing the seasonal alternation of winter and summer layers is reproduced in fig. 53. This specimen affords a good illustration of the varves described in Chapter IV: an alternation of coarser and finer layers is clearly shown in the photograph. The folding was no doubt caused by a mass of ice forced against the yielding sediment, possibly a stranded iceberg, or by the snout of a glacier pushing its way over the floor of a lake or shallow sea in which the summer and winter streams had deposited their heavier and lighter burden of mud and sand. Leaves of *Rhacopteris* (see fig. 57, A) and *Cardiopteris* (fig. 60, *C.*) casts of *Asterocalamites* (fig. 60, *As.*) and other plants bear witness to the spread of vegetation under a relatively low temperature and in an atmosphere charged with volcanic dust. The Kuttung plants agree much more closely with Lower Carboniferous types in the

[1] See also Süssmilch and David (20).
[2] Sahni (28). See also Sahni (30).

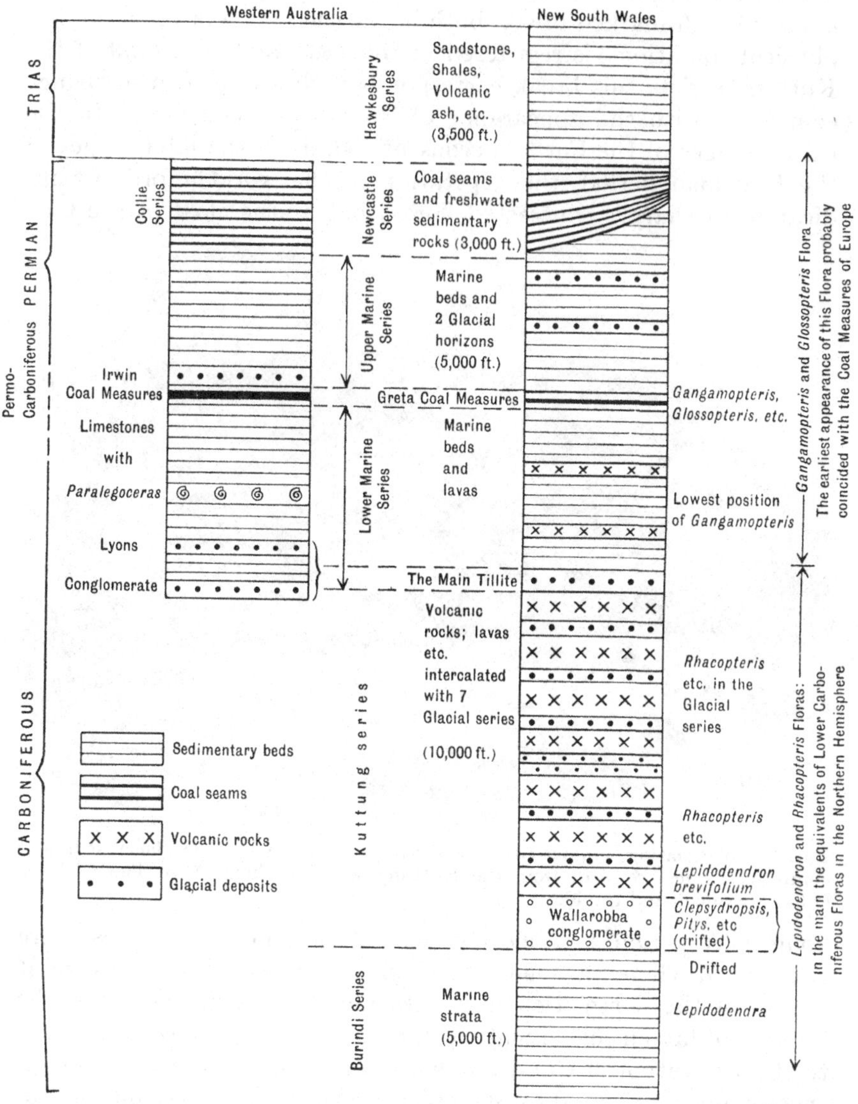

Fig. 52. Diagrammatic sections illustrating the sequence of events in Australia from the early part of the Carboniferous period to the Triassic period. (Simplified from David and Süssmilch.)

northern hemisphere than with members of the later Carboniferous floras. A well-marked break, both in the succession of fossils and in physical conditions, is registered in the rocks at the summit of the Kuttung series: this break corresponds with a crustal disturbance coincident with the appearance of the Glossopteris flora. In the earlier stages of the Carboniferous period, as in the later stages of the Devonian period, the vegetation of the world affords no indication of sharply contrasted floras; but at the close of the Carboniferous period two clearly defined botanical provinces were established. There was some commingling of genera, but the luxuriant forests of the north, rich in genera and species and bearing the impress of favourable climatic conditions, afford a marked contrast to the vegetation of Gondwanaland, which was relatively poor in species and dominated by such plants as *Gangamopteris* and *Glossopteris*—fern-like in habit though probably pteridosperms—and unknown in the contemporary floras north of the Tethys Sea. The earliest examples of this Glossopteris flora occur in sediments associated with glacial deposits. The conditions were no doubt very

FIG. 53. Alternating summer (dark) and winter (light) deposits of glacial sediment contorted by ice-pressure: from the Kuttung series, Seaham, New South Wales. (After Süssmilch and David.)

similar to those in Alaska at the present day and comparable with those in the western part of New Zealand (frontispiece): a vast continent on which a uniform vegetation flourished in proximity to glaciers and ice-sheets. The probability is that when the Upper Carboniferous forests of Europe and North America had hardly passed their prime, Gondwanaland was in the grip of an Ice Age; the ice-free areas clothed with associations of plants differing widely from those of the northern continents (cf. figs. 60, 74).

Returning to the Hunter River section, we find above the Kuttung series a thick glacial deposit at the base of the Lower Marine series, a series of marine strata with intercalated lava-flows. In Australia the oldest known examples of *Gangamopteris*, the earliest representatives of the Glossopteris flora, were obtained from the shales at approximately 200 ft. above the base of this series. From a bed near the base of the Lower Marine series were collected specimens of the bivalve shells of *Eurydesma* which indicate an Upper Carboniferous age. It is hardly possible precisely to correlate these rocks with equivalents in the northern hemisphere, but it is significant that all the recognizable plant remains from beds below those in which *Gangamopteris* occurs appear to be most closely akin to Lower Carboniferous species in the north. The broken line on the left-hand side of fig. 52 indicates the lack of data necessary to enable us to fix the boundary between the Carboniferous and the Permian periods. Above the Lower Marine series are the Greta Coal Measures, in all probability of Upper Carboniferous age, with *Gangamopteris* and its associates: then follows the Upper Marine series in which are included two more sets of glacial deposits demonstrating a recurrence of ice after a long interval. In the overlying Tomago and Newcastle series coal seams are the dominant feature. Finally the next series of sedimentary beds, known as the Hawkesbury series, records the history of the Triassic period.

Additional evidence bearing on the age of the Australian rocks is furnished by sections in other regions. In Western Australia there is a series of sedimentary beds and seams of coal known as the Collie Coal Measures, the equivalent of the Newcastle series of New South Wales; the latter series has yielded Upper Permian insects. Below the horizon of the Collie Coal Measures in another area occur deposits containing glacial boulders: these rest on the western Irwin Coal Measures which are equivalent in age to the Greta Coal Measures of New South Wales (fig. 52). If, as seems practically certain, the Greta Coal Measures are equivalent in age to the Irwin Coal Measures the glacial beds below them must be included within the Upper Carboniferous stage. Next in descending order are beds with marine fossils indicating an infra-Permian or Upper Carboniferous age. These beds rest on marine sediments rich in the shells of a cephalopod, *Paralegoceras Jacksoni*, which may be described as remotely connected with the living *Nautilus* and a species of a genus which ranges from Lower Carboniferous rocks to well into the Permian system. Dr Dighton Thomas, who is describing the Irwin River *Paralegoceras*, is convinced from a comparison of it with cephalopods from other regions and horizons that it is Upper Carboniferous in age. This piece of evidence is important as it lends support to the view that the Glossopteris flora began before the close of the Carboniferous period. In Queensland sedimentary beds with some volcanic ash and plant-remains—the Upper Bowen series—correspond to the Permian Newcastle series and the Collie series. Below this is the Middle Bowen series which includes some glacial deposits and, at the base, Coal Measures with *Glossopteris*, which are correlated with the Greta Coal Measures of New South Wales. It is clear from these and other geological data, from Tasmania and elsewhere, that the recurrence of glacial conditions and the existence of the Glossopteris flora are features characteristic of the Australian continent in the later stages of the Carboniferous period and in the succeeding Permian period.

So far attention has been concentrated on the eastern portion of Gondwanaland where the geological record demonstrates an apparently sudden change in the nature of the vegetation before the close of the Carboniferous period. It demonstrates also that this change, so far as we can tell, was coincident with a change in the environment which set the scene for a Glacial period. In the Hunter River section (fig. 52) there are ten different series of glacial deposits ranging in age from Lower Carboniferous to about the middle of the Permian period.[1] In other parts of the southern continent there is less definite evidence than in Australia of alternating glacial and interglacial phases, but in the main there is a similar sequence of events.

[1] In addition to references already given to Australian sources see David (96), (24).

CORRELATION TABLE. *Upper Palaeozoic rocks of Gondwanaland.*

	Falkland Islands	Brazil	South Africa	India: The peninsula	India: Salt Range	India: Kashmir	India: Spiti	Australia
PERMIAN	Sedimentary rocks	Marine beds	Sedimentary beds: coal and plants	Freshwater beds: coal and plants	Productus limestone	Zewan series	Productus shales Fenestella shales, etc.	Upper marine series
	Gangamopteris	Iraty beds						
CARBONIFEROUS	Glacial shales, etc.	White band		Karharbari beds: coal and plants	Speckled sandstone			Greta Coal Measures Lower marine series
		Rio Bonito beds	Dwyka shales					
	Tillite	Orleans tillite	Dwyka tillite	Talchir tillite	Tillite	Gangamopteris beds	Shales with *Rhacopteris* and other Lower Carboniferous plants	Tillite
								Volcanic and glacial beds
								Conglomerate
								Marine strata
	Devonian plants		Devonian plants					Devonian plants

The appended correlation table may serve as a guide to the order of the strata mentioned in the following brief descriptions.

India

In the peninsula of India the Talchir conglomerate forms the lowest member of the Gondwana system, which consists of sedimentary freshwater beds containing *Gangamopteris* and other genera. The Talchir tillite[1] (fig. 16, p. 38), which includes erratic blocks up to thirty tons in weight, is believed to have been deposited by glaciers which had their source in the Aravalli Hills in Rajputana. The only evidence of any submergence in the interior of the peninsula during the Gondwana period is afforded by a marine band resting on the tillite in the coalfield of Umaria[2] containing marine fossils. In the opinion of Dr Fermor, of the Indian Geological Survey, the fossils from the Umaria bed point to the presence of a Carboniferous sea over a region that is now the Rewah State; the marine shells are believed by Dr Dighton Thomas to support this view, and by another geologist they are stated to include both Upper Carboniferous and Lower Permian species. Resting on the Talchir tillite and the superposed plant-beds are the Karharbari beds with coal seams, *Gangamopteris* and other plants. In the Indian peninsula the Talchir beds rest on much older rocks, and there is no definite evidence of their geological age. On the other hand, in the Salt Range, to the north-west of the Himalayas, there is a tillite believed to be equivalent in age to that in the peninsula and overlain by beds containing Upper Carboniferous marine fossils, e.g. *Eurydesma*, the bivalve recorded also from Australia and South Africa.[3] The Salt Range strata, known as the Speckled Sandstone, are succeeded by beds of limestone which, from the abundance of the shells of *Productus*, an extinct marine type of bivalve shell, is known as the Productus Limestone, and by other beds with Permian marine shells. Dr Thomas, in discussing Prof. Schuchert's view that the Productus Limestone and the beds below it, which rest on the Talchir tillite, are Permian in age, points out that the fossil shells from the lower part of the limestone, above the boulder bed, do not

[1] See Chapter IV. [2] Reed, F. R. C. (28); Fermor (23).

[3] Since this was written my attention has been drawn to two important recent papers which give additional evidence from this area strongly supporting the view that the Palaeozoic glaciation was either Upper Carboniferous or Lower Permian. G. de P. Cotter (29); Reed, de Cotter and Lahiri (30).

favour a geological horizon later than Lower Permian. This being so, the still older Talchir tillite and the Speckled Sandstone containing *Eurydesma* should be assigned to the Carboniferous system.

The occurrence of *Gangamopteris* and other plants among marine sediments of the Tethys Sea in Kashmir throws additional light on the question of geological age. The plant-beds are overlain by the Zewan series, the lower part of which is probably Upper Carboniferous in age. In the district of Spiti (the River Spiti is a tributary of the Sutlej in the Punjab) a few plant fragments, recognized by Zeiller[1] as Lower Carboniferous species, were found in the Po series: these include a *Rhacopteris*, one of the genera recorded from New South Wales. The Spiti plants are the oldest recognizable vegetable remains from India. Above the Po series are the Fenestella and Productus shales, probably Upper Carboniferous and homotaxial with the lower part of the Zewan series in Kashmir.

South Africa and South America

The Dwyka tillite of South Africa, so called from the Dwyka River near Prince Albert, covers a wide area and reaches a thickness of 2000 ft. North of latitude 33° S. it rests on glaciated pavements,[2] a fact indicating ice on the land; in the south it rests on shales and was deposited in water, thus pointing to floating ice. At Vereeniging impressions of *Gangamopteris* were found by Dr T. N. Leslie between the base of the tillite and the underlying Pre-Devonian Dolomite. *Eurydesma* occurs in the Dwyka shales, also a crustacean, *Pygocephalus*, believed by Mr H. Woods[3] to be most closely allied to Upper Carboniferous forms of the genus. Dr Broom informs me that some recently discovered specimens of *Pygocephalus* do not in his opinion afford convincing evidence of a close affinity to European Carboniferous species: he does not regard the genus as an important index-fossil. At a higher level there is a bed, known as the white band, which has yielded bones of the reptile *Mesosaurus*. In the Ecca series occur *Glossopteris* and other members of the flora, also coal seams and stumps of trees.

Above Devonian rocks in the Falkland Islands are sedimentary beds known as the Lafonian series. At the base is a tillite deposited by glaciers from Pre-Devonian hills to the south of the present islands; above this are beds containing *Glossopteris*. Tillites occur

1 Zeiller in Hayden (04).
2 See Chapter IV, fig. 17.
3 Woods (22).

in South America[1] from the tropic of Capricorn to latitude 52° S. In Brazil the Orleans tillite is overlain by sedimentary beds and, at a higher level, there is a black carbonaceous shale (termed the white band because it weathers white) in which *Mesosaurus* was discovered. This band corresponds exactly with the white band of South Africa and is believed to be Upper Carboniferous. Fossils from the Rio Bonito plant-beds have been identified as remains of crustaceans (eurypterids) generically identical with fossils from the Witteberg beds of South Africa and closely allied to European Devonian species. The discovery a few years ago by Dr L. du Toit of specimens of *Cardiopteris* in Argentina "just above a tillite resting on a glaciated surface"[2] is of great importance; it supplies additional evidence in support of the Carboniferous age of the oldest phase of the Glossopteris flora.

This summary, though necessarily very incomplete, may suffice to enable us to reconstruct in their broad features the closing scenes of the Palaeozoic era on the continent of Gondwanaland. We see an enormous land-region comparable in its mantle of ice to Greenland at the present day: in some places glaciers piled up moraines and their streams deposited seasonally banded muds and sand; in other places from the cliffs of an ice barrier were detached icebergs carrying boulders that found a resting-place in the mud of a sea-floor. In the course of the latest phase of the Palaeozoic era ice-sheets and glaciers spread from the remote south beyond the equator: lands that are now in the tropical zone were then ice-bound. This striking contrast, as Dr Simpson[3] tells us, on meteorological grounds, is a physical impossibility. The facts are not in dispute: only the interpretation of them. Here we have an argument in favour of the Wegener hypothesis and of the assumption that the region that is now tropical India was in the latter part of the Palaeozoic era much farther south. The world was divided into three or perhaps more sharply contrasted regions; a northern region where rank vegetation covered thousands of square miles of swamp and low hills, and a vast southern continent where another flora flourished in proximity to retreating glaciers. We have next to compare the two botanical provinces. There is, however, another question to be considered: is there in the Upper Carboniferous rocks

[1] White, D. (08); Woodworth (12).
[2] From a letter, June, 1924.
[3] Simpson (29).

or in the early stages of the Permian period any evidence of glacial conditions in the northern province comparable to those of Gondwanaland? With a few exceptions the northern continents seem to have been ice-free. In Massachusetts an undoubted tillite[1] with banded glacial muds has been described as Permo-Carboniferous: its precise age has not been determined. There are a few other instances of local glaciation; but the important point is that during the period covered by the Coal Measures the climate of the northern continents differed very considerably from that of Gondwanaland.

[1] Sayles (14), (19).

CHAPTER XI

THE EARLIER CARBONIFEROUS VEGETATION

The youthful bloom of one age is the stately dignity of another. *H. Montagu Butler*

FROM a consideration of the oldest known terrestrial vegetation we pass to the much more abundant, more varied, and more legible botanical documents furnished by the rocks of the Carboniferous system. These documents enable us to follow in a relatively stable world the unfolding of new types from precursors in the later Devonian floras which had colonized wide stretches of the earth's surface. There is no sharp line of demarcation between the two periods: in the earlier Carboniferous floras we find many new plants which are recognizable as descendants of Devonian types: others come before us as strangers. We are impressed by the general uniformity in the character of the vegetation which has left traces in the rocks of Spitsbergen, North America, Europe, and Australia. This comparative uniformity is characteristic of the earlier stages of the Carboniferous period, and in sharp contrast to the striking difference between the later Carboniferous vegetation of the northern continents and that of Gondwanaland to the south of the Tethys Sea.

In the account of the Devonian vegetation a distinction was drawn between an earlier and a later phase within the limits of the period. Similarly a comparison of the plants preserved in the lower strata of the Carboniferous period with those from the upper beds reveals well-marked differences in the character, the luxuriance, and the geographical distribution of the floras. The late Dr Kidston[1] spoke of the sudden disappearance of an older flora and the equally sudden appearance of a later flora as the successive plant-beds of the British Carboniferous system are followed upwards, as one of the most remarkable facts in palaeobotany. It may be that this apparently abrupt contrast is merely the expression of geographical conditions: Prof. Bailey[2] points out that there were two main land areas in Europe during the Carboniferous period; a northern region, and to the south an island-festoon formed of the emerging hills

[1] Kidston (23–25).

[2] Bailey, E. B. (26).

of the Hercynian range. The early Carboniferous vegetation flourished over both of these regions which in England were separated from one another by a clear and comparatively deep sea: the southern flora continued to develop, as is shown by the succession of plants in Silesia, from the lower to the upper stages of the Carboniferous system. The flora of the northern land, as Prof. Bailey puts it, got out of date. Eventually, as the sundering sea in the British area became shallower and deltas stretched from shore to shore, the more progressive southern flora was able to invade the northern territory.

The earlier Carboniferous flora carried on the traditions established in the latter part of the preceding period. The later Carboniferous vegetation persisted with reduced numbers of genera and species into the earlier stages of the Permian period. As it is often impossible to draw a clear line between the rocks of the two systems, the term Permo-Carboniferous is occasionally applied to that chapter of geological history of which the records were preserved when the Palaeozoic era was drawing to its close. In many parts of the world, e.g. in North America, Europe, and China, conditions became much less favourable to the development of plant-life during the later stages of the Permian period. Arid lands and seas like the Caspian or the Great Salt Lake of Utah replaced the more genial environment of former days.

The physical conditions during the first half of the Carboniferous period were ill adapted to the preservation of samples of terrestrial vegetation as compared with those in the second half when sedimentary rocks and seams of coal were slowly accumulating in swamps, lagoons, and estuaries. We will first glance at a few of the records of plant-life furnished by the Lower Carboniferous limestones, rocks which consist mainly of calcareous material derived from the shells and skeletons of marine animals and were built up in clear water beyond reach of sediment and plant-débris carried by rivers into the sea. These limestones are rich in the remains of corals and with them have been found many other fossils some of which may confidently be assigned to algae. It is well known that the so-called coral reefs of modern seas are often composed of a mixture of corals and lime-secreting seaweeds, indeed in many instances algae are the more abundant. We cannot always express a definite opinion on the affinities to existing forms of the plants of the Carboniferous ocean, nor is it possible in certain instances to decide between an organic and an inorganic origin of calcareous structures simulating animal or plant bodies. We may picture ridges and mounds of limestone in course of formation by the ceaseless activity of lime-secreting animals and plants near the surface of the sea. Blocks detached from the living reef by the waves would mark the site of the submerged colonies as in coral atolls at the present day.

Calcareous, Reef-building Algae

Descriptions of a few examples of Calcareous Algae and of problematical structures discovered in Carboniferous limestones will suffice to convey an impression of the nature of some of the available material. Reference to original sources are given in the footnote.[1] In the Museum of the Geological Survey, London, and in the Natural History Museum, Brussels, may be seen transparent sections cut from the limestones of Namur which show a layered arrangement of particles suggesting either a regular precipitation by chemical action independent of life or, as is more probable, the work of lime-secreting organisms. In the nodules from which the sections were prepared there is no undoubted cellular framework: some specimens referred by Gürich[2] to the genus *Spongiostroma* are believed by some authors to be animal in origin, while others prefer to regard them as Calcareous Algae. *Spongiostroma* has been recorded also from Lower Carboniferous rocks of Westmorland, England, and from Silurian rocks of Gothland. The researches of Prof. Garwood have brought to light various forms of calcareous nodules in English limestones which, on microscopical examination, afford evidence of cellular structure. The nodules are sometimes etched into relief by a natural process of weathering.[3] Under the generic name *Ortonella* (from the village of Orton in Westmorland) Garwood described nodules "varying in size from that of a marble to that of a tangerine orange" made up of very narrow, unseptate branched tubes radiating from a common centre. The tubes remind one of *Girvanella*,[4] but in *Ortonella* they are radially and not concentrically disposed. *Ortonella* may be a lime-secreting Siphoneous Alga: it marks a definite horizon in the limestone of northern England and of the Avon gorge near Bristol. Another form is illustrated by *Mitcheldeania* which resembles *Ortonella* in its tubular structure, though differing from it in certain details. Nodules of a distinct

[1] Garwood (13), (14); Garwood and Goodyear (19); Pia in Hirmer (27).
[2] Gürich (06). [3] Reynolds (21). [4] See p. 105.

type are referred by Garwood to *Aphralysia*:[1] seen under the microscope this genus resembles a mass of small bubbles, plano-convex in section and arranged in alternating rows. No recognizable plant-cells have been discovered. The regular form of the units in the foam-like groups indicates some organic construction, but the actual nature of the fossils remains an open question. One of the most satisfactory examples of Calcareous Algae from Lower Carboniferous rocks is *Solenopora Garwoodi* (fig. 41, B): this genus was first recorded from Ordovician, Silurian, and Jurassic rocks, and subsequently Dr Hinde[2] recognized well-preserved examples in specimens of the Carboniferous Limestone from northern England. *Solenopora Garwoodi* is almost certainly an alga allied to *Lithothamnium* (fig. 25, B), one of the Rhodophyceae which has been a reef-builder since the Cretaceous period. The lack of reproductive organs in the Carboniferous Algae seriously increases the difficulty of precise comparison of extinct and living types. Geologically the various forms are valuable as indices of geological horizons within the Carboniferous system, and the occurrence of some of them over areas several hundred square miles in extent shows their importance as rock-builders.

Lycopodiales

From marine shallow-water sediments, deposited near enough to the coast to receive river-borne scraps of the vegetation, it has been possible to learn something of the nature of the contemporary flora, though our knowledge is mainly derived from fossils in fresh-water beds and from stems, branches, leaves, and seeds preserved in wonderful perfection in volcanic ash. We will first pass in review a few of the plants obtained from the volcanic rocks of southern Scotland, the majority of which are from Pettycur in Fifeshire on the northern shore of the Firth of Forth. In the early stages of the Carboniferous period southern Scotland was the scene of considerable volcanic activity:[3] trees and humbler plants were periodically overwhelmed by showers of ash and their tissues became impregnated by petrifying material derived by percolating water from freshly ejected mineral substances. One of the finest specimens of an early Carboniferous *Lepidodendron* is seen in fig. 20 which shows the polished surface of a cross-section of a stem 33 cm. in diameter.[4] This was found in a bed of ash at Dalmeny south of the Firth of

[1] ἀφρός, foam; ἅλυσις, a chain. [2] Hinde (13). [3] Geikie (97).
[4] For a description of this stem see Seward and Hill (00).

Forth. The dark outer cylinder consists of the resistant tissues of the fissured bark; the more delicate inner tissue has been replaced by a mass of volcanic ash, and the harder cylinder of wood, deprived of support, fell from its central position against the outer shell. The woody axis consists of a narrow cylinder of primary wood surrounding the pith and enclosed by a broader cylinder of secondary wood with no indication of rings of growth. Similar *Lepidodendron* stems were discovered in contemporary volcanic beds on the north-east coast of the Isle of Arran about sixty years ago.[1] The genus *Lepidodendron* played a conspicuous part in the earlier as in the later Carboniferous floras. Trees closely related to Carboniferous species existed in the Devonian period (see fig. 48), but it is the more abundant and better preserved specimens from Carboniferous rocks which enable us to reproduce not only the form of the living plant but also to reconstruct in detail its several parts. In habit *Lepidodendron* resembled *Cyclostigma*, a Devonian genus which persisted into the succeeding period. The paper coal of central Russia consists of matted layers of compressed and mummified cuticular skins of stems closely akin to those of the Carboniferous *Bothrodendron* and assigned by some authors to *Porodendron*.[2] More resistant to bacterial attack than the tissue beneath the surface, the sheets of cuticle, formed of the outermost part of the superficial layer of cells, were eventually carried by water to some tranquil spot where they remained practically unaltered for millions of years. One of the most completely known *Lepidodendra* is the species *L. Veltheimianum*: this characteristic and widely distributed member of the early Carboniferous floras is probably the species preserved as sandstone casts in Victoria Park, Glasgow, which are rooted in the rock by long forking arms originally clothed with absorbing rootlets. Another impressive example of the remains of a *Lepidodendron* forest has been found in Nova Scotia where ninety-six vertical trunks were found in an area 120 by 15 ft.[3] The surface of a *Lepidodendron*, except on the older portions of trunks on which the bark had been split by pressure from within as the stem increased in girth, was covered with spirally arranged, elongated, oval and rounded cushions[4] with a small triangular scar near the upper end

[1] Williamson (80). Good examples may be seen in the Manchester Museum.
[2] Zalessky (09); Walton (26); Bode (29).
[3] Pirsson and Schuchert (20), fig. 426.
[4] See fig. 54; also, for a photograph of the stem of *Lepidodendron Veltheimianum*, Seward (10).

of each where the short needle-like leaves were attached. On some of the larger stems, known as *Ulodendron*[1] (fig. 60, *U.*), there are vertical rows of cup-shaped depressions, a few centimetres in diameter, which mark the position of branches thrown off by a natural process of abscission: precisely similar scars are a familiar feature on stems of the Kauri Pine of New Zealand (*Agathis australis*). As in modern lycopods, the characteristic method of branching in the much greater and more imposing *Lepidodendra* was by equal bifurcations, though not infrequently the regular forking of the shoot-system was modified by the unequal development of some of the branches. Fertile branches bore pendulous cones similar in form though not in structure to those of the Spruce Fir (*Picea excelsa*). Cones of *Lepidodendron Veltheimianum* found at Pettycur illustrate the production of two kinds of spore, as in the diminutive cones of livings pecies of *Selaginella*. In this feature, as in the possession of a scale, known as the ligule, borne above the detachable portion of the leaf, *Lepidodendron* comes nearer to *Selaginella* and *Isoetes* than to *Lycopodium* which it most nearly resembles in general habit and in foliage. One of the largest cones of a *Lepidodendron* is the species *Lepidostrobus Brownii*:[2] it is customary to include the cones in a separate genus because they are usually found detached and not in organic union with stems. In height comparable to a forest tree; in its ability to add successive layers of wood to the central conducting cylinder and to the bark, *Lepidodendron* differed from all living species of *Lycopodium* and *Selaginella*. The genus *Isoetes* (quillworts), distinguished by its stumpy stem and the tuft of quill-like leaves from other members of the lycopod group, has in common with *Lepidodendron* and other extinct arborescent lycopods the habit of adding to the tissues of the stem by secondary growth in thickness, though in *Isoetes* the amount is small. Trees in modern forests owe their power of resilience and ability to meet bending strains to the strength of the wood which constitutes by far the greater part of the trunk. In *Lepidodendron* the wood was a small central rod insufficient to serve as an efficient mechanical support in times of stress: this defect was remedied by the development of a broad external cylinder of comparatively strong tissue which increased in amount as the tree grew. A comparison of figures E.–H., fig. 64, will help to make clear some of

[1] Renier (10), (26[2]). [2] Zeiller (11[2]).

the more obvious differences in the structure of the stems of living and extinct genera of the Lycopodiales.

The *Lepidodendra* are often spoken of as the ancestors of our lycopods and other surviving members of the Lycopodiales, but it is difficult[1] to visualize a series of steps by which the giants of the later Palaeozoic floras could be converted into the lycopods which now grow on our moors and in tropical forests. On the other hand, there is no doubt a direct relationship between certain herbaceous lycopodiaceous plants that have left traces in the sedimentary rocks of the Coal Measures and those which still exist; but while it may be convenient for purposes of popular exposition to quote the Palaeozoic *Lepidodendra* as instances of evolution, working through retrogression rather than by progressive development, we are on safer ground when we think of the arborescent lycopods as the moribund representatives of a separate line of evolution within the group Lycopodiales, as plants which illustrate a common tendency towards gigantism followed by extinction. There is not much satisfactory evidence of the persistence after the early days of the Permian period of the Palaeozoic *Lepidodendra* and *Sigillariae*. Reference should, however, be made to a solitary example of a large lycopod cone, *Lycostrobus Scotti*, closely resembling a cone of *Lepidodendron*, which was described by Prof. Nathorst[2] from the Rhaetic beds of southern Sweden, and to cones from Triassic rocks in Germany;[3] also to some imperfectly preserved pieces of stems and a fossil which resembles *Stigmaria* from Triassic rocks in France.[4]

The great majority of the arborescent lycopods bore cones on which crowded fertile leaves produced sporangia with spores of one or two sizes, but specimens from Pettycur and elsewhere have demonstrated the occurrence of a more specialized type in which some of the sporangia enclose a single large spore and are themselves enclosed in a protective envelope, an integument, which is open along the upper surface of the sporangium to admit the pollen. This more advanced form of reproductive shoot has been named *Lepidocarpon*[5] because of its essential difference from the usual *Lepidodendron* cones which are included in the genus *Lepidostrobus*. The difference between *Lepidocarpon* and *Lepidostrobus* is that the former had reached the level of a seed-producing apparatus, while

[1] Arber, A. (28). [2] Nathorst (08). See Chapter XIII.
[3] Frentzen (20). [4] Fliche (10). [5] Scott (01).

the latter retained the simpler methods of spore-production and dispersal that are still characteristic attributes of lycopodiaceous plants. The discovery of *Lepidocarpon* and of a few other fertile shoots of more or less similar construction affords a striking illustration of the advanced state of organization reached by some of the Carboniferous lycopods. In many respects, notably in the production of organs exhibiting some of the essential characters of seeds, the Palaeozoic lycopods had acquired a much greater degree of specialization than we find in any surviving members of the Lycopodiales.

In a species described from Lower Carboniferous rocks in Scotland as *Lycopodites Stockii*[1] we have an example of what is apparently an herbaceous lycopod agreeing closely with modern forms. Though the Lepidodendreae of the late Carboniferous floras differed, so far as we know, in no very important respects from the smaller number of species which flourished in the earlier days of the period, there are certain distinguishing features—notably the form of the leaf-cushions and the structure of the stem—which can be employed usefully as criteria of geological age. Among the Lower Carboniferous species are *Lepidodendron Veltheimianum*, *L. Rhodeanum*, *L. Volkmannianum*, *L. Nathorsti* and others. The long and slightly sinuous, tapered ends of the leaf-cushions shown in fig. 54 are a distinguishing feature of a few of the older species of the genus. The genus became cosmopolitan during the earlier stages of the Carboniferous period: specimens identical with or hardly distinguishable from species from Spitsbergen, Europe, and North America have been recorded from Australia; *Lepidodendron Veltheimianum* and closely related forms have been found in many parts of Europe, in Spitsbergen, in north-eastern Greenland (latitude 80–81° N.), Asia Minor, North Africa, the Sinai peninsula, and North and South America. Confirmatory evidence of the occurrence of this species in Australia has recently been furnished by a description of a petrified stem[2] which is apparently specifically identical with specimens previously recorded from Pettycur in Scotland.

The genus *Sigillaria*, though abundant in the late Carboniferous floras, seems to have been very rare in the early part of the period. The few specimens so far described have ribbed stems of a type foreshadowed on portions of the trunk of the Upper Devonian *Protolepidodendron primaevum* (see fig. 48, *Pn.*). Branches, known

[1] Kidston (84); Bower (08), p. 298.

[2] Barnard (28).

as species of *Archaeosigillaria*,[1] occasionally bearing short leaves and covered with contiguous, hexagonal leaf-scars, from Spitzbergen, Scotland, Wales, England, and North Africa, are probably closely allied to *Protolepidodendron*, which has recently been recorded from early Carboniferous strata in Australia.

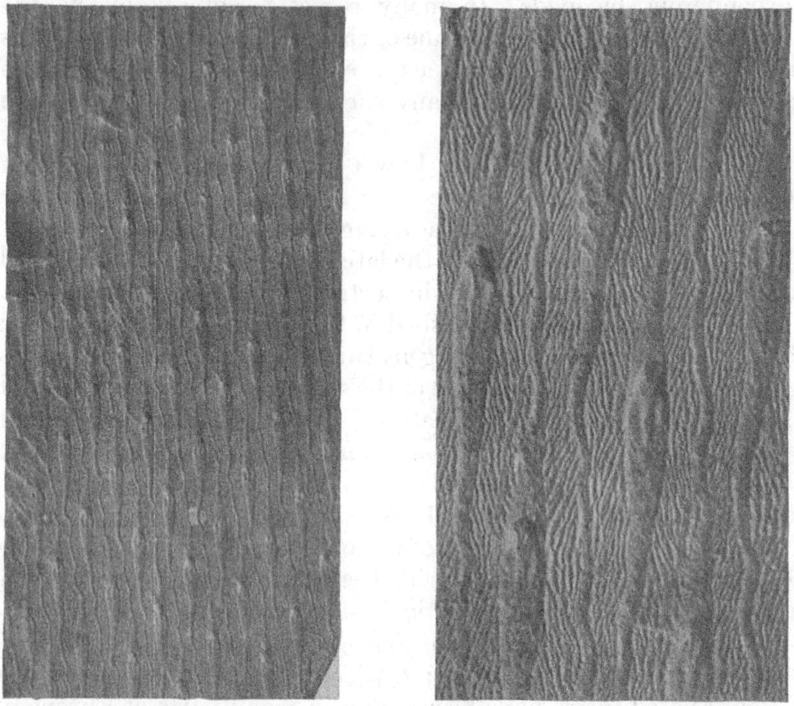

A B

FIG. 54. *Lepidodendron Nathorsti* from Linlithgowshire, Scotland, showing long and narrow leaf-cushions on the wrinkled stem. A. Nat. size. B. Enlarged. (Kidston Collection, Geological Survey Museum.)

Articulatae

We may next briefly consider some other plants obtained from southern Scotland. The genus *Sphenophyllum*, which is represented in Upper Devonian floras and continued into the Permian period, is recorded from Lower Carboniferous strata in north-eastern Green-

[1] White, D. (07); Carpentier (13); Walkom (28); Carpentier (30) who describes both Sigillarias and *Archaeosigillaria* from Morocco.

land, Pettycur, Newfoundland, and Europe. Another member of the Articulatae, plants with jointed stems and leaves borne in whorls, is the genus *Asterocalamites* (fig. 60, *As.*): this was a tall equisetaceous plant differing from *Calamites* and modern *Equiseta* in the narrower, more spreading and repeatedly forked leaves, and in the lack of alternation at the nodes of the vertical ribs which mark the position of the conducting strands. A few species of *Calamites* also occur in Lower Carboniferous rocks,[1] but the genus is much more characteristic of the later Carboniferous floras and of the earlier Permian vegetation.

Pteridosperms

One of the outstanding features of the early Carboniferous vegetation, as of the later Carboniferous floras, is the abundance and variety of fern-like fronds borne by pteridosperms, plants producing seeds and small pollen-bearing sporangia. One of the best known pteridosperms is the genus *Heterangium*. The oldest species was founded on petrified stems from Pettycur: the stem rarely exceeds 1·5 cm. in diameter; there is a single axial strand of primary wood which became surrounded by a cylinder of secondary wood (fig. 64, S) comparable in structure to that of cyads.[2] *Heterangium* and pteridosperms generally differed widely from ferns both in methods of reproduction and in the more highly specialized structure of the stem and other organs. The slender stem produced large, freely branched leaves of the form known as *Sphenopteris elegans*, which reminds one of some living species of *Davallia*, but the fronds of *Heterangium* are distinguished by a forked method of branching which is unusual in modern ferns. With the leaves, though not actually attached to them, occur seeds (*Sphaerostoma*), 3·5 mm. in length, which, despite their small size, bear a close resemblance in structure to the much larger seeds of living cyads. *Heterangium* flourished more abundantly in the late Carboniferous floras. Several examples of pteridosperms and fronds which may conveniently be called pteridophylls, because in the absence of reproductive organs it is often impossible to say whether a specimen of a leaf was borne by a fern or by a pteridosperm, were described by the late Prof. Nathorst from the rich Lower Carboniferous plant-beds of Spitsbergen.

[1] For photographs of the various species of *Calamites* see Kidston and Jongmans (15).
[2] For descriptions of different kinds of *Heterangium* see Scott (17); Kubart (14).

Emphasis was laid in Chapter IX on the richness of the flora preserved in the rocks of Bear Island in latitude 75° N. and of Ellesmere Land several degrees farther north. One of the largest collections of Lower Carboniferous plants is from Spitsbergen:[1] the size of the stems and fronds affords a remarkable contrast to the stunted herbage which now ekes out a precarious existence in the same region. It is worth while to glance at some of the members of this ancient Arctic flora. Fronds of *Sphenopteridium* and *Sphenopteris* are represented by several forms which differ only in a minor degree from Upper Devonian species. The frond of the species *Sphenopteris Nordbergi* with a bifurcating rachis and branches bearing deeply divided leaflets resembles those of some existing ferns as also the leaves of the fennel (*Foeniculum vulgare*). There is no absolute proof of the nature of the reproductive organs, though in all probability some associated seeds belong to the same plant: these seeds (fig. 76, F) have a long pointed beak clothed with stiff hairs and recall the fruits of the stork's bill (*Erodium*). Many other instances might be given of seeds of Palaeozoic pteridosperms which in form, and in the possession of appendages facilitating dispersal by wind or by animals, suggest comparison with types evolved after a lapse of millions of years by the fruits of flowering plants. In the course of ages similar results have been achieved by many diverse groups of plants: age after age there has been a repetition of unconscious effort towards the same end. Groups which had reached what might be regarded as an advanced state of efficiency became extinct; after a long interval new creations repeated with little or no change in plan the structural design produced by long-forgotten and possibly blindly ending lines of evolution. The pteridosperms of the later Palaeozoic floras, in the form of their leaves, in the construction of their seeds and male organs, as also in certain anatomical features, seem to foreshadow the flowering plants of later ages. Is it possible that these analogies may be something more than instances of parallel development?

Another type of pteridosperms is illustrated by species of *Adiantites* (fig. 60, *Ad.*; fig. 63, C),[2] fronds with leaflets resembling those of the maidenhair fern (*Adiantum*), and superficially not unlike the foliage of *Thalictrum*, a genus of flowering plants. One form, *Adiantites bellidulus*, described from Spitsbergen, is frequently associated with seeds and male organs, while in another species

[1] Nathorst (94), (14). [2] Kidston (89), (23), (23[2]).

from slightly higher Carboniferous rocks in America seeds have been found attached to the leaflets. There are also fronds known as *Cardiopteris* (fig. 60, *C.*) with an unbranched rachis bearing two rows of large, almost circular leaflets, and others named *Cardiopteridium* in which the rachis becomes forked in the upper part. *Cardiopteris* is a characteristic and widely distributed Lower Carboniferous genus which is recorded from many European localities, from North and South America, and elsewhere. Before leaving the fronds of pteridosperms, reference may be made to *Sphenopteris*[1] (or *Telangium*) *teiliana*, a species very similar to *S. Nordbergi*, from Lower Carboniferous rocks in North Wales. Prof. Walton, whose hypothetical restoration of the fertile frond is reproduced in fig. 55, has shown that the main axis of the large frond gives off two large and leafy divergent arms; the main axis at a slightly higher level forks into two equal branches, each of which again forks repeatedly into bush-like clusters of branchlets with groups of sporangia at their tips. This frond affords a striking contrast, in its more bushy habit and in the division of labour illustrated by the difference in position and in form between its sterile and fertile branches, to the fronds of true ferns: the stem is not known. Another form of pteridosperm frond, the genus *Rhodea*,[2] though not confined to Lower Carboniferous floras, is a characteristic member of them: the leaflets are deeply cut into narrow almost hair-like segments each with a single vein (fig. 63, K), but some species shade into the *Sphenopteris* type of frond (fig. 63, G).

Various kinds of petrified stems have been obtained from Lower Carboniferous rocks in Scotland and elsewhere which, from their resemblance in anatomical characters to stems known to have borne seed-bearing fronds, are also referred to the pteridosperms. Prof. Gordon[3] described a stem, about 2 cm. in diameter, from Pettycur as *Rhetinangium Arberi*: though not unlike certain ferns in the arrangement of the primary conducting strands it agrees with *Heterangium* in having a fairly broad zone of secondary wood. Another form of stem construction is illustrated by *Stenomyelon*[4] from southern Scotland: in its anatomical characters (fig. 64, T) it

[1] Prof. Walton proposes to institute for this type of frond a new generic name, *Diplopteridium*.

[2] Carpentier (29) describes the fructification of *Rhodea*. For illustrations of the fronds see Kidston (23–25); Oberste-Brink (14).

[3] Gordon, W. T. (12).

[4] Kidston and Gwynne-Vaughan (12).

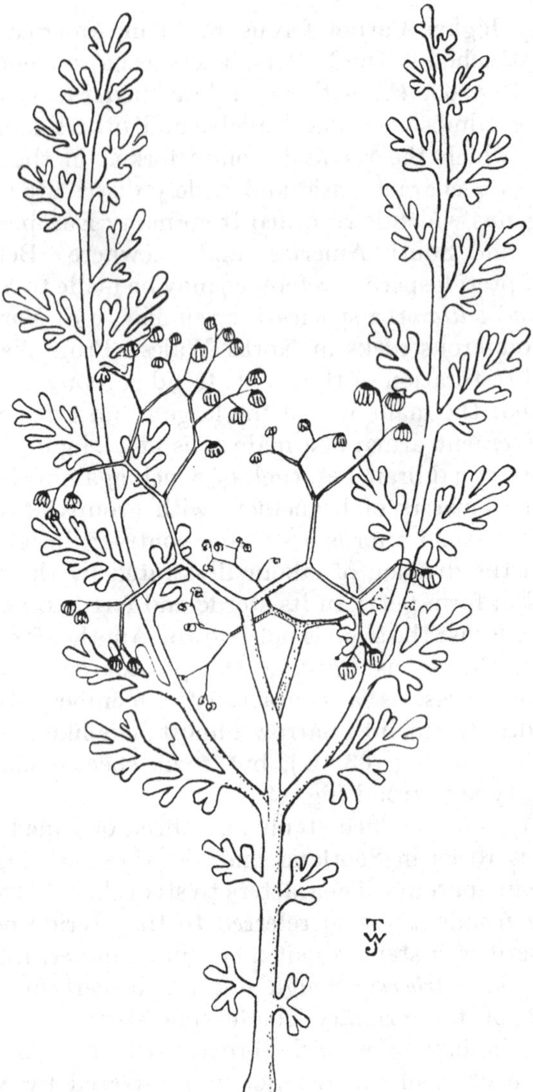

FIG. 55. Restoration by Prof. Walton of a fertile frond of *Diplopteridium teilianum* (formerly called *Sphenopteris teiliana*). The connexion of the spore-capsules with the axis of the frond has not been definitely proved. This reconstruction was first published in the *Encyclopaedia Britannica* [Walton (29³)].

differs from all known living plants. There are also species of *Calamopitys*[1] recorded from Britain, Germany, Hungary and other regions: these stems are a few centimetres in diameter; a fairly large pith, consisting either entirely of soft tissue or with an admixture of woody elements, is surrounded by a ring of primary conducting strands external to which is a cylinder of secondary wood. We know nothing of the reproductive organs; only the lower ends of leaf-stalks are preserved in union with the stem. The wood is similar in its finer structure to that of living araucarias. The genus *Cladoxylon*, already mentioned as a member of the Devonian vegetation, is recorded from Lower Carboniferous rocks in Germany, England, and Scotland. In the arrangement of the conducting tissue in the stem it differs in some respects from all living types. Instead of the single column of wood as in a pine or an oak there were separate vertical bands (fig. 64, D), some simple, others divided into radially disposed arms which in cross-section resemble an irregular star. Each plate consists of primary wood surrounded by a varying amount of secondary wood. These examples suffice to show the complexity and the diverse forms of Carboniferous stems which are believed to be pteridosperms. A larger stem, nearly 1 ft. in diameter and probably a pteridosperm, is represented by *Protopitys* (fig. 64, V). Nothing is known of the leaves or reproductive organs. The stem has a fairly large pith, elliptical in section, enclosed by a zone of primary wood to which the plant added by regular increments a cylinder of secondary wood. A peculiar feature, which was adopted after the lapse of many ages by the palm *Ravenala* (travellers' tree) is the emission of conducting strands from the ends of the long axis of the main cylinder to supply two opposite rows of leaves.

Ferns

We turn next to a few leaves the affinities of which are uncertain. Fig. 56 shows portions of two fronds of *Rhacopteris*:[2] one is attached to a piece of stem. The fronds of this genus are usually simple and not forked, but a few specimens are known in which the upper part forks into two arms bearing clusters of sporangia. The leaflets are wedge-shaped or almost circular (fig. 57, B) and traversed by

[1] For descriptions and illustrations of this and other genera reference must be made to text-books, e.g. Scott (23) or Seward (17).
[2] Kidston (23^2); Oberste Brink (14); Walton (27).

numerous veins: occasionally the leaflets are dissected into finger-like segments (fig. 57, A). This genus, though not entirely confined to the lower beds of the system, is characteristic of the earlier Car-

FIG. 56. Two fronds of *Rhacopteris* attached to a stem. Lower Carboniferous. Nat. size. (After Walton.)

boniferous floras; it occurs in many parts of the northern hemisphere and is recorded also from Australia. It may be a true fern.

An interesting feature of the early Carboniferous vegetation is the occurrence of several peculiar ferns. Several forms have been described from the Pettycur beds. The specimens consist of well-preserved pieces of leaf-stalks and occasional stems, also in some instances of spore-capsules. We know little of the external form of the leaves. Though these plants are no doubt correctly assigned to the ferns, none of them bear a close resemblance to existing

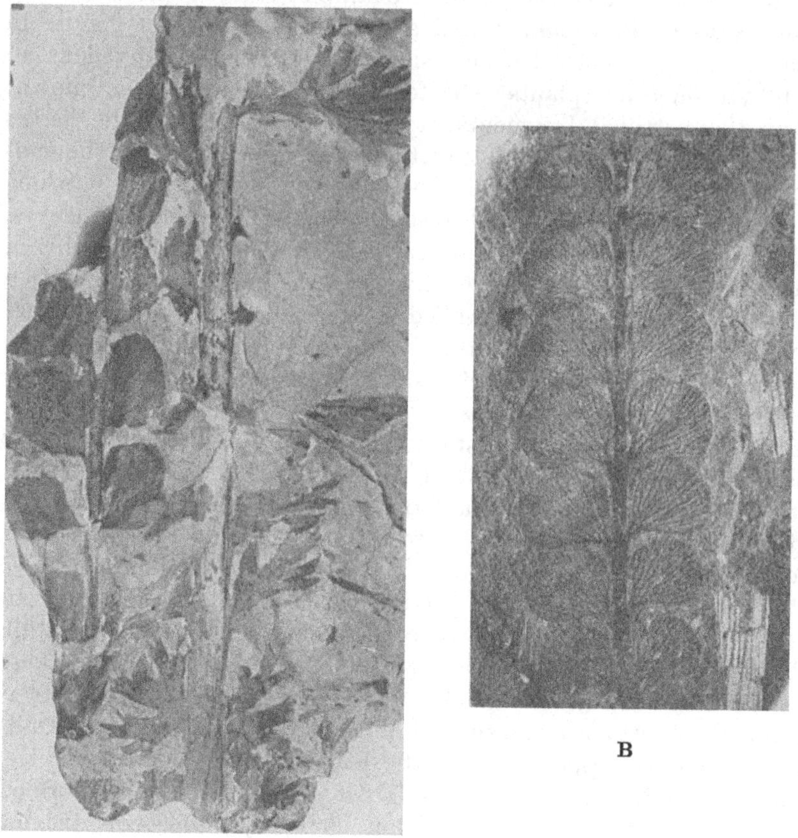

B

A

FIG. 57. A. *Rhacopteris ovata* and *R. intermedia*. Carboniferous, New South Wales. Manchester Museum. Nat. size. (Photo. E. Ashby. After J. Walton.) B. *Rhacopteris circularis*. Manchester Museum. Nat. size. (After J. Walton.)

members of the group. They are included with the Devonian *Asteropteris* in a family for which the name Coenopterideae was proposed because of the association within a single group of anatomical characters that are now distributed among several families. In some respects the Coenopterideae show points of contact with the

Osmundaceae, the Ophioglossaceae, and the Hymenophyllaceae; but these resemblances are not close enough to serve as direct connecting links between the oldest known ferns and those of the present day. One of the simplest examples is *Botryopteris antiqua* discovered at Pettycur: it had a thin stem, less than 2 mm. in diameter, and, as Dr Hirmer[1] suggests, it may have lived as an epiphyte on other plants. The form of the fronds is not known: from the petrified fragments of stems we learn that the leaves were branched and bore no leaflets. As seen in fig. 58, A, the conducting tissue forms a solid rod of simple structure from which large branches were given off to supply the bifurcate axes of leaves. A root is seen at *r*. The spore-capsules, hanging from the tips of branchlets, agree with those of existing ferns in the possession of an annulus, a band of thicker-walled cells which on contraction cause the rupture of the capsule-wall. The annulus consists of a band three to four cells broad and is not a single layer as in modern ferns. A less simple type is seen in *Metaclepsydropsis*.[2] Fig. 58, C, shows a transverse section of a leaf-stalk with an oval axial strand having two spaces, one at each end of its longest diameter, which were originally occupied by delicate tissue. The space surrounding the woody core was filled with less resistant tissue which had decayed before petrification. In fig. 58, B, the wood is seen on a larger scale: at one end are two strands passing out to a forked leaf-stalk. In fig. 58, D, the base of the two equal arms of a branched leaf-stalk are seen above the main axis: at a higher level on the stem these strands would form the core of each of the separate, divergent arms of the frond. At *a* are two smaller and imperfectly preserved strands which supplied appendages of the leaf-base known as aphlebiae. Such appendages, rare in modern ferns, are regarded as transformed leaflets and are relatively common on many Palaeozoic fronds. Though no complete fronds of *Metaclepsydropsis* have been found we know that the leaf-stalk bore four rows of branches, a series of two bifurcating arms on each side. This architectural feature, characteristic of several members of the Coenopterideae, is in striking contrast to that of existing ferns. In most ferns the leaf-stalk leaves the stem as a single axis and then gives off a series of branches (pinnae) on two opposite sides, the pinnae being usually so placed that the branches and leaflets lie in one plane: the whole branched frond is flat. An approach to the habit of many Coeno-

[1] Hirmer (27); Kidston (08). [2] Gordon, W. T. (11).

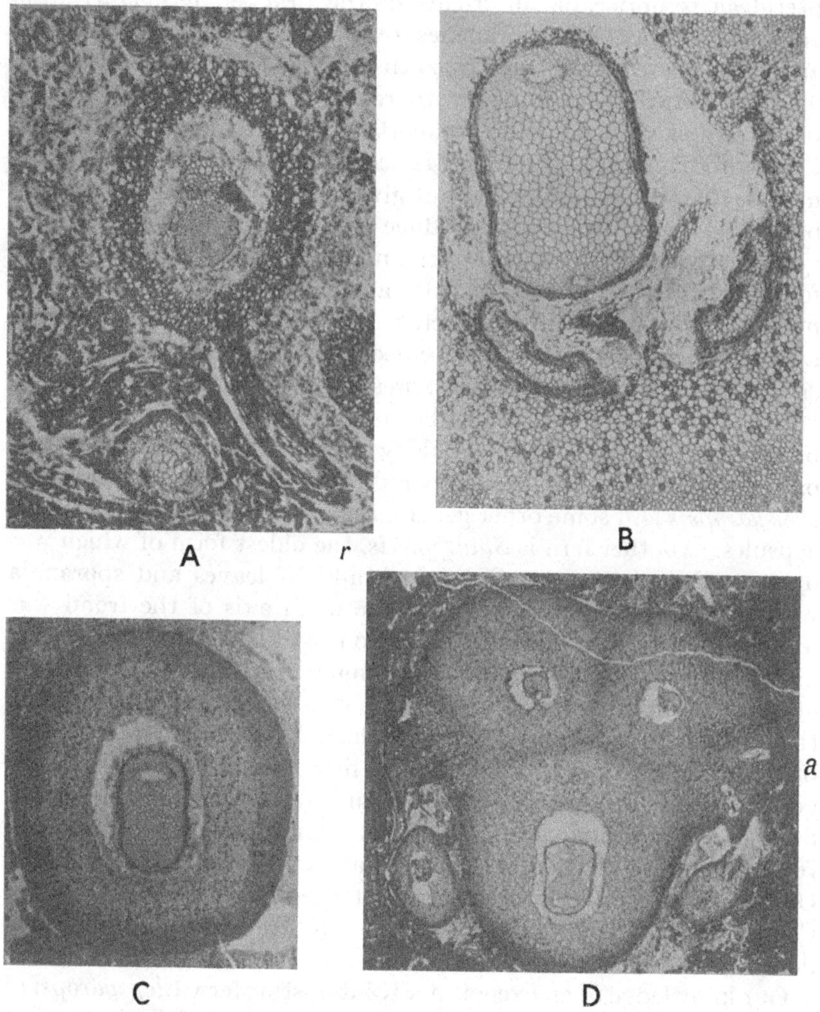

FIG. 58. Sections of stem (A) and leaf-stalks (B–D) of Lower Carboniferous ferns from Pettycur, Scotland. A. *Botryopteris antiqua* stem with root and petioles. B. *Metaclepsydropsis duplex*. Petiole (leaf-stalk) and, below, smaller strands of conducting tissue supplying the forked branches (pinna) of the frond. × 19. C. *Metaclepsydropsis duplex*. Complete transverse section of a petiole. × 83. D. *Metaclepsydropsis duplex*. Two pinnae being given off from the main axis of the frond: this section shows also, on the flanks, the smaller strands of aphlebiae seen on the right-hand side opposite *a* as black lines with clear spaces. *ca.* × 3. (Photograph by Prof. Gordon.)

pterideae is noticeable in fronds of the bracken fern (*Pteridium aquilinum*): the lateral branches tend to place their leaflet-clad arms in a plane at right angles to that of the main axis. Attention may be drawn to a tendency to revert to a forking method of branching in some varieties or sports of modern ferns. In several Carboniferous ferns, including *Metaclepsydropsis* (fig. 58, B–D) the main leaf-stalk is unbranched but gives off two series of repeatedly bifurcate branches which produce an almost bush-like frond-system. Another peculiarity of the majority of the Lower Carboniferous ferns is the absence of flat leaflets. The frond-branches were presumably green and the important function of absorbing carbonic acid gas from the air was performed by the plexus of the slender subdivisions of the leaf. An inflorescence of the commonly cultivated plant *Gypsophila*, a member of the carnation family (Caryophyllaceae), affords a rough analogy: the small scaly bracts at the base of the forks may be compared with the aphlebiae of *Metaclepsydropsis* and some other genera, and the flowers with the spore-capsules. Another fern is *Stauropteris*, the oldest form of which was discovered at Pettycur. Only the stalks of leaves and sporangia with spores have been described. The main axis of the frond was forked and by the repeated bifurcation of the branches the leaves acquired a bush-like habit. Sporangia with an annulus were attached to the tips of ultimate branchlets. In transverse section the conducting tissue of the rachis has the form of a cross with blunt arms, a pattern unlike that in living ferns. In *Diplolabis*, a genus allied to *Metaclepsydropsis* and recorded from Scotland, Silesia, and France, the stem had a cylindrical axial strand and from this were detached H-shaped strands to supply the leaf-stalks. On the main frond-axis were series of forked branches. Many of the Coenopterideae continued to flourish in the later Carboniferous floras.

Our knowledge of an exceptionally interesting fern, *Clepsydropsis*,[1] has recently been substantially increased by Prof. Sahni's account of some Australian specimens. Fig. 59 shows a transverse section of part of a large silicified block which was found in the Kuttung[2] (Lower Carboniferous) series of New South Wales. In the section are seen several small stems, *S*, recognizable by the five-armed form of the woody axis: these are scattered among a large number of

[1] From κλεψύδρα = a water-clock, from the hour-glass form of the conducting strand of the leaf-stalk. Sahni (28).

[2] See p. 171.

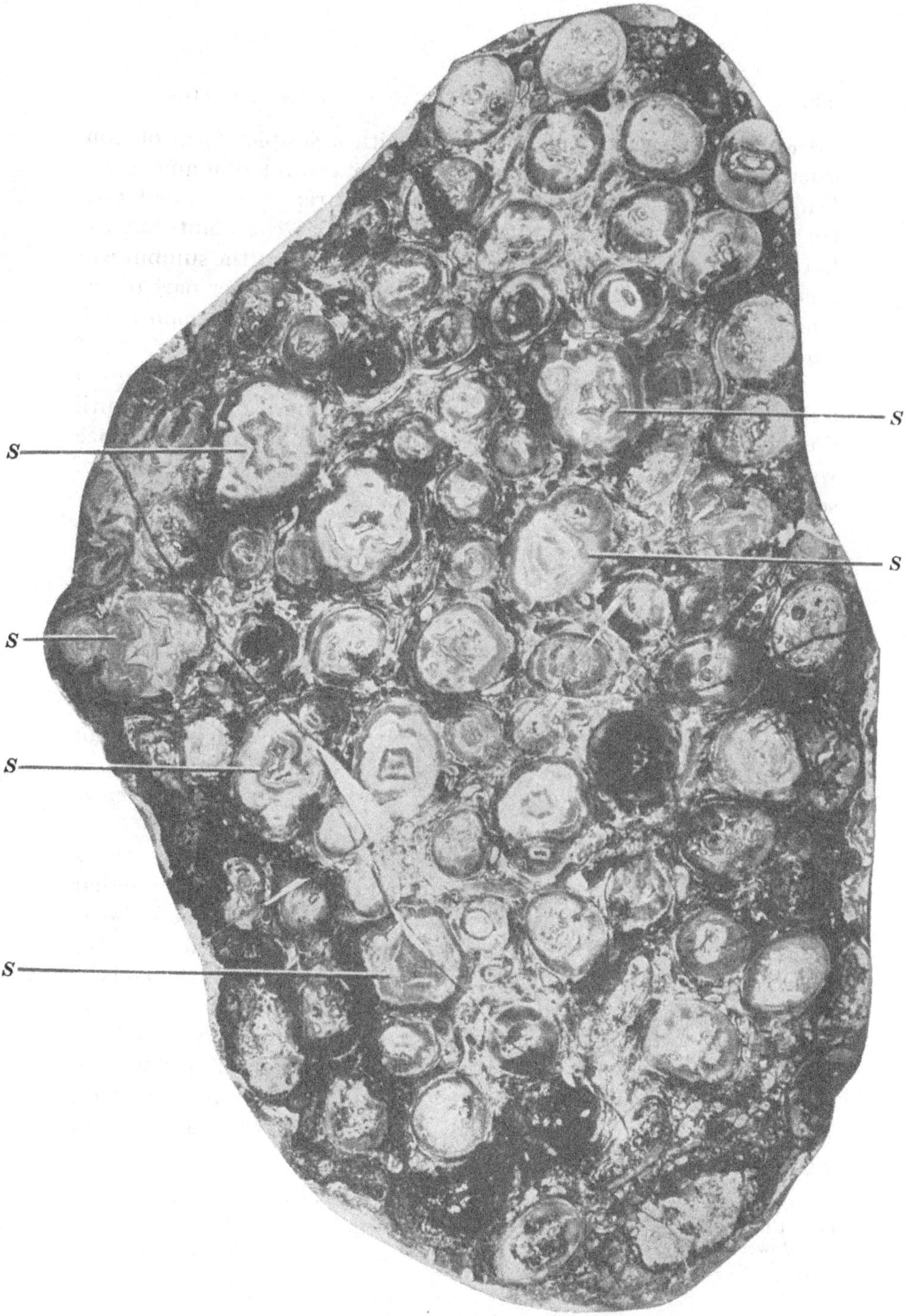

FIG. 59. A petrified *Clepsydropsis* from Lower Carboniferous rocks in New South Wales consisting of a mass of leaf-stalks and a few stems embedded in a plexus of roots. The stems of the fern are shown at *S*. Nat. size. (From a section sent to me by Prof. Sir T. W. Edgeworth David. A.C.S.)

rather smaller sections of leaf-stalks with a simpler form of conducting axis, and a mass of associated roots wandering among the leaf-stalks and stems. The section probably represents about one-third of the complete stem. Sahni thinks that the plant had the habit of a tree-fern and a height of 12 ft. or more. At the summit was a crown of leaves, but of these we know only the lower part of the stalks. A remarkable feature is the construction of the main trunk which is spoken of as a false stem because it is in reality a group of stems (fig. 59, *S*) formed by the repeated branching of an original axis. As the plant grew the stems became more numerous until finally a robust multi-stemmed column was produced. The surface was covered with a tangle of roots such as we see in the stems of some existing ferns. A similar form of construction is illustrated by the Lower Cretaceous fern *Tempskya*, and comparison may be made with the stem of the South African osmundaceous *Todea barbara*. *Clepsydropsis* is the only member of the Coenopterideae so far discovered in the southern hemisphere: it was originally described from Lower Carboniferous rocks in central Europe and from Siberia. Prof. Sahni, who recently examined the Siberian and German specimens, assigns the Siberian fern to a new genus, *Asterochlaenopsis*,[1] and speaks of it as possibly Permian in age: he is not convinced of the generic identity of the fossils from Thuringia, originally described as *Clepsydropsis*, and the Australian *Clepsydropsis*.

Looking at the early Carboniferous ferns as a whole, we are struck by the diversity in structure of stems and petioles, by the peculiar style of the usually leafless fronds, and by the occurrence of sporangia, not as in most modern forms on the upper surface of the leaflets but as single capsules or tassels pendulous from the ends of fine branches. In some respects, notably in the lack of leaflets and in the position of the sporangia, these extinct ferns may be said to be relatively primitive.[2] They are highly specialized products of evolution which, though possessing certain features in common with living genera, cannot be directly linked either with Mesozoic or with surviving members of the fern class.

[1] In a paper recently communicated to the Royal Society of London and published after this chapter was in type: Sahni (30).

[2] Bower (23–28).

Psygmophyllum *and* Pitys

Examples of the large wedge-shaped leaves of *Psygmophyllum*, a Devonian and Carboniferous genus of uncertain position, which were not less than 17 cm. long and 15 cm. broad, were described by Dr Arber[1] from Lower Carboniferous rocks in Newfoundland.

Several petrified stems from Lower Carboniferous rocks present a striking similarity in the structure of the secondary wood to the araucarian conifers *Araucaria* and *Agathis*, though in the character of the primary wood they form a group apart. The largest tree so far discovered in Britain may be seen in the grounds of the Natural History Museum, London: it was found at Craigleith near Edinburgh.[2] The incomplete trunk is nearly 50 ft. high and at the lower end the wood is 5 ft. across. This tree, *Pitys Withami*, is one of several Lower Carboniferous gymnosperms represented by petrified stems, but in one species, *Pitys Dayi*,[3] discovered by Prof. Gordon in East Lothian, Scotland, we know that the leaves were small and similar to those of *Araucaria excelsa* and allied conifers. Some trees of *Pitys* (*Pt.*) with a *Lepidodendron* are shown in the right-hand upper corner of fig. 60. Diagrammatic transverse sections of *Pitys* are shown in fig. 64, R and U. *Archaeopitys*, described from Lower Carboniferous rocks in Kentucky,[4] is an allied genus. These and other trees afford evidence of the further development both in the number of forms and in the range of structure of a group which was represented in the Devonian floras; they demonstrate the remote antiquity of the araucarian type of wood. The chief structural difference between these early Carboniferous trees and modern conifers is the greater separation in the extinct genera of the primary from the secondary wood. If we look at a transverse section of the stem of an *Araucaria* or a pine we see a single cylinder of wood encircling the pith, and on closer inspection we note a gradual shading off of the irregularly grouped elements of the primary wood at the edge of the pith into the regular rows of elements which make up the bulk of the tree. In *Pitys* (fig. 64, R, U), *Archaeopitys*, and other ancient forms the primary conducting tissue forms strands encircling or scattered through the pith and not merged into the subsequently developed secondary wood. The occurrence of separate strands of primary conducting tissue seems to be an archaic feature[5] which did not persist in later products of evolution.

[1] Arber, E. A. N. (12²). [2] Witham (31). [3] Scott (23).
[4] Scott and Jeffrey (14). [5] Scott (02²).

The restoration reproduced in fig. 60, though purporting to depict a scene in the latter part of the Carboniferous period, includes in the right-hand corner a group of plants characteristic of the earlier Carboniferous vegetation. In the middle distance are a few trees of *Pitys* (fig. 60, *Pt.*) with foliage-shoots at the ends of branches bearing small crowded leaves. A *Lepidodendron* is seen in company with the larger trees. In the right-hand lower corner at *Ad.* is a large frond of the pteridosperm *Adiantites*, with leaflets similar to those of a modern species of *Adiantum* but with small seeds on some of the branches. To the left of this is a curved stem of an *Asterocalamites* (*As.*) bearing regular whorls of bifurcating slender leaves, embraced by a *Sphenophyllum* (*Sp.*) with its whorls of deeply cut foliage. At *C.*, *C.* are fronds of *Cardiopteris*, and two species of *Rhacopteris* are seen at *R.*, *R.* At *S.* is the fern *Stauropteris burntislandica* with forked and spreading fronds bare of leaflets.

The geographical distribution of some of the more characteristic early Carboniferous plants is shown in Table B (p. 266).

CHAPTER XII

THE LATER CARBONIFEROUS VEGETATION WITH SOME ACCOUNT OF PERMIAN FLORAS

...a wide and melancholy waste
Of putrid marshes. *Shelley*

BEFORE describing selected examples of the more conspicuous or more interesting plants of the later Carboniferous and early Permian vegetation it is worth while to enquire what impression would be made upon a visitor from the modern world to the swamps and hills which were the site of the Coal Age forests. As an additional aid to an appreciation of the general aspect of the vegetation an attempt has been made to reproduce a typical scene (fig. 60) in which a few of the more abundant trees and herbaceous plants are represented. The restorations are for the most part based on substantial evidence, but it is freely acknowledged that in the interpretation of the available data imagination necessarily plays some part.

A Scene in the Coal Age

The reconstruction reproduced in fig. 60, though intended primarily to illustrate a scene in the Coal period in the northern hemisphere, includes on the right-hand side of the picture a few trees and herbaceous plants which flourished in the earlier part of the Carboniferous period (see Chapter XI). Marshes and sheets of water with a few low hills reaching to the horizon: in the middle distance and beyond swampy ground studded with *Calamites*; in the centre lying obliquely behind one of the herbaceous lycopods, which differs but little from some living members of the family, is a stranded trunk of a woody tree. In the left-hand corner one sees the spreading fronds of the pteridosperm *Neuropteris*, bearing rows of oblong, rounded leaflets and, in place of some of them, large tapered seeds (fig. 60, *N.*): near the fork of the frond on the edge of the picture (*N.*) two larger and more rounded leaflets are clearly shown: they are the leaflets known as *Cyclopteris*, so-called from their more or less circular shape, which are a characteristic feature of some *Neuropteris* fronds.[1] Lying obliquely across the corner and partially hidden by the fern-like foliage, is a dead stem of *Ulodendron* (*U.*), a

[1] For photograph and reconstruction of a *Neuropteris* frond see Carpentier (30^3).

type of arborescent Lycopod distinguished by rows of large scars which mark the position of cast-off branches. Behind the *Ulodendron* is a small plant of *Psygmophyllum* (*Pg.*) with wedge-shaped leaves. A tall calamite in front of the prone trunk is distinguished by its regular transverse nodal lines and whorled foliage-shoots and branches bearing stalked, long and narrow cones: a *Sphenophyllum* has twisted itself round the calamite stem (*Cal.* with *Sp.*). Farther back, to the left, is a group of plants having the semblance of tree-ferns, though actually pteridosperms. The trees in the upper left-hand corner are representatives of gymnosperms allied to certain existing conifers: the shorter tree with a columnar stem bearing a dense tuft of branches resembling those of some araucarias is a species of *Walchia* (*W.*), a late Carboniferous and Permian genus: the taller trees are examples of *Cordaites* (*Cd.*) one form with broader and another with narrower strap-like leaves. The nearest tree in the water is a *Lepidodendron* (*Ln.*) with a more open canopy of leafy branches: on each side of it is a *Sigillaria* (*Sg.*) with clusters of stalked, slender cones and at a higher level crowded needle-like leaves. Behind the larger *Sigillaria* are three kinds of *Calamites* differing from one another in the arrangement of the slender branches. In the middle of the foreground is the decaying stump of a *Lepidodendron* or a *Sigillaria* with its forked and partially submerged spreading "roots" (*Stigmaria, Sa.*). Farther to the right are two plants closely resembling one another in habit: the taller is a *Psaronius* (*Ps.*), a true tree-fern; the shorter, bearing a large forked frond and others not yet uncoiled, is a pteridosperm (*N.*). The two compound leaves with broad fleshy axes (rachises) and lateral arms belong to one of the true ferns, a species of *Etapteris* (*Et.*), one of the Coenopterideae, in which some fronds bore very small and deeply lobed leaflets and others tufts of spore-capsules.

Let us first look more closely at some of the trees: groves of *Calamites* (fig. 60, *Cal.*)—their columnar stems, bare below, where branches had been cast off; the bark fissured by the continued expansion of the wood within; the tapering upper portions hidden by closely set tiers of whorled branches bearing star-like clusters of leaves at short intervals—might suggest comparison with greatly enlarged horsetails (*Equisetum*) in which whorls of spreading foliage replaced the toothed leaf-sheaths. Some of the calamites, with stems less freely branched or with crowded whorls of foliage-shoots more widely separated from one another on the main axis, would

FIG. 60. A landscape in the Coal Age. On the right a group of older Carboniferous plants. (Drawn by Mr Edward Vulliamy.) *Ad. Adiantites*; *As. Asterocalamites*; *C. Cardiopteris*; *Cal. Calamites*; *Cd. Cordaites*; *Et. Etapteris*; *Lg. Lyginopteris*; *Ln. Lepidodendron*; *N. Neuropteris*; *Pg. Psygmophyllum*; *Ps. Psaronius*; *Pt. Pitys*; *R. Rachopteris*; *S. Stauropteris*; *Sa. Stigmaria*; *Sg. Sigillaria*; *Sp. Sphenophyllum*; *U. Ulodendron*; *W. Walchia*.

recall other forms of modern horsetails. Trees such as *Lepidodendron* (fig. 60, *Ln.*), with forked trunks and branches stretching upwards by repeated subdivision into a canopy of needle-clad shoots, would at a distance recall some existing conifers; though on closer inspection many peculiarities would become apparent. An even greater contrast to the ordinary type of forest tree would be furnished by the *Sigillariae* (*Sg.*) recognizable by their tall and bare stems, some unbranched, others with an occasional fork, soaring upwards into an elongated conical summit encased in a tuft of long pine-like needles. In habit the *Sigillariae* suggest comparison with well-grown plants of the Australian Liliaceous genera *Kingia* and *Xanthorrhoea* (grass-trees or black boys), or with some of the taller species of unbranched cacti. The handsome *Cordaites* (*Cd.*) would seem more familiar, the spreading branches clothed with long strap-shaped leaves like those of a *Yucca* or of the larger-leaved species of the conifer *Agathis* (kauri pine of New Zealand and other forms). A less common tree in the late Carboniferous and the early Permian forests was the genus *Walchia* (fig. 60, *W.*) with leaves resembling those of the Norfolk Island pine (*Araucaria excelsa*).

Here and there among the *Calamites* and *Lepidodendra* would be found tree-ferns superficially indistinguishable from modern species, the older part of the stem made thicker by a covering of sinuous roots and the apex embowered in a crown of spreading fronds (fig. 60, *Ps.*). On a nearer view the tendency of many of the fronds to branch by repeated forking would be noted as an interesting feature. In addition to the tree-ferns the forests contained several other ferns, much too small and inconspicuous to attract attention in a general view. These humbler members of the group would be found to agree closely with the genera already described in the account of the older Carboniferous flora.

An observer would be struck by the abundance and variety of plants which to him appeared to be ferns: some having stems like miniature tree-ferns, others of lower growth with fronds borne on creeping rhizomes, and possibly some living as epiphytes, their green leaves standing out against the more sombre coloured supporting trunks of forest trees. On a closer view he would discover that by far the greater number of these supposed ferns were plants—both small and large—bearing seeds and clusters of inconspicuous spore-capsules filled with pollen, members of a group which we now call pteridosperms. The pteridosperms were a dominant group; they had evolved a style of compound leaf (fig. 60, *N.*) constructed on the general plan of a fern frond. The position attained by these plants with their amazing range in the design of the leaflets and in the structure of the reproductive organs (fig. 61) is one of the outstanding features of the later Palaeozoic floras. They are products of evolution which it might almost seem were created to serve as a warning to botanists against placing trust in superficial resemblances as guides to affinity.

Restorations of the late Carboniferous vegetation enable us to form a general idea of the plant associations or at least they represent interpretations, reinforced by imagination, of such fragmentary documents as are available. They tell us nothing of the flora composed of plants too small to be shown in a comprehensive view. The vegetable kingdom in the Permo-Carboniferous age as in other ages included many representatives of fungi, algae, mosses and liverworts: our knowledge of them is meagre but there are a few points of general interest worth attention.

The Lower Plants

An examination under the microscope of thin sections of wood and other tissues prepared from the petrified litter of the late Palaeozoic forests often reveals the occurrence of very delicate threads, with here and there a spherical swelling, which are in all respects identical with the cells and reproductive organs of modern fungi.[1] Palaeozoic trees, like the trees of to-day, were susceptible to the attacks of fungi: some lived as parasites in the tissues of their food-producing hosts; others lived in the soil or on the organic substances in the miscellaneous débris of broken branches, cast off leaves, and other scraps of dead vegetation. In the absence of well-preserved reproductive organs it is usually impossible to compare with any precision fossil and living forms, and it is rare to find among Palaeozoic specimens or indeed specimens of any age satisfactory criteria of affinity. Fungi are not always harmful to the higher plants in which they live: in the roots of many forest trees there is an intimate association of fungi with the cells of the host; moreover the invaders are in some instances essential for the well-being of the tree. In the "villainous and rascally heaths", in salt marshes,[2] and in other places fungi live as beneficial partners with

[1] For a recent, general account of fossil Fungi see Pia in Hirmer (27).
[2] Mason, E. (28).

flowering plants. There are indications of similar associations in Carboniferous forests. In the roots of *Cordaites*, an extinct gymnosperm, Prof. Osborn[1] found cells of a fungus in circumstances which suggest a similar co-operation. Another instance is described by Prof. Weiss[2] in specimens of an undetermined Carboniferous plant. The little we know of Permo-Carboniferous Fungi, though it gives no clue to the ancestry or interrelationship of existing forms, is sufficient to demonstrate the remote antiquity of the construction of the plant-body and of the manner of life which we now associate with the class. Sections prepared from a disintegrating fragment of a tree which lived in the Coal Age might, so far as the fungus is concerned, have been cut from a recent plant.

Reference has already been made to reef-building algae in the early Carboniferous seas. A few algae, including examples of calcareous Siphoneae, are recorded from some of the Upper Carboniferous marine deposits, but they are of no special botanical interest. It is the freshwater algae for which we search in the débris of the swamps and pools of the latter part of the Carboniferous period. We cannot expect to find recognizable traces of such delicate filaments or other cellular aggregates as form the bodies of algae which now live in lakes and meres. Unlike the fungi, which are often preserved because of their occurrence within the protective and stronger bodies of other plants, the algae were free-living and independent.

Much has been written on the structure and origin of certain kinds of coal known as boghead, torbanite, and kerosene shale from Carboniferous or Permian horizons in Scotland, France, North America, Brazil, New South Wales, and other parts of the world. It is believed that these close-grained brown rocks, which are rich in oil, represent a precipitate containing the bodies of innumerable gelatinous plants related to existing Blue Green Algae or to colonial forms of Green Algae. As in modern lakes so also in Palaeozoic pools and lakes certain kinds of freshwater, microscopical algae flourished abundantly and at certain seasons caused the phenomenon which we now call the bursting of the meres or "water-flowers", caused by the rapid multiplication of the plants and their accumulation as a dense green scum on the surface of the water. A full consideration of the evidence on which this view is based would involve a critical

[1] Osborn, T. G. B. (09).
[2] Weiss, F. E. (04).

discussion of a highly controversial subject.[1] On magnification a thin section of a piece of kerosene shale[2] from New South Wales shows a dark ground with groups of golden yellow bodies consisting of microscopical colonies of algae, more or less spherical or cuboidal in form. The French authors, MM. Bertrand and Renault, regarded these bodies as algae and referred some of them to the genus *Reinschia,* comparable in the hollow nature of the colony to the living alga *Volvox.* In the bogheads of Scotland and France the commonest alga is the genus *Pila,* a simpler type made up of colonies that are solid and not hollow (fig. 60 *a*).

The chief opponent of the algal nature of these fossils is Prof. Jeffrey of Harvard, who believes that the supposed algae are spores of higher plants with patterns on their resistant outer membranes simulating cell-structure. Jeffrey speaks of the algal origin of the bogheads and similar rocks as "a superfluous hypothesis based on erroneous data": he advances many arguments in support of his opinion and draws special attention to the difficulty of reconciling the wonderful preservation of delicate algae with the occurrence in the same matrix of woody fragments which have retained much less of their more resistant framework. In this connexion it may be noted that certain living algae with which some of the fossil forms are compared are rich in oil, and this may have rendered them less liable to decay. It is true that in many coals spores of *Lepidodendron* and other plants are common objects: as clouds of pollen are blown on to lakes and ponds from pine forests, so the pollen and other spores of Palaeozoic trees were showered into the stagnant water of lagoons and the forest-studded swamps. Spores have no doubt been mistaken for algae; on the other hand, it is equally true that the great majority of the boghead fossils are not spores but algae. One naturally turns to present-day conditions for guidance in the interpretation of problematical fossils. Dr Zalessky[3] found an alga, *Botryococcus,* in Lake Balkash in Turkestan which is hardly distinguishable from the Palaeozoic *Pila*: he compares masses of the minute bodies of the *Botryococcus* which had been washed ashore with sheets of a rubber-like material. Sections of the rubbery material on magnification show a structure resembling that seen

[1] Thiessen, with references (25); Jeffrey (09), (10), (24), (25); Zalessky (28^3). For an account of plant-tissues in coal, see Seyler and Edwards (29).
[2] Bertrand, C. E. and Renault (94); Bertrand, C. E. (96); Arber, E. A. N. (11).
[3] Zalessky (26).

in the bogheads. Salt-water lakes and lagoons in South Australia contain quantities of an organism which has been named *Elaeophyton* and is apparently a Blue-Green Alga: it forms a green scum on the water and, when blown on shore, a rubber-like mass locally known as Coorogonite, from the district where it occurs. Similar

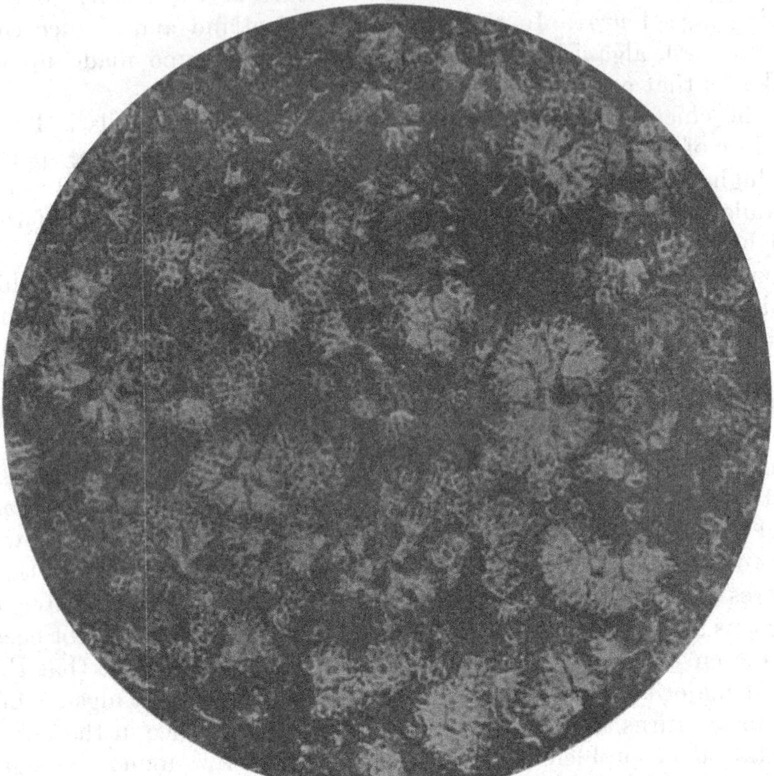

FIG. 60*a*. Section of a piece of boghead from a pit near Bathgate, Scotland, showing a mass of *Pila* (*Pila minor*) believed to be practically identical with the living *Botryococcus*. × 300. (Photograph supplied by Prof. P. Bertrand.)

organic material from Portuguese East Africa examined by Dr Boodle was found to be composed of a Blue-Green Alga resembling the living genus *Coelosphaerium*, and very similar in structure to the boghead fossils. In a full and well-illustrated account of the various Palaeozoic beds containing these microscopical structures Dr Thiessen favours their algal origin.

The researches of Prof. C. E. Bertrand[1] which first threw light on the nature of the boghead fossils are being continued by his son, Prof. Paul Bertrand of Lille, whose account of *Pila* and *Reinschia* at the meeting at Cambridge of the Fifth International Botanical Congress made out a good case in favour of his father's main conclusions. Fig. 60 *a*, for which I am indebted to Prof. P. Bertrand, shows a horizontal section of a piece of boghead from Scotland: in the dark matrix of the boghead are countless, highly magnified bodies of the alga *Pila*: the colonies consist of small cells with white (yellow in the actual specimen) swollen walls of gelatinous material more or less radially arranged or forming a reticulum, the arrangement depending upon the direction of the section of the colony. The dark patches between the walls are the cell-cavities. The conclusion is that the bogheads and similar deposits in Scotland, France, Kentucky, Alaska, Russia, and New South Wales are in great part composed of bituminous material rich in the remains of algae which lived in freshwater lakes or in brackish lagoons. The algae are sometimes associated with spores of vascular plants, fragments of wood, and other plant fragments. The important point is that in the latter part of the Palaeozoic era there existed many forms of Blue-Green Algae and colonial types of Green Algae differing but little from living genera. Despite the enormous gap in time between then and now, we can best visualize the conditions in some of the quiet waters of the Carboniferous swamps by observing what is happening in many lakes at the present day.[2]

Mosses and liverworts, though as yet evidence of their abundance is lacking, were undoubtedly present in the forests of the Coal Age; some living in damp places in the undergrowth, some with their light green flattened bodies clinging to the trunks and branches of stronger plants. Until a few years ago there was a tendency on the part of many botanists to regard the Bryophyta (mosses and liverworts) as a comparatively modern group unrepresented in the Palaeozoic rocks. This view illustrates the risk of attaching too much weight to negative evidence. The absence of any remains of these plants among the petrified material preserved in coal-balls and the lack of carbonized impressions were most easily explained by the

[1] Bertrand, C. E. (94); Bertrand, C. E. and Renault (92).
[2] For a description of some Permian Algae see Florin (29^2).

assumption that mosses and liverworts did not exist. By the application of modern methods to the investigation of fossils the occurrence of different forms of liverworts has been demonstrated in the plant-bearing shales of the Coal Measures. One species, *Hepaticites metzgerioides*, bears a close resemblance to members of the living genus *Metzgeria*, a small plant with a body about 1 mm. broad, which occurs in almost all parts of the world. The fossil form consists of a delicate, dichotomously branched ribbon, one cell thick, with a median rib marking the position of elongated conducting cells. Another species, also from English Carboniferous shale, had two kinds of leafy lobes attached to a slender axis reminding one of modern examples of *Fossombronia* and the genus *Treubia*.[1] *Fossombronia* is now almost cosmopolitan in its distribution; *Treubia* is a Javanese genus. No reproductive organs have been found in organic union with the Palaeozoic liverworts, but it is probable that some small masses of associated spores were produced by species of *Hepaticites*. The discovery of liverworts in the Coal Measures by Dr John Walton[2] justifies an opinion expressed more than twenty years ago by Prof. Campbell, that a more thorough examination of Palaeozoic fossils by modern methods might be expected to add materially to the small number of species that have already been referred to the Bryophyta.

We know only the vegetative parts of a few of these ancient hepatics, but such facts as are available afford an impressive instance of the persistence (fig. 136) through a long succession of ages, of a delicate organism undisturbed by the influences which transformed again and again the character of the world's vegetation.

Evidence of the occurrence of Carboniferous mosses is less satisfactory. A re-examination of an impression described several years ago as *Muscites polytricaceus*, from Upper Carboniferous beds of Central France, strengthens the probability that a comparison of the tufted group of leafy shoots with the existing genus *Polytrichum* was fully justified. Another moss of the same geological age, *Muscites Bertrandi*, is represented by a petrified and incompletely preserved stem bearing on its surface long hair-like filaments crossed at intervals by oblique septa like those characteristic of the absorbing threads of living mosses.

[1] For an account of present-day liverworts see Macvicar (26).
[2] Walton (25), (28^4).

Pteridosperms

Reference has already been made to the pteridosperms, a class which had become prominent before the end of the Devonian period. In the early Carboniferous floras they played a still more conspicuous part and reached the culminating point of their progress in the forests of the Coal period. Some of them, e.g. *Lyginopteris* (the plants like tree ferns, *Lg.* fig. 60), had fairly tall, occasionally

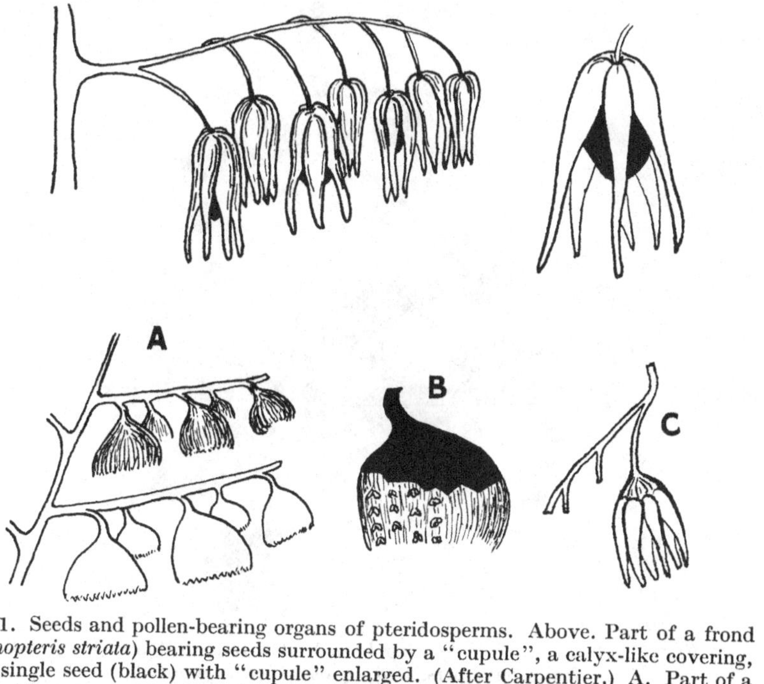

FIG. 61. Seeds and pollen-bearing organs of pteridosperms. Above. Part of a frond (*Sphenopteris striata*) bearing seeds surrounded by a "cupule", a calyx-like covering, and a single seed (black) with "cupule" enlarged. (After Carpentier.) A. Part of a frond with male organs (*Potoniea*) replacing the ordinary leaflets. B. A single male organ enlarged showing pollen-sacs exposed on removal of a portion of the outer covering. (After P. Bertrand.) C. A group of pollen-sacs of *Sphenopteris striata*. Enlarged. (After Bertrand.)

branched stems, bearing large fronds with small leaflets; others had fronds with larger leaflets similar to those of *Osmunda regalis*; some were much humbler plants with finely cut leaflets resembling those of filmy ferns. Examples of different forms of leaflets, distinguished by their venation and relation to the supporting axis, are shown in fig. 63. Pteridosperms bore two kinds of spore, large and small; the larger spore (megaspore) after fertilization became the seed and was enclosed in a closely fitting covering such as we are familiar with in the integument of the ovule of a conifer: this covering (integument) was free from the body of the sporangium at the apex where the pollen had access to the tip of the spore containing the egg-cell. The sporangia, or ovules, after fertilization by the male cells, which were developed in the smaller spores formed in the less conspicuous capsules, became the seeds. In some species

FIG. 62. Leaf and fruit-bearing branch of a tomato plant. The small fruits have persistent sepals. Slightly reduced.

the young seeds were surrounded by a second envelope, a kind of loose and often lobed sheath (fig. 61) similar to the husk of a hazel nut or to the spreading star-like calyx which forms a fringe at the base of a small tomato fruit (fig. 62).[1] In others the seeds were borne directly on the ordinary leaflets.[2]

[1] Bertrand, P. (26); Carpentier (29[2]).

[2] Since this chapter was written a paper by Prof. Halle was received in which he gives restorations of the following seed-bearing pteridosperm fronds based on material from Shansi, China: *Sphenopteris tenuis*, *Pecopteris Wongii*, *Emplectopteris triangularis*, *Nystroemia pectiniformis*. This paper [Halle (29[2])] is an exceptionally important contribution to our knowledge of pteridosperms.

The fronds of some pteridosperms must have looked very like those of the living North American fern, *Cystopteris bulbifera*, on which small, green, oval bodies about a quarter of an inch in length are borne at the point of junction of the lateral branches (pinnae) with the main axis.[1] These green bodies, known as bulbils, fall off and produce new plants: they consist of fleshy scales enclosing a small bud, and are in fact buds provided with a store of reserve food which serve the same purpose as seeds in the life-history of the pteridosperm, though they are independent of any sexual process of reproduction and belong to an entirely different category of organs. In many respects more advanced and more complex than true ferns, several pteridosperms had stems which grew in thickness by regular additions of secondary conducting tissue resembling, anatomically, the wood of existing cycads. In the possession of two kinds of spore a pteridosperm agrees with *Selaginella* and *Isoetes* among the lycopods and with the water ferns (Hydropterideae) *Pilularia* and *Azolla*; but the larger spores of a pteridosperm were not shed from the plant before fertilization as, with very rare exceptions, they are in existing Selaginellas and water ferns which have two kinds of spores. The seeds of a pteridosperm, though not enclosed in anything which can legitimately be called a carpel, were in many instances partially enclosed in a protective investment (fig. 61) which is analogous to the ovary-wall of a flowering plant.[2] The seeds often replaced leaflets, or were borne on them, and were not completely enclosed in an ovary. We may best picture the state of affairs by thinking of pteridosperm seeds and their accompanying cup-like envelopes (cupule) as instances of evolutionary tendencies which failed to reach the standard eventually attained by the flowering plants. Another comparison with the higher plants may be based on similarity of leaf-form: the branched, compound leaves of many members of the Umbelliferae and several other families are, superficially, very close to those of some pteridosperms.

The numerous forms of pteridosperm fronds are described under different generic names having reference to the shape and veining of the leaflets (fig. 63), the method of branching of the whole leaf, and in some fertile fronds to the structure and arrangement of the spore-capsules. Superficially the fronds of many of these extinct

[1] For an illustration see Bower (23–28), vol. I, fig. 66.

[2] In this connexion see description of abnormal leaves of *Ginkgo* with seeds on the margin by Sakisaka (29).

plants appear to be almost identical with those of living ferns. There are, however, certain differences in habit: a characteristic feature of the Palaeozoic fronds is their tendency to branch dichotomously, whereas in living ferns the large leaves usually consist of a central axis (rachis) bearing two rows of lateral arms (pinnae) set with leaflets (pinnules). Fig. 63, B, is a diagrammatic representation of one type of regular forking characteristic of a pteridosperm frond which is common in Upper Carboniferous floras. The fronds are pinnate, bearing two rows of leaflets attached directly to the rachis as in the familiar *Polypodium vulgare*; but as a rule they are bipinnate—the leaflets of the pinnate type being replaced by pinnate lateral members; or tripinnate—each lateral pinna bearing still smaller pinnae with leaflets; or quadripinnate. In the great majority of pteridosperms the main axis of the frond and frequently the pinnae are regularly forked. Anatomically these fronds differ in certain respects from those of true ferns. In *Pecopteris*, *Alethopteris*, *Lonchopteris*, *Callipteris*, *Odontopteris*, *Callipteridium* (fig. 63, O, I, N, A, F, P) each leaflet is attached by the whole breadth of the base and this may be extended a short distance along the supporting axis (decurrent) as in *Alethopteris* and *Odontopteris* (I, F). In *Neuropteris* (L) the leaflets are attached by the central portion of the lamina, the edges of the base being free and rounded. In *Sphenopteris* (G) the pinnules are more or less deeply lobed or rounded, and there is a central vein. In most of the genera a strong central vein (midrib) runs up the centre and from it spring simple or forked smaller veins which pass outwards to the edge of the leaflet at various angles. Leaflets of *Lonchopteris* (N) resemble those of *Alethopteris* (I) except in the reticulate arrangement of the lateral veins: a similar venation distinguishes *Linopteris*[1] from *Neuropteris* (M and L).

In some of the later Carboniferous and early Permian pteridosperms the occurrence of leaflets not only on the branches of the frond but also on the main axis (rachis), as in such Palaeozoic genera as *Callipteris* (fig. 63, A) and *Callipteridium* (P), recalls the distribution of leaflets on the compound leaves of a potato or a tomato plant (fig. 62). The pollen was conveyed to the female organ by wind, and probably insects had a share in pollination. It may be that the sticky glands on the ragged cup of the seeds of *Lyginopteris* were attractive to dragon flies and other winged insects which are

[1] For description of *Linopteris* seeds see Dix (28).

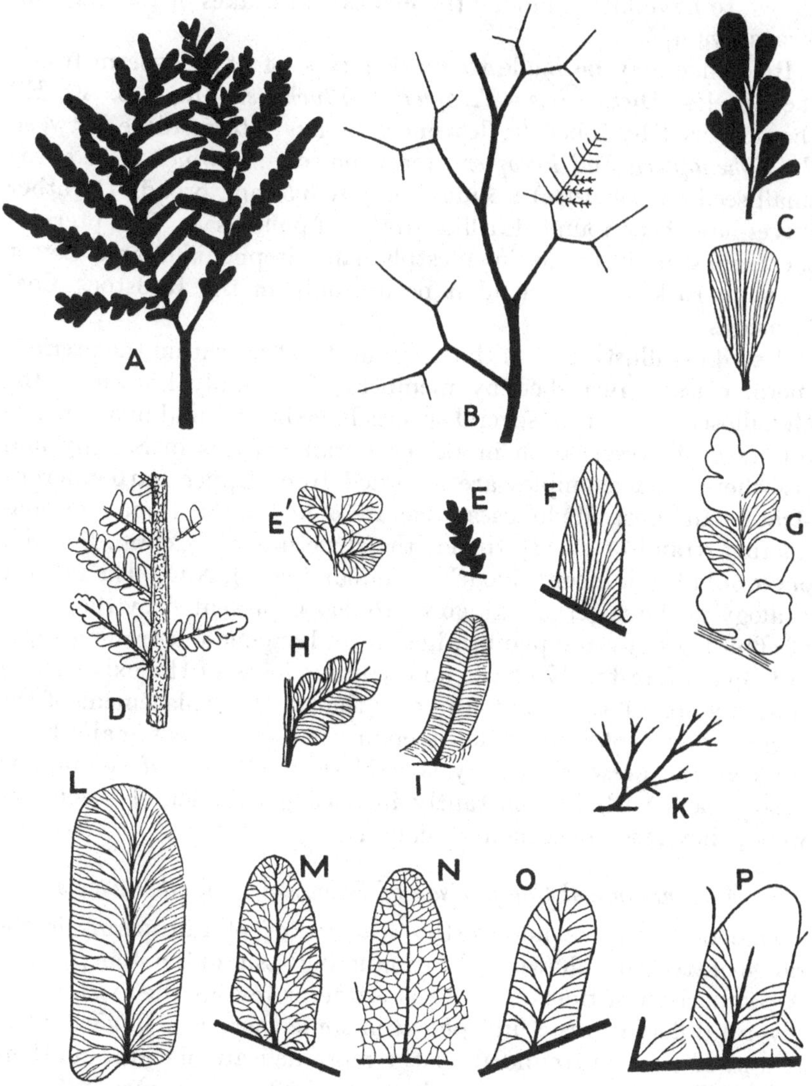

FIG. 63. Fronds and leaflets of pteridosperms. A. *Callipteris Pellati*. Permian frond from Lodève, France. B. Diagram of the branching of a frond of *Mariopteris nervosa*. C. Pinna and enlarged leaflet of *Adiantites antiqua*. D. *Lepidopteris Ottonis* (Rhaetic). E, E′. *Mariopteris nervosa*, part of a pinna (E) and enlarged leaflets (E′). F. Leaflet of *Odontopteris reichiana*. G. Lobed leaflet of *Sphenopteris obtusiloba*. H. Part of a pinna of *Thinnfeldia odontopteroides* (Triassic). I–P. Leaflets (pinnules) of I, *Alethopteris*; K, *Rhodea*; L, *Neuropteris*; M, *Linopteris*; N, *Lonchopteris*; O, *Pecopteris*; P, *Callipteridium*. A, E, M, N, after Zeiller; B, C, G, K, L, O, P, after Kidston; D, after Antevs; F, after Potonié.

known to have flitted among the glades and brakes of the Carboniferous swamps.

Reference may be made to another type of pteridosperm frond, the species *Dicksonites* (*Pecopteris*) *Pluckeneti* (fig. 76, E, E′) characterized by lobed leaflets more or less intermediate between the *Sphenopteris* and *Pecopteris* forms on some of which were borne small seeds[1] (*Leptotesta*) 4·5 mm. long by 3·5 mm. broad. On other leaves have been found star-like groups of pollen-sacs. This pteridosperm has been found in Westphalian, Stephanian, and Lower Permian rocks: in England it occurs only in the Radstock Coal Measures.

A striking illustration of the range in structure within the pteridosperm class is furnished by members of a family known as the Medulloseae. Stems of several species have been found in a wonderful state of preservation in Lower Permian rocks in Saxony and elsewhere, and examples are recorded from Upper Carboniferous strata. In most Medulloseae the stem had more than one conducting strand of woody tissue: the structure was extraordinarily, one would think inconveniently, complex (fig. 64, N,O). The nearest analogy in the vegetable kingdom to-day is presented by the stems of climbing flowering plants which form hanging and trailing ropes in tropical forests. We have no direct evidence of the existence of climbing Medulloseae; but we know that in the replacement of the single woody column, which is common to the great majority of trees at the present day, by several stout strands of conducting tissue, each with its constantly increasing cylinder of secondary wood, they resembled modern lianes.

An Anatomical Comparison of Palaeozoic and other Stems

One of the more noteworthy features of pteridosperms is the range of structure shown in the conducting tissue of the stem (fig. 64). A comparison of the stems of modern ferns reveals the existence of several different structural plans: in some there is a single central strand of wood surrounded by the more delicate phloem; in others the conducting tissue forms a hollow cylinder, or in rare instances two or even three concentric cylinders: in the more recently evolved ferns the conducting tissue is a hollow cylinder pierced by large, spirally arranged openings as in a lattice-work bent into tubular

[1] Loubière (29). See Halle (27), (29[2]) for description of another species of *Pecopteris* frond with seeds; also Scott (29).

form. In living ferns, with one trifling exception, there is no zone of secondary conducting tissue formed by a cambium. In the stems of pteridosperms, on the other hand, an actively growing layer of cells (cambium) produced secondary tissue in radially disposed series, forming a ring of wood gradually increasing in breadth from year to year, as in modern conifers and most flowering plants. The primary conducting tissue assumed various patterns. By the primary conducting tissue is meant the wood and phloem, concerned with the transport of water and salts and the distribution of food manufactured in the cells of the green leaves; it is the tissue which is developed at the apex of the stem and has not been added to by the action of a cambium: the component elements are not arranged in regular radial series as are the elements produced by a cambium. In *Heterangium*,[1] for example, the primary wood forms a solid central core (fig. 64, S): in *Lyginopteris* a ring of relatively small strands on the inner edge of the secondary wood (fig. 64, P); in *Medullosa* it consists of separate strands, each with its own enclosing cylinder of secondary wood, or in some species (N, O) it assumes the form of complete hollow cylinders. In the possession of secondary conducting tissue developed from a cambium pteridosperm stems agree much more closely with existing gymnosperms and angiosperms than with ferns, though in complexity and in the varied architectural plans they give the impression of types in which the structural features were less stabilized and more diverse than in any one group of existing plants. This unusual range in the form of the conducting tissue would seem to be a characteristic feature not only of pteridosperms but of some other Palaeozoic groups. Reference should be made to the diagrams of the pteridosperm stems and of the Lower Carboniferous gymnosperms *Pitys*, *Callixylon* and *Protopitys* shown in fig. 64.

In the scattered distribution and in certain structural features of the conducting strands (vascular bundles) the leaf-stalks of Medullosean fronds such as *Alethopteris* closely resemble in transverse section the stems of many monocotyledons such as the sugar cane, Indian corn (*Zea*), and numerous others. This similarity led an author to describe a specimen from the American Coal Measures as a Palaeozoic angiosperm:[2] it was subsequently pointed out[3] that the fossil is anatomically identical with European Medullosean leaf-stalks and can be compared with monocotyledons only in so far

[1] Scott (17); Kubart (14). [2] Noé (23). [3] Seward (23[2]).

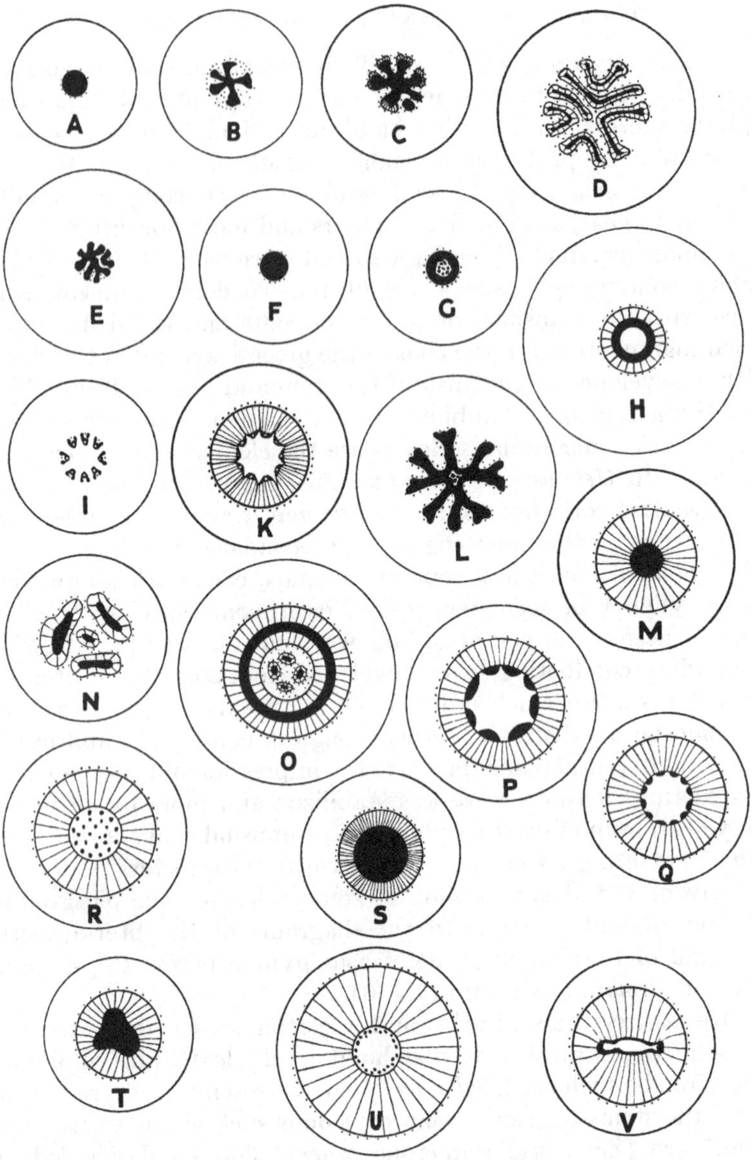

FIG. 64. Diagrammatic drawings of transverse sections of various types of stem in some extinct and living genera, both pteridophytes and gymnosperms. The black represents primary wood, the radial lines secondary wood, and the dots phloem. A. *Rhynia*: Devonian. B. *Asteroxylon*: Devonian. C. *Schizopodium*: Devonian. D. *Cladoxylon*: Devonian and Carboniferous. E. *Lycopodium*: recent. F–H. Three types of *Lepidodendron*: Carboniferous. I. *Equisetum*: recent. K. *Calamites*: Carboniferous, cf. fig. 65. L. *Asterochlaena*: Permian. M. *Botrychioxylon*: Carboniferous. N, O. *Medullosa*: Carboniferous and Permian. P. *Lyginopteris*: Carboniferous. Q. *Callixylon*: Devonian. R. *Pitys Dayi*: Carboniferous. S. *Heterangium*: Carboniferous. T. *Stenomyelon*: Carboniferous. U. *Pitys*: Carboniferous. V. *Protopitys*: Carboniferous.

as there is a general analogy in some anatomical characters. While it is possible that there may be some connexion between pteridosperms and plants in the line of evolution of the angiosperms, it is unfortunate that the reference of a petiole of *Medullosa* to an angiosperm has been quoted by more than one writer as evidence of the existence of Palaeozoic flowering plants.

The ancestry of pteridosperms is unknown: there are no adequate grounds for assuming a direct relationship to ferns, and we can only speculate on their connexion with the seed-bearing plants which ultimately replaced them. There is evidence of affinity of the Medulloseae to the cycads, and it may be that the pteridosperms were highly specialized members of an evolutionary series which includes also the cycads. The maximum development of the pteridosperms was in the Permo-Carboniferous age, but, as we shall see later, there is reason to believe that the class was represented, though in reduced numbers and in less variety in the earlier part of the Mesozoic era.

Articulate Plants: Calamites *and others*

The group Equisetales was well represented in the later Carboniferous forests. The equisetalean genus *Asterocalamites*, which had wandered far during the earlier stage of the period, was associated with a few pioneer species of *Calamites*. The latter genus rose rapidly to a prominent position in the course of the later stages of the Carboniferous period and was represented by trees very diverse in form: it persisted, though in greatly reduced numbers, into the Permian period. The existence of many kinds of *Calamites* has already been mentioned. It is customary to classify the various forms of calamite stems on the position and frequency of the branches: species in which whorls of branches occur only at certain nodes and are separated by comparatively long branchless portions of stem are included in a section of *Calamites* known as *Calamitina*; stems on which branches occur in all or nearly all the nodal regions are referred to *Eucalamites*, and those with very few or no branches to the section *Stylocalamites*. The cones also show a considerable range in structure; they were not as simple as the fertile shoots of an *Equisetum*, and some of them produced two sizes of spore, a not uncommon attribute of extinct plants belonging to groups in which heterospory is now very rare or absent. Anatomically the young stem of a *Calamites* resembled the mature stem of a modern *Equisetum* (fig. 64, I), at least in its main characters: a large pith, which soon became hollow, was surrounded by a ring of small and separate primary conducting strands each having on its inner margin a vertical canal. As the plant grew, new wood was added to the primary bundles which formed a thick cylinder of regularly disposed conducting tubes traversed by vertical plates of shorter cells

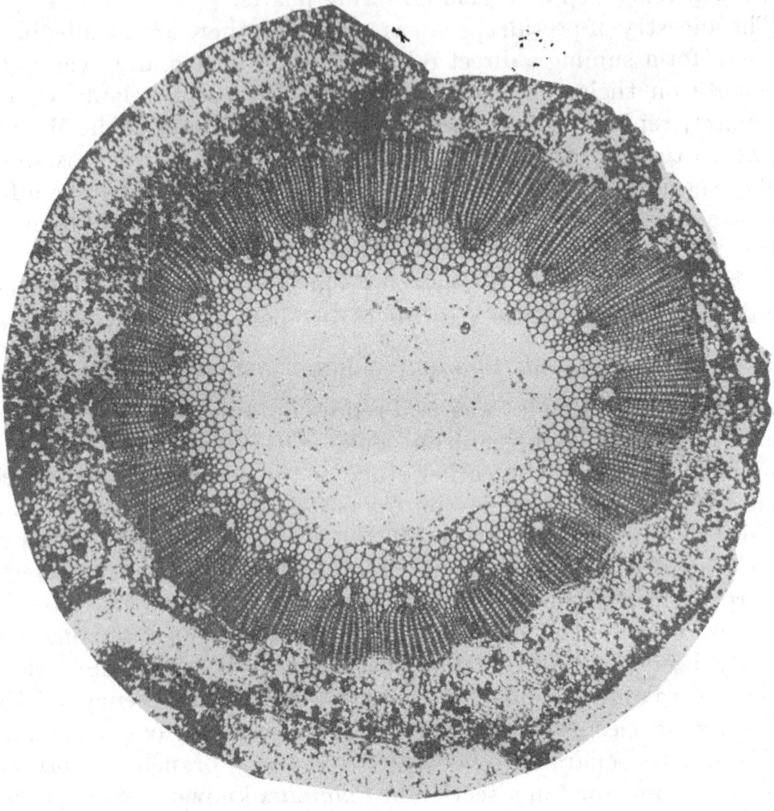

FIG. 65. Transverse section of a fairly young stem of *Calamites* showing the hollow pith; a cylinder of wood, most of which is secondary, with a carinal canal at the apex of each wedge of wood; and the bark. × 18. (Kidston Collection, Glasgow University.)

(medullary rays). Fig. 65 shows a transverse section of a comparatively young *Calamites* with a narrow zone of secondary wood, the apices of the primary strands being indicated by vertical canals (see also fig. 64, K). In older stems the wood reached a considerable thickness and *pari passu* with its growth a protective cylinder of bark was developed by regular increments. In their method of secondary thickening calamites behaved as conifers or dicotyledons and differed from all existing Equiseta in which the stem never becomes woody. In size and in structural plan the Palaeozoic *Calamites* was superior to *Equisetum*: it is indeed difficult to regard the extinct genus as a direct ancestor of the modern representatives of the Equisetales. Some large *Calamites* stems are known from Permian rocks, but most of the Mesozoic members of the group are of a different type and much more closely allied to the surviving Equiseta. It is probable that a few Upper Carboniferous plants which have been described as species of *Equisetites* and bore leaf-sheaths and cones similar to those of *Equisetum*, are Palaeozoic members of a long line of evolution which is still in being. The foliage-shoots,[1] because of their frequent occurrence independently of the stems which bore them, are described under separate generic names: those with circles of narrow leaves united basally into a narrow collar clasping the branch and spread out at right-angles to the axis are known as *Annularia*; those in which the circles of narrow and tapered leaves are composed of separately attached members are included in the genus *Asterophyllites*.

Another type of plant with articulated stems is illustrated by *Sphenophyllum*, a genus which, like *Calamites*, reached its zenith in the later Carboniferous floras and apparently became extinct before the beginning of the Mesozoic era. *Sphenophyllum* was in habit similar to *Galium aparine* (cleavers); its wedge-shaped leaves, frayed or toothed at the ends or deeply cleft into almost hair-like segments, were borne at regular intervals on a sparsely branched stem which was always comparatively slender and may have depended for support upon more robust plants. Some authors prefer to think of *Sphenophyllum* as living in water, though there is no indication of this in the structure of the stem. On the other hand, the discovery of a species in China with spinous stems[2] favours a scrambling habit. In the anatomy of the stem, which grew in diameter by the activity of a cambial layer and in the structure of the cones, *Sphenophyllum* cannot be closely matched with any living plants. It is one of many genera in the later Palaeozoic forests which impress us by the complexity of the mechanism of the plant-body.

The more one realizes the wealth of form and structure revealed by the fossils scattered through the peat beds and the sedimentary deposits of the closing stages of the Palaeozoic era, the greater

[1] Thomas, H. H. (12[2]). [2] Yabe and Ôishi (28).

becomes the difficulty of reconstructing genealogical trees. Development proceeded along many diverse lines and some of them have been traced through only a small fraction of geological history. A few have been followed from the Devonian period to the present day, but they are exceptional instances of successful adjustment of plant mechanism and construction to the physical environment.

Lycopodiales

The Lycopodiales were represented by many different forms of *Lepidodendron*, *Sigillaria*, *Bothrodendron*, and other genera which attained the stature of trees. From the base of the stem long forking arms bearing spirally attached rootlets spread a long distance through the mud or sand and, like the underground organs of some modern fen plants, avoided the deeper layers of the soil that were poor in oxygen. It is these spreading and forked roots known as *Stigmaria* (fig. 60, *Sa.*) which are a characteristic feature of the rock below each seam of coal. It is interesting to find from a recent account of a new species of *Stigmaria*[1] that this genus affords an additional instance of diversity in structural plan: in most examples there was a pith in the centre, but in one form a solid axial strand of primary wood. The cones of *Lepidodendron* reached a considerable size and many of them had spores of two kinds. The frequent occurrence of spores, some large, some small, in countless thousands in seams of coal and scattered over the surface of slabs of hardened mud and clay from the later Carboniferous estuaries and lagoons, enables us to picture the cones of *Lepidodendra* and *Sigillariae* swinging like censers and scattering clouds of pollen as the yellow dust of pine trees is wastefully scattered by the wind.

There were also herbaceous lycopodiaceous plants in the undergrowth of the forests and some may have lived as epiphytes on the boughs of trees, a method of existence now adopted by several species of tropical lycopods. It is these humbler plants, rather than the giant *Lepidodendra*, which afford the most satisfactory evidence of the persistence through the ages (fig. 136) of a group of plants some of which still flourish abundantly in tropical, temperate, and even in arctic regions. There is evidence of the production of seeds in place of the large spores (megaspores) by a few of the later Palaeozoic lycopods, both the arborescent and herbaceous forms.

[1] Leclercq (28).

Gymnosperms: Conifers, Cycadophytes, Ginkgoales, etc.

Attention has been called to the occurrence of trees in the late Carboniferous forests which may be compared with living species of *Agathis* and *Araucaria* in the nature of the secondary wood, though in the character of the primary tissue the Palaeozoic types differed from all existing gymnosperms. One of the few Palaeozoic fossils assigned to the genus *Araucarites* was discovered in Lower Permian beds near Autun in France:[1] the plant bore single-seeded scales very similar to those on the cones of living species of *Araucaria* and *Agathis*. A contrast between the ancient and modern tree is seen in the organization of the reproductive shoots. The seeds of the extinct genera on the whole resembled those of cycads and *Ginkgo* rather than the seeds of conifers; they were not as a rule produced on cones as in pines and firs. In *Cordaites*, which is characterized by an araucarian type of wood, the seeds were borne on small bud-like branches[2] and not on the scales of a cone. The pollen-grains were contained in sacs borne on long stalks among a cluster of protective leaves and not in a compact group as in existing conifers. *Cordaites* and many other extinct gymnosperms, though resembling modern conifers in habit, are included in a special group of naked-seeded plants, from which in later ages there may have evolved the direct ancestors of the araucarian family.

In illustration of the variety in leaf-form exhibited by the plants of the Coal Age, reference may be made to two genera, *Dolerophyllum*[3] and *Titanophyllum*.[4] The latter name is applied to some remarkable, short, strap-like and comparatively thick leaves from the Stephanian series of the Commentry coalfield in France reaching a length of 75 cm. and 25 cm. broad; and traversed by several parallel veins. Those of *Dolerophyllum* are relatively broader, and almost circular, 7 by 9 cm.; the best known specimens are preserved as buds consisting of a rolled up group of petrified leaves: a recent examination of the anatomical features has led Dr Florin to assign the genus to a family of its own and to connect it with the Cordaitales. *Dolerophyllum* is recorded from Westphalian rocks in Germany and from the Permian Kusnezk flora (see map, fig. 78).

The genus *Psygmophyllum*, mentioned in the account of Devonian floras, was also a member of the earlier[5] and later Carboniferous

[1] Zeiller (06). [2] Schoute (25), (25[2]). [3] Florin (25); Seward (17).
[4] Renault and Zeiller (88–90). [5] Arber, E. A. N. (12[2]).

vegetation. Its large cuneate leaves, closely resembling in shape the foliage of *Ginkgo*, occasionally show a peculiar feature in the anastomosing of the veins, which are more numerous and finer than those of the surviving genus. Though most leaves of *Psygmophyllum* are similar in shape and lobing to the foliage of *Ginkgo*, some specimens included in the genus are much larger and distinguished by the division of the leaf into long and narrow segments. An example of this latter form is *Psygmophyllum cuneifolium*[1] from the Permian plant-beds of Russia (fig. 66). Nothing is known of the reproductive organs nor of the anatomical structure of the stem: we cannot therefore express a definite opinion on the precise position of the plant. It is almost certain that fossils from different horizons in the later Palaeozoic periods, and from widely separated localities, which are included by authors in *Psygmophyllum*, do not all belong to one genus or even to one family or group of plants. There is a marked difference between the wedge-shaped leaves from European Coal Measures[2] and from the late Carboniferous or possibly early Permian beds of South Africa[3] and the forms such as that shown in fig. 66 from Upper Permian rocks in Russia. Some Russian leaves of similar habit have recently been described by Zalessky[4] under a new name, *Idelopteris*. There are other leaves from Permo-Carboniferous beds which agree even more closely with those of *Ginkgo*, notably the leaves included in the genus *Saportaea* (fig. 67) recorded from the Lower Permian rocks of North America and China. It is difficult to resist the conclusion, though it is based on leaves alone, that this type is a Palaeozoic forerunner of the group Ginkgoales, a group which produced many different forms destined to colonize widely scattered regions in the Triassic and Jurassic periods and is to-day represented by the solitary but venerable tree *Ginkgo biloba* (fig. 31). Two other genera may be mentioned as examples of gymnosperms that are probably conifers. *Dicranophyllum*,[5] characterized by the repeatedly forked, narrow, linear leaves nearly 20 cm. long, would seem to be a type remote in habit from any living genus. It is represented by one British species from Staffordshire and is recorded from Upper Carboniferous beds in Belgium, Portugal, France, also from Lower Permian beds in China and Europe. Another tree, *Walchia*, with foliage-shoots similar to those of *Araucaria excelsa*, bore two forms of cones, cylindrical cones with

[1] Zalessky (18), (27), (29^2).
[2] Seward (19).
[3] Seward (03).
[4] Zalessky (29^2).
[5] For references to illustration see Seward (19).

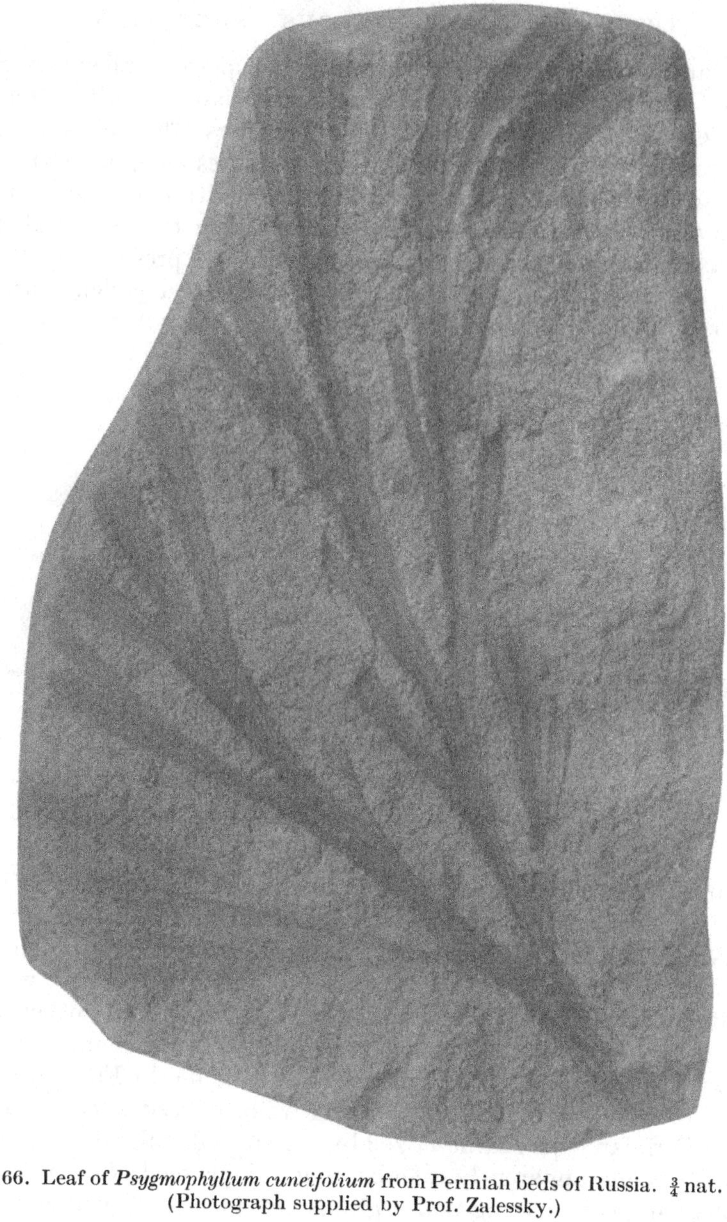

FIG. 66. Leaf of *Psygmophyllum cuneifolium* from Permian beds of Russia. ¾ nat. size. (Photograph supplied by Prof. Zalessky.)

seeds and others consisting of spirally disposed pollen-bearing scales. *Walchia* (fig. 60, *W.*), though araucarian in habit, differed in its reproductive organs from all living conifers. This genus, which is characteristic of the uppermost Coal Measures (Stephanian) and Permian floras, bore cylindrical female cones with lanceolate scales in the axil of which (that is, the angle between the scale and the cone-axis) was a short shoot with scales which presumably bore seeds, also large male cones with spirally disposed pollen-bearing members: the spores were provided with a ring-shaped air-chamber[1]

FIG. 67. Leaf of *Saportaea nervosa* from Shansi, China. Slightly reduced. (Photograph supplied by Prof. Halle.)

which, like the bladders of the pine pollen, was an aid to dispersal by wind. This is one of many instances of the occurrence of appendages or inflated membranes on spores of extinct plants demonstrating persistence through the ages of similar features which facilitated the distribution of reproductive cells. Among other late Palaeozoic representatives of the conifers[2] are the Stephanian genus *Lecrosia* (so called from Le Cros, near St Étienne in France) with needle-like leaves and cones bearing winged seeds; *Ernestia*, a Lower Permian type characterized by fan-shaped scales divided into three lobes and apparently each bearing a single seed; *Pityanthus*, a Stephanian genus having pollen-grains provided with bladders like those of the pollen of a pine; *Pityospermum*, a Permian genus with winged seeds, and a few other very imperfectly known forms. Over

[1] Lück (13). [2] Florin (27), (29).

and over again, as knowledge increases, we find that resemblances to modern plants exhibited by the foliage and the anatomical structure of the stem are discounted as evidence of affinity by the contrasts shown in other and more important characters.

Though no petrified fossils having the anatomical features of

FIG. 68. Fronds of *Plagiozamites* from Shansi, China. Nat. size. (Photograph supplied by Prof. Halle.)

existing cycads or of the Mesozoic Cycadophyta, which occupied a dominant position in the late Triassic and in the Jurassic floras, have been discovered in Permo-Carboniferous strata, it is noteworthy that the wood of many pteridosperms shows a striking resemblance to that of the cycadean type. There are, however, a few fronds from Upper Carboniferous and Lower Permian beds which are cycadean in form. The genus *Plagiozamites* (fig. 68), described from Lower

Permian rocks in France, Germany, Virginia,[1] and China, has usually been regarded as a form of cycadean frond: it has, however, recently been shown that specimens formerly believed to be fronds with two ranks of leaflets are branches bearing spirally disposed leaves twisted into one plane as in shoots of *Taxus* and other conifers. More satisfactory examples of cycadean fronds are furnished by specimens assigned to *Pterophyllum* and *Dioonites*. The former genus is known from France, Germany, and elsewhere. Dr Hamshaw Thomas[2] states that an examination of the structure of the superficial layer of cells on the fossil leaflet of a French specimen shows that the characters agree generally with those of certain Mesozoic cycadean fronds. Good specimens of *Dioonites* have been found in Permian rocks in China.[3] Both genera became abundant in the Mesozoic era. It would seem highly probable that the class Cycadophyta came into existence before the close of the Palaeozoic era, though it was not until the latter part of the Triassic period that it became numerically important.

Ferns

The genus *Psaronius* is represented in many museums by attractive specimens of petrified stems which were originally described by a German author as "Staarsteine" (starling stones), from a slight resemblance of polished sections to the mottling on a starling's breast. In habit *Psaronius* must have closely resembled many living tree ferns, but in certain anatomical features it differs from any modern ferns. The fronds were borne in two or four vertical rows (fig. 60, *Ps.*) or, in some species, in crowded spirals: this is clearly shown both in the structure of petrified stems and in casts of structureless stems on which large leaf-scars are prominent. The central part of the stem was occupied by band-like strands of conducting tissue and occasionally by plates of strengthening cells: these strands form a great variety of patterns. In the peripheral region are seen numerous roots embedded in a mass of tissue formed from the matted hairs of roots and stem; and between this and the centre are the bases of leaf-stalks in process of detachment from the parent axis. In the simplest structural type, which is also the oldest and was discovered in the Lower Coal Measures of England, the conducting tissues form a hollow cylinder and not, as in the other species, a complex net-

[1] Bassler (16) with references.
[2] Thomas, H. H. (30).
[3] Halle (27).

work. *Psaronius*[1] is recorded from Upper Carboniferous or Lower Permian rocks in many parts of Europe, in North America, and Brazil. In its structure and in the nature of the fertile fronds it resembles living members of the Marattiaceae, but the peculiarities of the extinct genus are too well marked to permit its inclusion in any of the families instituted for modern representatives of the ferns.

In addition to a certain number of true ferns included in the genus *Psaronius* (fig. 60, *Ps.*), there were other plants, relatively inconspicuous in comparison with the pteridosperms, that can confidently be called ferns. *Psaronius* was especially abundant in the forests of central France in the closing stage of the Carboniferous period; the stems reached a height of five metres. Several of the early Carboniferous Coenopterideae persisted into the latter part of the period and some of them into the early Permian floras; but the group, so far as we know, is essentially Palaeozoic. The Mesozoic ferns are much more closely related to existing types and their precise relationship to the older genera is by no means clear. The question arises, Can we recognize in the comparatively small number of Permo-Carboniferous ferns any undoubted representatives of present-day families? Some fertile fronds from Upper Carboniferous and Lower Permian strata have been assigned to the genus *Asterotheca*, a name suggested by the star-like groups of sporangia on the lower face of the leaflets. But the sporangia lack the annulus which is a distinguishing feature of most living ferns. One of the best known examples of this genus is *Asterotheca* (*Pecopteris*) *Miltoni* (fig. 63, O), a common species in the Coal Measures. *Asterotheca*, though possibly an extinct member of the Marattiaceae, is hardly entitled to be included in that family without some qualification. Although the spore-capsules of *Asterotheca* and some other Palaeozoic genera agree closely with those of modern ferns, it does not follow that plants which bore them did not also bear seeds. Another Permo-Carboniferous genus, *Ptychocarpus*, had large, repeatedly branched fronds bearing groups of six to eight sporangia in compact circular clusters opening by an apical pore, a method of dehiscence like that in the sporangia of *Christensenia*, a tropical genus of Marattiaceae. There is evidence of connexion both of *Asterotheca* and *Ptychocarpus* fronds with *Psaronius* stems. There are a few other genera with sporangia recalling those of the Marattiaceae; but even if we include some of the petrified *Psaronius* stems, we can

[1] Schuster (11[3]) describes a Permian fern with secondary tissues as *Xylopsaronius*.

only say of these late Palaeozoic plants—some of which, it seems certain, are true ferns—that they agree more closely with the Marattiaceae than with any other surviving family.

Another type of spore-producing apparatus is characteristic of *Senftenbergia*, a genus rare in Britain and recorded from Belgium and France: this consists of sporangia, each of which has two to five rows of thick-walled cells at the apex, a form that is now found in the family Schizaeaceae. There are, however, differences in detail: the fossil sporangia have an annulus composed of several cell-rows; in the modern ferns it is very rare to find more than a single row. Moreover, in the living ferns there is a plate of thinner cells on the actual tip of the sporangium, and this is not present in the sporangia of *Senftenbergia*. The Palaeozoic genus undoubtedly foreshadows the schizaeaceous type: whether this is an instance of parallel development or a mark of direct relationship remains to be determined. Similarly the genus *Oligocarpia*, with sporangia in small circular groups, has been classed with the Gleicheniaceae because of the position of the annulus. A recent examination of the fossil sporangia has, however, revealed peculiar characters which throw doubt on the close affinity of the genus to members of the Gleicheniaceae, a family which in the latter part of the Mesozoic era was spread over the world and is to-day abundantly represented in the southern hemisphere.

Some beautifully petrified stems from Permian rocks of Russia furnish definite evidence of affinity to the Osmundaceae, one of the most ancient families of ferns. The two Russian genera, *Zalesskya* and *Thamnopteris*,[1] are examples of Palaeozoic Osmundaceae which differ from the Mesozoic and the existing members of the family in the structure of the stem: the wood is not arranged in more or less separate strands enclosing a central region of soft tissue as in the more modern types, but as a continuous mass which in *Thamnopteris* forms a solid axial core. A similar concentration of the conducting tissue in the centre of the stem is a feature of several Palaeozoic plants which are now represented by genera with the corresponding tissue encircling a central pith.

Attention is called to a few other Palaeozoic fern-stems which exhibit a marked contrast to living forms in their anatomical characters. The genus *Asterochlaena*[2] (fig. 64, L) of Lower Permian age is represented by a stem a few centimetres in diameter bearing

[1] Kidston and Gwynne-Vaughan (08), (09); Zalessky (24).
[2] Bertrand, P. (11).

several roots and enclosed in a mass of leaf-stalks: the centre is occupied by a deeply fluted column of conducting tissue which in transverse section has the form of a strand with radially disposed arms. This is one of the most complex types in the Coenopterideae and, anatomically, is unlike any living fern. The habit of the fronds is unknown. Another interesting fern-stem discovered in the Lower Coal Measures of England is the genus *Botrychioxylon* (fig. 64, M)[1] in which a solid axial strand of primary wood is surrounded by a cylinder of secondary wood. This genus, known only as a petrified, dichotomously branched stem, is named after the living fern *Botrychium* (moonwort), a member of the Ophioglossaceae, because *Botrychium* is peculiar among ferns of the present day in having secondary wood, though the amount is very much less than in the fossil: we cannot say definitely whether or not there is any real affinity between the Palaeozoic and the recent plant.

The general position may be stated as follows: the oldest known representatives of the ferns of which we possess the fullest knowledge cannot confidently be assigned without reservation to any of the families based on living genera. Some of the extinct genera had certain anatomical features which subsequently became characteristic of more modern ferns. Others had fronds bearing sporangia without a well-developed annulus, and in this respect they are comparable to some living Marattiaceae. Some had sporangia with a mechanism for the scattering of the spores which clearly foreshadowed that which is now characteristic of the Schizaeaceae and Gleicheniaceae. The families which have the strongest claim to direct ancestors in the late Carboniferous and early Permian floras are the Marattiaceae, the Osmundaceae, the Ophioglossaceae; possibly also the Schizaeaceae and Gleicheniaceae. The ancient ferns may be described as plants having a wide range in vegetative and reproductive characters: this would seem to be an attribute of an early stage of evolution when characters that are now segregated into different families were often found in a single genus or in a collection of genera. In the Palaeozoic era evolutionary tendencies along various lines were still in the experimental stage, and it was not until the Mesozoic era that patterns became stabilized and have persisted with no radical change to the present day.

Several forms of isolated sporangia agreeing more or less closely with those of living ferns have been described from Carboniferous

[1] Scott, D. H. (12), (20).

beds, but the difficulty is to discover whether they are strictly comparable to the sporangia of ferns and were the only spore-bearing organs of the extinct plants; or whether they belong to plants which bore two kinds of spores, pollen-grains (small spores) and ovules or undeveloped seeds (large spores of a special kind). Similar differences are presented by specimens of fertile fronds. Dr Kidston[1] described under the generic name *Zeilleria* pieces of a small fertile frond on which some of the delicate leaflets have a very slight development of the lamina, and the projecting veins bear a minute globular body which on ripening opened into 4–5 valves: he noticed a scar in the middle of the open cup which he believed to mark the presence of a small seed and therefore assigned this Westphalian genus to the pteridosperms. More recently a French author[2] states that the leaflets bear groups of sporangia indicating affinity to the marattiaceous ferns.

Botanical Provinces: Floras (see figs. 78, 79)

The forests which spread over a large area in Europe were in their salient features similar to those on the great continent on the western side of the northern hemisphere: *Calamites*, *Lepidodendron*, *Sigillaria*, pteridosperms, *Cordaites*, and many other trees flourished abundantly from west to east (fig. 78, *A–C*). Some of them reached the Arctic regions. There were differences between the two botanical provinces, though these seem to us insignificant as compared with the general uniformity in the composition of the vegetation as a whole. It must be remembered that only a small proportion of the available material has been thoroughly investigated. There are still large collections, for example in North America, which have not been described on modern lines. It has recently been shown that some of the northern plants wandered as far south as Sumatra[3] and the Malay Peninsula[4] before the closing stages of the Permian period. The Sumatran plants, or at least most of them, indicate a late Carboniferous or an early Permian age of the beds. These remote outliers are of special interest because over the greater part of the southern hemisphere the vegetation had a different facies from that of the northern lands; moreover, the physical conditions of the two regions were in striking contrast.

[1] Kidston (24).
[2] Corsin (27).
[3] Jongmans and Gothan (25); Posthumus (27), (29).
[4] Edwards (26^2).

A Far Eastern Flora

We may now turn to the Far East and to the rich coalfields of China and Korea. The best material is that which has been described from central Shansi by Prof. Halle[1] in a model memoir. The rocks of the region with which we are now concerned are classified as follows: at the base, a series of coal-bearing beds resting on older Palaeozoic (Ordovician) rocks, known as the Yuehmenkou series; above them a set of delta sediments rich in plants, the Shihhotse series; resting on these occur barren sandstones intercalated with beds of gypsum which indicate desert conditions, the Shihchienfeng series. The conclusion reached by Halle is that the plant-bearing Yuehmenkou and Shihhotse series are Permian in age; but in view of the discovery of several species in the lowest series which occur also in European and American Upper Carboniferous strata we may refer to the oldest flora as Permo-Carboniferous. The Shihhotse plants are also believed to be Lower Permian in age, and if this surmise is correct the uppermost and barren Shihchienfeng series should be correlated with the Upper Permian rocks of Europe which, like those in Shansi, were deposited under more or less arid conditions. The vegetation, of which samples are preserved in the plant-beds of central Shansi, affords clear evidence of a close agreement with the Permo-Carboniferous floras of North America and Europe. Out of a hundred and three species obtained from China seventy are new, but many of them belong to genera which are abundant in the northern floras: twenty species appear to be specifically identical with European forms, and fourteen species are North American.

It is worth while briefly to consider some of the Shansi genera. A type of equisetaceous plant common in the Far East and not so far found in Europe or America is the genus *Lobatannularia* (fig. 69) represented by specimens bearing fan-shaped, whorled clusters of linear leaves on branches similar to the foliage-shoots of calamitean plants described under the name *Annularia* from many northern localities. The name *Lobatannularia* was proposed by Kawasaki[2] a short time before the publication by Halle of the generic name *Annularites*, which it replaces. The eastern plant is distinguished by its more copious branching, and by a constant difference in the size of the leaves of a single whorl. Moreover, the star-like groups of leaves, instead of being radially disposed round the branch as in

[1] Halle (27). [2] Kawasaki (27).

calamitean shoots, and on the stem of a *Galium* or an *Asperula*, lie in one plane so that a whole branch-system may be compared in its flat expansion of green surfaces with a large frond of a fern.

FIG. 69. *Lobatannularia ensifolius* from Shansi, China. Nat. size. (From Prof. Halle.)

The foliage of *Lobatannularia* in the form of the individual leaves resembles that of a lupin (*Lupinus*), though in the latter genus each set of leaflets is borne on a long stalk, whereas in the fossil plant each excentric stellate group of leaves encircles the branch by a narrow sheath. The regular variation in size of the individual leaves is interpreted by Halle[1] as a correlation of form with efficiency in function, a plan by which overlapping of successive leaf-circles was minimized and every leaf received its fair share of sunlight. The genus *Asterophyllites*, another form of calamitean foliage-shoot common in Europe and America, occurs also in the Shansi beds. In view of the occurrence of *Asterophyllites*, *Annularia*, and *Lobatannularia*, all of

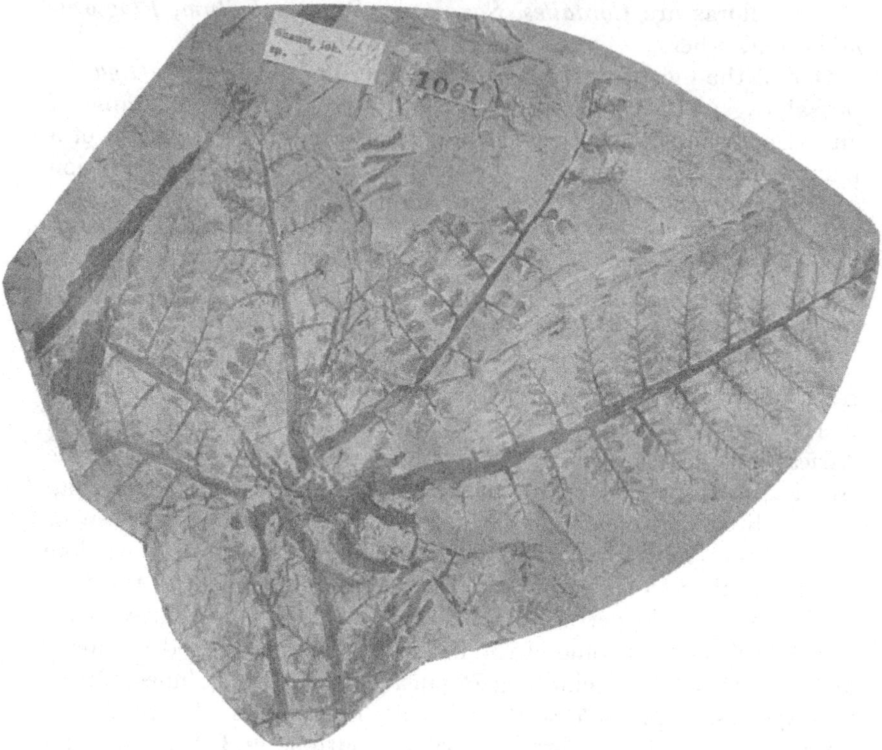

FIG. 70. *Oligocarpia* (*Sphenopteris*) *Gothani*. Leaves attached to a stem or branch, from Shansi, China. Nat. size. (From Prof. Halle.)

which were probably borne by calamitean plants, the great rarity of stems of calamites in the Chinese floras is difficult to understand: only one species, *Calamites Suckowi*, which is one of the most widely distributed forms in the northern floras, is recorded from Shansi. *Sphenophyllum* is represented by several species. There are also many fronds of pteridosperms and some that are probably ferns. An unusually complete specimen of *Oligocarpia* (*Sphenopteris*)

[1] Halle (28).

Gothani is shown in fig. 70. The leaves, which may belong either to a pteridosperm or a fern, are attached to a piece of stem, and on some of the leaflets there are imperfectly preserved sporangia similar to those of *Asterotheca* or *Oligocarpia*. There are only two species of *Lepidodendron*, both of which are European. *Sigillaria* has not been found. Other genera common to the northern and Far Eastern floras are *Cordaites*, *Saportaea*, *Psygmophyllum*, *Plagiozamites*, and others.

One of the most interesting plants described by Halle is *Gigantopteris*[1] (figs. 71, 72). *Gigantopteris* affords a striking example of an extinct genus with a discontinuous geographical range and of a plant which differs widely from any existing type. Its distribution is shown in fig. 78, *Gi.*: it was first discovered in Lower Permian beds in China, subsequently in Korea, North America, and Texas. It is unknown in Europe. Pieces of a frond of one of the Chinese species, *Gigantopteris nicotianaefolia*, so called from a superficial resemblance of the fragments that were first described to a leaf of a tobacco plant (*Nicotiana*), are shown in fig. 71. There is a general resemblance between the leaf reproduced in fig. 71, A, and that of the leguminous plant *Erythrina caffra*, the Kaffir-boom of South Africa. From subsequently discovered specimens we know that the fronds probably exceeded a metre in length and were more than 20 cm. broad; they bore large lateral leaflets with an uneven or toothed margin and a terminal segment. The veins form a net-like system similar to the simpler meshwork in the leaflets of the Carboniferous pteridosperm *Lonchopteris* (fig. 63, N). There is evidence that in this species some of the flat leaflets were replaced by hooks and formed efficient climbing organs.[2] Among the Chinese forms is a species, *Gigantopteris Whitei*, with long ribbon-shaped fronds having a straight (entire) or waved outline and traversed by parallel ribs giving off series of finer veins (fig. 72). A very closely related species, *Gigantopteris americana* from the Lower Permian strata in Oklahoma and Texas has similar ribbon-like and forked fronds. The genus is probably a pteridosperm: its wide distribution in the Far East has led to the employment of the term Gigantopteris flora for an early Permian association of plants. The American species has recently been found in Lower Permian beds in Sumatra (fig. 79, *Gi.*).

The Shansi flora is worthy of attention from another point of view:

[1] Halle (27); Yabe (17); White, D. (12).

[2] Halle (29).

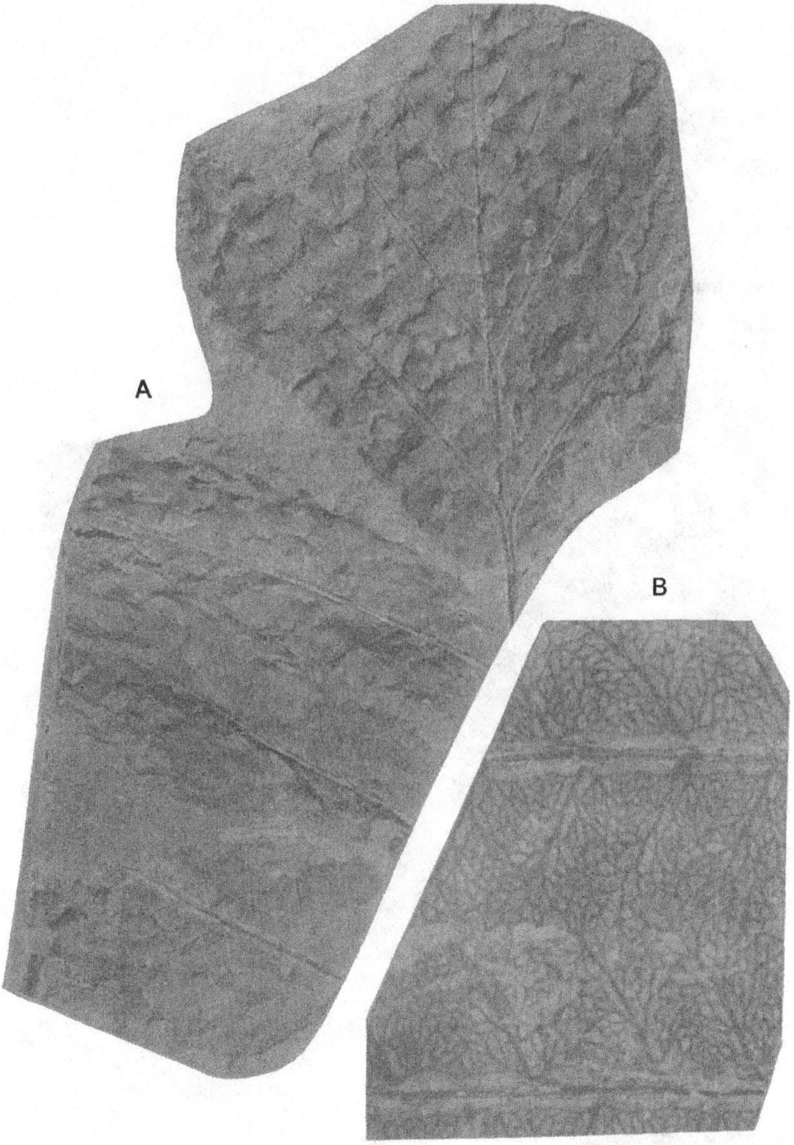

FIG. 71. *Gigantopteris nicotianaefolia* from Shansi, China. A, apical portion of frond, nat. size; B, piece enlarged to show venation. (From Prof. Halle.)

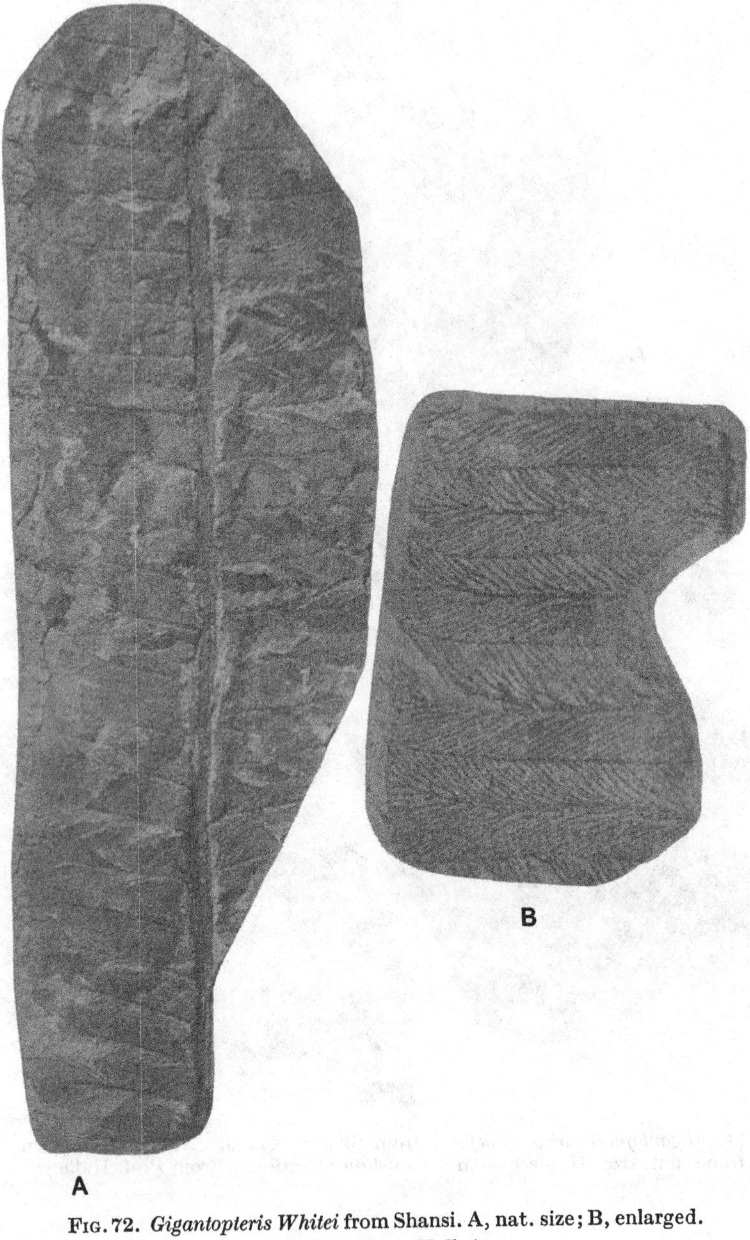

Fig. 72. *Gigantopteris Whitei* from Shansi. A, nat. size; B, enlarged. (From Prof. Halle.)

it includes not only widely distributed and typical Permo-Carboniferous plants, but also genera which furnish some of the comparatively few satisfactory links so far discovered between the late Palaeozoic and the early Mesozoic floras. If we compare on broad lines the Permo-Carboniferous vegetation of North America and Europe with that of the earlier part of the Mesozoic era, we cannot fail to be struck by the contrasts not only in detail but in general facies presented by these two epochs in the history of the plant-world. The flora of Shansi, though agreeing in the main with the North American and European floras, provides more points of contact with the Mesozoic vegetation. The genus *Baiera*, an extinct member of the Ginkgoales, is represented by large leaves cut into narrow segments which agree very closely with an Upper Triassic species, *Baiera multifida*, from North America.[1] Other connecting links are supplied by a species of *Cladophlebis* (fig. 88), a genus of ferns founded on the form of the leaflets and characteristic of Mesozoic floras throughout the world. *Cladophlebis Nystrœmii* has been recorded by Halle from Lower Permian rocks in China. The occurrence of some cycadean fronds, belonging to genera which are particularly abundant in the earlier Mesozoic floras, in Permian floras of China, and in rare instances in North America and Europe, has already been mentioned. Another genus that was fairly common in the Shansi floras is *Taeniopteris*: this name is given to simple linear fronds agreeing in form and venation with the leaves of the existing ferns *Scolopendrium*, *Oleandra* and others. *Taeniopteris* was a cosmopolitan genus in Mesozoic floras: it is recorded also from Permian beds in the northern continent and China. The numerous species include examples of more than one group of plants: some are almost certainly leaves of cycads; others probably are the foliage of pteridosperms, and some are ferns. A species, *T. multinervis*, from Permian rocks in France, Morocco, and North America (Kansas) occurs also in China. Other Chinese forms are compared by Halle with Upper Triassic species from Virginia. Similarly the genus *Chiropteris*, founded on reniform, net-veined leaves of unknown affinity, is represented both in China and in the older Mesozoic floras in the southern hemisphere. Several of the Shansi plants have been recorded from Korea.

Recent contributions to our knowledge of the late Permian floras of the Far East show that the vegetation which was spread over a

[1] Fontaine (83).

wide area in North America and Europe had invaded southern Asia; it had succeeded in colonizing a strip of territory as far south as Sumatra and the Malay Peninsula. This vigorous company of plants from the north had intruded itself, probably with the help of mountain chains as lines of migration (see fig. 78), into the continent known as Gondwanaland where the Permo-Carboniferous vegetation was dominated by *Glossopteris* and other genera which are not represented in the floras so far discussed.

The Glossopteris Flora

It has long been known that the climatic conditions in Gondwanaland during the closing stages of the Palaeozoic era were in sharp contrast to those north of the Tethys Sea. Reference has been made in a former chapter to the convincing evidence of geology in support of a Permo-Carboniferous Ice Age which affected vast areas in South America, South Africa, India, and Australia. Accurate correlation of the ancient boulder clays of Gondwanaland with Permo-Carboniferous strata in the northern hemisphere is extremely difficult and has given rise to much discussion. One view, to which prominence is given in Chapter IV, is that the stage was set for the development of glaciers, and eventually ice-sheets, in Australia and some other regions of the southern continent before the end of the Carboniferous period, that is, before the disappearance of the luxuriant forests of the Coal Age. We may illustrate the contrast between the vegetation of the northern continent, Eria, with tentacles reaching to southern China and Malaya, and the contemporary vegetation of by far the larger part of Gondwanaland, by comparing the former with the Dismal Swamp forest of Virginia and the latter with the Arctic floras of southern Greenland and Alaska. The contrast, though exaggerated in this rough analogy, was unquestionably striking.

The term Glossopteris flora[1] is usually applied to the southern vegetation because of the great abundance and wide range of tongue-shaped leaves with a median rib and a network of lateral veins. The leaves (fig. 74, *G.*) probably belonged to pteridosperms. None have been found bearing sporangia like those of ferns, but several have been described in intimate association with seeds and with scales which may have enveloped seeds. No seeds have been

[1] For illustrations of *Glossopteris* and other members of the flora see Arber, E. A. N. (05), also the publications mentioned on p. 288.

found organically attached to fronds, though seeds and leaves are often associated.[1] It is very rare to find *Glossopteris* leaves connected with a supporting axis, though the frequent association with them of branched stems known as species of *Vertebraria* renders it practically certain that leaves and stem were originally united. *Glossopteris* is often accompanied by leaves of a similar form—*Gangamopteris* (fig. 74, *Gg.*)—though these are generally rather larger and distinguished by the lack or feeble development of a central rib. It is not always easy to draw a distinction between the two forms: *Gangamopteris* is probably the more primitive; it occurs in the lowest strata containing members of the Glossopteris flora, and it did not persist as long as *Glossopteris*. Both genera occur in Gondwanaland in beds which are believed to be Upper Carboniferous, and in some places, notably in South Africa and the Argentine, specimens have been found in close relation to the boulder beds. Neither genus has been found in the Upper Carboniferous or Lower Permian beds of North America and Europe, nor in the coalfields of China; but *Glossopteris* and some other members of the Glossopteris flora are recorded from strata in northern Russia (fig. 78, *D*) that are of Upper Permian age and at localities farther south in Siberia. In 1912 fragments of *Glossopteris* were collected by Dr Wilson and Lieut. Bowers from rocks exposed on the slopes of Buckley Island on the Beardmore glacier 300 miles from the south pole[2] (fig. 79). We have no means of forming a very definite idea of the habit of *Glossopteris* or of *Gangamopteris*. In the reconstruction (fig. 74) they are shown as shrubs. Another widely spread member of the Glossopteris flora is a plant, represented only by fronds, which has recently been re-christened *Gondwanidium*[3] and was formerly called *Neuropteridium*. The fronds are long and bear two rows of large and more or less deeply lobed leaflets; they are always sterile and the assumption is that they are the leaves of a pteridosperm. The name *Neuropteridium* has long been applied to specimens from Gondwanaland of Permo-Carboniferous age and to others from Triassic strata in Europe: but it is probable that the two sets of leaves are not generically identical, hence the change in nomenclature recently suggested by Dr Gothan. In general appearance *Gondwanidium* agrees closely with fronds known as *Cardiopteris* (fig. 60, *C.*) from Lower Carboniferous beds in the northern hemisphere.

[1] Walkom (21), (28²); Seward (17), p. 140; Thomas, H. H. (21).
[2] Seward (14).
[3] Gothan (27).

No true *Calamites* are recorded from Gondwanaland, but articulate plants are represented by species of *Schizoneura*, a genus differing from *Calamites* and *Equisetum* in having longer and broader leaves. Some forms with narrow, linear leaves approach the Chinese and Korean *Lobatannularia* (fig. 69), though most of them are distinguished by the frequent occurrence of two sizes of leaf spreading from each joint of the branch, some relatively broad with several ribs and others with only a single median vein. *Schizoneura* occurs also in Triassic floras of both hemispheres: it is shown in figs. 74, 84, *Sz*. This genus, though one of the most widely spread in the southern floras, did not, as far as we know, exist in the Coal forests north of the equator. There are, however, several genera common to the northern and southern botanical provinces: a few sigillarias and *Lepidodendra* have been found in South America and South Africa. The tree fern *Psaronius* (fig. 74, *Ps.*) is represented by an almost perfectly preserved stem from Brazil, and another fern, *Tietea*,[1] which appears to be an allied genus, is recorded from the same country. A few pteridosperm fronds, similar in habit to northern species, have been found in India and other regions in company with *Glossopteris*.

Psygmophyllum,[2] a genus previously described, is another type common to both floras. An additional connecting link is supplied by long strap-like leaves usually described from Gondwanaland as species of *Noeggerathiopsis*, but very closely resembling the foliage of the northern *Cordaites*. The beautiful transverse section of a gymnosperm, *Dadoxylon indicum*,[3] shown in fig. 73, illustrates an interesting feature characteristic of several stems from different parts of Gondwanaland, namely the occurrence of well-marked annual rings. Other examples of stems from beds of similar age are described from Australia[4] and South America. It is noteworthy that in stems of closely allied plants from Carboniferous and Lower Permian rocks in the regions north of the Tethys Sea the wood is much more uniform in structure: this contrast may, in part at least, be due to the differences in climate, though it would be rash to assert that the degree of development of rings can invariably be taken as a measure of climatic conditions. The Indian specimens agree generally in their anatomy with stems from Europe referred to the Cordaitales: in all probability these Indian plants and other

[1] Solms-Laubach (13); Derby (15).
[2] Seward (03).
[3] Holden, R. (17).
[4] Sahni and Singh (26); Walkom (28^3).

closely allied species from Australia, South Africa, and South America bore the leaves which are usually spoken of as *Noeggerathiopsis* or *Cordaites*. Whether or not all the leaves described under these two names are foliage of a single genus, there can be no doubt

FIG. 73. Transverse section of a stem of *Dadoxylon indicum*, Barakar, India, showing rings of growth in the wood. × 6. (From a section in the Botany School Collection, Cambridge.)

of the near relationship of the cordaitalean trees of the northern and southern continents in the Permo-Carboniferous Age. Examples of branched shoots, of doubtful affinity, from Permo-Carboniferous beds in India, which were formerly believed to be generically identical with the European Triassic *Voltzia*, have been transferred to a new genus, *Buriadia* (fig. 74, *B.*),[1] because of the frequent, apical forking of the small leaves.

The vegetation of Gondwanaland was much poorer in genera and species than the contemporary forests in the northern hemisphere. Its most characteristic plants are unknown in the late Carboniferous and early Permian floras of North America, Europe, and southern Asia. With the exception of a few arborescent lycopods, a few pteridosperm fronds, *Psaronius*, *Sphenopteris*, *Sphenophyllum*, *Annularia* and *Psygmophyllum*, the great majority of the plants which played a conspicuous part in the northern forests have not been discovered in the Glossopteris flora.

The most striking example of a mixture in the Gondwanaland province of plants characteristic of the northern and eastern, Permo-Carboniferous vegetation with members of the Glossopteris flora is afforded by a collection recently described by Dr Walton[2] from the Wankie Coalfield in Southern Rhodesia. *Glossopteris* is represented by several species; there is also a *Phyllotheca*, the common Gondwanaland leaves *Cordaites* (*Noeggerathiopsis*) *Hislopi* and *Sphenophyllum speciosum*: with these were found *Pecopteris arborescens*, a type of frond common in northern Stephanian and Lower Permian floras, and recorded also from Sumatra; *Pecopteris* (*Ptychocarpus*) *unita*, a northern Stephanian and early Permian species; *Sphenophyllum oblongifolium*, abundant in Upper Carboniferous and Lower Permian floras in Europe and recorded from China and Sumatra. Dr Walton regards the Wankie beds as Lower Permian or possibly uppermost Carboniferous in age.

A scene in Gondwanaland is represented in fig. 74. The dark cliff in the middle distance is the exposed edge of a mass of boulder clay: behind this are glaciers and drifting icebergs, and in the background a range of hills with patches of snow in the ravines and on the gentler slopes. On the right-hand side is a *Lepidodendron* (*Ln.*); to the left of it a *Schizoneura* (*Sz.*), and to the right *Glossopteris* with a tuft of *Gangamopteris* leaves in the right-hand corner. The small plant with jointed stem and cup-like whorls of leaf-sheaths cut into short linear segments is a species of *Phyllotheca* (fig. 74, *P.*), which, with *Schizoneura*, represented the Equisetales in the vegetation of Gondwanaland. *Phyllotheca* occurs also in Upper Carboniferous rocks in Asia Minor and in Permian rocks in Siberia. Farther to the left is a plant of *Gondwanidium* (*Gd.*) and beyond it some *Glosso-*

[1] Seward and Sahni (20); Sahni (28²); Florin (29). [2] Walton (29²).

FIG. 74. A scene in Gondwanaland during the late Palaeozoic Ice Age. (Drawn by Mr Edward Vulliamy.) *B. Buriadia. C. Cordaites* (*Noeggerathiopsis*); *G. Glossopteris*; *Gd. Gondwanidium*; *Gg. Gangamopteris*; *Ln. Lepidodendron*; *P. Phyllotheca*; *Pg. Psygmophyllum*; *Ps. Psaronius*; *Sg. Sigillaria*; *Sp. Sphenophyllum*; *Sz. Schizoneura*.

pteris shrubs. The small tree at *Pg.* is a *Psygmophyllum* with its divided *Ginkgo*-like leaves. The tree-fern to the left is *Psaronius* (*Ps.*) characterized by the attachment of fronds in two opposite series.[1] Farther to the left is a tall *Sigillaria* (*Sg.*). A tree of *Cordaites* (*Noeggerathiopsis*) (*C.* with *Sp.*) is seen on the extreme left with a *Sphenophyllum* twisted round the stem.

The Kusnezk Flora

It remains to call attention to a northern province occupied in the latter part of the Permian period by an association of plants differing in several respects from the Permo-Carboniferous vegetation of North America and Europe. Our knowledge of this Kusnezk flora (so called from the place Kusnezk in Siberia; fig. 78, *Ku.*) is based on descriptions of plants at many localities, from Vladivostock on the Pacific coast to northern Russia: its age is probably Upper Permian. Though occupying a territory separated by a short distance from the Shansi region, the Kusnezk flora has little in common with most of the floras to the south and west; and it was not strictly contemporary with them. In an admirable atlas compiled by Prof. Zalessky[2] fifty-four species are illustrated, and of these only one, *Pecopteris anthriscifolia*, a pteridosperm frond recorded also from Lower Permian beds in France and Portugal, is common to the Kusnezk flora and the Permo-Carboniferous floras of Europe. One of the typical Permian genera of the Kusnezk flora is the pteridosperm *Callipteris* (fig. 63, A): the species *Callipteris conferta*, not represented in the Kusnezk flora, is among the more widely distributed and characteristic fronds in Permian floras: it was most abundant in the older stages, but persisted into the latter part of the period. With it occur the conifers *Ullmannia*, *Walchia*, and *Dicranophyllum*, also *Psygmophyllum* (fig. 74, *Pg.*), and a plant with large, deeply lobed leaves similar to those of *Psygmophyllum*. The Kusnezk flora differs from the slightly younger Shansi flora in the presence of *Gangamopteris* and some other Gondwanaland genera. It contains also species of *Czekanowskia*, *Ginkgoites*, and *Phoenicopsis*, plants allied to *Ginkgo* which were widespread in the earlier Mesozoic floras; together with *Podozamites*, *Cladophlebis*, *Baiera*, and a few other genera which supply additional evidence of the approaching transformation of the ancient vegetation of the closing

[1] Solms-Laubach (04[2]); Derby (13), (14).
[2] Zalessky (18), (27), (28[2]), (29), (29[2]), (29[3]).

stage of the Palaeozoic era into the more modern, Mesozoic type. The province occupied by the Kusnezk flora is known as Angaraland, from the Angara River in Siberia. Significant features of the Angaraland vegetation are the admixture of Gondwanaland forms and plants characteristic of early Mesozoic floras, also its contrast to the Upper Carboniferous and Lower Permian floras of North America and Europe.

A Permian Flora in Arizona

It has recently been shown that some members of the Kusnezk flora were represented in the Permian flora of Arizona. This flora, which is of exceptional interest, has recently been described by Dr David White[1] from plant-beds in the Grand Canyon of Arizona. As previously pointed out in the description (Chapter II) of the section of the canyon reproduced in fig. 4, a considerable thickness of late Palaeozoic strata is exposed on the walls of the cliff. In the locality from which the fossil plants were obtained Permian beds are well developed: these include about 300 ft. of red shales with some sandstones, which are known as the Hermit shale and referred to the upper part of the Lower Permian period. On the exposed surface of the shale intersecting cracks or fissures, occasional moulds of salt crystals and small pits in the rock afford evidence of strong desiccating sunlight, the deposition of salt, and the impact of hailstones: the occurrence of drifted sand in the sun-cracks furnishes additional proof of a semi-arid and warm climate. Both the nature of the rocks and the character of the plants point to relatively dry conditions. The flora differs in certain respects from the typical, richer and older Permian floras of Kansas, Oklahoma, Colorado, New Mexico, Texas, and the Appalachian region in North America, and in Europe. Such genera as *Calamites, Cordaites, Odontopteris, Callipteridium, Neuropteris* and *Alethopteris*—all of which are common survivals from the late Carboniferous vegetation in Lower Permian floras—are absent. This difference, though partly due to a slightly younger age of the Arizona beds, may also be a consequence of climatic conditions unfavourable to many of the older genera.

The widely distributed pteridosperm *Callipteris* (fig. 63, A) is well represented, including the characteristic Permian species *Callipteris conferta*; there are also many fronds, no doubt pteridosperms, similar in habit to *Callipteris*, which Dr White assigns to a new genus

[1] White, D. (29).

Supaia (from an old Indian name), using the term in a comprehensive sense as a group-designation: he includes in it fronds that are forked into two equal branches and bear relatively thick leaflets agreeing in venation with *Alethopteris* or *Neuropteris* and *Callipteris*. In the *Supaia* group are included fronds which it is customary to refer to the genus *Thinnfeldia* and by some authors to *Dicroidium*, also the fronds originally described as *Danaeopsis Hughesi* from India and other parts of Gondwanaland. Fronds similar in habit to those of *Supaia* are represented by species of *Callipteris*, *Callipteridium*, and *Glenopteris*, a name used for pieces of Permian leaves from Kansas. It would be out of place to discuss questions of nomenclature raised by the institution of the genus *Supaia*: the point is that there is a strong family likeness between many fronds ranging in age from Upper Carboniferous to Triassic, Rhaetic and even Jurassic strata; and it is probable that such genera as *Thinnfeldia*, *Glenopteris*, *Callipteris*, *Dichopteris* and others are all pteridosperm fronds. The species *Danaeopsis Hughesi*, characterized by large forked fronds bearing broadly linear leaflets, which is recorded from the Upper Permian Kusnezk flora, from Upper Permian or Triassic rocks in India and Australia, and from Rhaetic beds in the Argentine, is represented in the Permian flora of China by a similar type referred after considerable hesitation by Prof. Halle[1] to a previously instituted genus *Protoblechnum*, a name suggesting affinity to the fern *Blechnum*, an affinity of which there is no evidence. Dr White would include *Danaeopsis Hughesi* in *Supaia*, and the Chinese *Protoblechnum* in *Glenopteris*.

The important facts which emerge from the account of the Hermit flora are: that it includes several plants not previously recorded from typical early Permian floras, many of which are closely allied to or identical with species from the Upper Permian Kusnezk flora of Angaraland, also species characteristic of Gondwanaland, and only a few which are typical Lower Permian plants in other parts of North America and in Europe. To take the last first: *Sphenophyllum*, *Taeniopteris*, *Callipteris*, *Walchia*, and a few other genera carry on the northern Permian tradition. A Hermit species of *Sphenophyllum* appears to be very closely related to a species from the Ural Mountains. Dr White regards *Sphenophyllum* as a water plant, but the anatomical structure of all the petrified examples of the genus does not favour an aquatic environment: moreover, a species recently

[1] Halle (27). See also Yabe and Oishi (28).

described from China has a spinous stem suggestive of a scrambling habit. The Hermit flora helps us to link the later Palaeozoic vegetation with that of the early part of the Mesozoic era. In the abundance of *Thinnfeldia* (*Supaia*) fronds similar to those characteristic of India and the southern hemisphere and in the occurrence of *Ullmannia*, *Voltzia*, and some other conifers of doubtful affinity, we have links both with the Triassic and latest Permian floras and with the vegetation of Gondwanaland.

It is noteworthy that, as in other Permian localities in many widely separated parts of the world, there is evidence in the Arizona rocks of arid or semi-arid conditions. The great regions of salt beds "distributed through hundreds of feet", which separated the Grand Canyon region from the great North American plains, and are slightly younger in age than the Hermit shale, afford a striking example of the change in climate which was probably the chief factor in causing the transformation of the late Palaeozoic vegetation into the very different vegetation of the opening stages of the Mesozoic era.

A comparative analysis of the Permo-Carboniferous floras of the world reveals to us for the first time in geological history a well-marked differentiation of plant-communities: there were at least three botanical provinces. We are apt to exaggerate the uniformity of past vegetations when we compare lists of fossil plants compiled from the incomplete documents scattered through the rocks of widely separated regions with the infinitely richer material available at the present day. It is much easier to be impressed by the occurrence of what we believe to be species of the same genus in places geographically remote than to remember how little we know of the characters of fossil forms which mark true specific differences. In the earlier part of the Carboniferous period, as in the later stages of the Devonian period, there were no botanical provinces with well-defined distinctive characters. This at least is the conclusion based on such data as we possess. Some Lower Carboniferous genera flourished in regions as remote from one another as North-East Greenland, Spitsbergen, Siberia, and Australia. Were our information less incomplete, we should doubtless find that although some early Carboniferous genera and species were cosmopolitan, others had a much more restricted range.[1] When we compare existing

[1] Gothan (15).

floras of countries far apart, the fact that certain species occur in almost all of them is often forgotten. But making due allowance for imperfect data and a proneness to interpret resemblances which may be misleading as proof of identity, we are compelled to admit that the vegetation of the world in the early days of the Carboniferous period was surprisingly uniform in composition in comparison with the later Carboniferous floras, and with the more diversified communities in the modern world.

Passing to the Upper Carboniferous period and including in our retrospect plants from Lower Permian rocks, we find a few genera common to Eria, China, and Gondwanaland, but such are comparatively rare and much more limited in their geographical range on the continent of Gondwanaland than are *Glossopteris* and its associates. Moreover, apart from differences in generic representation, the northern vegetation was luxuriant and varied; the Glossopteris flora monotonous and relatively meagre. The flora of Angaraland differed from that of Gondwanaland in the greater range in genera and species and in its closer connexion with the Mesozoic phase of plant evolution. The occurrence of *Glossopteris* in Upper Permian rocks in Russia,[1] near the mouth of the Dwina River, is evidence of the northward migration of a genus, which had colonized enormous areas in Gondwanaland, from Antarctica to north-west India and tropical Africa. There must have been some land-bridge or island stepping-stones across the Tethys Sea. The failure of the Glossopteris flora to establish itself in China may be in consequence of physical barriers or of unfavourable climatic conditions.

Glossopteris began its successful career before the end of the Carboniferous period; it had wandered as far as northern Russia and other parts of Angaraland before the beginning of the Mesozoic era. Its coexistence with a Rhaetic flora was first demonstrated by Prof. Zeiller[2] in his account of the rich plant-beds of Tongking: more recently Dr Harris obtained leaves from Rhaetic strata in East Greenland, which, with good reason, he compares to *Glossopteris*. The Greenland leaves, which reached a considerable size, are examples of a type originally described from Rhaetic beds in southern Sweden as *Anthrophyopsis*: whatever their true nature may be, the Greenland and Swedish leaves do not appear to differ in any essential features from *Glossopteris*. *Schizoneura* is another instance of a Palaeozoic genus which established itself in Europe in the early

[1] Amalitzky (01). [2] Zeiller (03).

part of the Mesozoic era. It is probable that discoveries in Antarctica will furnish clearer evidence than we at present possess of the origin of some members of the Glossopteris flora in the far south, whence they gradually spread to the northern confines of Gondwana-land and crossed the Tethys Sea as invaders of northern lands.

Climate

Much has been written on the Coal Age climate.[1] Some authors create forests shrouded in mist, an atmosphere rich in carbonic acid gas, and stifling heat. Others lay stress on the fact that peat—which in some respects is the modern representative of coal—is now characteristic of temperate countries, though by no means unknown in the Tropics. As a matter of fact, it is impossible to estimate with any degree of accuracy the temperature of the northern forests. Arguments advanced in favour of a tropical or subtropical climate are based on very insecure foundations: the occurrence of ferns believed to belong to the Marattiaceae, a family that is now mainly tropical in its range; the habit of some trees, notably *Sigillaria,* to bear cones directly on the older part of the stem rather than on slender shoots, a feature which is quoted as an example of the phenomenon of cauliflory (i.e. the production of flowers on the older wood) in some modern tropical trees. Another argument of no real value is based on the supposed rapid growth of the stems of *Cordaites,* suggested by the breaking up of the pith into detached and parallel diaphragms such as we see if we split longitudinally a twig of a walnut tree (*Juglans*). The occasional occurrence of fronds of the cycadean pattern is adduced in favour of a comparison of the late Palaeozoic vegetation of the northern hemisphere with that of the warmer countries to-day. But the tropical cycads of the modern world are too remote geologically and botanically from the Palaeozoic members of the class to serve as a guide to climatic conditions in the Coal Age. The absence of regular rings of growth in the secondary wood of the Carboniferous trees is one of the most consistent pieces of evidence in support of a climate without sharply contrasted seasons. The section of a Malayan species of *Agathis* reproduced in fig. 75 affords a good example of an existing conifer in which there are no obvious rhythmic variations in the diameter of the wood elements: the wood is relatively uniform as compared with that of a conifer grown in temperate countries.

[1] See Köppen and Wegener (24); Brockmann-Jerosch (14).

Though the two photographs are not on the same scale, a comparison of the stem of *Agathis* with that of the extinct gymnosperm *Dadoxylon indicum* (fig. 73), which in the structure of the secondary wood closely resembles the living tree, reveals obvious differences: wood with little or no trace of rings, and wood with clearly defined

FIG. 75. Section of a stem of *Agathis alba* showing the absence or feeble development of annual rings. About ⅓ nat. size. (Photograph supplied by Mr F. W. Foxworthy, the Forest Research Office, Kuala Lumpur, F.M.S.)

boundaries between the spring and the late summer zones. There is a close connexion between rings of growth and climate, and the absence of any recurring zones of spring and late summer wood is a significant fact. On the other hand, existing trees show considerable differences in the sharpness of contrast between the seasonal growths: there are probably other factors than those connected with the environment which have an influence. The plants themselves may behave differently: moreover, we must remember that the trees of the Carboniferous forests are for the most part distantly related to modern types.[1] Conclusions based on the supposed relationship of extinct to recent plants are of doubtful value, particularly when comparisons are made between such genera as *Lepidodendron* and *Calamites* with modern lycopods and *Equiseta* or between Palaeozoic and recent ferns. The luxuriance of the Coal-period vegetation and the extraordinary range in plant-form, also the height reached by many of the trees, do not in themselves necessitate a tropical climate. Although beds of peat are now being formed in tropical Sumatra,[2] Ceylon, tropical Africa, and the Amazon valley, it is none the less true that we can more readily picture vast accumulations of plant débris either on swampy ground or on the floor of a lake[3] in a temperate rather than in a tropical country. On the other hand, the absence of annual rings is a point in favour of a climate more equable than that in temperate regions.

On the analogy of modern plants it might be urged that the protection of the stomatal pores in deep grooves on the lower face of a *Lepidodendron* leaf is evidence of dry conditions: in the leaves of living plants the position and number of stomata are often quoted as an example of correlation of structure and habitat, but the same structural features are found under different physical conditions. Plants growing in places which are physically dry and plants on shingle beaches furnish equally striking illustrations of devices which have the effect of reducing loss of water from the leaves. It is very probable that many of the Carboniferous trees grew in salt or brackish water and developed certain characters, such as protection of the stomata, which were not the result of dryness in the soil but, in some way not yet fully understood, a response to a particular environment. We have no means of ascertaining the percentage of carbonic acid gas in the atmosphere of a period separated from the present by many millions of years: it may have been higher than now; we cannot tell. The science of ecology, the study of plants in relation to environment, is still young: the wide range in leaf-form and in the habit of plants growing side by side raises many problems concerning the relations of the diverse units of an association to the several factors conditioning existence. These problems can be solved only by intensive study, and though good progress has been made

[1] Antevs (17) on rings of growth.
[2] Potonié (09).
[3] Griffiths (27). On climatic conditions see also Dachnowski (11).

by the experimental investigation of living plants, we are still unable to supply answers to the questions presented to us when we search for reasons underlying the relation between cause and effect.

The occurrence of large air-spaces in the cortex of a calamite root is one of several pieces of evidence in support of the view that some of the Carboniferous plants grew either in water or in wet marshland. On the other hand, it is equally probable that some of the trees were confined to higher and drier ground. There were mixed forests occupying wide reaches of low ground; there were stretches of muddy flats monopolized by a few kinds of trees. There were other trees on the slopes of undulating ground overlooking the watery waste of shallow estuaries or brackish lagoons. The wide geographical distribution of many Carboniferous plants, and indeed of plants of all ages, implies efficient means of dispersal and efficiency as colonizers. Dr David White,[1] in his account of the Carboniferous Missouri flora, goes so far as to say that the conditions favourable for plant dispersion and the consequent homogeneous distribution of the successive floras of the northern hemisphere during the first half of the Carboniferous period "have never been equalled". As we shall see later, the world-wide occurrence of certain Jurassic plants affords another instance, no less striking, of adaptability to a wide range of climate and the possession of seeds fitted for dispersal. A few examples chosen from a large number of seeds of Carboniferous plants are shown in fig. 76. The broad-winged *Samaropsis* (A) from the Newcastle Coal Measures of New South Wales recalling the seeds of the *Bignonia* and *Welwitschia*, spinning as it fell like the fruits of a maple, is admirably fitted for dispersal by wind. Similarly, *Samaropsis bicaudata* (B) from the Lower Carboniferous beds of Scotland, one of many Palaeozoic seeds simulating winged fruits of living plants, such as the elm (*Ulmus*), illustrates another and a slightly different device which increases buoyancy and allows more time for aerial transport. The flanged *Diplopterotesta* (D), a Lower Carboniferous genus, may be an example of a well-protected seed able to resist the action of sea water or possibly fitted for dispersal by animal agency. *Aetheotesta elliptica* (C), a Lower Permian, French seed, with its cycad-like pollen-chamber (*pc.*) and a thick outer coat lightened by large air-spaces (*s.*), suggests journeys with ocean currents. The seeds on the fern-like leaflet of a late Carboniferous pteridosperm frond (E, E′)

[1] White (99).

are small enough to be scattered by the wind; while the Lower Carboniferous seed *Thysanotesta* (F), closely resembling the fruit of an *Erodium*, its long tapered beak thickly set with hairs, was probably dispersed by animals in days long before the existence of mammals.[1]

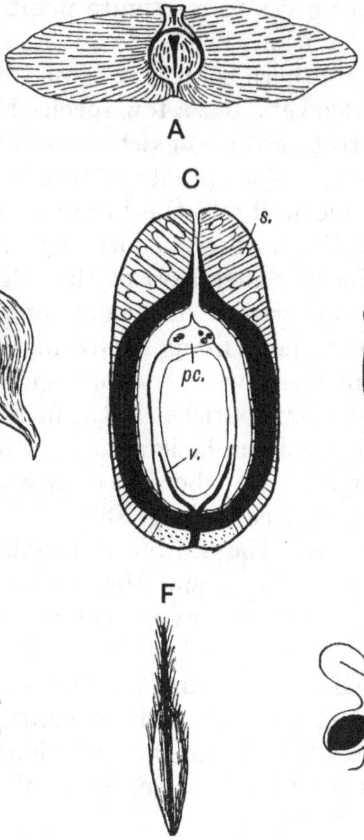

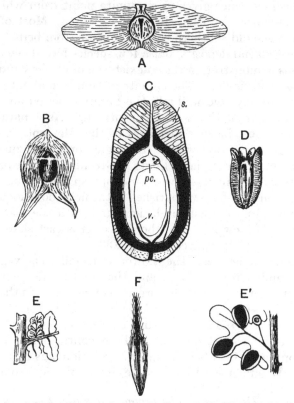

FIG. 76. Seeds of Carboniferous plants. A. *Samaropsis Pincombei.* 1¾ nat. size. (After Walkom.) B. *Samaropsis bicaudata.* Slightly enlarged. (After Kidston.) C. *Aetheotesta elliptica* in longitudinal section. *pc.* pollen-chambers; *s.* cavities in the fleshy coat; *v.* conducting (vascular) strands. Slightly enlarged. (After Renault.) D. *Diplopterotesta spetsbergensis.* Twice nat. size. (After Nathorst.) E, E′. Sterile leaflets and a fertile, seed-bearing, leaflet of *Dicksonites Pluckeneti.* E, nat. size; E′ × 3. (After Kidston and Zeiller.) F. *Thysanotesta sagittula.* Slightly enlarged. (After Nathorst.)

There were insects with a wing-spread of over 2 ft. and giant cockroaches which may have played some part as agents of dispersal; also, as Huxley said, "the stupid Salamander-like Laby-

[1] For other examples of Palaeozoic seeds see Seward (19).

rinthodonts which pottered with much belly and little leg like Falstaff in his old age among the Coal forests".

The Climate of Gondwanaland

Geological evidence gives us a definite point from which to consider the climatic conditions in Gondwanaland. Most of the fossil plants from the old southern continent come from beds lying above the thickest glacial deposits, but a few species have been discovered in positions demonstrating the co-existence of glaciers and members of the Glossopteris flora. The climate of Gondwanaland was doubtless comparatively cold well into the Permian period and much less genial than that of the northern continent. As a parallel to the existence of some hardier members of the Glossopteris flora even when the glacial period was at its height, one may quote the 400 odd species of vascular plants living in Greenland at the present day. As conditions became less severe, one may suggest comparison with the vegetation which now flourishes in the neighbourhood, and even on the moraines, of Alaskan glaciers, and with the glaciers of the south island of New Zealand where tree ferns cast shadows on the ice below them (frontispiece and fig. 138).

The contrast between the Lower Carboniferous vegetation of Australia and South America and that of the succeeding Permo-Carboniferous Age is no doubt directly correlated with the geological revolution of which there is striking evidence in the rocks of New South Wales. The older genera such as *Rhacopteris*, *Cardiopteris* and others ceased to exist on the southern continent: a new type of vegetation came into being, and its evolution was coincident with and directly influenced by a change in the physical conditions.

Geological Distribution of Carboniferous and Permian Plants

Having given a general account of the vegetation, we may next consider the distribution, both in space and time, of the earlier and later Carboniferous floras and of certain representative genera. One of our aims is to follow the development of the successive floras of the northern hemisphere immediately before and after the stage which marked the culminating phase of the Palaeozoic plant-world. In order to do this it is necessary to supplement the brief reference already made to such geological subdivisions as are essential to a fuller understanding of the floras. The correlation of Carboniferous rocks is too complex and controversial a subject to be discussed at length in this volume. Complete accord among geologists and palaeobotanists is an ideal seldom achieved; some measure of agreement was, however, reached at a Congress held at Heerlen in 1927.[1] Both fossil plants and the remains of freshwater and marine molluscs have been used as guides to geological horizons. These two classes of records do not always tell the same story; the animal fossils may indicate boundary-lines at certain levels in a pile of strata inconsistent with the evidence of the plants: for our purpose plants are the most valuable guides.

There is a general consensus of opinion in favour of speaking of the earlier portions of the Carboniferous period as Dinantian (from Dinant in Belgium). In the United States of America the marine phase of the system is termed Mississippian (from the valley of the river where rocks of this age are exceptionally well developed): the Upper Carboniferous is spoken of as Pennsylvanian. There has been much discussion on the position at which the most natural boundary line should be drawn between Lower and Upper Carboniferous. Dr Kidston[2] showed that the series of coarse sandstones known as the Millstone Grit contains in its lower beds plants agreeing generally with those typical of the oldest, or Dinantian, flora, while the upper strata are linked botanically with the Upper Carboniferous epoch. It is noteworthy that the term Millstone Grit has not always been consistently used; also that the break in the succession of plants characteristic of the British area is not universal. One result of the Heerlen Conference was the adoption of the name Namurian (from Namur in Belgium), previously employed by some authors for a series of beds between the Dinantian and the richest coal-bearing rocks in the upper part of the Carboniferous system. The higher limit of the Namurian state is defined by the occurrence of the Cephalopod *Gastrioceras subcrenatum*. This subdivision is omitted from the correlation table given below partly for the sake of simplicity and partly because the limits of the Namurian, as defined by *Gastrioceras subcrenatum*, do not coincide with any clearly marked botanical breaks. The lowest subdivision of the Upper Carboniferous, the Westphalian, includes some beds assigned by geologists to the Namurian. For present purposes the term Westphalian is adopted for coal-bearing rocks immediately following the Dinantian; the still higher coal seams and associated sedimentary

[1] Jongmans (28) and other papers on classification contained in the same volume.
[2] Kidston (23–25).

deposits are referred to the Stephanian, a name suggested by the development of rocks, containing plants regarded as characteristic of this age, in the district of St Étienne in France. Kidston used the term Westphalian in a narrower sense, and as this led to confusion Prof. Watts[1] proposed as an alternative the term Yorkian. Under Westphalian in the wider sense are included strata which in Britain are often termed the Lower and Middle Coal Measures. It should, however, be noted that the lower part of the Westphalian is equivalent in age and in the composition of the flora to Kidston's Lanarkian series, which he regarded as equivalent to the Lower Coal Measures and the upper part of the Millstone Grit. Kidston also instituted another subdivision based on plant evidence, the Staffordian or Transition series, intermediate between the Middle and Upper Coal Measures of England and Scotland. Finally the Stephanian stage includes part at least of Kidston's Radstockian (from the Radstock coalfield in south-west England), and a set of beds known as the Keele series (from Keele in Staffordshire).

The accompanying table is far from complete: it shows a classification of Carboniferous rocks based primarily on the plants; also the relation in geological age of some of the better known continental coalfields to those of Britain.[2]

There are practically no species common to the Dinantian and Westphalian floras: *Lepidodendron Veltheimianum* crosses the line between the lower and upper divisions of the Carboniferous system, but it is essentially a Dinantian plant and one of the index-fossils for that stage. The early Carboniferous vegetation was not only more cosmopolitan (see Table B); it was also much poorer in genera and species than Westphalian and Stephanian floras. It was during the Westphalian phase that the forests excelled in quality and in wealth of forms. The superiority of the Westphalian floras in numbers of species and in diversity of composition is shown in Table B. In the Stephanian flora we miss several of the older plants; there are many additions, but more losses. In the early part of the Permian period the decline becomes still more apparent, and we are

[1] Watts (22).

[2] For further information on the classification and correlation of Carboniferous plant-bearing strata see papers by Bertrand, Gothan, Jongmans, Trueman, and Zalessky in the Report of the *Congrès de Stratigraphie Carbonifère* held at Heerlen in June 1927, published at Liége, 1928; also Kidston (94), (01), (02), (23–25); Potonié (96); Watts (22); Gothan (23); Renier (26); Jongmans (28–28³); Kendall in Evans and Stubblefield (29).

Correlation Table of Carboniferous and Permian Strata.

	Great Britain	France	Holland and Belgium	Central Europe	North America	Other Regions
PERMIAN	Magnesian Limestone Marl Slate	Lodève plant-beds		Zechstein Kupfer-schiefer	Hermit flora; Arizona (see p. 249)	Kusnezk flora (see p. 248)
	Lower Permian	Autunian: Brive, Blanzy, etc.		Rothliegendes	Mid-continent floras. Dunkard group: Virginia, etc. Washington, New Brunswick and Nova Scotia	Sumatra; China (Shansi) and Korea: Donetz basin (South Russia)
Stephanian	Poorly represented: Radstock series and Keele group	St Étienne, Commentry, etc.	Poorly represented	Ottweiler series (Saar coal-field)	PENNSYLVANIAN: Upper beds of Rhode Island and Nova Scotia. Allegheny formation. Illinois; Indiana. Upper Coal Measures of New Brunswick and Nova Scotia; Lower beds of Rhode Island and Missouri	Italy, Bohemia, Donetz basin. Possibly Chinese and Korean plant-beds in part
Westphalian	Staffordian (Transition series) Middle and Lower Coal Measures	Valencienne coalfield	Upper Westphalian	Westphalian coalfields: Aachen, Ruhr, Schatzlarer series (Silesia)	PENNSYLVANIAN: Pottsville	Eregli (Asia Minor), Poland, Donetz basin, North Africa
	Part of Mill-stone Grit		Namurian	Ostrau and Waldenburg beds		
LOWER CARBONIFEROUS	Part of Mill-stone Grit Carboniferous Limestone Calciferous Sandstone of Scotland	Vosges, etc.	Dinantian	Culm series: Saalfeld beds (Thuringia)	MISSISSIPPIAN	Spitsbergen, Greenland, Urals, Donetz basin, Siberia, Eregli, Poland, Bulgaria, Pyrenees, North Africa, Argentina, Australia, Spiti (India)

prepared for the still more meagre records furnished by the relatively barren rocks of the latter part of the Permian period and the early part of the Triassic period. This change, from the luxuriant forests of the Coal Age to the depauperated oasis-vegetation of Triassic deserts, illustrates the disastrous consequences of the geological revolution which reached its climax at the end of the Carboniferous period. We interpret the apparently sudden impoverishment of the vegetation, indicated by the scarcity of fossils, as evidence of direct correlation of a changed physical environment with a transformed plant-world. In this instance the conclusion is probably correct, but there is always the possibility of being misled by negative evidence. We cannot assume that the contemporary vegetation was exceptionally meagre or locally unrepresented merely because no fossils have been found; it may be because the conditions under which the barren sediments were formed were not such as were favourable to the preservation of plant fragments. The richest floras have been obtained from strata accumulated in river deltas or in estuaries, and it is important to remember that in many instances sediments were deposited in places where the area of their deposition was beyond the reach of forest-clad banks and valley slopes which are the sources of most of the material in the herbaria of the rocks.

We may next consider another point: though the Westphalian vegetation may be regarded as the expression of a certain phase in evolution of the plant-world, it is not to be supposed that at this, or at any of the stages into which geological epochs are more or less arbitrarily divided, there were no fluctuations in the proportional representation of classes or in the component units of the plant associations. A careful analysis of the Westphalian plant-beds demonstrates a shifting of the balance from one group to another, and frequent changes in the nature of the forests symptomatic of changes in the environment. There is no stability either of the earth's crust or of the plant communities on its surface. To our eyes the present inorganic and organic worlds may appear relatively static; it is only by viewing them foreshortened and in comparative disregard of the lapse of time that we see them in their true relation.

An exceptionally good example of an intensive study of a series of plant-bearing beds is furnished by results obtained by Mr David Davies,[1] who made a most laborious and thorough investigation

[1] Davies, D. (21), (29).

of the rocks associated with coal seams over part of the South Wales coalfield. He collected samples of nearly 400,000 plants from strata within the Westphalian stage (using this term in the more comprehensive sense) associated with twenty-nine different coal seams over an area of 30 square miles, and through a thickness of rock of about 1150 yards. Many statistical tables are given in the papers to which readers are referred for a full presentation of the evidence. The most interesting conclusions may be briefly summarized: certain classes of plants were found to be dominant at particular horizons; and in each class certain genera occupied a dominant position. He also shows that when a genus became common the number of species generally decreased: this, as Mr Davies suggests, may indicate a diminution in developmental activity under an unpropitious or unstimulating environment. The annexed diagram (fig. 77), part of a larger one given by Mr Davies, illustrates by the relative breadth of the black columns the varying fortunes of some of the groups. When, for example, the Lycopodiales were dominant, the ferns and pteridosperms diminished in quantity; conversely, the lycopods fell off when the ferns and pteridosperms were in the ascendant. The calamites, as represented by stems and foliage, and the cordaitean trees were found to maintain a more uniform proportion, as though less susceptible than other groups to varying circumstances. Recurrent shifting of the balance among classes, groups, and genera is interpreted as evidence of fluctuations in the relative height of the land above sea-level: the site of the forests was not always unmitigated swamp and morass; from time to time the land rose and drier conditions prevailed. Mr Davies regards an increase in the number of *Lepidodendra*, *Sigillariae*, and allied trees as an index of wetter conditions, while the dominance of ferns and pteridosperms, he believes, points to higher and drier ground. Whether or not these conclusions are correct, it has been definitely proved that in the course of the Westphalian stage the physical environment was by no means uniform; the land rose and sank, and the shifting of the background, as we learn from the compressed and concentrated records, seems to have been rapid. Forests, which had grown for thousands of years on the low, swampy ground of a delta, were eventually destroyed by incursions of the sea registered in the hardened beds of mud associated with the coal. Recurrent periods of undisturbed possession of an area by rank vegetation were followed by periodic floodings: the land and water oscillated up and

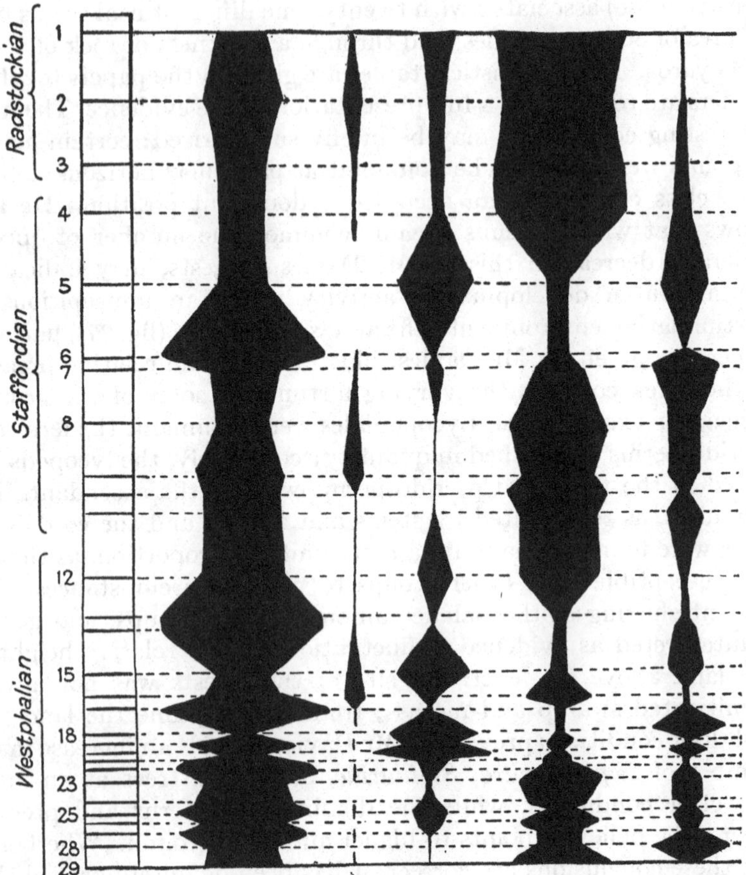

Fig. 77. Diagram showing the proportional representation of groups of plants at different stages of the Coal Measures of South Wales. (After Mr D. Davies. From *Discovery*.)

down; the rise of the land was not always of the same degree, and the conditions were not uniformly suitable for the same class of plants.

In order fully to appreciate the significance of the results of the researches into the relation between generations of Westphalian forests and the factors conditioning their growth, it is necessary to consult the original sources and to study the statistical tables compiled by Mr Davies. It is by patient and critical investigations in the field, such as Mr Davies has successfully carried out, that we may hope to obtain data enabling us to visualize in the rocks and their fossils the diverse scenes which formed the background to successive companies of plants.

The Upper Carboniferous floras differ more widely from the Dinantian flora than the Dinantian vegetation differed from that of the latter part of the Devonian period. On the other hand, the differences between the Westphalian and Stephanian floras are not greater than we should expect to find in tracing the development of the plant-world through a long series of years during which the physical conditions, though by no means constant, varied within comparatively narrow limits.

Another question is the possibility of utilizing fossil plants as sign-posts guiding us to geological horizons within the Carboniferous system: this is a matter of vital importance in the application of palaeobotany to coal-mining and other commercial enterprises. The literature on the classification of the Coal Measures is both voluminous and scattered and the lists of plants given by different authors are by no means free from discrepancies. A species may have persisted longer in one region than in another: some genera and groups of species have been found in practice to furnish much more valuable information than others. There are plants called by a French author "vagabond species",[1] which have been traced through several thousand feet of strata and were evidently able to hold their own as conservative types through a relatively long period: these are of little value as aids to the subdivision of a series of beds into narrower time-limits. Pteridosperms and some fronds which may have belonged to ferns afford the best clues. There were regional peculiarities in the Carboniferous forests as there are in the forests of to-day; for example, the fronds known as *Lonchopteris*[2] (fig. 63, N) mark a definite substage in the Westphalian series of Europe, but the genus has not been found in some of the European coalfields and is unknown in North America.

[1] Bertrand, P. (28).

[2] Gothan (07^2), (25^2).

TABLE B. *The Geological and Geographical Distribution of some Carboniferous and Permian Plants*

	Devonian	Lower Carboniferous	Upper Carboniferous: Westphalian	Upper Carboniferous: Stephanian	Permian: Lower	Permian: Upper	Triassic		N. America	Arctic regions	Europe	N. Africa	Asia Minor: South Russia	Kusnezk flora	China: Korea	Malaya	S. America: S. Africa: Australia	India	Notes and References
Equisetales																			
Asterocalamites		○	—						×	×	×	×	×	.	.	.	×	.	Jongmans (11)
Calamites		—	○	—	—	—			×	.	×	.	×	.	×	×	×	.	Kidston and Jongmans (15–17); Yabe and Endo (21) record the genus from Japan
Calamites Suckowi		—	○	—	—				×	.	×	.	×	.	×	×	.	.	
C. ramosus		—	○	—					×	.	×	.	×	.	.	.	.	.	
C. gigas				—	○				.	.	×	.	.	.	.	.	.	.	
Asterophyllites equisetiformis			—	—	—				×	.	×	.	×	.	×	.	.	.	
A. longifolius			○	—	—				×	.	×	.	.	.	×	.	.	.	
Annularia stellata			—	○	—				×	.	×	.	×	.	×	.	×	.	Walkom (16)
A. sphenophylloides			—	○	—				×	.	×	.	.	.	×	.	.	.	
Lobatannularia				—	—	- - -	- - -	→	.	.	.	.	.	.	×	.	.	.	Kawasaki (27); Halle (27). Cf. *Annulariopsis* from the Rhaetic of Tongking: Zeiller (03); Halle (28)
Schizoneura				—	—	—	—		.	.	.	.	.	×	.	.	×	×	
Phyllotheca			—	—	—	—	—	→	×	.	.	.	×	×	×	.	×	×	
Equisetites			—	—	—	—	—	→	.	.	.	.	.	.	.	.	.	.	The later forms are indistinguishable from *Equisetum*
Sphenophyllales																			
Sphenophyllum	—	—	—	—	—	—			×	×	×	.	×	×	×	×	×	×	
S. tenerrimum	—	○							×	×	×	.	.	.	.	.	.	.	
S. cuneifolium		—	○	—					×	.	×	.	×	.	.	.	.	.	
S. oblongifolium				—	—				×	.	×	×	×	.	.	.	×	.	Lundquist (19). Cf. *S. sino-coreanum*; Halle (27)
S. Thoni				—	○				×	.	×	.	.	×	×	.	×	.	Halle (27); Walton (29[2])
S. speciosum				—	—				.	.	.	.	.	.	×	.	×	×	
Cheirostrobus		—							.	.	×	.	.	.	.	.	.	.	Scott, D. H. (97)

Lycopodiales																		
Lepidodendron	—	—	○	—	—				×	×	×	×	×	·	×	×	×	·	
L. veltheimianum		○	—						×	×	×	×	×	·	×	×	×	·	
L. spetsbergense		—							×	×	×	·	·	·	·	·	×	·	Walkom (28). Cf. *L. Osbornei* of New South Wales
L. rimosum		—	○						×	·	×	·	×	·	·	·	·	·	
L. aculeatum		—	○	—					·	·	×	×	×	·	·	·	·	·	
L. Gaudryi			—	—					·	·	×	·	·	·	×	·	·	·	
L. oculus felis				—	—				·	·	·	·	·	·	×	·	·	·	Halle (27)
Lepidophloios laricinus		—	○	—					×	·	×	·	·	·	·	·	×	·	Cf. *L. scoticus*; Kidston (93); Lundquist (19)
Sigillaria		—	○	—	—				×	·	×	·	×	·	×	·	×	·	
Sigillaria Youngiana		—							·	·	×	·	·	·	·	·	·	·	Kidston (94). Cf. *Lepidodendron spetsbergense*
S. mammillaris			—						×	·	×	·	×	·	·	·	·	·	
S. tessellata			○	—					×	·	×	·	×	·	·	·	·	·	
S. Brardi			—	—	—				×	·	×	×	·	·	×	×	×	·	Seward (97[2]); Lundquist (19)
Stigmaria	—	—	—	—	—	—	...		×	×	×	×	×	·	×	×	×	·	Walkom (28[3])
Archaeosigillaria	—	—							×	×	×	×	·	·	×	·	×	·	Cf. *Protolepidodendron* (see Table A) Krasser (00)
Bothrodendron minutifolium		—	○						×	·	×	·	·	·	·	·	·	·	
Asolanus			○	—					×	·	×	·	×	·	·	·	×	·	Hirmer (27)
LEPIDOSPERMEAE																			
Lepidocarpon		—	—						·	·	×	·	·	·	·	·	·	·	Scott (01)
Miadesmia			—						·	·	×	·	·	·	·	·	·	·	
Lycopodites		—	—	—	—	—	—	→	·	·	×	·	·	·	·	·	·	·	
Selaginellites			—	—	—	—	—	→	·	·	×	·	·	·	·	·	·	·	Zeiller (06); Halle (07)
Filicales																			
COENOPTERIDEAE		—	—	—					·	·	·	·	·	·	·	·	·	·	
Botryopteris		—	—	—	—				·	·	×	·	·	·	·	·	·	·	
Metaclepsydropsis		—							·	·	×	·	·	·	·	·	·	·	
Stauropteris		—	—						·	·	×	·	·	·	·	·	·	·	
Diplolabis		—							·	·	×	·	·	·	·	·	·	·	
Botrychioxylon			—						·	·	×	·	·	·	·	·	·	·	
Etapteris		—	—	—	—				·	·	×	·	·	·	·	·	·	·	
Clepsydropsis		—							·	·	×?	·	·	·	·	·	×	·	Sahni (28), Prof. Sahni thinks that the Australian and European ferns may not be generically identical
Ankyropteris			—	—	—				·	·	×	·	·	·	·	·	·	·	Holden, H. S. (30)
Asterochlaena					—				·	·	×	·	·	·	·	·	·	·	Bertrand, P. (11)
OTHER FERNS																			
Thamnopteris					—	—			·	·	·	·	·	×	·	·	·	·	Kidston and Gwynne-Vaughan (08). For other Osmundaceae, see Zalessky (24)
Psaronius			—	—	○				×	·	×	·	·	·	·	·	×	·	Zeiller (90). For an account of another South American fern, see Solms-Laubach (13)
Rhacopteris (*Anisopteris*)		○							×	×	×	·	·	·	·	·	×	·	Kidston (23[2]); Walton (27)
Oligocarpia			○	—	—				×	·	×	·	×	·	×	·	·	·	
Senftenbergia			—						·	·	×	·	·	·	·	·	·	·	Kidston (24)

Table B. *The Geological and Geographical Distribution of some Carboniferous and Permian Plants* (cont.)

	Devonian	Lower Carboniferous	Upper Carboniferous: Westphalian	Upper Carboniferous: Stephanian	Permian: Lower	Permian: Upper	Triassic		N. America	Arctic regions	Europe	N. Africa	Asia Minor: S. Russia	Kusnezk Flora	China: Korea	Malaya	S. America: S. Africa: Australia	India	Notes and References
PTERIDOSPERMEAE and SOME FERNS	—	—	—	—	—	—	—	→	.	.	.	.	.	.	.	.	.	.	
I. Fronds																			
Pecopteris			—	○	—	—	—		×	.	×	×	×	×	×	×	×	.	For illustrations of Triassic fronds of similar form see Fontaine (83)
Pecopteris (*Asterotheca*) *Miltoni*			○	—	—				×	.	×	×	×	.	×	×	.	.	Halle (27) describes a closely allied form (*P. Wongii*) with seeds
P. (*Asterotheca*) *arborescens*			—	○	—				×	.	×	.	.	.	.	×	×	.	Walton (29[2]); recorded from Turkestan by Zalessky (28)
P. (*Ptychocarpus*) *unita*			○	—					×	.	×	.	×	.	×	.	.	.	
P. (*Dicksonites*) *Pluckeneti*			—	○					×	.	×	.	×	×	.	.	.	.	Loubière (29); Kidston (24)
Pecopteris anthriscifolia					—	—			.	.	×	.	.	×	×	.	.	.	Zalessky (27); Halle (27)
Cladophlebis (fern fronds)					—	—	—	→	.	.	.	.	.	×	×	.	×	×	A very common Mesozoic form of frond
Alethopteris			○	—	—				×	.	×	.	×	×	×	.	×	×	
A. lonchitica			—						×	.	×	.	×	.	.	.	.	.	
A. Serlii			○	—					×	.	×	.	×	.	.	.	.	.	
A. Grandini				○	—				×	.	×	.	×	.	.	.	.	.	
Lonchopteris			○		- - -	- - -	- - -		.	.	×	.	×	.	.	.	.	.	Cf. *Emplectopteris*; Halle (27). For Triassic fronds of similar habit see Fontaine (83)
Callipteris					—	—			×	.	×	×	.	×	.	×	.	.	The characteristic Permian species, *C. conferta*, is abundant, though not universal, in Europe. See White (29)
Callipteridium				—	○				×	.	×	×	.	.	×	.	×	.	
Odontopteris			—	—	○				×	.	×	×	×	×	×	.	.	.	For a good illustration see Zeiller (00[3]) and a copy in Seward (10)
Neuropteris		—	○						×	.	×	×	×	.	×	.	.	.	
N. antecedens		—							.	.	×	.	.	.	.	.	.	.	Stur (75–77); Walton (28[5])
N. heterophylla			—						×	.	×	.	×	.	.	.	.	.	
N. Scheuchzeri			○	—					×	.	×	.	×	.	.	.	.	.	
Neuropteridium					—	—	—		.	.	.	.	.	×	×	.	.	.	Halle (27). For illustrations of Triassic examples, see Blanckenhorn (86)

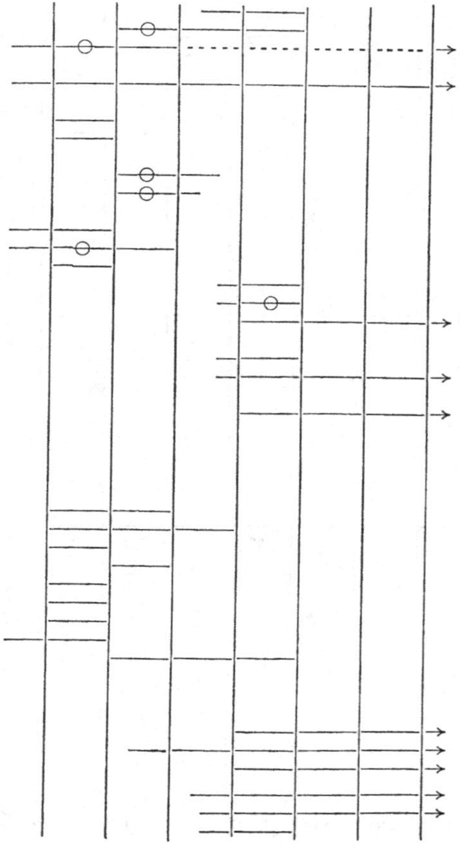

Gondwanidium	.	.	.	.	.	.	.	.	×	×	Gothan (27)
Linopteris	×	.	×	×	×	.	.	.	.	.	Zeiller (06); Dix (28)
Adiantites (the Mesozoic and later species are ferns)	×	×	×	.	×	.	.	.	×	.	Kidston (23)
Sphenopteris (the Mesozoic and later species are ferns)	×	×	×	.	×	.	×	.	×	×	
Sphenopteris (*Telangium*) *affinis*	.	×	×	.	.	.	.	.	.	.	Kidston (24); Zalessky (09), *S. bifida*
Diplopteridium (= *Sphenopteris teiliana*)	.	.	×	.	.	.	.	.	.	.	Walton (27), (29[3])
Sphenopteris obtusiloba	×	.	×	.	×	.	.	.	.	.	
Mariopteris nervosa and *M. muricata*	×	.	×	.	×	.	.	.	.	.	
Sphenopteridium	×	×	×	×	.	.	.	.	.	.	
Rhodea	.	.	×	.	×	.	.	.	.	.	
Cardiopteris	.	×	×	.	×	.	.	.	×	.	
Gigantopteris	×	.	.	.	.	.	×	×	.	.	Halle (29); Posthumus (29)
Taeniopteris multinervis	×	.	×	×	.	.	×	×	.	.	
Thinnfeldia (*Supaia*) *group*	×	.	×	.	.	×	×	.	×	×	White (29); Sellards (00) describes the genus *Glenopteris* of this group
Gangamopteris	.	.	.	.	.	×	.	.	×	×	
Glossopteris	.	.	.	.	.	×	.	.	×	×	Sahni (23) describes the structure of the epidermis
Chiropteris	.	.	.	.	.	.	×	.	×	.	Halle (27). The Gondwanaland examples are from early Mesozoic strata: see Du Toit (27[2])
II. Stems, etc. Pteridospermeae and genera of uncertain affinity											
Lyginopteris	.	.	×	.	×	.	.	.	.	.	
Heterangium	.	.	×	.	.	.	.	.	.	.	Scott (17); Kubart (14)
Rhetinangium	.	.	×	.	.	.	.	.	.	.	Gordon (12)
Megaloxylon	.	.	×	.	.	.	.	.	.	.	Seward (99)
Calamopitys	×	.	×	.	.	.	.	.	.	.	
Stenomyelon	.	.	×	.	.	.	.	.	.	.	Kidston and Gwynne-Vaughan (12)
Protopitys	.	.	.	.	.	.	.	.	.	.	Solms-Laubach (93)
Cladoxylon	.	.	×	.	.	.	.	.	.	.	
Medullosa	×	.	×	.	.	.	.	.	.	.	A leaf-stalk of *Medullosa* described by Noé (23) as a Palaeozoic monocotyledonous stem: Seward (23[2])
Cycadophyta											
Dioonites	.	.	.	.	.	.	×	.	.	.	Halle (27)
Pterophyllum	.	.	×	.	.	×	.	.	.	.	
Sphenozamites	.	.	×	.	.	.	.	.	.	.	
Ginkgoales	.	.	.	.	.	.	.	.	.	.	
Baiera	.	.	×	×?	.	×	×	.	.	.	Fritel (25)
Saportaea	×	.	.	.	.	.	×	.	.	.	

Table B. *The Geological and Geographical Distribution of some Carboniferous and Permian Plants* (cont.)

	Devonian	Lower Carboniferous	Upper Carboniferous: Westphalian	Upper Carboniferous: Stephanian	Permian: Lower	Permian: Upper	Triassic		N. America	Arctic Regions	Europe	N. Africa	Asia Minor: S. Russia	Kusnezk flora	China: Korea	Malaya	S. America: S. Africa: Australia	India	Notes and References
Cordaitales																			
Cordaites		—	—	—	—	- - -	- - -	→	×	×	×	×	×	×	×	×	×	×	For descriptions of cordaitalean wood (*Dadoxylon*, etc.) see Halle (11); Walkom (28[3]); Seward and Walton (23); Holden, R. (17); Sahni and Singh (26)
Pitys: Archaeopitys		—							×	·	×	·	·	·	·	×	·	·	Scott and Jeffrey (14); Walkom (28[3])
Coenoxylon					—	—			·	·	·	·	·	×	·	·	·	·	Zalessky (11)
Dolerophyllum			—	—	—				×	·	×	·	·	×	·	·	·	·	Florin (25)
Coniferales																			
Araucarites					—	—	—	→	·	·	×	·	·	·	·	·	·	×	Zeiller (06)
Walchia			—	—	○	—			×	·	×	×	·	·	·	·	×	·	Florin (27) for Palaeozoic conifers; White (29)
Ernestia (*Walchia filiciformis*)					—				×	·	×	·	·	·	·	·	·	·	
Voltzia					—	—	—		×	·	×	·	·	·	·	·	·	×	Walton (28[2]). For an account of a similar conifer from India (*Buriadia*) see Seward and Sahni (20); Sahni (28[2])
Pityanthus				—					·	·	×	·	·	·	·	·	·	·	Florin (27)
Pityospermum					—	—			·	·	×	·	·	·	·	·	·	·	
Dicranophyllum		—	—	—	○	—			×	·	×	·	·	×	×	·	·	·	
Ullmannia					—	—			×	·	×	×	·	·	·	·	·	·	Solms-Laubach (84); White (29)
Gomphostrobus					—	—			·	·	×	·	·	·	·	·	·	·	Gothan (28[3])
Plants of uncertain position																			
Psygmophyllum	—	—	—	—	—	—			×	×	×	·	·	×	×	·	×	·	
Plagiozamites				—	—				×	·	×	·	·	·	×	·	·	·	Bassler (16); Halle (27)
Tingia				—	—				·	·	·	·	·	·	×	·	·	·	Halle (25), (27); Kon'no (28)
Pelourdea				—	—	—	—		·	·	·	·	·	·	×	·	·	·	Seward (17); Halle (27). The generic name was proposed primarily for Mesozoic fossils resembling leaves of *Cordaites*

Table B is intended to serve two purposes: on the left-hand side is shown the range in time, from the Devonian period to the end of the Palaeozoic era, of several genera and species selected from much longer lists that are available. The maximum development, as measured by relative frequency of specimens, is indicated in some instances by a small circle on the horizontal distribution-line: a genus or species, though it may be common to more than one stage or formation, is usually found to be most abundant at one geological horizon or stage of which it is regarded as a characteristic or an index-fossil. On the right-hand side is shown the range in space of genera or groups of species, and a few references to original sources are given in the last column.

We will deal first with range in time. The table is in the main self-explanatory, but it may be helpful to draw attention to a few facts bearing on the vertical distribution of plants through the rocks of the Carboniferous system. There are certain genera which are generally regarded as characteristic of Dinantian floras, though species of some of them are recorded from higher levels, e.g. *Asterocalamites*, *Adiantites*, *Rhodea* (fig. 63, K), *Rhacopteris* and *Cardiopteris*. Among the ferns there are a few genera not as yet traced beyond the Dinantian; but as these cannot be identified in the absence of petrified material and occur only in exceptionally favoured localities, they are of secondary importance as index-fossils. Similarly, genera such as *Cheirostrobus* (Sphenophyllales), *Rhetinangium*, *Stenomyelon*, *Pitys* and others, founded on unusually well-preserved specimens—mostly from southern Scotland—have a limited value as chronological criteria because of their rarity or absence in most Carboniferous plant-bearing strata.

Calamites, though essentially an Upper Carboniferous genus, is represented by a few species in the older floras which in some respects are intermediate between *Asterocalamites*, with the ribs and grooves on the pith-casts continuous from one internode to the next, and the typical Westphalian *Calamites* with a regular alternation of the ribs at each node: the Dinantian species, *Calamites Haueri*, *C. ramifer*, and others have been assigned by Hirmer[1] to a special genus, *Mesocalamites*. *Sphenophyllum*, a legacy from the Devonian period, was widely spread in the early part of the Carboniferous rocks, but it did not reach its full development until the latter part of the Coal Age; it persisted well into the Permian

[1] Hirmer (27).

period. Attention has already been called to the wide geographical distribution of certain species of *Lepidodendron* in the Dinantian stage: prominent in the earlier Carboniferous forests, the genus was more prolific, in an evolutionary sense, in the latter part of the period; it was rare and apparently moribund in Permian floras (e.g. *Lepidodendron Gaudryi*) and evidence of its survival into the Triassic period is unconvincing. *Sigillaria*, having played a very subordinate part in Dinantian floras, reached its maximum in the Westphalian: it is noteworthy that species with ribbed stems, on which the vertical arrangement of the leaf-scars on the vertical ribs is more obvious than the spiral disposition (e.g. *Sigillaria elegans*, *S. scutellata*, etc.), are characteristic of the Westphalian and relatively rare in the Stephanian; while, in the Stephanian and Permian floras, species in which the leaf-scars are spirally scattered on a smooth stem (e.g. *S. Brardi*, *S. McMurtriei*) are more abundant. The genera *Bothrodendron* (e.g. *B. minutifolium*), *Asolanus*, another lycopodialean stem with small spirally disposed leaf-scars and recognizable by a certain form of sculpturing on its surface, and *Lepidophloios*, closely related to *Lepidodendron*, are characteristic of Westphalian floras.

As shown in Table B, pteridosperm fronds are among the more abundant Westphalian and Stephanian fossils: *Alethopteris* (e.g. *A. Serlii* and *A. lonchitica*) reached its fullest development in Westphalian floras: *Alethopteris Grandini* is one of the few Stephanian species. Fronds of the *Neuropteris* type (e.g. *N. Scheuchzeri*, *N. rarinervis*, etc.) and their associated seeds and spore-bearing organs are among the most characteristic Westphalian fossils, the genus *Mixoneura* (e.g. *M. ovata*),[1] differing but slightly from *Neuropteris*, is also a useful index-fossil in the uppermost beds of the Westphalian series, with *Linopteris* (fig. 63, M), *Sphenopteris obtusiloba*, *S. Hoeninghausi*, and other species. *Lonchopteris*, which is characteristic of a definite level in the Westphalian stage, had a much more restricted geographical range than many of the other pteridosperms; it persisted into Stephanian floras and it occurs, though under a somewhat different form—described by Halle as *Emplectopteris*—in the Lower Permian, or possibly in part Stephanian, vegetation of China. Fronds with leaflets closely resembling those of *Lonchopteris* are recorded from Upper Triassic rocks. The genus *Pecopteris* affords an example of a type of pteridosperm frond which was most abun-

[1] Bertrand, P. (26[2]).

dant in Stephanian floras: *Pecopteris arborescens* is one of the characteristic Stephanian and Lower Permian species in North America and Europe and it has recently been found in the Wankie coalfield of Rhodesia.[1] The Stephanian stage is well developed in the coalfields of central France, in the Saar basin and in other parts of Germany, and the uppermost beds pass without a break into the older strata of the Permian period.

Stephanian time coincides with the entrance on to the world's stage of several new types destined to reach their highest development in later ages. Fronds of the genera *Odontopteris* and *Callipteridium*, which became more abundant in early Permian floras, make their appearance; leaves of *Cordaites*, though common in some localities, are for the most part specifically distinct from the older forms. The band-like leaves of *Taeniopteris* are first recorded from the Stephanian series, also the pinnate fronds of *Pterophyllum*, the superficially similar foliage-shoots of *Plagiozamites*, the conifer *Walchia* and some other genera which are pioneers of the Mesozoic vegetation. *Calamites cruciatus*, *Annularia stellata*, *Asterophyllites equisetiformis* are characteristic representatives of the Equisetales. The most striking features of this latest stage in the Carboniferous system are the absence or rarity of many of the Westphalian plants[2] and the incoming of several new types destined to play a more prominent part in later periods.

For other examples of differences in the composition of the floras reference should be made to Table B, which shows an apparently sudden manifestation of evolutionary energy as a striking attribute of the Westphalian stage. The much greater number and variety of plants in the Upper Carboniferous floras is clearly established, but this should not lead us to underrate the high level of specialization already reached by some of the Dinantian plants. Such Dinantian genera as *Cheirostrobus*, *Clepsydropsis*, *Pitys*, *Calamopitys* and many others afford convincing evidence of a high degree of organization which compares favourably with that shown by members of the later floras. It is none the less true that under the exceptionally favourable conditions in the northern hemisphere subsequent to the Dinantian stage the Palaeozoic vegetation reached its fullest development. To take a rough analogy: the American "aloe" (*Agave americana*) dies after the supreme effort of producing its amazing inflorescence; the plant-world of the Palaeozoic

[1] Walton (29^2). [2] Bertrand, P. (28^2), (28^3).

era reached its blossoming period in the Westphalian stage, and it is no exaggeration to say that this was followed by the extinction of a considerable number of its most characteristic components. In this connexion it is interesting to notice that the decrease in variety among the calamites subsequent to the Westphalian stage is associated with a marked increase in the size of stem, indicated by the large pith-casts of the Stephanian and Lower and Middle Permian species *Calamites gigas*. Pith-casts of this species are described by Grand'Eury[1] from Stephanian beds at St Étienne as 5 metres long with little change in diameter. It has been suggested that this may be an instance of what has been called gigantism, a phase which appears to be a premonition of approaching extinction.

Reference has already been made to the general continuity between the later Carboniferous vegetation and that of the earlier part of the Permian period. In England the older Permian rocks are usually separated by an unconformity from the uppermost Carboniferous, though the boundary in some places is by no means clear. It is, however, relevant to point out that strata of Stephanian age are only very partially represented in Britain. In central France, in many parts of Germany, in Italy and in North America, as also in Gondwanaland, the Carboniferous and Permian succession is more gradual and indeed almost imperceptible. The term Autunian is often applied to Lower Permian rocks because of the occurrence of rich plant-bearing beds of this age in the neighbourhood of Autun, in France: beds of corresponding age, but marine in origin, are termed Artinskian, from their development in Russia. The Autunian, or Lower Permian series, has furnished many more plants than have been obtained from the Upper Permian strata, which include the Magnesian Limestone (Zechstein) and Marl Slate or Copper Slate (Kupferschiefer). Most of our knowledge of the vegetation has been gained from France and Germany; in central France from the Autun district, the basins of Blanzy and Brive; in Germany from Thuringia, Baden, Saxony, particularly in the Chemnitz district, where magnificent collections of petrified stems and other fossils—many of them displayed in the Chemnitz Museum—have been obtained. In North America Permian beds are well developed in Kansas, Colorado, Arizona, western Virginia, and in other States; also in Nova Scotia and New Brunswick. It is significant that both in North America and Europe the Permian strata,

[1] Grand'Eury (12).

particularly those belonging to the upper part of the system, are frequently barren and include layers of gypsum and salt which bear witness to a serious disturbance in the factors controlling the plant-world.

The differences between the Stephanian and the Lower Permian vegetation are comparatively slight. *Calamites Suckowi* persisted from the Westphalian to the Autunian stage, also certain types of calamitean foliage. The genus *Lobatannularia*, described by Halle from early Permian beds in China, illustrates a new introduction, though it is one which shows clear indications of relationship to *Annularia*, as also to the genus *Schizoneura*, a form characteristic of the early Glossopteris flora, which has been found in the early Permian Kusnezk flora and is a characteristic member of the oldest Triassic flora of Europe. Another equisetalean genus, *Phyllotheca*, is a link between the Carboniferous, Permian, and Mesozoic floras. The chief difference between the Carboniferous and Permian Equisetales is the relatively poor representation of the arborescent calamitean species in the later period. Reference has already been made to *Calamites gigas* and to *Lepidodendron*. *Sphenophyllum* was still vigorous in the first half of the Permian period: *S. Thoni* is one of the more widely distributed Permian species. It is noteworthy that in several genera of fronds characteristic of some of the Permian floras the leaflets are relatively thick or leathery and probably more resistant than the thinner leaflets of other pteridosperms to adverse conditions.

Passing to the true ferns, we find some members of the Coenopterideae, an exclusively Palaeozoic family, in early Permian floras: *Etapteris*, first recorded from Lower Carboniferous rocks, survived into the Permian period; *Tubicaulis*, represented by petrified stems a metre long and several centimetres in diameter, and sporangia of the *Botryopteris* type, lived from Westphalian into early Permian time; a Siberian fern stem, recently described by Prof. Sahni[1] under a new name, *Asterochlaenopsis*, which resembles the older *Clepsydropsis* (fig. 59), and *Asterochlaena* (fig. 64, L). Another genus of Coenopterideae common to both periods is *Ankyropteris*[2] with a stem in which secondary conducting tissue was spasmodically developed. Some fronds of the *Pecopteris* form, included in the genus

[1] Sahni (30) speaks of the Siberian *Asterochlaenopsis* as a link between *Asterochlaena* and *Ankyropteris* and as affording an additional piece of evidence in favour of a common origin of the Zygopterideae and Osmundaceae.

[2] Holden, H. S. (30).

Asterotheca because of the stellate arrangement of the spore-capsules, and which may be the foliage of ferns allied to modern Marattiaceae, have been described from Upper Carboniferous and Lower Permian rocks: similar fronds are recorded from Upper Triassic strata.

It is in the Upper Permian Kusnezk flora (fig. 78, K) that the oldest petrified examples of osmundaceous fern stems, *Thamnopteris* and *Zalesskya*, were discovered: the conducting tissue of these genera is of a more primitive type than in any of the later members of the *Osmunda* family.[1] Among examples of fronds, almost certainly the foliage of pteridosperms, which are characteristic of Permian floras are *Odontopteris*—also a member of Westphalian and Stephanian floras; *Callipteridium*—also a Stephanian type; species of *Taeniopteris*—particularly *T. multinervis*; and fronds with comparatively thick rounded leaflets known as *Callipteris*—one of the most trustworthy index-fossils of Autunian strata; the species *Pecopteris anthriscifolia* and others. Some additional reference should be made to some of these Stephanian and Permian fronds which are known to be the foliage of pteridosperms. The two types of fronds, *Callipteridium* and *Odontopteris*, resemble *Callipteris* in the forking of the main axis and in some other respects, such as venation characters and the occurrence of leaflets on the rachis. All the fronds, though undoubtedly borne by plants closely related to those which had foliage of the *Alethopteris* type or leaves with *Pecopteris* leaflets, exhibit certain distinctive characters marking them off from most other forms. *Callipteridium* (fig. 63, P) differs from *Callipteris* in the absence of any backward spreading of the leaflet base, and in the greater degree of bifurcation of the frond; *Odontopteris* is distinguished by the absence of a midrib in the leaflets (fig. 63, F). Both *Callipteridium* and *Odontopteris* occur in Upper Carboniferous and in Lower Permian floras. The type of frond characteristic of these genera is seen also in those of the Lower Permian *Glenopteris*[2] with handsome leaves bearing leaflets reaching a length of about 15 cm., also in many forms included in the genus *Thinnfeldia*, which was widely spread in early Mesozoic floras, as in other Mesozoic genera such as *Lomatopteris*. It is clear that *Callipteris* is one of several genera which may reasonably be regarded as

[1] For an account of an older Permian fern believed to be allied to *Thamnopteris* see Beck (20) on *Protothamnopteris*.

[2] Sellards (00).

members of a group of pteridosperms which bridge the gap between the Palaeozoic and Mesozoic eras.

Petrified stems of *Medullosa* are among the most striking fossils in the Permian flora of Chemnitz. A good example of discontinuous geographical distribution is furnished by the large, reticulately veined, fronds of *Gigantopteris* (figs. 71, 72) discovered in North America, China and Sumatra (figs. 78, 79 *Gi.*). Fronds resembling in habit those of some living cycads, though doubtless borne by very remotely related genera, occur in some Stephanian floras: additional examples of similar Mesozoic types made their first appearance in the early part of the Permian period, e.g. *Sphenozamites* and *Dioonites*. The genus *Walchia* is represented both in the Stephanian and early Permian vegetation by the species *W. piniformis*: another species of this Palaeozoic conifer, *W. filiciformis*, which has recently been assigned to a new genus, *Ernestia*,[1] is exclusively Permian. Another genus, *Lecrosia*, characterized by oval cones bearing scales with winged seeds and needle-shaped leaves, is a Stephanian forerunner of Permian conifers. The genus *Ullmannia*, which is characteristic of the later Permian floras, has been found in Arizona in beds at the upper limit of the lower part of the Permian system. Foliage-shoots of similar habit have recently been described from an Upper Permian flora in Germany as *Archaeopodocarpus*[2] on the ground that their fertile branches resemble those of living species of the southern hemisphere conifer *Podocarpus;* but the evidence of relationship hardly amounts to proof. Reference may also be made to an imperfectly known species, *Araucarites Delafondi*, described from the Autunian of Blanzy in France. *Rhipidopsis*, *Baiera* and *Saportaea*, genera founded on leaves, are probably the oldest representatives of that extraordinarily vigorous Mesozoic class, the Ginkgoales. *Saportaea* is recorded from Lower Permian rocks in North America and China; like *Gigantopteris*, it affords an example of discontinuous geographical range. *Rhipidopsis*,[3] characterized by its large, deeply cleft *Ginkgo*-like leaves, is another Permian genus. This genus has been found in a much later Rhaetic flora of China. The oldest examples of *Baiera*, which flourished in both hemispheres in the Triassic and Jurassic periods, are from Lower Permian beds. The genus *Tingia*, originally described as *Pterophyllum* and believed to be cycadean, has recently been more fully described by Halle[4]

[1] Florin (27).
[2] Weigelt (28).
[3] For illustrations and references see Seward (19).
[4] Halle (25), (27).

from ample material discovered in the Lower Permian beds of China: it is an interesting Permian type characterized by four rows of relatively large, wedge-shaped or broadly linear leaves traversed by many veins; two rows are attached to the upper face of the stout axis and spread out in one plane and two rows spring from the lower face, a habit similar to that of *Selaginella*. Halle, on the evidence then available, was inclined to regard *Tingia* as a possible member of the class Ginkgoales. Subsequently Prof. Kon'no described additional species of *Tingia* from Korea, and in close association with the foliage-shoots were discovered cones which throw additional light on the nature of this extinct type. The cones, borne at the end of a branch, consist of four vertical series of crowded fertile leaves (sporophylls), each with a large spore-capsule on its upper face divided into four compartments. Prof. Kon'no,[1] to whose full description reference should be made, agrees with Prof. Halle that the foliage-shoots resemble most closely those of existing species of *Selaginella*; the cones, referred to a separate genus *Tingiostachya*, in his opinion indicate a nearer affinity to the family Psilotaceae, now represented by *Psilotum* (fig. 29) and *Tmesipteris*. All we can say is that *Tingia* is an example of an extinct family which lived during one of the most critical periods in the history of the plant-kingdom and in a region of the world exceptionally favourable for the study of the transitional stages between the Palaeozoic and Mesozoic eras: it is one of many plants of which we know enough to make us long for more information.

The foliage-shoots known as *Plagiozamites*, recorded from Stephanian and early Permian rocks in Europe and North America, are believed to belong to plants closely akin to if not generically identical with *Tingia*.

Some members of the Cordaitales, a class characteristic of the later Carboniferous floras, continued to flourish in the Permian period, and there is some evidence pointing to the survival of closely related plants into the early part of the Mesozoic era.

We have seen that as the Palaeozoic era drew to its close the character of the vegetation changed, though only in part: the majority of the plants in the Upper Carboniferous and early Permian forests were members of a fairly homogeneous company, but an impending transformation, actually accomplished before the end of the Triassic period, and faintly foreshadowed in the Stephanian, became

[1] Kon'no (29).

more apparent in the Permian period. It may be that this mingling of Palaeozoic and Mesozoic plants was most pronounced in the Siberian Kusnezk flora, which is probably Upper rather than Lower Permian in age: but it has been suggested by Prof. Obrutschew[1] that the unusual number of plants of very pronounced Mesozoic aspect, recorded as associates of typical Permian genera, may be partially explained by our imperfect knowledge of the provenance of the collections. The rich plant-beds of Siberia, which include Carboniferous, Permian, and Mesozoic strata, are still very imperfectly known. The more we know of the plants preserved in the older Permian rocks and in the beds classed as Permo-Triassic in China and Korea, the more promise there seems to be of discoveries which will enable us to reconstruct some of the missing links in the chain of plant-life connecting the Palaeozoic and Mesozoic floras. There are indications that the original home of many genera characteristic of the earlier Mesozoic vegetation may have been in the continent of Angara, a region which has been pre-eminently immune from the disastrous consequences of geological revolutions. The recently described flora from the Arizona Canyon[2] throws fresh light on the vegetation of the northern continent as it was in the middle of the Permian period. We know comparatively little of the plant-world of Upper Permian time. The richest Upper Permian flora is that which has already been mentioned as the Kusnezk flora (fig. 78, K): between this and the Arizona flora there are many connecting links and in both floras are several new elements that are obviously direct ancestors of Mesozoic species. In other parts of the world Upper Permian floras are rare, and most of them are much poorer in the variety and number of plants than those which preceded them. The Upper Permian flora of Lodève in France[3] is one of the richest.

The Transition from one Era to another

The important consideration for our present purpose is the comparison of the latest phase of the Palaeozoic plant-world, as revealed by a study of the relics of vegetation preserved in the rocks deposited at that stage of geological history which is recognized as the last act in the Palaeozoic era, and the phase represented by the earliest known Triassic floras. We will first draw attention to the most striking difference between the later Palaeozoic and the oldest Mesozoic plant-world. By the beginning of the Stephanian—that

[1] Obrutschew (26). [2] White, D. (29). [3] Zeiller (98).

is the uppermost stage of the Carboniferous period—the luxuriant forests of the Westphalian stage had already lost some of their old-established trees; the decline was beginning. Through the earlier stages of the Permian period many of the old genera still maintained their ground though in diminished numbers: some new types, which were better fitted for the altered circumstances of the new environment, replaced the old and furnished evidence of continuity between a depauperated vegetation of the past and a very different vegetation of the future. Before the end of the Permian period the *Lepidodendra* had disappeared, though the discovery of the cone *Lycostrobus Scotti* in the Rhaetic flora of southern Sweden and the subsequent discovery of its spores in the Greenland flora of the same age suggest the possibility of the continued existence of some of the lycopodialean trees after the close of the Palaeozoic era. No Sigillarias have been found in strata higher than Lower Permian: the much smaller Triassic *Pleuromeia* (fig. 84, *Pa*) may be a direct descendant of the arborescent lycopods of the Coal Age; but the glory of the race has departed. The true calamites seem to have become extinct by the end of the Permian period. The same fate befell *Sphenophyllum*. Through *Callipteris* and its allies the Palaeozoic pteridosperms passed the boundary between the Palaeozoic and the Mesozoic era: *Neuropteris*, *Alethopteris*, *Medullosa*, and very many other vigorous Palaeozoic members of this class failed to survive. It may be that in spite of the apparent failure of the pteridosperms to retain their dominance beyond the limits of the Palaeozoic era they are still represented in a different guise in the present flowering plants: this is a heterodox suggestion founded on fancy which, one is tempted to think, may be promoted to the realm of fact as knowledge increases. The Palaeozoic ferns afford a striking contrast to those which became widely spread before the end of the Triassic period; the Coenopterideae were replaced by genera of much more modern aspect; the Osmundaceae, well established in the latter part of the Permian period in Angaraland, assumed an increasing importance as the Mesozoic era advanced. Fronds of the form known as *Cladophlebis*, some of which are undoubtedly Osmundaceae and are among the more cosmopolitan of Upper Triassic, Rhaetic and Jurassic fossils, are recorded from the Lower Permian flora of northern China. Reference has already been made to Carboniferous ferns foreshadowing features characteristic of the Gleicheniaceae and Schizaeaceae. Connecting links between the Permian and

Triassic vegetation are afforded by a few conifers already mentioned: it is this line of evolution and the line followed by the Ginkgoales which can be traced more clearly than some of the others from the Permian to the early part of the Mesozoic era. In this connexion it is interesting to find in the Kusnezk flora the earliest examples of such members of the Ginkgoales as *Phoenicopsis*, *Czekanowskia* and *Ginkgoites* which are essentially Mesozoic genera. Before the end of the Permian period there existed conifers, such as *Voltzia* and *Pseudovoltzia*, *Pityospermum* and *Pityanthus* which show evidence of relationship to the Abietineae, and there is good evidence of the occurrence of the araucarian family before the end of the Palaeozoic era. There is no proof of the survival of *Cordaites* and its allies into the Mesozoic era, but the occurrence of leaves, hardly distinguishable from the Carboniferous species, in Triassic and even later floras and the discovery of other pieces of evidence, suggest the possibility that the line of evolution represented by the Cordaitales stretched unbroken into later ages.

As already pointed out, it is fairly certain that the Cycadophyta were represented in Stephanian floras. The probability is that these plants appeared before the end of the Carboniferous period, but it was not until the latter part of the Triassic period that they definitely began their rapid progress towards a position of dominance.

The Geographical Distribution of Carboniferous and Permian Plants

We will next glance at the geographical distribution of Carboniferous and Permian plants. As seen in Table B, certain characteristic Dinantian genera and species, or groups of closely related species, were cosmopolitan. Such evidence as we have justifies the statement that the older Carboniferous plants from all parts of the world afford no indication of any clearly marked contrast between the vegetation to the north of the Equator and that of Gondwanaland. A selected number of localities where plants of Dinantian age have been found are shown on the maps (figs. 78, 79) as red spots. In North America two regions only are marked: Kentucky is one of the few districts in the New World where petrified specimens of Lower Carboniferous plants have been discovered. These include leafstalks of ferns and other fossils agreeing generally with species from Saalfeld in Thuringia, also a large *Lepidodendron* cone closely allied to the western European species *Lepidostrobus Brownii*.[1] Some

[1] Zeiller (11^2).

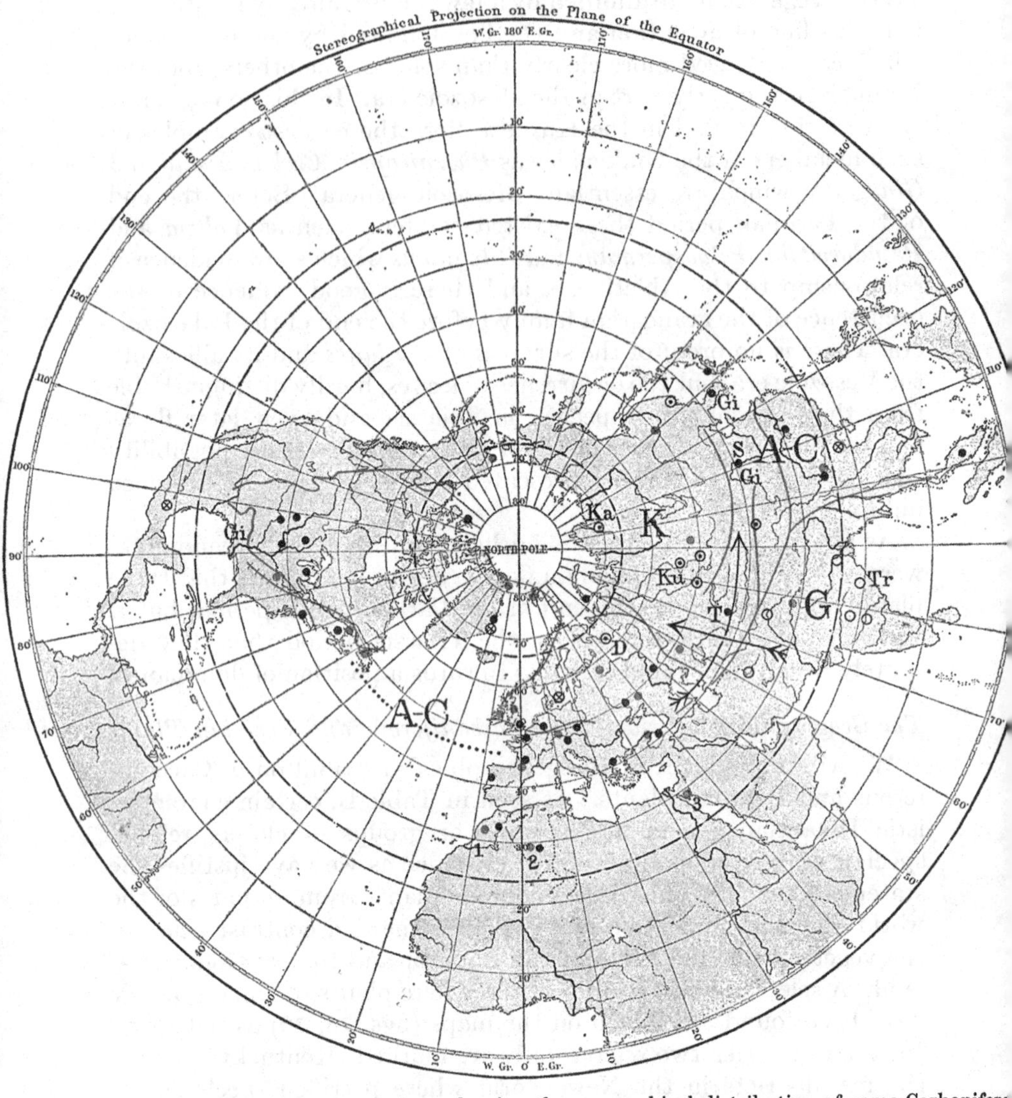

FIG. 78. Map of the northern hemisphere showing the geographical distribution of some Carbonifer[ous] and Permian floras. For explanation of map, see pp. 281–287.

Note. Locality 3 (red) should perhaps be represented by a black dot, as Coal Measures and not Low[er] Carboniferous image. See p. 283, footnote 1.

Lower Carboniferous species have been obtained from Pennsylvania, Virginia and other States. Plants compared with European species have been found in Newfoundland, and both Lower and Upper (Westphalian) plant-beds occur in Nova Scotia and New Brunswick. Early Carboniferous plants are recorded from Alaska and north-eastern Greenland. A particularly rich Dinantian flora was described by Nathorst from Spitsbergen: this includes not only certain cosmopolitan species, but some which have not been discovered elsewhere. To avoid confusion, only a few localities are shown in Europe: Pettycur and other places in southern Scotland and the adjoining district in northern England; Saalfeld in Germany (the red spot in central Europe) is one of several districts where specimens have been collected. Some Lower Carboniferous plants, which probably grew on the southern shore of the Tethys Sea, have been described from Morocco together with relics of a Westphalian flora. *Asterocalamites* and Dinantian species of *Lepidodendron* are recorded from several places (a few of which are shown in fig. 78) in Siberia and along the Ural chain. In the Donetz basin, north of the Black Sea, there is a succession of Lower and Upper Carboniferous beds passing into Permian strata, also at Eregli to the south of the Black Sea. Specimens of *Lepidodendron* apparently identical with European Dinantian species were found some years ago by Mr Ball of the Egyptian Geological Survey in the Sinai Peninsula.[1] The only Indian specimens agreeing with early Carboniferous species were discovered at Spiti (the red spot near the letter G).

Turning to the southern hemisphere (fig. 79), characteristic Dinantian genera, and some species which seem to be indistinguishable from northern forms, have been described from New South Wales and Queensland, also from other localities not marked on the map. Two South American localities are shown, one in Argentina and one (P) in Peru. The plants from Peru were assigned by one author[2] to an Upper Carboniferous age; by two other authors[3] to the Lower part of the system. No undoubted examples of early Carboniferous plants are known from South Africa.

The geographical distribution of Upper Carboniferous and of Permian plants is shown by the black spots in figs. 78 and 79. Only a comparatively few localities have been selected. Our knowledge

[1] Since this was written the specimens have been more carefully examined by Dr. Stockmans in my laboratory, and we are inclined to regard them as possibly Westphalian in age. [2] Berry (22^3). [3] Gothan (28); Seward (22^4).

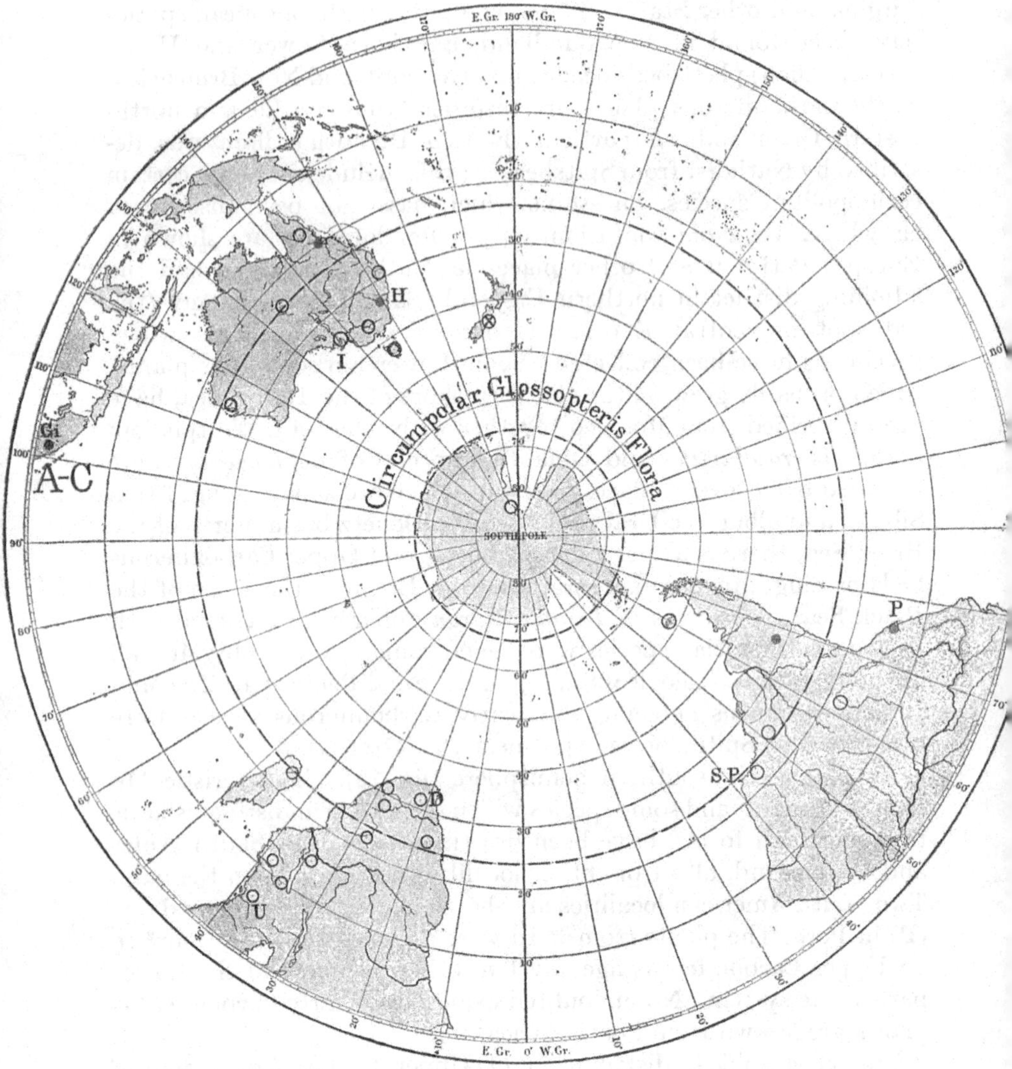

Fig. 79. Map of the southern hemisphere showing the geographical distribution of some Permo-Carboniferous floras, etc., see pp. 281–287.

Note. Lower Carboniferous localities are shown in red.

of Arctic floras of this age is unfortunately very meagre: fragments brought long ago by Arctic explorers from Banks Island and Bathurst Island on long. 120° N. suggest comparison with Carboniferous species, but it must have been in a moment of enthusiasm that Oswald Heer spoke of the specimens as evidence of a "luxuriant and stately forest" in the "inhospitable hyperborean region". A few fossils obtained from the southern part of Novaya Zemlya were described by Nathorst[1] as possibly of Middle Carboniferous age. The coal forests of North America occupied an area of at least 250,000 square miles: from Rhode Island in the east and the Appalachian coalfield through Michigan and Illinois to Missouri, Iowa and western Kentucky to Nebraska and Texas. In its general composition the vegetation closely resembled that of Europe (fig. 78, A-C). The letters A-C stand for Atlantic-Chinese flora. Dr White,[2] in his Memoir on the Upper Carboniferous Coal Flora of Missouri, states that of sixty-nine species thirty-four are identical with British species and thirty-five closely allied to European types. The letters *Gi.* (fig. 78) to the west of Florida mark the occurrence in Lower Permian beds of Texas of *Gigantopteris*, a genus unknown in Europe but common in the early Permian flora of China and Korea and recorded also from Sumatra (fig. 79, *Gi.*). Turning to Europe, Westphalian or Stephanian floras are known from Portugal, Spain, Britain, France, Belgium, Holland, and Switzerland through central Europe, to Italy, Bulgaria and southern Russia.

Passing farther east, the circles with a central dot indicate the approximate position of localities in the province of the Kusnezk flora (K) extending from the Dwina River (D) to Vladivostock (V) on the Pacific coast and reaching the Khatanga River (Ka) in northern Siberia. The locality Kusnezk is shown at Ku. The distinguishing feature of this flora, which is probably of Upper Permian age, is the association of typical genera of European and American Permian genera with some elements derived from the Permo-Carboniferous vegetation of Gondwanaland. From lat. 40° N. to the circle of Capricorn, in China and Korea, there was another botanical province with a vegetation closely linked by many species with the Stephanian and early Permian floras of Europe and North America. The genus *Gigantopteris* (*Gi.*) was a widely spread member of this Chinese vegetation. Still farther south, in the Malay Peninsula (fig. 78) and more especially in Sumatra (fig. 79) evidence has

[1] Nathorst (94). [2] White, D. (99).

been found of the invasion of Gondwanaland by far-travelled members of the American and European Permo-Carboniferous forests. The circles in the Indian peninsula and to the north-west of the Himalayas indicate some of the localities of the typical Glossopteris flora, which in the latter part of the Carboniferous period had become established over almost the whole of Gondwanaland (figs. 78, 79). Although no typical representatives of the Glossopteris flora have so far been found in New Zealand (circle with cross), it is by no means unlikely that leaves recorded from beds of Rhaetic age under the name *Linguifolium* are closely allied to *Glossopteris*.[1] This is a debatable point: the late Dr Arber[2] was strongly opposed to the suggestion of any direct affinity between the New Zealand fossils and *Glossopteris*, but both Dr Gothan[3] and myself are of opinion that the striking resemblance in form, despite the lack of regular anastomoses of the lateral veins of *Linguifolium*, is more than an accident of parallel development. None the less the fact remains that no evidence is available of the existence of the older Glossopteris flora in New Zealand. Returning to the northern hemisphere, the circles enclosing a cross at Tongking, in southern Sweden, eastern Greenland, and Mexico indicate the occurrence of the genus *Glossopteris*, or some plant with leaves of the *Glossopteris* type, in Rhaetic floras.

The red lines with cross-bars show the approximate trend of the pre-Carboniferous Caledonian mountain ranges in western Europe and in North America: the other red lines mark the position of some of the later Palaeozoic Alps, the Cordillera ranges on the west, the Armorican range through Ireland, Wales, and southern England, the Variscan Alps through France, the Vosges, and eastward to Bohemia and other ranges still farther east. It is probable that the Palaeozoic Ural chain served as a partial barrier to the eastward spread of the European vegetation. Similarly the east and west range in China may have served as a partial obstacle to the northward migration of the Gigantopteris flora. It may also be suggested that the branched Malayan arm of the transverse range in southern China was the route by which members of the Permo-Carboniferous flora of Shansi were able to establish an outpost of the great northern flora in the southern hemisphere. It would be premature to attempt to correlate more closely the physical features, as determined by the uplifting of Palaeozoic mountain chains, and the boundaries of

[1] Seward (14). [2] Arber, E. A. N. (17). [3] Gothan (27); Seward (14).

the Carboniferous and Permian botanical provinces; but it is worth while to draw attention to a line of enquiry which, with fuller knowledge, may be productive of important results.

The arrows are intended to suggest possible routes of migration. *Glossopteris* and some other Gondwanaland genera crossed the Tethys Sea and reached northern Russia before the end of the Permian period; it is probable that the northern forests of the Coal Age occupied in China and Malaya an area greater than that indicated by the records so far discovered in the rocks: at one time there must have been a connecting stretch of country between the original home of the late Carboniferous vegetation and the remote southern regions.

Notes on Literature

Additional information on Carboniferous and Permian floras may be obtained from the following sources, which are a few selected from a voluminous literature as examples of descriptive accounts of some of the better known and richer floras or as compilations likely to be useful to students. Many important early memoirs are omitted because of their inclusion in published bibliographies.

For illustrations and descriptions of Carboniferous plants students should consult palaeobotanical text-books in which references are given to original sources. The most important descriptive memoirs are those by the late Prof. W. C. Williamson published in the *Philosophical Transactions* of the Royal Society from 1871 to 1892 and the later memoirs under the joint authorship of Prof. Williamson and Dr D. H. Scott, also published by the Royal Society.

For a general account, with references to palaeobotanical papers, see Kendall and Versey in Evans and Stubblefield (29); geological text-books, such as Geikie (03), Chamberlin and Salisbury (04–06), Pirsson and Schuchert (20), Haug (20), Kayser (23–24); also Gothan (23).

Lower Carboniferous

Great Britain: Kidston (82), (89), (94), (03), (23–25); Arber, E. A. N. (04^2), (12); Scott (02); Walton (27), (28^5). *France and Belgium*: Bureau (14); Carpentier (13); Renier (10^2). *Central Europe*: Weiss, E. (85); Solms-Laubach (96); Stur (75–77); Gothan (27^2), (27^3) with references; Jongmans (09); Potonié (01) who describes as Silurian plants that are almost certainly Lower Carboniferous in age; Rydzewski (19). *Russia and Siberia*: Zalessky (07), (09), (28^4); Obrutschew (26). *Asia Minor*: Zeiller (99). *China*: Krasser (00); Halle (27^2); Grabau (23–28). *Arctic Regions*: Nathorst (94), (02), (11^2), (14), (20). *North Africa*: Flamand (07); Douvillé and Zeiller (08); Fritel (25); Carpentier (30). *North America*: White, D. (95); Arber, E. A. N. (12^2); Scott and Jeffrey (14). *South America*: Szajnocha (91); Kurtz (94); Berry (22^3), (28) Upper Carboniferous species; Seward (22^4); Gothan (28); du Toit (27). *Australia*: Süssmilch (22); Süssmilch and David (20); Howchin (29); Walkom (19), (28), (28^3); Benson and others (20–21); Sahni (26).

Upper Carboniferous, including some early Permian Floras

For a general account of the Coal Measures see Gibson (20), North (26), Crookall (29), in addition to the text-books mentioned above. A useful list of references up to 1905 is given by Arber, E. A. N. (06[2]).

Great Britain: Kidston (01), (02) general; Kidston (86) Radstock; (88), (91), (14) Staffordshire; (96), (97) Yorkshire; (17) Wyre. For other references to literature see the synonyms given by Kidston in his monographs (23–25), (23). Arber, E. A. N. (09), (12[3]), (14) Kent coalfield; (14[2]) Wyre; (12[4]) Forest of Dean. Crookall (25) Bristol and Somerset. *France, Belgium and Holland*: Zeiller (88); Renault and Zeiller (88–90); Grand'Eury (12), (13); Carpentier (13), (29[2]) and other papers; Bertrand, P. (28[2]), (28[3]); Renier (10[3]), (26); Kidston (11); Kidston and Jongmans (15–17); Jongmans (09), (11), (17), (28[2]), (28[3]); Koopmans (28); Leclercq (28). *Germany and other countries*: Von Bubnoff (26); Stur (85); Sterzel (07); Jongmans (09); Gothan (07[2]), (13), (23), (25[2]); Potonié (96), (21); Bode (27). *Asia Minor*: Zeiller (99). *Poland*: Rydzewski (13). *Russia*: Zalessky (07), (28[4]). *North America*: Lesquereux (79); White, D. (99), (04); Stopes (14); Willis, B. (22). *Siberia, China, Japan*, etc. (plant-beds probably Lower Permian in age, but it is not always possible to draw a definite line between the Stephanian and early Permian floras): Zalessky (28); Halle (27) with good bibliography, (27[2]); Yabe and Endo (21); Schenk (83); Yokoyama (08); Colani (19); Mathieu (21), (22); Kawasaki (27); Yabe and Oishi (28); Kon'no (29). For information on the Geology of China see Grabau (23–28), Gregory (25), Berkey and Morris (27), Barbour (29). *Sumatra and the Malay Peninsula*: Jongmans and Gothan (25); Posthumus (27), (29); Edwards (26[2]).

Permian

France: Zeiller (90), (92), (98), (06); Renault (96). *Germany*: Solms-Laubach (84); Potonié (93), (96); Sterzel (07); Lück (13); Gothan (23); Gothan and Nagalhard (22); Weigelt (28); Augusta (29); Florin (29[2]). *North America*: Sellards (00); White, D. (04); Schuchert (28) with references. *North Africa*: Carpentier (30[2]), (30[3]). *Kusnezk Flora* (fig. 78, K): Zeiller (96); Amalitzky (01); Zalessky (18), (27), (28[2]), (29), (29[2]), (29[3]). *China*, etc.: see above.

Gondwanaland

For a general account of the Glossopteris flora see Arber, E. A. N. (05), Seward (23), (29), (29[3]), du Toit (21), (26), Bertrand, P. (09).

India: Sahni (22), (26); Seward and Sahni (20); Seward (07[2]); Seward and Woodward (05); Zeiller (02); Wadia (19); Feistmantel (79[2]), (80), (81), (81[2]), (86). *Australia*: Feistmantel (90); Süssmilch and David (20); Walkom (22), (28[2]), (28[3]); Arber, E. A. N. (02). *Antarctica*: Seward (14); Edwards (28). *Africa*: du Toit (26); Arber, E. A. N. (10); Potonié (00); Gothan (14[2]); Zeiller (96[2]); v. Brehmer (14); Seward and Leslie (08); Seward (97[2]), (03); Kräusel and Range (28); Dixey (26); Walton (29[2]). *Madagascar*: Boule (08). *South America*: du Toit (27); White, D. (08); Zeiller (95); Lundquist (19); Keidel (22); Gothan (27); Kurtz (94). *Falkland Islands*: Halle (11); Baker (22); Seward and Walton (23).

CHAPTER XIII

THE FIRST PHASE OF THE MESOZOIC ERA: THE TRIASSIC PERIOD

And the desert shall rejoice, and blossom as the rose. Isaiah, xxxv, 1

THE Triassic period, which derives its name from the three well-marked series of strata assigned to it by German geologists, may be described as the opening chapter of a new world epoch.

Between the end of the Palaeozoic era, as represented by the Permian rocks, and the beginning of the Mesozoic era there is no well-defined interval in the succession of strata: the beds of the younger system are usually superposed in regular sequence on those of the older. During the earlier stages of the Triassic period the surface in some parts of the world was slowly rising: seas gave place to sandy plains studded with salt lakes and pools. There was no great geological revolution coincident in time with the boundary adopted by geologists between the two periods.

A New Era

Why then do we regard the Triassic period as the first act of a new drama? The history of the earth is in the main a connected story, not a record of recurrent cataclysms: in the organic as in the inorganic world changes occurred gradually. If, however, we compare the two eras from a broad point of view the differences between them, both in the geological conditions and in the organic kingdoms, are seen to be great enough to call for special emphasis. Mesozoic history is a record of relatively stable conditions; the strata of the successive periods, Triassic, Jurassic and Cretaceous, are for the most part sedimentary beds: we miss the recurrent manifestations of volcanic activity recorded in the widespread lavas and other igneous rocks of the Palaeozoic era. Many of the Palaeozoic rocks can be traced with little or no change over a wide area: the strata of the Mesozoic systems are less homogeneous and less persistent, and indicate rapidly shifting scenes within restricted areas. There have always been crustal oscillations varying within wide limits in duration and intensity. The most striking contrast between the two eras is presented by the floras and faunas: we are concerned only with the floras, and it is no exaggeration to describe the Mesozoic vegetation as poles asunder from that of the Palaeozoic era. If we compare in their broader features the forests of the Coal Age with those which it is possible to reconstruct from the fossils found in Upper Triassic and in Rhaetic beds, we find that they have little in common. The late Palaeozoic vegetation derived its characteristic features from floras composed in great part of extinct types, whereas among the plants of the late Triassic floras we at once recognize the precursors of more familiar and more modern forms. The preliminary steps in this transformation were taken in the latter part of the Palaeozoic era. We have already seen that far-reaching crustal disturbances began before the end of the Carboniferous period: these were the factors more directly concerned in setting the stage for the new era. Geological revolutions created new environments: the plant-world reacted to the play of forces in the inorganic world. The transition from the humid swamps of the Coal Age to the gradual desiccation that is clearly indicated by the history of the Permian and Triassic periods began in Permian time and reached its highest expression in the Triassic period. Geologists have fixed the upper limit of the Palaeozoic era at the end of the Permian period; but botanically the more natural position would seem to be nearer the middle of the Permian system. The meagre collections of plants obtained from the latest Palaeozoic rocks reveal the existence of a vegetation more closely linked with that of the earliest stages of the Mesozoic era than with the plant-world as it was in the first half of the Permian period.

The Triassic record resembles that of the Devonian period in being written in two sets of documents; one set proving the prevalence over wide areas in North America and Europe of salt-laden inland seas and deserts: another set recording a wide expanse of open water from southern Europe to the Himalayas, and on the Pacific coast of the western continent. Examples of the latter, or marine, phase are seen in upraised calcareous deposits which now form part of the Tyrolese Alps (fig. 82); from them we have discovered the existence in the Triassic sea of banks of Calcareous Algae. It is from the other set of documents that we follow the development of the new plant-world. The map reproduced as fig. 80 purports to give a general idea of the distribution of continents and oceans in the latter part of the Triassic period. It will be noticed that as in the Palaeozoic era the Tethys Sea stretched across the world: on

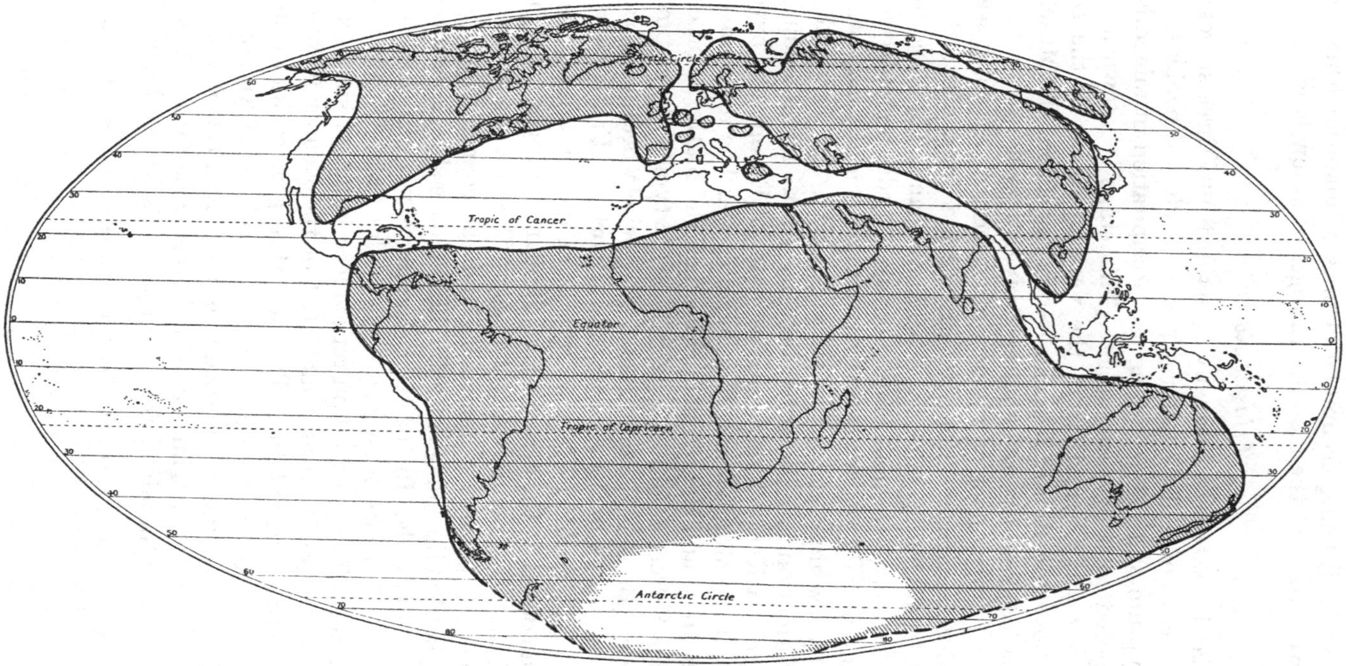

FIG. 80. Map of the world at the end of the Triassic period. The shaded areas are land. (Based on Arldt's map.)

its floor were still being accumulated piles of material which in a later era were to form the mountain ranges of a new land.

Triassic Rocks

The Triassic system in Germany is divided into three stages: at the base a series of shallow-water deposits, sands and pebbles, with beds of gypsum and salt recording the precipitation of minerals in water exposed to a dry atmosphere. This stage is known as the Bunter (*bunt* = variegated) from the varying colour of the sediments. Above the Bunter is a calcareous formation, the Muschelkalk, so-called from the abundance of bivalve shells: this material accumulated on the floor of a large, impoverished sea which was a northward extension of the more open Tethys from which it was separated from time to time by a barrier of land. The uppermost stage, or Keuper (a German mining term), records the replacement of the Muschelkalk sea by scattered salt lakes and sandy wastes relieved by swamps clothed with vegetation. In England the Triassic formation is represented only by strata corresponding to the Bunter and Keuper: the western border of Europe was beyond the limits of the Mid-Triassic (Muschelkalk) sea. The red sandstone of St Bees Head and the red pebble beds on the south Devon coast at Budleigh Salterton and at other places are typical developments of the older (Bunter) Triassic rocks. The pebble beds were transported by rivers from the ancient hills of northern France, and the finer red sands enable us to picture a scene comparable to that in the African desert at the present day. Similar conditions are recalled by the Keuper sediments: the salt-beds of Cheshire and Worcestershire suggest a Dead Sea landscape, and the coarser piles on the flanks of the Mendip and Quantock Hills tell us of shingle-beaches on the shores of islands. In the thick Keuper Marls of Leicestershire we have the sediments of an inland sea studded with islands of Pre-Cambrian rock. The photograph reproduced in fig. 81 shows a crag of Mountsorrel granite from which part of the Keuper mantle has been removed by denudation. The smoothed grooves and rounded ridges on the granite afford unmistakable evidence of wind-erosion, of sand storms lashing an exposed rock-face. As Prof. Watts[1] says, the picture gives us a glimpse of a "Triassic landscape". A few plants have been obtained from Keuper beds in the Eden valley

[1] Watts (03); Bosworth, T. O. (12). On Triassic desert conditions in England see also Goodchild (00); Greenly (94).

FIG. 81. Granitic rocks in Charnwood Forest, England, grooved and polished by Triassic sand storms. (Photograph by Mr P. W. Wright, supplied by Prof. W. W. Watts.)

and from Worcestershire: but it is mainly from extra-British localities that specimens of Triassic vegetation have been collected.

From eastern North America, as from portions of Europe, the open sea was excluded in the Triassic period. A region of salt lakes and arid scenery is illustrated by the red Newark series of conglomerates and sandstones which form a discontinuous band from Nova Scotia to South Carolina. There is abundant proof that over large areas in the northern hemisphere the climatic conditions were similar to those in the present steppes of Asia. But the abundance of Upper Triassic plants in Lower Austria, in the coalfield of Richmond in Virginia,[1] and in other places points to conditions favourable to the spread of forests: it is from these plant-beds that we discover the magnitude of the change in the facies of the vegetation subsequent to the closing stages of the Palaeozoic era. One of the most impressive proofs of the former existence of a Triassic forest of giant trees is afforded by the innumerable coniferous logs—the wood changed into agate, jasper and chalcedony—strewn over an area of many square miles in Arizona. Some of the trunks reach a length of 200 ft. and a diameter of 10 ft.[2]

The sedimentary rocks of the Keuper are overlain by the Rhaetic series which takes its name from the Rhaetian Alps. Rhaetic beds stretch across England from the Yorkshire coast to the coast of Dorset. The composition of the sediments and their fossils suggest comparison with muds that are now forming on the floor of the Black Sea. Strata of this age are widely developed in Germany and in southern Sweden and have furnished abundant samples of a vigorous vegetation. A rich flora almost identical with that from Swedish rocks has been described from Scoresby Sound in east Greenland. The Rhaetic stage[3] is intermediate between the Triassic and Jurassic systems: its sediments point to the invasion of the inland Keuper sea by the fresher water of the southern ocean, a marine transgression which in the early days of the succeeding Jurassic period gained in extent and duration.

On the continent of Gondwanaland freshwater Triassic strata in South America, South Africa, India, and Australia have furnished

[1] Fontaine (83); Ward (05).

[2] Ward, L. F. (00).

[3] By some geologists the Rhaetic stage is included in the Jurassic system; by others in the Triassic. After the marine transgression at the close of the Keuper stage, which is indicated by the Lower Rhaetic marine beds, there was a temporary return to the lagoons of the Keuper: it is the Rhaetic floras and rocks which illustrate the initiation of Jurassic vegetation and of Jurassic geographical conditions. Gignoux (26).

abundant plant remains. In some of the southern floras, notably the exceptionally rich flora of Tongking, residual elements from the Palaeozoic era are associated with the more modern types which make up the bulk of the vegetation. In the southern hemisphere the Palaeozoic and Mesozoic eras shade more gradually into one another than on the northern continents; there is not the same evidence of marked contrasts either in the inorganic or the organic worlds. The outstanding feature of the Keuper and Rhaetic floras to the north of the Tethys Sea is the eloquent testimony they bear to a world-wide change in the character of the vegetation.

Triassic Algae

The occurrence of two very different types of rock indicating sharply contrasted conditions of deposition is one of the outstanding features of the Triassic period in Europe and Asia. This contrast is reflected in the fossil plants. From the limestones and dolomites of the Tyrolese Alps and the Himalayas it is possible to form a general idea of the algal flora of the Triassic sea. Scattered through the uplifted rocks carved into the peaks and precipitous walls of the Dolomites are shattered masses of old coral reefs, of reefs made of calcareous seaweeds, and of algae which passed their lives attached to stones and shells near enough to the surface of the water to receive the radiant energy of the sun. The sloping strata of the Schlern dolomite on the face of the Fermeda Turm,[1] seen on the right-hand of fig. 82, are rich in the calcareous casings of *Diplopora* and other algae. Similarly on the steep flanks of the mountains on the left are exposed other layers of upraised masses of débris telling us of the plant-communities in the channels and bays of a Mid-Triassic ocean.

An entirely different scene is recalled by the comparatively barren sandstones in the Vosges and in other parts of Europe which have furnished fragmentary relics of terrestrial plants from the oases of an early Triassic desert. Passing to the higher members of the Triassic system we find in some regions beds of gypsum and salt and other evidence of the persistence of desert conditions, while in other places rocks of similar age demonstrate the occurrence of a vegetation far surpassing in luxuriance and variety that of the earlier part of the Triassic period.

It was during the latter part of the Permian period, particularly

[1] Gordon, M. Ogilvie (28).

in the northern hemisphere, that the plant-world entered upon a new phase. Reference has already been made to the geological revolution which began before the close of the Carboniferous period as an influential factor in creating a new environment. The forests of the Coal Age were replaced by arid lands and scattered patches of a sparse vegetation. At the end of the Westphalian stage some new

FIG. 82. The northern face of the Gersterospitzen in the Dolomite mountains. (Photograph supplied by Dr Ogilvie Gordon.)

genera appeared in the Stephanian flora and these, unlike the majority of their older associates, survived the hardships of the transition period and formed part of a small group of early Triassic plants. Before giving fuller consideration to the transition from the Palaeozoic to the oldest Mesozoic vegetation we will first pass in review some of the plants discovered in Triassic rocks.

The abundance of lime-encrusted algae in Triassic marine deposits is due to the enormously increased chance of preservation given to them by the resistant, calcified parts of their bodies: very little is known of contemporary seaweeds which were not thus protected. In the warmer seas of the present age there are about ten genera of a group of Calcareous Algae known as the Dasycladaceae, a family to which reference was made in Chapter VII. An example of an existing Mediterranean representative is shown in fig. 83. In this alga, *Dasycladus clavaeformis*, the main axis is an undivided tube, seldom more than 5 cm. tall, attached by thread-

FIG. 83. *Dasycladus clavaeformis*, a Calcareous alga from the Gulf of Naples. × 3. (Photograph of a specimen received from the Director of the Naples Marine Biological Station.)

like holdfasts to a rock or shell on the floor of a sea shallow enough to receive the necessary amount of light. The upper, light green, part is covered with a felt of narrow, branched tubes springing in regular whorls from the central axis: reproductive organs are borne at the ends of the main lateral branches. Another living genus of similar habit is *Neomeris*,[1] a tropical alga resembling a miniature sausage: some species grow between blocks of coral and are protected by a thick coating of lime against the buffeting of the waves. Members of the Dasycladaceae have already been described in the account of Devonian and Carboniferous floras; and still older repre-

[1] Church (95).

sentatives of this and allied families flourished in Ordovician seas. Dr Pia of Vienna, the leading authority on this group, has described twelve Triassic genera, and it is safe to assume that the actual number was much greater. Attention may be called to the following points: in Triassic seas the Dasycladaceae were more abundant and more varied than in the warm seas of to-day; they lived farther from the Tropics than their modern descendants. It is worthy of note that these Mid-Triassic Algae are valuable guides to the geologist[1] in the determination of the relative age of rocks: the species have a limited range in time and do not pass from one geological stage to another; moreover, they often occur in rocks which are poor in fossils. One of the commonest Triassic genera is *Diplopora*; the plants are usually rod-like, a few centimetres long; in some forms the central axis swells into a club-shaped apex, and in others the central stalk is contracted at regular intervals. As in modern genera, the main axis bore crowded clusters of slender branches: these are not often preserved, but rings of small holes on the calcareous stem indicate their position. In the living relatives the branches are freely subdivided, while in the Triassic species they are unbranched. *Diplopora* is abundant not only in the southern European Alps, but it is distributed through masses of limestone in central Germany, in Siberia, and other regions. A Tertiary specimen of an allied genus, *Dactylopora*, is shown in fig. 113, C. The rows of holes on the inner face of the broken tubular axis mark the position of the main lateral branches, while the smaller holes and cavities in the calcareous wall were left on the decay of finer branches and spore-capsules. *Dactylopora*, using the name in a strict sense, is known only from the older Tertiary floras of France and Belgium.[2]

Bearing in mind the delicate nature of many siphoneous algae and the disastrous consequences inseparable from the conversion of the marine oozes and reefs into the crystalline rocks of the Alps, we may confidently assert that the number of genera so far recorded is but a fraction of the whole algal flora of Triassic seas. The early Mesozoic forms far outnumbered the existing species. There are certain differences between the extinct and the living types, but the resemblances are more striking than the contrasts. These peculiarly constructed algae were probably evolved long before the advent of

[1] Pia (25^2).

[2] For references to descriptions of Triassic Algae see Pia in Hirmer (27); Pia (24), (25), (26).

the oldest terrestrial vegetation: species changed with comparative rapidity, some few persisted with little modification from one age to another, and the group as a whole seems to have reached its zenith, both in variety and geographical distribution, in the Triassic period (fig. 136) when many of the genera played a prominent part as rock-builders.

Diplopora may serve as an example of a company of several algae characterized by relatively large cylindrical stems of different form. Another type is represented by the genus *Sphaerocodium* (fig. 41, E), which cannot be positively assigned to a position in the Algae: it occurs as small, spherical or oval pieces of limestone up to 2 cm. in diameter in the alpine Trias, and consists of a mass of very small, forked tubes.

An Early Triassic Scene

As an introduction to the terrestrial vegetation of the earlier stages of the Triassic period we turn to the reconstruction reproduced in fig. 84. This attempt to picture a desert scene is based on material obtained from the sandstones of the Vosges country.[1] On the left, the broad and smooth grooves on the weathered granite scarp, suggested by the granite rocks of Charnwood Forest seen in fig. 81, show the effect of wind-driven sand from the billowy waste of dunes stretching to the horizon. The trees on the cliff, resembling in habit living araucarias and other familiar conifers, are reconstructed from branches and foliage-shoots of the extinct genera *Voltzia* and *Albertia*. The broad, yucca-like leaves (fig. 84, *Pl.*)[2] above the left-hand corner were probably borne on a tree comparable to *Cordaites* which reached its full development before the end of the Palaeozoic era. Below are fertile and sterile fronds of the fern *Neuropteridium* (*Nd.*); to the right the long pinnate fronds of a cycad (*Cy.*), and farther to the right another species of *Neuropteridium*. The two plants with cylindrical stems bearing a terminal tuft of narrow, spiky leaves are examples of the imperfectly known *Pleuromeia* (*Pa.*), a Triassic plant believed to be a descendant of the arborescent Palaeozoic lycopods:[3] the specimen higher on the

[1] Schimper and Mougeot (44); Fliche (10); see also Blanckenhorn (86); Frentzen (15), (20), (22).

[2] For an account of leaves of this type see Wills (10); Arber, E. A. N. (09^2); Schlüter and Schmidt (27).

[3] The reconstruction of *Pleuromeia* is based on a drawing by Walther (12).

FIG. 84. An early Triassic landscape. (Drawn by Mr Edward Vulliamy.) *Cy.* A cycad; *Eq. Equisetites*; *Nd. Neuropteridium*; *Pa. Pleuromeia*; *Pl. Pelourdea* (*Yuccites*); *Sz. Schizoneura*.

sandy slope shows a fertile apex above the drooping and moribund foliage. Bordering the lake on the left,

"A still salt pool, lock'd in with bars of sand,"

are a few plants of *Schizoneura* (*Sz.*), in habit recalling bamboos, but in their whorled leaves and jointed stems comparable with the Palaeozoic *Calamites* and the modern *Equiseta.* To the right is a group of large *Equiseta* (*Eq.*), apparently identical, except in their greater size and more robust stems, with the modern horsetails. A log of some conifer projects from the sand in the right-hand corner.

The earliest Triassic vegetation, as one would expect from the nature of the environment in which it existed, is represented by a small number of specimens collected in Europe from the Bunter beds of the Vosges, the Black Forest, and other localities.

For general purposes it is unnecessary to draw a distinction between the plants furnished by the lowest beds of the Triassic system—the Bunter series—and the few that are recorded from the Muschelkalk—the middle stage of the Triassic period. As will be shown later, this early Triassic flora differed widely from the much richer floras obtained from Upper Triassic, or Keuper, rocks. There is a marked contrast between the vegetation represented by the latest Permian and the earlier Triassic floras and that of the Keuper stage. On the other hand, there is little difference in the general features of the vegetation between the Keuper and Rhaetic floras. The plant-world changed in the course of the Triassic period: there was an apparently sudden increase in the development of fresh forms in the later stages which led to the production of a new type of vegetation which, with relatively minor fluctuations, persisted until the latter part of the Jurassic period.

It is worth while emphasizing the fact that the most natural boundary lines between periods in the history of floras do not necessarily correspond with those adopted by geologists in their classification of the rocks: it is, however, convenient to make use of the accepted time-divisions in our retrospect of the plant-world.

A. The Earlier Triassic Vegetation

Equisetales

One of the notable features of the earlier Triassic vegetation is the presence of large equisetaceous plants agreeing closely with existing horsetails in the external characters of the stem: sheaths, presumably formed by the fusion of whorls of narrow leaves, clasp the main axis and branches at regular intervals and each leaf-sheath is divided at the upper edge into pointed or blunt ends representing the free apical portions of the individual segments. Similarly there have been found in the Triassic strata cone-like bodies consisting of crowded appendages bearing spore-capsules and spores constructed on a plan essentially like that of the corresponding organs in living *Equiseta.* The great difference between the Triassic plants, spoken of as species of *Equisetites*, and their modern descendants is the much greater size of the extinct species. Of their internal structure practically nothing is known, but the probability is that the early Mesozoic forms had harder and more woody stems. Additional features of the Triassic plants which remind us of modern horsetails are the production by a few of the ancient species of underground tubers, closely resembling except in their larger size the food-storing tubers of some of our common species of *Equisetum*; also the possession of horizontal plates of tissue (diaphragms) at each joint (node) of the stem. The geographical distribution of some of the more characteristic species of early Triassic *Equisetites* is shown in Table C. (p. 332). We have already noticed the occurrence of stems generically identical with these early Mesozoic plants in the Westphalian stage of the Carboniferous period (p. 266).

Another member of the Equisetales characteristic of the earlier Triassic floras is *Schizoneura* (fig. 84, *Sz.*) which, as we have seen, flourished in Gondwanaland before the close of the Palaeozoic era and subsequently migrated far into the northern hemisphere. *Schizoneura* differs in no very important respects from the genus *Neocalamites*, a plant which became abundant in the later Triassic and in Rhaetic floras: the two are very closely allied and may perhaps be regarded as modified descendants of the Palaeozoic *Calamites.* A fuller comparison of these and other Triassic plants with members of the Palaeozoic floras is made in the latter part of this chapter.

Lycopodiales

Among the fossils from the Bunter and Muschelkalk series there are a few incomplete and unconvincing specimens preserved as casts and without any trace of internal structure, which have been described as survivals of the Palaeozoic *Lepidodendra* and allied genera. There are some cylindrical stems about 11 cm. in diameter, from Middle Triassic strata, the surface of which is covered with rhomboidal or almost square scars recalling the small pieces of a tessellated pavement: these were originally regarded as tree-ferns and called *Caulopteris tessellata*, but subsequently renamed *Lepidodendrites* in the belief that they are related to *Lepidodendron*. With them were found casts bearing scars similar to those on the ubiquitous stigmarias of the Coal Age. An early Triassic genus, *Lesangeana*, which is recorded also from Rhaetic rocks, may perhaps be another relic of the Palaeozoic *Lepidodendra*. An obscure fossil, named by Prof. Fliche[1] *Pœcilitostachys*, which appears to be part of a large cone, is most probably lycopodiaceous and allied to the cones (*Lepidostrobus*) of the older arborescent lycopods. Another early Triassic fossil which resembles the Palaeozoic *Lepidodendron* cones is *Lepidostrobus palaeotriassicus*.[2] One of the most characteristic and widely spread early Triassic plants is *Pleuromeia* (fig. 84, *Pa*), a genus of which little is known beyond the surface features of the stem and the forms of some of the reproductive appendages. It was originally assigned to *Sigillaria* and regarded as the solitary Mesozoic descendant of that genus. The first specimens were found in a piece of Triassic sandstone dislodged during a storm from the tower of Magdeburg cathedral.[3] The transversely elongated scars, resembling in shape and arrangement those of some sigillarias, are a distinctive character of the slightly tapered stems; but the resemblance is not complete. The lower end of the stem passes into blunt arms arranged cross-wise and bearing numerous scars of rootlets, a character suggesting comparison with the much shorter and more stumpy stems of a modern *Isoetes* (quillwort). No leaves have been found; only large sporangia which were attached to the upper part of the stem. *Pleuromeia* is exclusively Triassic and characteristic of the early Triassic floras: it may be described as one of the very few Mesozoic plants which form a link—how complete a link is not known—with the arborescent lycopodiaceous genera of the Palaeozoic era: it may also be related to the early Cretaceous genus *Nathorstiana*.[4]

Ferns

Among the few ferns discovered in Lower Triassic sediments *Neuropteridium* is the most characteristic genus. Both sterile and

[1] Fliche (10).
[2] Frentzen (20).
[3] Potonié (04); Walther (12); Kryshtofovich (23).
[4] Richter (09).

fertile fronds are known: the sterile leaves appear to have been simply pinnate, that is, they had an axis reaching a length of over 3 ft. which was unbranched and bore two rows of linear leaflets similar in form and venation to those of the Palaeozoic pteridosperm *Neuropteris* and the living *Osmunda regalis*. The fertile fronds had longer and much narrower leaflets bearing numerous sporangia, too imperfectly preserved to afford evidence of their affinity to those of living ferns. This genus is usually believed to be a true fern, comparable in its two kinds of fronds with the modern European and North American fern *Blechnum Spicant* (hard fern), but we do not know enough about it to refer it to a family position. Some pieces of petrified roots and a portion of the stem of a fern from Bunter beds of Germany have been described as *Psaronius triassicus*,[1] in the belief that they belong to a species of the genus which is characteristic of the later Carboniferous and early Permian floras. Our knowledge of early Triassic ferns is very meagre: impressions of branched fronds bearing leaflets of the *Cladophlebis* type are recorded from Bunter beds. The *Cladophlebis* form of leaflet, attached by the whole base, often slightly curved, with a central main vein giving off arched or almost straight, forked lateral veins, is shown in the Rhaetic species represented in fig. 88. There are also examples of a genus described as *Asterotheca*, or *Asterocarpus*, which is characterized by small groups of sporangia on its leaflets: its affinities are by no means certain. Some of the fern-like fronds from the older Triassic rocks may well belong to pteridosperms. Among probable representatives of the pteridosperms in the early Triassic floras of both the northern and southern hemispheres are fronds described as species of *Callipteridium*, a Stephanian and Lower Permian genus, and *Odontopteris*. A petrified stem from the Muschelkalk of Silesia described as *Knorripteris Jutieri*[2] is an example of a fern, represented by the stem, which differs too widely from any living genus to be assigned to a family position.

Cycads and Conifers

A few cycadean fronds, some referred to the genus *Zamites* because of the superficial resemblance of the comparatively long and pointed leaflets (pinnae) to those of the living genus *Zamia*, and others assigned to *Pterophyllum*, indicate the occurrence in the earlier Triassic flora of plants agreeing in their foliage with the few

[1] Frentzen (20). [2] Hirmer (27); Hörich (10).

Palaeozoic representatives of a group which became much more prominent in the later floras of the Triassic period.

Among Triassic conifers the genus *Voltzia* is the commonest and best known: its foliage-shoots are among the most abundant and widespread fossils in beds of the Bunter series. In habit *Voltzia* resembled some existing araucarias, such as *Araucaria excelsa* (Norfolk Island pine); the more slender branches bore short, spirally disposed leaves like those of the shorter-leaved species of *Araucaria* and others that are straighter and longer, a difference in leaf-form that is characteristic of the genus. Both male and female cones are known, the male only as imperfectly preserved fossils. The female shoots are rather lax cones made up of broadly triangular scales with, usually, a five-lobed upper margin: each scale is double, a lower narrow scale and an upper broader scale bearing three seeds.[1] This double nature of the cone-scales is significant as indicating a possible relationship to conifers of the family Abietineae, such as pines, larches, firs, and cedars. On the other hand, the occurrence of three seeds on the *Voltzia* scales is a distinguishing feature, and the structure of the wood of the Triassic genus is of the araucarian and not the abietinean type. Like many other extinct genera, *Voltzia* seems to be a generalized type, that is, it exhibits points of contact with more than one living plant. Another and more imperfectly known genus is *Albertia*, distinguished by its broader leaves similar to those of some modern species of *Agathis*, e.g. *Agathis australis*, the kauri pine of New Zealand. Reference has already been made to a remarkable occurrence of the remains of a Triassic forest in Arizona, though in age it is probably rather later than the European rocks assigned to the Bunter series. Among the trunks scattered over the Arizona desert are some with well-marked rings of growth and a structure practically identical with that of the araucarias and *Agathis* of the present day.

There is abundant evidence in early Triassic floras of the northern hemisphere of the existence of trees with araucarian wood, though it must be admitted no typical araucarian cones have so far been discovered north of the equator in Triassic rocks. On the other hand, a cone-scale bearing a single seed, in size and form practically identical with the cone-scales of some living araucarias, has been described from Permian rocks of France. All we can say is that trees closely resembling the southern hemisphere araucarias in the

[1] Walton (28[2]). See also Wills (10); Schlüter and Schmidt (27).

microscopical characters of the wood were abundant in Triassic floras: some of them differ from existing conifers in the structure of their cones and appear to be extinct types combining in one single individual characters which suggest that they may be generalized forms whose descendants developed along separate, and possibly divergent lines. There have recently been described petrified stems and cones, both characterized by araucarian features, from a fossil forest in Patagonia[1] said to be Middle Triassic in age. The seed-bearing cones, which closely resemble some Jurassic species, are similar to those of the living *Araucaria Cookii* and allied species. The seeds are described as having structural features intermediate between those of *Araucaria* and abietineous conifers. Petrified wood from Lower Triassic rocks has been recorded under the name *Palaeotaxodioxylon*; it differs in certain microscopical features from araucarian wood, but its affinities are uncertain. Other conifers characteristic of later Triassic floras are briefly described later.

B. The Later Triassic and the Rhaetic Floras

Representatives of the Ginkgoales are recorded from many parts of the world in Upper Triassic and Rhaetic rocks, also from South Africa and Australia in beds which are probably equivalent in age to Lower Triassic strata in Europe. There is no doubt of the existence of plants in the still older Permian floras which are usually regarded as pioneers of the class, which by the latter part of the Triassic period had become prominent in the vegetation of both hemispheres.

The vegetation in the northern hemisphere entered upon a new phase in the latter part of the Triassic period: the scanty floras represented by the material collected from the Bunter and Muschelkalk stages seem suddenly to become much more varied: many new types are recognized; and though desert conditions persisted in some regions, there were tracts of country tenanted by a vegetation comparable in the variety of genera and in its general richness with the luxuriant forests of the Coal Age. One of the richer Keuper floras is that of Virginia in the eastern part of the United States, but the descriptions of the plants, published many years ago, are now inadequate, and it is hoped that the abundant material that has been collected from the Richmond coalfield will be re-examined

[1] Wieland (29); Gothan (25).

by modern methods. Another Keuper flora has been obtained from Switzerland and others from Germany, Austria and elsewhere. Passing to the Rhaetic series, we find still more widely spread and more thoroughly investigated floras which may conveniently be considered with the closely allied Keuper floras.

Rhaetic Greenland

The reconstruction shown in fig. 85 purporting to represent a scene in east Greenland is based on fossils collected by Dr Hartz of Copenhagen and in greater abundance by Dr T. M. Harris of Cambridge from the shales and sandstones of a regular series of Rhaetic sediments reaching a thickness of about 90 metres. The beds were probably deposited by a river from the interior of Greenland in the neighbourhood of Scoresby Sound between lat. 70° and 71° N. Some of the specimens were collected from rocks exposed on the face of the cliff shown in fig. 86. In the foreground of the estuary on the left the tall tree is the conifer *Stachyotaxus* (*St.*) with foliage-shoots similar to those of the yew, but differing from any known genus in the rows of seeds attached to the relatively slender fertile branches. Behind the *Stachyotaxus*, near the upper part of the trunk, is a branch of *Podozamites* (*Pz.*), probably an extinct conifer, with much larger leaves in two ranks and clusters of smaller, fertile leaves. Other conifers, resembling in habit the bald cypress of North America (cf. fig. 115), are represented as growing in shallow water on the right. Across the left-hand corner is a stem of *Neocalamites*[1] (*Nc.*), an equisetaceous genus which was widely distributed in Rhaetic floras. Other equisetaceous plants, closely resembling living *Equiseta*, are seen in the middle of the picture by the water's edge. Prominent among the trees were species of *Ginkgoites* (*Gk.*), one of which is shown at the right-hand upper corner: another member of the same class, *Baiera* (*B.*), with large fan-shaped leaves, is seen in the left-hand corner. The cone hanging from a branch over one of the boughs of the Ginkgo-like tree in the right-hand upper corner is *Lycostrobus* (*Ls.*), belonging to a Rhaetic lycopodiaceous plant probably allied to the Palaeozoic *Lepidodendron*. In the middle of the picture is a group of cycadophytes in habit resembling palms and including *Nilssonia*[2], *Pterophyllum*, and other genera. To the right of the central group is another member of the Cycadophyta, the genus *Wielandiella* (below *Wa.*), characterized

[1] For reconstruction, see Berry (18^2): see also Berry (12). [2] Nathorst (09).

FIG. 85. A scene in east Greenland in the Rhaetic Age. Based in part on material obtained from the rocks exposed in the cliff shown in fig. 86. (Drawn by Mr Edward Vulliamy.) *B. Baiera*; *D. Dictyophyllum*; *Gk. Ginkgoites*; *Gs. Glossopteris*; *Lc. Laccopteris*; [illegible] *Pz. Podozamites*; *Ss. Sagenopteris*; *St. Stachyo-*

by its forked branches and smaller fronds. The smaller trees on the right in the middle distance, with relatively large fronds that are superficially fern-like, are examples of *Lepidopteris* (*Lp.*), a pteridosperm and one of the few well-authenticated survivals in Mesozoic floras of that vigorous Palaeozoic class. Another probable survival from the Palaeozoic era is shown as a branch below the *Ginkgoites* on the right, a plant (*Pa.*) bearing long ribbon-like leaves like those of *Cordaites*.

Special attention is called to the plant in the foreground stretching obliquely over the water and distinguished by its four-lobed leaves and leafless fertile branches bearing berry-like fruits: this is the genus *Sagenopteris* (fig. 85, *Ss.*). The name *Sagenopteris* is applied to small leaves usually with four, narrow or oval leaflets resembling the leaves of *Marsilia*, one of the existing water-ferns, and, superficially, not unlike those of an *Oxalis*, though three or four times as large. Each leaflet has a more or less prominent central vein and a network of finer veins (a character implied by the generic name).[1] The position in the plant-kingdom of this widely distributed Upper Triassic or Jurassic genus has long been in doubt: it has, however, recently been shown by Dr Hamshaw Thomas[2] that *Sagenopteris* leaves are in all probability the foliage of plants which bore reproductive organs indicating affinity to the flowering plants. A description of the fertile shoots is given in the next chapter (p. 366). At the lower right-hand corner are the large and reticulately veined leaves of a plant which may be a late survival of the Palaeozoic genus *Glossopteris* (*Gs.*). Prominent among the ferns are those with tall fronds divided at the summit of a long stalk into spreading arms: these include *Laccopteris* (*Lc.*) and allied genera: others, such as *Dictyophyllum* (*D.*), have deeply cut leaves resembling those of the living tropical fern *Dipteris* (fig. 95). These are but a few of the plants which have been found in the Rhaetic beds of Greenland. This partial reconstruction of a flora, the remains of which include at least 200 species, enables us to realize the amazing contrast between the past and the present: the cliffs seen in fig. 86 form the edge of a plateau on which a comparatively small number of Arctic plants now complete their year's growth in the course of a short summer season, the shortness of which is in some degree compensated by a continuous supply of solar energy. In the Rhaetic period the vegetation which left scattered samples in the sedimentary

[1] σαγήνη, a net.

[2] Thomas, H. H. (25).

FIG. 86. Neill's Cliffs. Scoresby Sound, Greenland. The foreground is frozen sea with a covering of snow. (From a photograph, taken in April, 1927, by Dr T. M. Harris.)

strata of which the cliffs are partially made was in every respect as luxuriant and vigorous as that which flourished at the same time 1000 miles farther south in the region that is now southern Sweden. Ferns seem to have been the most abundant plants and the flora was very rich in members of the Ginkgoales. We will now look more closely at some of the Keuper and Rhaetic plants.

Equisetales, Lycopods, and Ferns

The genus *Neocalamites*[1] (fig. 85, *Nc.*), its jointed stems bearing whorls of long and narrow leaves, separately attached and not laterally united, recalls the Palaeozoic *Calamites*, and the Far Eastern type *Lobatannularia*, which is probably generically identical with a plant described from the Rhaetic beds of Tongking as *Annulariopsis*: in *Neocalamites* we have a link between the Palaeozoic Equisetales and those of the Mesozoic era, possibly also a Rhaetic successor of the early Triassic and late Palaeozoic *Schizoneura*. With *Neocalamites* were several species of *Equisetites*, some larger than any existing horsetails.

The most interesting representative of the Lycopodiales is the cone *Lycostrobus* (fig. 85, *Ls.*), the only example among Mesozoic plants of a fertile shoot, which in its structure and large size is strongly suggestive of a direct connexion with the arborescent lycopods of the Palaeozoic forests. *Lycostrobus*[2] was discovered in Rhaetic beds in Scania and well-preserved spores have since been found in east Greenland. Nothing is known of the stem or foliage-shoots. Species of *Lycopodites* from Sweden afford a connecting-link between the herbaceous lycopods of the late Palaeozoic floras and those of Jurassic and even present-day species.

A noteworthy feature of the Keuper-Rhaetic floras is the introduction into the world's vegetation of several new ferns: these are the earliest examples of a race which continued to flourish in many parts of the world during the Jurassic period and subsequently became reduced in numbers and restricted to tropical and subtropical countries south of the equator. These ferns afford no evidence of any direct connexion with Palaeozoic types; they are a striking illustration of the modern aspect of some of the characteristic members of the early Mesozoic vegetation. Among the new introductions are *Dictyophyllum*, *Thaumatopteris*, *Hausmannia*, *Clathropteris* and *Laccopteris*. The oldest is *Dictyophyllum*[3] (fig. 85,

[1] Halle (08); Berry (18[2]). [2] Nathorst (08). [3] Nathorst (06).

D.), which is recorded from Keuper beds and is said to have been found in Yunnan in association with Permian plants: the fronds are similar in form and size to those of *Dipteris*, a genus which now occurs in northern India, Malaya, New Guinea, the Philippines, and some other southern lands. The sporangia have a well-developed annulus and differ but little except in their larger size from those of the living genus. Fronds of similar habit, in which the radially divergent branches reach a length of over 30 cm. and are distinguished by the more rectangular meshes formed by the veins, are included in the genus *Clathropteris*,[1] which is represented in Keuper or Rhaetic floras in Virginia, Sweden, Tongking, Chile, and elsewhere. A smaller fern, *Hausmannia*, which occurs in Rhaetic and later floras, agrees even more closely than the other genera with living species of *Dipteris*. In *Laccopteris*, which first appears in Keuper rocks, we have a widely spread Mesozoic genus nearly related to the Malayan *Matonia* (fig. 94). Some of the distinguishing features of *Laccopteris* are illustrated by the drawings (fig. 85, *Lc.*; fig. 87) of *Laccopteris groenlandica*, a Rhaetic species recently discovered in Greenland by Dr Harris. The frond is similar in habit to *Matonia*, but the leaflets are rather longer and narrower in the extinct fern. The venation of sterile and fertile leaflets is seen in fig. 87, A and E: the fertile segments bear circular groups of sporangia (C) with a clearly defined annulus (B). A single spore is shown in fig. 87, D. This species differs only in minor characters from *Laccopteris Muensteri*, a well-known Rhaetic form, and from the common Jurassic species, *L. elegans*.

It is a remarkable fact that these and other Keuper and Rhaetic ferns are representatives of what has been called the *Matonia-Dipteris* alliance,[2] a branch of the fern group which reached its maximum development as an almost cosmopolitan company in the earlier floras of the Mesozoic era. At the present day the two genera *Matonia* and *Dipteris* and a few other allied ferns, such as the Malayan *Cheiropleuria* (fig. 96), are represented by very few species and have a restricted geographical range in the southern Tropics, thus affording a marked contrast to the former occurrence and variety of closely allied genera within the Arctic Circle. A similar contrast between the past and the present is furnished by the genus *Gleichenites*, recorded from some of the Keuper floras of Europe and from the Rhaetic rocks of east Greenland: this fern is almost

[1] Nathorst (06^2). [2] Bower (23–28).

identical in the form of its fronds with some species of *Gleichenia* (fig. 26) which is now one of the commonest members of tropical floras south of the equator. Reference has previously been made to the occurrence in the late Palaeozoic flora of Russia of petrified stems affording evidence of osmundaceous affinities: the family Osmundaceae, of which the royal fern, *Osmunda regalis*, is an almost cosmopolitan and the best known member, was widely distributed in the later Triassic and Rhaetic floras. The species *Todites Williamsoni* and allied forms, characterized by large bipinnate fronds

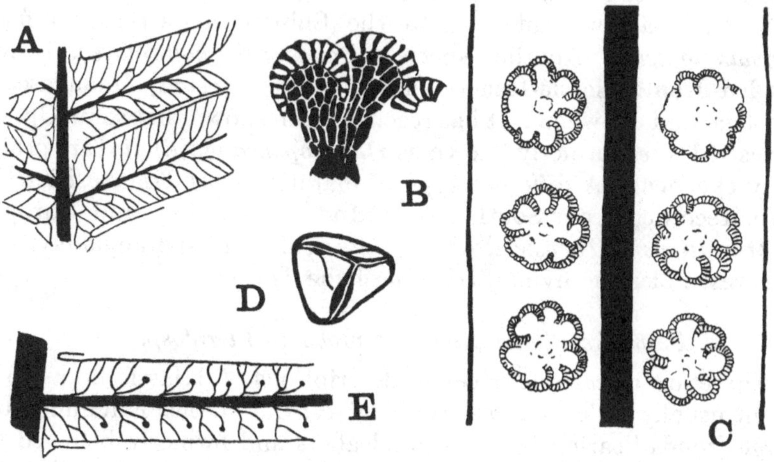

FIG. 87. Leaflets and sporangia of *Laccopteris groenlandica*; Rhaetic beds of Scoresby Sound. A. Pieces of leaflets (pinnules) showing venation. × 4. B. Sporangia showing the annulus and some of the other cells of the wall. × 30. C. Groups of sporangia (sori) on the lower face of a pinnule. × 10. D. A spore. × 250. E. Venation of a fertile leaflet. × 4. (After T. M. Harris.)

bearing closely set branches (pinnae) on a thick main axis with short slightly curved leaflets, often covered with sporangia like those of *Osmunda* and its companion genus *Todea*, was a member of Upper Triassic or Rhaetic floras in many parts of the world as far apart as Tongking, Australia and Greenland. Other ferns of the same geological age, which may be regarded as near relatives of the existing Osmundaceae, are often assigned to the genus *Cladophlebis*, e.g. *Cladophlebis nebbensis* and *C. Roesserti*. The latter species, which had a very wide geographical range, is well represented in the Rhaetic flora of east Greenland: the habit of the frond is illustrated in fig. 88 A, and the typical form and venation of *Cladophlebis* leaflets are shown in fig. 88 B, C and D. A Keuper-Rhaetic fern, distinguished by its partially coherent leaflets on lobed linear pinnae, though not referable on such data as are available to any family, is the genus *Bernoullia*.[1] One of the oldest families of living ferns, the Marattiaceae, which is definitely foreshadowed in some of the later Palaeozoic genera, is represented in Keuper and Rhaetic floras by species included in the genus *Danaeopsis*, so called because they were believed to be most nearly allied to the living *Danaea*. We now know that one of these fossil species, *Danaeopsis fecunda*,[2] agrees most closely in the structure of its sporangia, borne on the long and narrow leaflets, with the Chinese marattiaceous fern *Archangiopteris*. Another species, *Danaeopsis marantacea*, is recorded from Triassic-Rhaetic floras in Switzerland, Madagascar, Austria, and elsewhere. It has recently been shown that some Upper Triassic leaves formerly known as *Danaeopsis angustifolia*[3] are probably cycadean. A different type of marattiaceous Triassic-Rhaetic fern, recorded from South Africa, India, Tongking, and Europe, is *Marattiopsis Muensteri*, which in its coalescent sporangia (synangia) resembles the living *Marattia* of the Tropics.

Fern-like Plants which are probably Pteridosperms

There are many references in descriptions of Triassic floras to a plant usually called *Danaeopsis Hughesi*: this species is founded on large fronds bearing broad linear leaflets and formerly believed to be allied to *Danaea*. It has already been stated that Prof. Halle substituted for *Danaeopsis* the generic name *Protoblechnum* and that, more recently, Dr White suggested as more appropriate the designation *Glenopteris*. This Triassic plant, of which we know only the fronds, is no doubt a very near relative of the widely spread early Mesozoic genus *Thinnfeldia*; and both are almost certainly pteridosperms.

We may next consider certain early Mesozoic plants represented by fronds agreeing in external form with those of ferns but differing essentially in some other respects. The genus *Thinnfeldia* is the commonest of these: in it for present purposes may be included species from Gondwanaland which some authors prefer to regard, and perhaps with good reason, as a distinct genus *Dicroidium*.[4] *Thinnfeldia*, in the wide sense, stands for fronds varying much in size, characterized by a stout axis which is either simple or forked,

[1] Frentzen (26).
[2] Halle (21).
[3] von Lilienstein (28).
[4] Antevs (14).

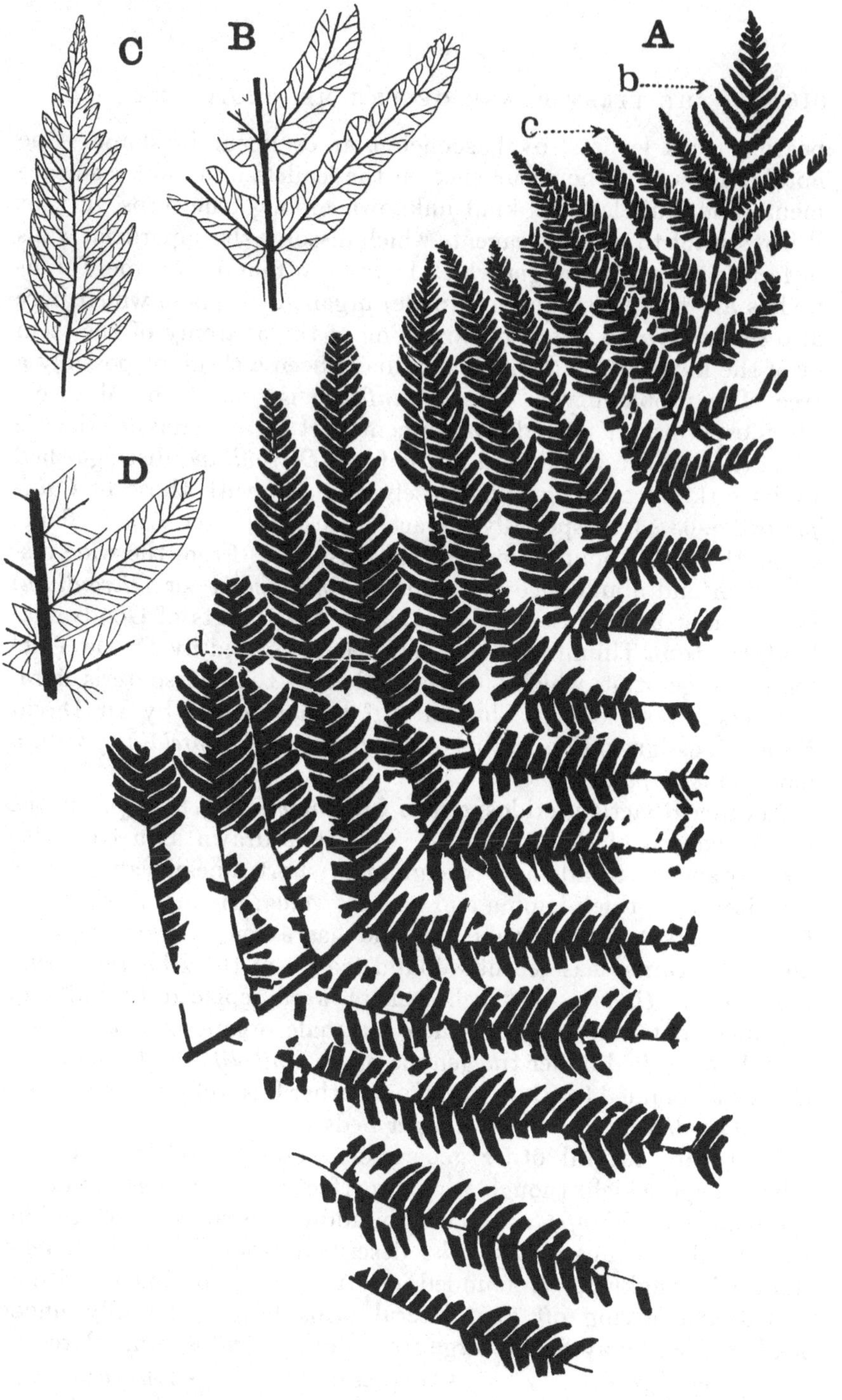

FIG. 88. Part of a frond of *Cladophlebis Roesserti* from the Rhaetic beds of Scoresby Sound. A. Upper part of a bipinnate frond. ½ nat. size. B–D. Leaflets enlarged to show the veins. × 2. (After T. M. Harris.)

bearing thick leaflets, oval, semicircular or linear in shape. One notable feature is the occurrence on the leaflets of an outer surface membrane (cuticle) of a kind unknown among true ferns, in that it is resistant to certain reagents which dissolve the superficial walls of fern leaflets. Sporangia which have been found on some Thinnfeldias are the pollen-bearing (male) organs of a plant which bore also seeds. We do not know anything of the anatomy of the stem or of the habit of the plant; it may have been a shrub or possibly a tree. The probability is that *Thinnfeldia* is one of the Mesozoic pteridosperms. It played a prominent part in the Triassic-Rhaetic vegetation of the world, and in the form *Dicroidium*, distinguished by its forked fronds agreeing closely in habit with those of *Callipteris* (fig. 63 A), is especially characteristic of the later Triassic floras of South Africa, South America and Australia. From the wide distribution and abundance of fronds of *Thinnfeldia* (or *Dicroidium*) in the plant-beds of South Africa, as in other parts of Gondwana-land, the term Thinnfeldia flora has been proposed by Dr Du Toit[1] for the vegetation which succeeded that of the Glossopteris flora. Reference was made in Chapter XII[2] to a proposal by Dr David White to assign various forms of frond of the *Thinnfeldia* type to a new genus *Supaia*.

Another Rhaetic pteridosperm is *Lepidopteris* (fig. 85, *Lp.*); in the reconstruction of the Greenland scene it is drawn as a tree with large branched fronds and triangular or short, linear leaflets characterized by a thick lamina and obscure veins: the main axis of the fronds bears numerous tubercles and has a rough, warty appearance. Dr Harris has recently found fronds of the Rhaetic species *Lepidopteris Ottonis* in which leaflets were replaced by tufts of capsules containing spores which are male organs of a type previously described under the generic name *Antholithus*. *Lepidopteris Ottonis* is confined to Rhaetic rocks: another less well-known species, *L. stuttgardiensis*, occurs in Keuper beds.

There are several other genera characteristic of the Keuper-Rhaetic age which, though very imperfectly known, bear some resemblance to *Thinnfeldia* and may be additional examples of pteridosperms. The genus *Ctenopteris*, described from the rich Rhaetic flora of Tongking,[3] is founded on frond-like specimens with a broad axis giving off long lateral arms bearing broadly linear leaflets: these may be the large fronds of a pteridosperm related to

[1] Du Toit (27²). [2] See p. 250. [3] Zeiller (03).

Thinnfeldia and to other Mesozoic genera, all of which have thick leaflets and none of them afford any convincing evidence of relationship to ferns. A simple type of frond, long, narrow and unbranched, with short and often broad leaflets, is illustrated by the Rhaetic genus *Ptilozamites*, which was particularly abundant in the flora of southern Sweden. Fronds generally resembling *Thinnfeldia* in habit and in the microscopical structure of the superficial layer, described under the name *Stenopteris*[1] and widely distributed in Upper Triassic and Rhaetic beds, may be included in the Thinnfeldia alliance.

Cycadophyta; Ginkgoales, etc.

Passing from the older to the younger Triassic floras and into the Rhaetic age we notice a large increase in the number and variety of cycadean plants. The few Stephanian, Permian, and Lower Triassic fronds which are usually considered to be cycadean are succeeded in Keuper floras by several new forms. The genus *Sphenozamites*, so called because of the wedge-shaped form of the pinnae which in size remind one of those on the leaves of some living species of the American cycad *Zamia*, occurs first in an early Permian flora of France; it peristed through the Triassic to the latter part of the Jurassic period. Some of the largest species are examples from Upper Triassic (Keuper) beds of Virginia: in one species the frond was at least a metre in length and the elongate wedge-shaped segments reached a length of nearly 30 cm. and at the rounded apex a breadth of 12 cm. Some of the cycadean fronds have long and narrow segments similar in shape to those of *Zamites* and *Pterophyllum* from Palaeozoic and Bunter floras, but they differ in venation and are assigned to the genus *Pseudoctenis*; others conform in the characters of the frond to the Palaeozoic *Pterophyllum*. A different type of leaf is illustrated by the genus *Otozamites* with its two-ranked segments attached to the upper face of the axis and more or less auriculate at the base: this form of frond became much more abundant in the Jurassic period. Similarly the genus *Nilssonia*, distinguished by its relatively broad simple fronds, sometimes with a lamina practically entire and not cut into segments, in other species with the leaf-surface divided into broad or narrow pieces, appears for the first time in Rhaetic floras.

Our knowledge of Triassic cycads is based mainly on fronds which

[1] Du Toit (27[2]); Walkom (17).

agree generally in habit with those of living cycads. The Rhaetic genus *Wielandiella* is a notable exception: it was probably a shrubby plant (shown in fig. 85 below *Wa.* to the right of the group of taller plants in the middle of the picture); the stem was regularly forked into divergent arms, bearing relatively small leaves similar in form to those of some of the smaller species of *Nilssonia* and referred to the genus *Anomozamites*. In the angles of the bifurcate stem were short, oval reproductive shoots producing near the base a circle of small male organs and on the more swollen upper part of the axis a layer of short appendages, some consisting of contiguous sterile scales, and others, fewer in number, of stalked female organs (ovules). These reproductive shoots, or flowers, agree in essentials with those of the numerous examples of the Cycadophyta which were among the most conspicuous plants in the late Jurassic and early Cretaceous floras. A fuller consideration of these Mesozoic plants, usually spoken of as cycads, though in their reproductive organs poles asunder from the modern genera, is reserved for the next chapter.

Among the few fossil plants discovered in early Mesozoic rocks which resemble in their reproductive organs existing members of the Cycadaceae (i.e. the family of the Cycadophyta to which the living cycads belong) is the genus *Dioonitocarpidium* recently described by Dr von Lilienstein[1] from Keuper beds in Germany. The fertile leaf consists of an axis bearing two rows of narrow and slightly sickle-shaped appendages, or small leaflets, and near the base of the axis two oval bodies about 1 cm. long which are presumably seeds. These seeds are similar to some described by Nathorst from Rhaetic beds in southern Sweden. The sterile leaves are of the *Taeniopteris* form, that is, single and ribbon-like, and were originally named *Danaeopsis angustifolia*. In the restoration given in the paper from which these facts are taken the plant is represented as bearing sterile leaves resembling the foliage of the Jurassic and Wealden *Williamsoniella*, and higher on the stem a cluster of pinnate seed-bearing leaves. The Keuper genus may be compared with the gymnosperm (conifer) *Podozamites* (fig. 85, *Pz.*) characterized by its linear leaves and lax cones composed of fertile leafy scales unlike those of any living plant; it would seem to be an extinct member of the Cycadophyta, resembling in its female organs the existing genus *Cycas* much more closely than it resembles the usual Mesozoic

[1] von Lilienstein (28).

cycadophytes of the group Bennettitales. Some obscure fertile organs similar in shape to the seed-bearing leaves of *Cycas* have been described under the name *Haitingeria*[1] from Keuper beds near Lunz; but these and other fossils referred to the genus *Cycadospadix* do not furnish proof of close relationship to the fertile leaves of living cycads.

The important point is that the Triassic-Rhaetic fronds described as cycads afford little evidence of the degree of relationship of the parent plants to the small band of cycads, most of which are tropical, that now represent the class Cycadophyta. Such information as is available on the reproductive organs of the earliest Mesozoic cycads, particularly those of *Wielandiella*, enables us definitely to state that although the various fronds—such as *Zamites*, *Pterophyllum*, *Nilssonia*, and others—agree generally in external form with those of existing genera, the plants which bore them differed widely in their methods of reproduction and in certain instances, e.g. *Wielandiella*, in a sharply contrasted form of stem. As we shall see later, the fossil plants which it is customary to call cycads cannot be included with living genera in one family; they are examples of an extinct group, the Bennettitales, which played a conspicuous part in the vegetation of the world in the Upper Triassic, the Rhaetic and the Jurassic periods, and in the early part of the Cretaceous period.

It was in the Triassic-Rhaetic floras that the class Ginkgoales first gained a strong position. Now there is one solitary survivor, the maidenhair tree (fig. 31): in the older Mesozoic floras there were trees with leaves agreeing closely in form and in the structure of their superficial membranes (cuticles) with those of *Ginkgo biloba*: they resemble also the foliage of the Permian genus *Saportaea* (fig. 67). These Ginkgo-like leaves, which are included in the genus *Ginkgoites* (fig. 85, *Gk.*), occur in Upper Triassic or Rhaetic strata in South Africa, Australia, South America, Sweden and east Greenland. Leaves included in the genus *Baiera* (fig. 85, *B.*), a type already in existence in the early Permian vegetation, afford additional evidence of the former abundance of the class: they differ from *Ginkgoites* in attaining, on the average, larger dimensions, in having a lamina deeply cut into long linear segments and in some, if not in all species in having no well-defined stalk. It is not always easy to draw a definite line between *Ginkgoites* and *Baiera*; one type of leaf shades into another. In some Baieras, notably in *Baiera specta-*

[1] Krasser (19).

bilis, a Rhaetic species, the lamina reaches a length of nearly 30 cm. and is without a stalk. In *Czekanowskia*, characterized by much narrower, simple, or repeatedly branched leaves borne in tufts clasped at the base by several small overlapping scales, we have another illustration of the range in form displayed by the extinct ginkgoalean plants. *Czekanowskia* is recorded from Rhaetic beds in Greenland and Sweden, but it did not become prominent until the Jurassic period.

Reference has already been made to some early Triassic conifers such as *Voltzia* and others: in the Keuper and Rhaetic floras the class was represented by several new forms, practically all of which differ in important respects, except in many instances in the structure of the wood, from all living conifers. The genus *Stachyotaxus*[1] (fig. 85, *St.*) is of special interest; it appears to be confined to Rhaetic floras and played a prominent part in the forests of east Greenland and southern Sweden. In its two kinds of foliage shoots, some with two-ranked leaves like those of the yew, others with small scale-leaves, it resembles some living species of the southern tropical conifer *Dacrydium*. The reproductive shoots bore small fertile organs in spikes, each consisting of a short stalk expanded into a triangular scale bearing two seeds attached to a basal cup. *Stachyotaxus* may be an extinct member of the *Dacrydium* family (Podocarpineae), which includes the commonest conifers in the southern hemisphere; or it may be more closely allied to the existing genus *Cephalotaxus*. In this connexion it is noteworthy that evidence has recently been obtained of the occurrence in Permian rocks of a conifer, renamed *Archaeopodocarpus*,[2] which is considered to be a still older representative of the Podocarpineae. Another Rhaetic genus, *Palissya*, with foliage-shoots of the yew form, is distinguished from *Stachyotaxus* by the structure of the narrow, cylindrical cones made up of cone-scales bearing 5–6 seeds in a cup: its affinity to modern genera is not clear, but it too may be related to the Podocarpineae.

Another gymnosperm—probably a member of the Coniferales—which first appeared in the Triassic period and became world-wide in range during the Jurassic period is *Podozamites* (fig. 85, *Pz.*): this is recognized by the foliage-shoot bearing long linear leaves very like those of some living species of the southern conifer *Agathis*. The fertile shoots bore clusters, or loose cones, of much shorter leaves

[1] Nathorst (08[2]).

[2] See p. 277.

with two seeds attached to a small scale near the base. It is a common and inevitable practice to describe foliage-shoots and associated reproductive organs under different names if there is no actual proof of original connexion: the seed-bearing scales found in company with *Podozamites* branches are known as *Cycadocarpidium.* Though in its long and narrow leaves, spirally attached to the stem, *Podozamites* agrees closely with existing species of the conifer *Agathis*, the nature of the seed-bearing organs (*Cycadocarpidium*) suggests a possible relationship to modern cycads. On the whole this common Mesozoic gymnosperm would seem to be more closely allied to conifers than to cycads; it may be one of many generalized types which, if we knew more about them, might enable us to trace to a common stock groups that now seem far apart.

Rhexoxylon, *a Triassic Stem with the Structural Features of a Modern Liane*

One of the most remarkable Triassic genera is *Rhexoxylon,*[1] known only as large petrified stems from several localities in South Africa, Nyasaland, and in Antarctica where it was found in a block of sandstone on the Beardmore Glacier. The adult stem has a large pith, containing small conducting strands and secretory canals, surrounded by separate masses of wood (fig. 89, A) instead of the continuous homogeneous woody cylinder of modern conifers. Some of the stems reach a diameter of 25 cm. and a length of 10 ft. *Rhexoxylon* stems bear a striking resemblance in the arrangement of the wood in detached masses embedded in softer tissue to those of some tropical climbing plants, e.g. *Tetrapteris* (fig. 89, B), belonging to different families of flowering plants; but this resemblance does not indicate actual affinity. Fig. 89, A, is a diagrammatic sketch of a cross-section of a *Rhexoxylon* stem showing the breaking up of the wood into separate masses by the growth of the softer tissue which forms the ground mass: a similar discontinuity in the wood of a modern tropical climber is seen in fig. 89, B. The genus is characteristic of the earlier Triassic floras of Africa and is important as evidence of the occurrence in an extinct gymnosperm of a type of construction that is now a distinguishing feature of tropical, dicotyledonous lianes. It is very unlikely that *Rhexoxylon* was a climbing plant: the substitution of separate wedge-shaped plates of wood for the usual continuous cylinder which is now characteristic

[1] Walton (23[2]).

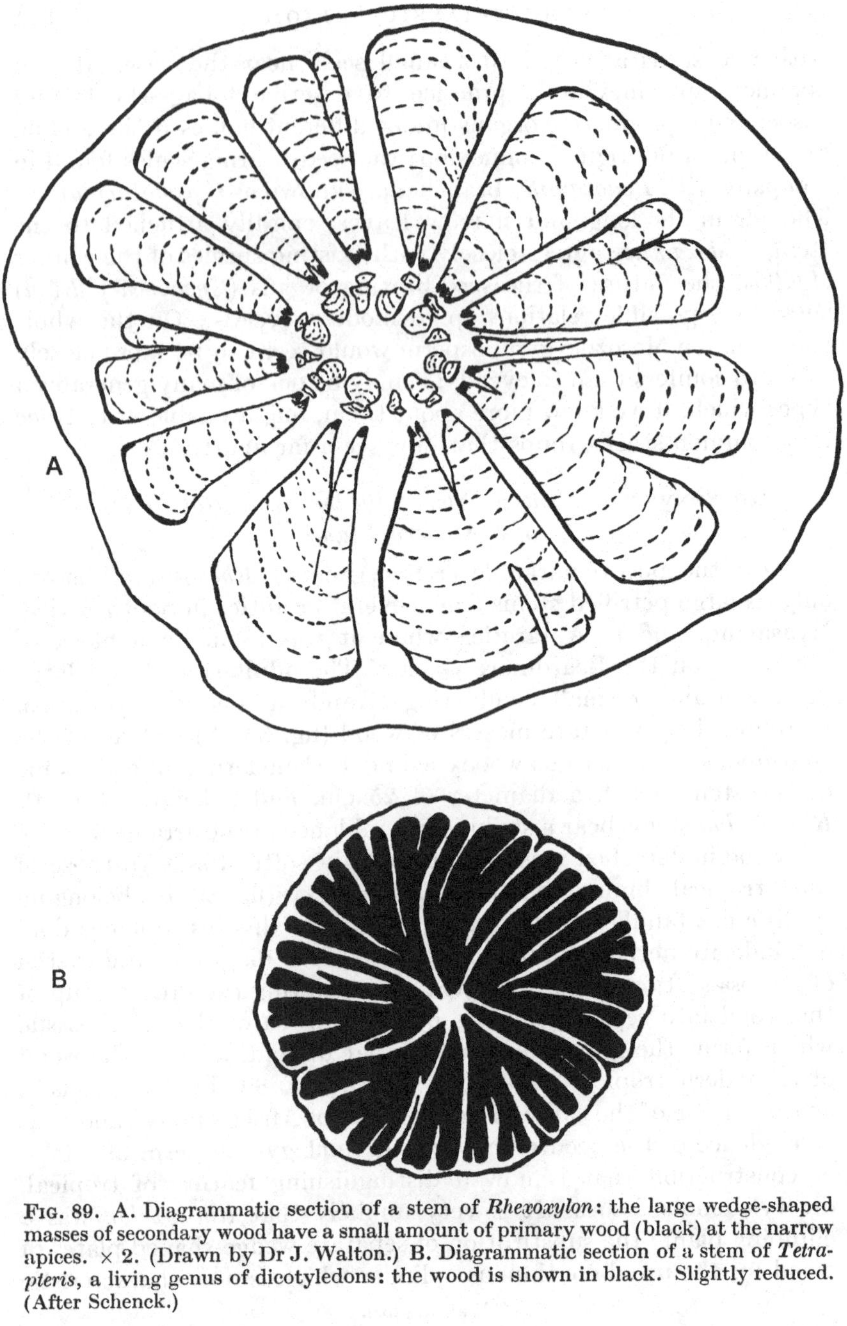

Fig. 89. A. Diagrammatic section of a stem of *Rhexoxylon*: the large wedge-shaped masses of secondary wood have a small amount of primary wood (black) at the narrow apices. × 2. (Drawn by Dr J. Walton.) B. Diagrammatic section of a stem of *Tetrapteris*, a living genus of dicotyledons: the wood is shown in black. Slightly reduced. (After Schenck.)

of lianes probably rendered the stems more pliable and better adapted to resist forces acting upon them, just as ropes made up of intertwined strands are able to withstand strain. The Triassic *Rhexoxylon*, though probably not a liane, foreshadows in its anatomy a type which in later ages became associated with stems belonging to a different class of plants.

Links with the Palaeozoic Era

Reference has already been made to several early Mesozoic plants which are of special interest as links with the vegetation of the latter part of the Palaeozoic era. Another example is furnished by a genus, originally called by such names as *Yuccites* or *Bambusium*—from the resemblance of the long strap-like leaves to those of *Yucca* or bamboos—and more recently renamed *Pelourdea*.[1] Leaves of similar form have been described as species of *Phyllotenia* (fig. 85, *Pa.*). These leaves, recorded from Lower Triassic rocks in the Vosges and from rather younger rocks in England, occur also in the Keuper beds of Switzerland, in the Rhaetic beds of Sweden and Tongking. Some of them agree closely with the Palaeozoic *Cordaites*[2] and it is possible that they belong to plants related to that extinct gymnosperm.

Though nothing definite is known of its systematic position the genus *Chiropteris* is worthy of mention: it is widely distributed in Upper Triassic rocks and occurs also in a few Rhaetic floras. *Chiropteris* is the name given to leaves which in form and size resemble those of some species of *Ginkgoites*, but the venation is a distinguishing feature. The lamina is broad, either complete or more or less deeply cleft: the veins anastomose and form meshes. This genus may be allied to the Palaeozoic *Psygmophyllum*, the affinity of which is also unknown, as in some species of that genus from the Coal Measures of England the veins occasionally anastomose.[3]

It must be admitted that the comparison of the leaves referred by authors to *Yuccites* with those of the Palaeozoic *Cordaites* rests on very uncertain foundations; we have no definite knowledge of the reproductive organs of the early Mesozoic plants which bore the Cordaites-like leaves. A good example of a Triassic specimen of *Yuccites vogesiacus* has recently been described from Lower Triassic rocks near Göttingen:[4] an axis 60 cm. long bears spirally

[1] Seward (17): see also Halle (10), (27).
[2] Kräusel (28).
[3] Seward (19).
[4] Schlüter and Schmidt (27).

attached leaves 30 cm. in length and reaching 4 cm. in diameter. Associated with the foliage-shoot were imperfectly preserved male cones of a type usually assigned to *Voltzia*, but occurring in a bed where branches of that genus are unrepresented. It is suggested that both these male cones and some scales, which seem to have borne a single seed, may belong to *Yuccites*; but proof of organic connexion with the foliage-shoots is lacking.

		India	South Africa	South America	Australia
Jurassic	Upper	Jabalpur series		Patagonian beds	
	Lower		Stormberg series: Volcanic rocks (Basalts)	Basalts	Walloon series (Queensland)
Trias-Rhaetic	Rhaetic, Upper Trias	Rajmahal series			
		Parsora stage	Stormberg series: Molteno beds	Cacheuta beds (Argentine); La Ternara beds (Chile)	Ipswich series (Queensland); Hawkesbury series (N.S.W): Wianamatta beds, Hawkesbury Sandstone, Narrabeen stage
	Lower Trias	Panchet series	Beaufort series: Beaufort (Upper)		
Permian		Damuda series: Raniganj stage	Beaufort series: Beaufort (Lower)		Newcastle Coal Measures
				Lower Lafonian	
		Damuda series: Barakar stage	Ecca beds		Upper marine series

NOTES ON THE TABLE

The table includes only plant-bearing beds in Gondwanaland: the precise correlation of Indian and southern hemisphere strata with those in Europe, North America, and northern Asia is often impossible. Reference should be made to a table given by Du Toit for additional information on South American beds and their correlation with South African strata; also to tables showing the Indian succession by Sahni (22), (26) and to a paper by Das-Gupta (29). Triassic plant-beds occur in east tropical Africa, but our knowledge of the flora is as yet very incomplete: Seward (22[3]). For Australian beds see Walkom (18[2]).

The Vegetation of Gondwanaland

We may now consider some of the distinguishing features of the Triassic-Rhaetic vegetation as revealed by the plants of India and the more southern parts of Gondwanaland. In these regions of the world there is a greater continuity in the sequence of Permian and

Triassic strata than we find in the northern hemisphere. In Europe there is a gap between the beds containing Permian—for the most part Lower Permian—plants and those from which we learn the composition of the oldest Mesozoic floras. It is in the more stable parts of the earth's crust, which have not been violently disturbed by geological revolutions, that we may expect to find links connecting the Palaeozoic and the Mesozoic plant-worlds. In the table given on p. 324 an attempt is made to correlate, so far as it is possible, the southern and northern plant-bearing rocks referred to the Triassic and Rhaetic periods.

The rocks of the Gondwana system of India form piles of sediment, with some seams of coal, deposited in lakes occupying depressions on the surface of a much more ancient foundation. From the scraps of vegetation found in these freshwater beds we are able to follow the development of successive floras from the time of the Palaeozoic Ice Age to the early days of the Cretaceous period. Resting on the Talchir and Karharbari series, believed by many of us to be of Upper Carboniferous age and by all students to be not younger than Permian, is a thick mass of sediments included by geologists in the Damuda series and subdivided into three stages. The flora is evidently Palaeozoic and Permian. The Damuda series is followed in ascending order by the Panchet series, representing the first chapter of Mesozoic history. This relatively poor flora presents a marked contrast to that of the upper beds of the Damuda series, and evidence of adverse climatic conditions furnished by the meagreness of the vegetation is supplemented by the nature of the rocks. In the still younger Parsora stage there is additional evidence of arid or semi-desert conditions. Both the Panchet and Parsora floras may be assigned to the Triassic period while the uppermost beds may be Rhaetic in age. *Glossopteris* and *Schizoneura* persisted into the Parsora stage, and the occurrence of leaves similar to those of *Cordaites* is another notable feature. With the older genera there were a few plants of a Jurassic type. A contrast between this flora and floras of the same age in Australia is the almost complete, or it may be complete, absence in India of ginkgoalean plants. Passing to the next higher Rajmahal series we find a flora of a much more modern aspect very similar to European Jurassic and in some respects to Rhaetic floras.

Abundant remains of Mesozoic vegetation have been found in many parts of Australia. During the last few years Dr Walkom has greatly extended our knowledge of the Australian floras: his work has made it possible to obtain a much clearer view of their relationship to floras in other parts of Gondwanaland and in the northern hemisphere. For our present purpose the most important set of beds is that known as the Ipswich series from Ipswich in Queensland. The plants indicate an Upper Triassic age: *Thinnfeldia* is represented by several species, also *Baiera* and *Ginkgoites*. The plant-beds included in the Walloon series are probably Lower Jurassic in age. Several well-known Rhaetic plants are recorded from New Zealand.

The geological age of the South African rocks containing remains of the *Glossopteris* flora is discussed in a previous chapter. We are now concerned with the plant-bearing strata assigned to the earlier part of the Mesozoic era, namely Triassic and Rhaetic. It is impossible to draw a very definite line between the Palaeozoic and Mesozoic strata: while the beds classed as Upper Carboniferous or as Permian contain members of the Glossopteris flora, the overlying beds have furnished abundant remains of *Thinnfeldia*. The Glossopteris flora was followed by a Thinnfeldia flora. The lowest beds of the Beaufort series (see table, p. 324) are referred to the latest stage of the Permian period; they are rich in *Glossopteris*. *Gangamopteris*, *Sigillaria* and *Psygmophyllum* which occur in the older Ecca series have disappeared. Passing to a higher level in the Beaufort series the flora becomes poorer: in the upper beds a few new genera appear and we are conscious of a change in the vegetation; a resemblance to Triassic and Rhaetic floras of other parts of the world becomes apparent. The upper Beaufort flora includes *Glenopteris* (*Danaeopsis*) *Hughesi*, *Thinnfeldia*, some cycads, and other plants characteristic of Mesozoic floras. Above the Beaufort series are beds assigned to the Molteno stage, the lowest member of the Stormberg series: these are regarded by Dr Du Toit, whose work has greatly advanced our knowledge, as Upper Triassic rather than Rhaetic. The Molteno flora is rich in *Thinnfeldia*: there are also species of *Ginkgoites* and *Baiera*, the fern *Marattiopsis*, *Rhexoxylon* and other genera. *Glossopteris* is recorded from these beds, but it is not certain that the leaves are correctly identified. The Molteno series is followed by some red beds and the Cave Sandstone which are overlain by 5000 ft. of basalt, probably the result of early Jurassic volcanic activity, which brought to a close the chapter of African geological history represented by the Karroo system.

A petrified stem of *Rhexoxylon*[1] from a moraine on the Beardmore Glacier is evidence of the existence of a South African Triassic genus in Antarctica. It should, however, be noted that the *Rhexoxylon* was not found in its original position and we do not know how far it may have been transported as a floating log before becoming embedded in sand.

Our knowledge of the South American plant-beds, though still very incomplete, has been extended by the recent work of Dr Du Toit.[2] There are two great geological systems which include both Palaeozoic and Mesozoic plant-beds: the Santa Catharina system of Brazil and the Paganzo system of the Argentine. The lower part of the Santa Catharina system is regarded as Upper Carboniferous and includes a basal series of glacial beds; resting on these beds is the Estrada Nova stage, the lower part being Permian while the upper part, separated from the lower by an unconformity, is referred to the Triassic period. Here there is a gap between the Palaeozoic and the Mesozoic eras, which it is suggested may represent the whole of the uppermost Permian and "perhaps much of the Lower Triassic". It is interesting to find that some of the South American Triassic rocks afford evidence of desert conditions: it would seem that in this period of the earth's history an arid climate was surprisingly widespread; similar evidence is furnished by rocks of approximately the same age in South Africa and in India as well as in the northern hemisphere and in China. In the other South American system, the Paganzo system, there is also an uncomformity in the succession similar to that in the Santa Catharina system. The Thinnfeldia flora is well developed, and Australian and South African species are recorded from the upper part of the Paganzo system in Argentina, which demonstrate the wide distribution and uniform character in Gondwanaland of the later Triassic, or Keuper, vegetation. Dr Wieland[3] has given an account of some problematical fossils from a Rhaetic flora in Argentina which he thinks may be winged fruits of a dicotyledon. One of these is reproduced as a tail-piece (fig. 90); a seed or possibly a fruit (if the former we are dealing with a gymnosperm; if the latter we have the earliest known angiosperm) is attached to a large wing with several veins. Dr Wieland adopted the name *Fraxinopsis* because of a resemblance to the winged fruits of an ash tree (*Fraxi-*

[1] Originally described as *Antarcticoxylon*: see Seward (14); Walton (23[2]).
[2] Du Toit (27). [3] Wieland (29[2]).

nus). There is, however, no definite evidence that the winged organ is a fruit; its nature can be ascertained only when specimens are discovered which reveal the internal structure. Meanwhile the comparison made with *Cycadocarpidium*[1] is most in accord with the evidence at present available.

A Rhaetic flora discovered in Chile contains, in addition to *Thinnfeldia*, genera such as *Dictyophyllum* and *Clathropteris* which do not occur in the Keuper flora of Argentina. From Madagascar a few plants have been described, proving the existence of a Triassic vegetation, probably of Keuper age.

The best collection of Rhaetic plants in the tropical zone is from Tongking: it may be that the lower plant-beds in this region are equivalent in age to the Keuper-Molteno series in South Africa, but the flora as a whole is definitely Rhaetic. Of the fifty-four plants described by Zeiller[2] only five are identified with Swedish species, but there is a general agreement in the genera: the Ginkgoales are very poorly represented; there are no Nilssonias and no *Thinnfeldia*, nor the fern *Laccopteris*. On the other hand, the occurrence of *Glossopteris* is a noteworthy feature.

In the vegetation of the world during the Keuper-Rhaetic age there were many cosmopolitan genera ranging from South Africa, Australia, and Tongking to Sweden, east Greenland and North America: ferns were abundant, also cycadean plants and members of the Ginkgoales, with representatives of the Equisetales. But a comparison of the several floras reveals certain well-marked differences between the North American and European vegetation and that which was spread over a large part of Gondwanaland. The distribution table (Table C, p. 332) illustrates the wide geographical range of genera and species; it shows that certain species appear to be confined respectively to Keuper and Rhaetic floras; it also illustrates regional peculiarities in the composition of the vegetation.

Between Two Eras: Connecting Links

In conclusion we may now briefly reconsider the relation of the latest phase of the Palaeozoic plant-world to the earlier Mesozoic vegetation. The question is: What are the connecting links between the floras of the two eras? The poverty of the earlier Triassic floras is a striking feature of the northern hemisphere vegetation. The

[1] Nathorst (114). [2] Zeiller (03).

history of the plant-world during the transition stage, incomplete though it is, presents to the student of evolution a problem of special interest. The Kusnezk flora of northern Asia already described in Chapter XII is exceptional among later Permian floras in the number and variety of its plants. With far-travelled wanderers from Gondwanaland there are associated several representatives of early Permian genera, and it is legitimate to assume that their survival was due to the prevalence in northern Russia and Siberia of climatic conditions more favourable than those in many other parts of the world where no such evidence of an equally diversified flora of Upper Permian age has been found. As Prof. Halle says, there are only one or two Kusnezk species which can be traced back to the closing stages of the Carboniferous period: there are several well-known early Permian types such as the almost cosmopolitan *Callipteris*, species of *Psygmophyllum*, *Odontopteris*, *Walchia*, *Pecopteris anthriscifolia*, and others. With these, in addition to *Gangamopteris*, there are other immigrants from the Glossopteris flora: there is the Upper Permian genus *Ullmannia*, also *Voltzia*, a characteristic early Triassic and late Permian genus. There are moreover several genera which are essentially Mesozoic: among these are *Czekanowskia*, not elsewhere recorded from rocks older than Rhaetic, *Ginkgoites* and *Phoenicopsis*, both members of the Ginkgoales, with *Podozamites* and some others. There are also *Cladophlebis*, *Dioonites*, *Neuropteridium* and *Chiropteris*, genera mainly, though not exclusively, characteristic of early Mesozoic floras: these and certain other genera occur in the plant-beds of central Shansi which have furnished remains of a flora closely related to the Stephanian and early Permian floras of North America and Europe, but differing from the great majority of contemporary floras in the larger number of plants of Mesozoic type.

Attention was called in Chapter XII to several Shansi plants which appear to be connecting-links between Palaeozoic and Mesozoic floras: such are *Lobatannularia*, *Cladophlebis Nystroemii*, *Neuropteridium polymorphum*, *Taeniopteris multinervis*—a late Palaeozoic species in Europe, which differs in no essential respect from some Upper Triassic species from North America. An equally good link between Permian and Mesozoic floras is afforded by *Baiera* which, beginning in the Permian period, became a cosmopolitan constituent of the late Triassic and of the Jurassic floras. It is clear therefore that in the Lower Permian, or possibly in part Stephanian flora, of central Shansi as well as in the younger Kusnezk flora there was a larger proportion of Mesozoic genera in a vegetation which was mainly Palaeozoic in character than we find in other regions.

In the northern hemisphere there are other examples of recently evolved types in Permian floras which became prominent in succeeding ages. The genus *Glenopteris* discovered in Permian beds of Kansas may be closely related to the widely distributed Thinnfeldias of Triassic, Rhaetic and Jurassic floras. Reference has been made to the insignificant position occupied by true ferns in the Palaeozoic floras: the great majority of fern-like fronds—and the number is considerable—obtained from Carboniferous and Permian strata undoubtedly belong to pteridosperms, the "flowering plants" of the later Palaeozoic forests. The Palaeozoic Coenopterideae, which are true ferns, are remote from any living genera and there are no direct descendants of them, so far as we know, in Mesozoic floras. Leaving out of account *Senftenbergia*, *Oligocarpia*, *Ptychocarpus* and a few other Palaeozoic genera—which in the characters of their sporangia closely resemble, but are not identical with, any living genera, nor do they appear to be represented by any Mesozoic species—the Palaeozoic ferns which survived the transitional period are very few in number. Reference has been made to connecting links among the ferns of the Kusnezk and Shansi floras between the Palaeozoic and Mesozoic floras. Mention may also be made of the existence of the osmundaceous fern stems *Thamnopteris* and *Zalesskya* in the Kusnezk flora which are some of the earliest members of a family destined to become more prominent in the earlier Mesozoic floras.

Among the Lycopodiales the presumably herbaceous *Lycopodites* affords an example of a conservative and very old type which bridged the gap between the two eras. As already stated, there is evidence that certain Lower Triassic plants carried on for a time the moribund line of arborescent Palaeozoic lycopods. The Rhaetic cone, *Lycostrobus*, may be the last relic of the old Lepidodendron stock.

Recent additions to our knowledge of the fertile organs of *Thinnfeldia* point to the inclusion of this and other Triassic-Rhaetic genera in the pteridosperms, and it would seem highly probable that this class of plant which reached its zenith in the forests of the Coal Age continued to play a conspicuous part in the earlier Mesozoic vegetation, though under a different guise, a difference possibly correlated with the altered climatic conditions. If we contrast the Keuper and Rhaetic floras with those which preceded them we become conscious of very great differences; the plant-world in the Permian period was dominated by *Calamites*, *Lepidodendra*, sigillarias, a host of pteridosperms and several kinds of gymnosperm trees such as *Cordaites*; here and there were a few plants of another type, the advance guard of a Mesozoic host. Before the end of the Triassic period the vegetation had put on a new dress; the dominant genera were of a much more modern type and most of them bore no obvious marks of direct relationship to the trees in the Palaeozoic forests. The contrast in the plant-world affords evidence of some far-reaching change in the organic world directly connected with a revolution in the inorganic world which caused a shifting of the scenes and the creation of a different set of external conditions.

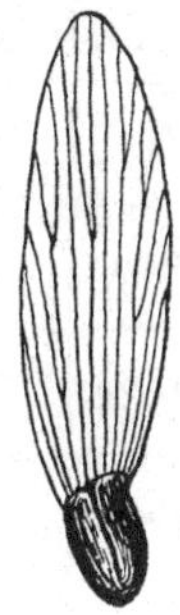

FIG. 90. *Fraxinopsis*. (After Wieland.)

TABLE C. *Geological and Geographical Distribution of some Triassic and Rhaetic Plants*

For Notes on the table, see p. 334.

	Palaeozoic	Trias–Rhaetic			Jurassic		Greenland 7	Europe VIII	Caucasia, etc. XI	Turkestan, etc. XII	India XIII	China; Tongking, etc. XIV	China; Japan; Korea, etc. XVI	W. North America XX	Mexico; Honduras 21, 22	Virginia, New Brunswick, etc. 23, 24	Africa; Madagascar	South America	New Zealand	Australia
		Bunter	*Keuper*	*Rhaetic*																
Equisetales																				
Equisetites	—	—	—	—	—	→	×	×	×	×	×	×	·	·	·	×	×	·	·	×
Schizoneura: *Neocalamites*	—	—	—	—	—		×	×	×	×	×	×	×	·	·	×	×	×	·	×
Lycopodiales																				
Lycopodites	—	—	—	—	—	→	×	×	·	·	×	·	·	·	·	·	·	·	·	·
Lycostrobus: *Lepidostrobus*	—	—	- - -	—			×	×	·	·	·	·	·	·	·	·	·	·	·	·
Pleuromeia		—					·	×	·	·	·	·	×	·	·	·	·	·	·	·
Filicales																				
Marattiopsis			—	—	—		×	×	·	×	×	×	×	·	·	·	×	·	·	·
Danaeopsis			—	—			·	×	·	·	·	·	·	·	·	×	×	·	·	·
Todites			—	—	—		×	×	×	×	·	×	·	·	·	×	×	·	·	×
Cladophlebis nebbensis, etc.	—	—	—	—			×	×	·	·	×	×	×	·	·	×	×	×	·	×
Gleichenites			—	—	—	→	×	×	×	×	×	·	·	·	·	·	·	·	×	·
Hausmannia				—	—	→	×	×	×	·	·	·	×	·	·	·	·	·	·	·
Dictyophyllum	—	—	—	—	—		×	×	×	×	·	×	×	·	·	·	·	×	×	×
Clathropteris			—	—	—		×	×	×	×	·	×	×	·	·	×	·	×	·	·
Laccopteris: *Andriania*			—	—	—		×	×	×	·	·	·	·	·	×	·	·	·	·	·
Asterotheca	—	—	—	—			·	×	·	×	·	×	·	·	·	×	·	·	·	×
Neuropteridium		—	—				·	×	·	·	·	·	×	·	·	·	·	·	·	·

Pteridospermeae (some uncertain)																				
Glossopteris	—	—	—	—			×	×	.	.	.	×	.	.	×	.	×	.	×?	.
Thinnfeldia and allied forms	—	—	—	—	—		×	×	.	.	×	×	×	.	.	×	×	×	×	×
Lepidopteris			—	—			×	×	.	.	.	.	.	.	.	.	×	.	.	.
Callipteridium	—	—	—				.	.	.	.	.	.	.	.	.	.	×	.	.	.
Cycadophyta																				
Wielandiella: *Williamsonia*				—	—		×	×	.	.	.	.	.	.	×	.	.	.	.	.
Pterophyllum	—	—	—	—	—		×	×	×	×	×	×	×	.	×	×	×	.	×	×
Nilssonia			—	—	—	→	×	×	×	×	×	×	×	.	.	.	×	.	.	.
Sphenozamites	—	—	—	—	—		.	.	.	.	.	.	.	.	×	×	.	×	.	.
Pseudoctenis: *Ctenis*			—	—	—		×	×	.	.	×	.	×	×	.	×	×	.	.	×
Otozamites		—	—	—	—	→	×	×	×	×	×	×	.	.	.	.	.	.	.	×
Ginkgoales																				
Ginkgoites			—	—	—	→	×	×	×	×	.	.	×	.	.	×	×	×	.	×
Baiera	—	—	—	—	—	→	×	×	×	×	.	×	×	.	.	×	×	×	×	×
Czekanowskia	—	—	—	—			×	×	.	.	.	.	.	.	.	.	.	.	.	.
Rhipidopsis	—	—	—	—			.	.	.	.	.	.	×	.	.	.	.	.	.	.
Coniferales																				
Araucarites	—	—	—	—	—	→	×	×	.	.	.	×	×	×	.	×	.	×	.	×
Voltzia	—	—	—				.	×	.	.	.	×	.	.	×	×	×	.	.	.
Stachyotaxus				—			×	×	.	.	.	.	.	.	.	.	.	.	×	.
Palissya				—			×	×	.	.	×	.	.	.	.	.	.	.	.	.
Podozamites				—	—	→	×	×	×	×	.	×	×	.	.	×	.	×	.	×
Plantae incertae sedis																				
Rhexoxylon[1]		—	—				.	.	.	.	.	.	.	.	.	.	×	.	.	.
Chiropteris	—	—	—	—			.	×	.	.	.	.	.	.	.	.	×	×	×	×
Pelourdea	—	—	—	—	—		.	×	.	.	×	×	.	.	×	.	.	×	.	.
Caytoniales																				
Sagenopteris			—	—	—	→	×	×	.	.	.	.	×	.	×	×	.	.	.	×

[1] *Rhexoxylon* occurs also in Antarctica.

Notes on Distribution Table C and on Figs. 103, 104

As already pointed out the period of geological history dealt with in this chapter covers the interval between the end of the Palaeozoic era and the beginning of the Jurassic period. It is important to bear in mind that the boundaries drawn by geologists do not represent any pronounced or widespread change in the vegetation. There is comparatively little difference between the meagre floras of the latter part of the Permian period and those of the earlier part of the Triassic period: on the other hand, the vegetation of the Keuper or latest stage of the Triassic period differed greatly from that of the Bunter and Upper Permian stages. It is customary to draw a distinction between Keuper and Rhaetic floras and between Rhaetic and Lower Jurassic (Liassic) floras; but such distinction is often arbitrary and based on differences that are slight and unimportant. The essential point is that though the employment of the term Rhaetic implies a stage in geological history between the Keuper and the oldest Jurassic strata, it does not connote any clearly defined and distinctive phase in the development of the plant-world.

Table C

The table is intended to give a general impression of the geographical distribution of genera and in a few instances of species or groups of species. On the left-hand side is shown the range in time, within certain limits, of each genus. My aim is both to illustrate the wide dispersal over the world of many extinct Mesozoic plants, and to induce students of palaeobotany to undertake more intensive and detailed research in a field of work—Palaeogeography—which becomes more and more promising as knowledge is increased by the publication of critical lists of fossils.

There are many deliberate omissions in the list of selected genera: such genera as *Sphenopteris* and *Pecopteris* among the ferns; *Elatocladus*, *Widdringtonites* and some others among the conifers, because they are in most instances applied to specimens which cannot be referred with confidence to a precise family-position. A few genera are treated as a single unit, e.g. *Neocalamites* and *Schizoneura*, because it is often impossible to detect any well-defined distinguishing characters in specimens included in one or other genus.

Readers should refer to the maps (Figs. 103 and 104), pp. 372, 373, which show the approximate position of most of the regions and localities where Triassic-Rhaetic and Jurassic plants have been found. For convenience of reference circles or arcs have been drawn to indicate areas enclosing localities shown as black dots, numbers, or letters. References are given in the table (pages 374–377) at the end of Chapter XIV to sources where fuller information may be obtained: the list is intentionally incomplete and students should consult the bibliographies at the end of several of the papers and books which are cited.

The broken circles mark the boundaries of Arctic, Antarctic and tropical regions: it will be noticed that records within the Tropics are relatively few.

CHAPTER XIV

THE JURASSIC PERIOD

The estuaries of rivers appeal strongly to an adventurous imagination. *Conrad*

THE name Jurassic was chosen because rocks recording the events of this stage in the history of the earth form the Jura mountains between France and Switzerland. Many of the muds and sands of Jurassic estuaries are exceptionally rich in fossil plants. A comparison of the remains of vegetation collected from widely scattered regions shows that the plant-world was more uniform than it is at the present day: moreover the discovery of a varied and abundant flora in the rocks of Graham Land on the borders of Antarctica (fig. 104) reveals an amazing contrast between the past and the present. The uniformity of the Jurassic vegetation was not as great as it appears: we are familiar only with fragmentary remains of plants which grew on marshes and in other situations favourable to their preservation in sediments: we know comparatively little of the vegetation which flourished on higher ground. In leaves and logs of wood, drifted perhaps for a considerable distance from their original homes, we have waifs and strays from forests and fern-covered banks which afford tantalising glimpses of scenes which we can never completely recapture. Our general impression of the composition of the scattered floras is based on very imperfect data: we cannot apply to the collections made by nature the methods of analysis that are used in the comparison of existing floras. The statement that the vegetation was uniform must not be taken too literally; it means that the available evidence indicates much less difference between floras from different zones of latitude than our knowledge of plant-distribution at the present day would lead us to expect. Another reason which has led to an exaggerated idea of the monotonous character of the vegetation is that floras from different geological stages within the Jurassic system are often treated as though they were contemporaneous.

Jurassic Rocks

A reconstruction of the Jurassic world is shown in fig. 91. A characteristic feature of Jurassic rocks is the frequent alternation of sand, mud, and calcareous sediments, both in a series of beds in

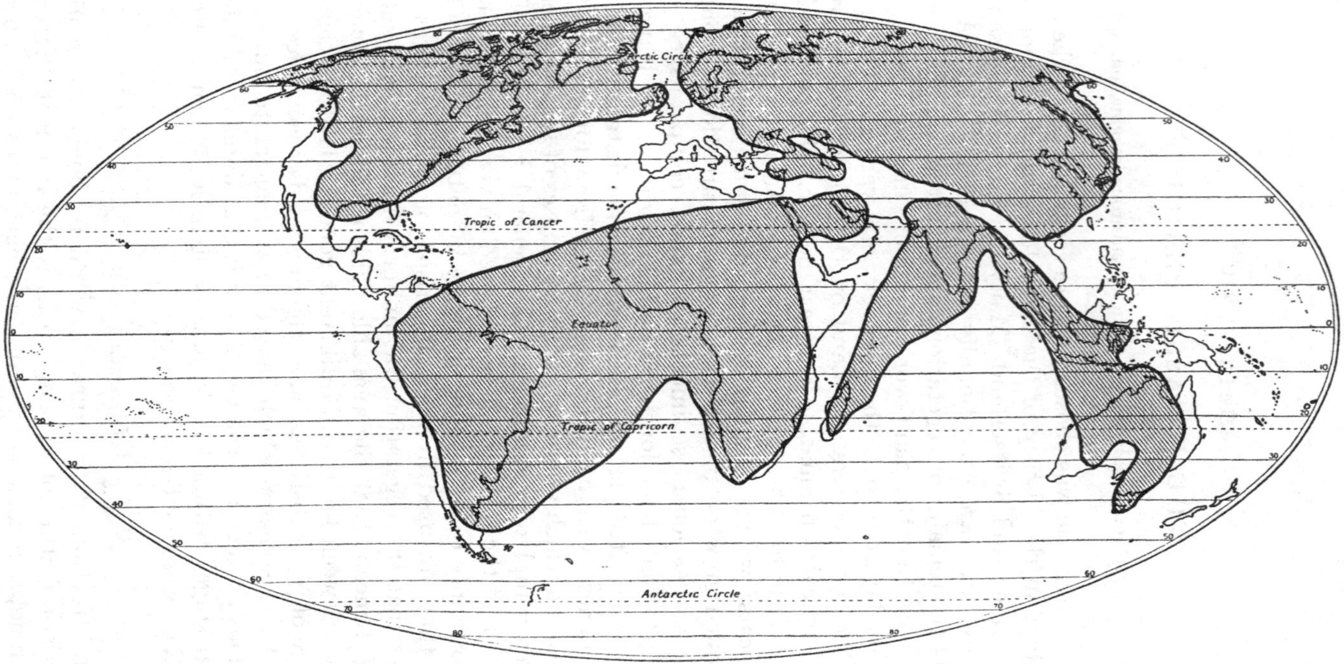

Fig. 91. Map of the world at the middle of the Jurassic period. The shaded areas are land. (Based on a reconstruction by Arldt.)

a single locality and when the strata are followed from one part of the country to another. The period was one of relative stability; the deposition of sediments on sea-floors and in inland basins was not suddenly interrupted by any serious crustal disturbance. Fluctuations in sea-level, gentle up-and-down movements of the earth's surface, were the accompaniments of a regular process of rock-building and of intermittent invasions of land by sea.

In England, with the exception of a few isolated patches or outliers, Jurassic sediments occur as a band running from east Yorkshire to the coast of Dorset. The differences between strata of the same age in the south, in the Midlands, and in Yorkshire are the result of greater or less proximity to land, of deeper water replacing shallower water, and of plant-covered tracts taking the place of a submerged coast. The following classification is intended to serve as a guide to the position of beds mentioned in descriptions of the plants.

OOLITE	Upper Oolite	Portlandian beds	Purbeck beds Portland beds
		Kimmeridge clay	
	Middle Oolite	Corallian: grits, limestones, and some clay (Oxford Clay)	
	Lower Oolite	Upper Estuarine series	
		Limestone	
		Middle Estuarine series. [Cloughton on the Yorkshire coast]	
		Marine bed	
		Lower Estuarine series. [Roseberry Topping and Hayburn Wyke plant-beds]	
		Dogger	
LIAS	Upper Lias	Shales, etc., with jet	
	Middle Lias	Ironstones and sandstones	
	Lower Lias	Marine sediments	

RHAETIC rocks: summit of the Triassic system

The two main divisions are the Lias and Oolite:[1] the term Lias, used by quarrymen for layers of rock, was adopted long ago by William Smith who from a study of Jurassic strata laid the foundations of our modern methods of reading geological history. The name Oolite refers to the abundance of small globular grains like the hard roe of a fish in many of the Jurassic limestones. The Liassic marine strata, limestones, clays, and sands rest on the deposits from the floor of the Rhaetic sea which was invaded by the fresher waters of the Tethys Ocean, occupying a large area in southern Europe including the regions now dominated by the Alps and Carpathians.

[1] For a general account of the Yorkshire Jurassic rocks see Fox-Strangways (92); Kendall and Wroot (24); Black (29).

Stems of conifers, some converted into jet,[1] in the Upper Liassic beds near Whitby and well-preserved impressions of leaves and other plant-remains in rocks on the Dorset coast afford glimpses of the older Jurassic vegetation. The Oolitic series consists of shallow-water limestones and sand: deeper-water marine deposits in southern England; and in Yorkshire a considerable development of

Fig. 92. Hayburn Wyke, east Yorkshire, England. The plant-bearing bed is across the bay just above the shingle, opposite the bush in the foreground. (After Kendall and Wroot: Phot. Mr Godfrey Bingley.)

freshwater estuarine or deltaic beds with the occasional intercalation of material deposited during incursions of the sea.

It is from the Estuarine series of east Yorkshire, represented by about 1000 ft. of strata, that the most important botanical records have been obtained. These plant-beds mark the site of deltas and estuaries to which rivers transported scourings from the rocks of a northern continent. The headland crowned by the Abbey of Saint Hilda at Whitby is built of Lower Estuarine beds on a foundation of Upper Liassic rocks: the modern hotel on the high cliff at Ravenscar stands on the site of a Roman fort that recalls the raids

[1] Seward (04).

of Viking pirates; the material of which the cliff is made (mainly Lower Estuarine beds) takes us back to an infinitely more ancient sea where the rocks of the Yorkshire coast were being accumulated as marine sediments.[1] At Hayburn Wyke (fig. 92), a few miles farther south, Lower and Middle Estuarine beds have supplied much information on the composition of Jurassic plant-communities: the Middle Estuarine sediments at Gristhorpe Bay, a short distance south of Scarborough, have furnished exceptionally valuable records of vegetation preserved where it grew on the marshes of a delta. The plant-beds of Gristhorpe, Cloughton, Hayburn Wyke, Roseberry Topping and other places have furnished a rich flora: it is interesting to note that there are considerable differences in the collections of plants obtained from the various beds of the series, due not to any difference in age but to the derivation of the samples from a variety of localities each with its own set of factors conditioning plant-growth. The limestones and clays of the Scarborough Castle hill enable us to follow successive stages in the deposition of later Jurassic strata.

The famous Portland stone, a calcareous rock containing marine shells, and the associated sands on the Isle of Portland and in the cliffs at Swanage are the deposits from the floor of an Upper Jurassic sea. Overlying the Portland series are the beds of the Purbeck series of the Isle of Purbeck, the district of Dorset which juts out into the English Channel at St Alban's Head and on the north is traversed by the narrow ridge of the Purbeck hills. At the base, resting on the Portland stone, is the lowest dirt bed, an old soil containing cycadean stems: this is succeeded at a higher level by a second dirt bed. The dark band shown in the photograph (fig. 93), below the head of the hammer, is one of the old surface-soils overlain by a calcareous rock which was deposited from water supersaturated with carbonate of lime: this material often encases stems belonging to the underlying forest. Some of the trunks of conifers are still rooted in the soil; others occur as prostrate logs which occasionally reach 20 ft. in length. The sudden change from the marine Portland series to the land-surface represented by the lower Purbeck dirt bed is an indication of a time-interval and a break in the sequence. The sea was converted into land or, in some places, into shallow estuaries. The dirt bed marks the site of the oldest forest which colonized the newly elevated surface.

[1] For a good account of the Yorkshire rocks, see Kendall and Wroot (24).

Resting on the higher Purbeck beds is a series of lacustrine sediments known as the Wealden series: these rocks are usually regarded as the base of the Cretaceous system. An argument in favour of including the Wealden beds in the Jurassic system is afforded by the fact that the uppermost, freshwater sediments of the Purbeckian are overlain in regular order by the lagoon or lacustrine strata formed on the floor of the Wealden lake. Whether or not we include the Wealden series in the Jurassic period is a matter

Fig. 93. A Jurassic surface-soil (dirt-bed) at Lulworth, Dorsetshire, England. (Photograph supplied by Prof. S. R. Reynolds.)

of secondary importance: of greater importance from our point of view is the general agreement of the plants preserved in Wealden deposits with those of the older Jurassic floras.

Jurassic rocks occur at a few localities in Scotland: of these the most interesting botanically are sediments which were deposited in a bay or estuary bordered by Devonian hills on the coast of Sutherland between Brora and Helmsdale.[1] From these Upper Jurassic beds exposed along the foreshore several plants have been collected and it is noteworthy that they bear a very close resemblance to species described from Wealden rocks.

[1] Stopes (07); Seward (11); Seward and Bancroft (13).

During part of the Jurassic period much of Europe was submerged and the sea spread far into the Arctic regions. There are many localities in France and Germany, Italy, Sardinia, and elsewhere which have yielded a rich harvest of Jurassic plants. On the Angara continent, a northern land deriving its name from the river in Siberia which comes from Lake Baikal and flows into the Yenissei, the conditions were specially favourable for the preservation of plants and the formation of coal as the region escaped any extensive marine transgression. Eastern North America was exposed to long-continued erosion during the Jurassic period, and the river-borne débris was deposited on the inaccessible floor of the present Atlantic Ocean. On the Pacific coast, in California, Oregon, and farther north, marine plant-bearing beds indicate a transgression of the sea and demonstrate the wide range of a vegetation which is more abundantly represented in European strata.

Our knowledge of the vegetation of the southern continent is mainly derived from the Upper Gondwana rocks of India, which contain abundant plant remains and form part of the thick series of freshwater and terrestrial deposits recording the changing fortunes of the plant-world through a succession of ages. Other floras have been described from Graham Land, from South Africa, Australia and New Zealand.

The Jurassic Plant-world

The Jurassic period as a whole presents us with a picture of a relatively uniform world: we see a long succession of floras from those of Rhaetic time to the floras which have been pieced together from specimens scattered through the beds of the Wealden series which it is customary to regard as the basal members of the Cretaceous system. Between the Rhaetic and the Wealden vegetation the differences are comparatively slight: the more prominent classes were conifers and ferns and members of an extinct group of cycadean plants. Though the stage was occupied by relays of changing genera and species the geological background varied but little, and the general uniformity in the conditions is reflected in the conservative character of the successive floras.

The Rhaetic vegetation persisted without any striking change into the earliest stages of the Jurassic period: as we follow the progress of evolution through the middle and upper series we are not conscious of any far-reaching transformation in the general aspect of the plant-world. Genera disappear and others take their place: Rhaetic species are followed by Jurassic species, the slight differences between them being often difficult to recognize. There was an almost imperceptible passage from one individual type to another in a vegetation which, in its broader features, was comparatively stable. In the account of Jurassic floras no attempt is made to follow in detail the relatively minor modifications within the limits of the period. The majority of the richest collections of plants come from sedimentary rocks assigned to a middle Jurassic age.

If we envisage the vegetation as a whole certain characteristic attributes become apparent. A comparison of Triassic and Rhaetic floras in the northern and southern hemispheres shows differences in composition which enable us to detect, in a more or less marked degree, differences in the predominant elements indicative of botanical provinces. When we make a similar comparison of Jurassic floras we are struck by the greater uniformity in associations of plants based on collections from localities as remote as northern Europe and the borders of Antarctica, California and India. Certain genera and families flourished more abundantly in some regions than in others; some plants, though common over wide areas, were by no means cosmopolitan. There were unquestionably regional peculiarities among widely scattered floras deeper and greater than the incomplete and fragmentary records appear to indicate; but such data as we possess lead us to conclude that the Jurassic vegetation was less diversified and less affected by geographical position than that of any other stage in geological history. This fact will be easier to appreciate after a short description has been given of the commoner and more characteristic constituents of Jurassic floras. A reference to Table D (p. 343) will serve as a guide to the geographical range of the commoner genera.

Another feature of Jurassic vegetation is the closer approximation of the plants to those of our own time; as Prof. Sahni says: "It is satisfactory to note that the hostile ranks of the species *incertae sedis* [the plants we cannot assign to a definite position] have suffered heavy losses".[1] Moribund survivors from the Palaeozoic world disappear and more modern and familiar types become increasingly abundant: we get into closer touch with lines of evolution leading directly to genera and species which are still with us. The scenes follow one another without any drastic change in the setting:

[1] Sahni (22).

See maps; figs. 103, 104, pp. 372, 373.

	Trias-Rhaetic	Jurassic	Cretaceous		Alaska I	Siberia; New Siberian Islands III	Arctic (Europe) IV	Europe: N. Africa VIII	Caucasia, etc. XI	Afghanistan, etc. XII	India: Ceylon XIII	Japan: China, etc. XVI	W. N. America XX	Mexico 21	Graham Land: Patagonia	Madagascar	Australia: New Zealand
Equisetales																	
Equisetites columnaris and allied forms		—			×	×	×	×	×	×	×	×	×	·	×	×	×
Neocalamites	—	—			·	×	·	×	×	×	·	×	·	·	·	·	×
Filicales																	
Marattiopsis	—	—	—		·	·	·	×	×	·	×	×	×	·	·	·	·
Todites } *Osmundites* }	—	—			·	×	·	×	×	×	×	×	·	·	×	·	×
Cladophlebis denticulata and allied forms	—	○	—		×	×	×	×	×	×	×	×	×	·	×	·	×
Gleichenites	—	—	—	→	·	·	×	×	×	·	×	·	·	·	·	·	×
Dictyophyllum	—	—	—		×	·	·	×	×	×	·	·	×	·	×	·	×
Laccopteris	—	—	—		·	·	·	×	×	·	·	×	·	×	·	·	×
Matonidium		—	—		·	·	·	×	·	·	·	·	·	·	·	·	·
Klukia		—			·	·	·	×	×	×	·	×	·	·	×	×	×
Eboracia	—	○			·	×	·	×	×	×	×	×	×	·	×	·	·
Coniopteris hymenophylloides and allied forms		—			×	×	·	×	×	×	×	×	×	×	×	·	×
Pteridospermeae																	
Thinnfeldia and allied genera	—	—	—		·	·	·	×	×	×	·	·	·	·	×	·	×
Cycadophyta																	
Cycadeoidea	—	—	○		·	·	·	×	×	·	·	·	×	×	·	·	·
Williamsonia	—	○	—		·	·	·	×	·	×	×	·	×	×	×	·	·
Wielandiella	—	—			·	·	·	×	·	·	·	·	·	×	·	·	·
Ptilophyllum		—			·	×	·	×	×	·	×	×	×	×	×	·	×
Otozamites	—	—	—		×	·	·	×	×	·	×	·	·	×	×	·	×
Ctenis } *Pseudoctenis* }	—	—			·	×	·	×	×	×	×	×	×	·	×	·	×
Dictyozamites		—			·	·	·	×	·	·	×	×	·	·	×	·	·
Nilssonia	—	—	—		×	×	·	×	×	×	×	×	×	·	×	·	×
Ginkgoales																	
Ginkgoites	—	—	—		×	×	×	×	×	×	×	×	×	·	·	·	×
Baiera	—	—	—		·	×	×	×	×	×	·	×	×	·	·	·	×
Czekanowskia	—	—			·	×	×	×	×	×	·	×	·	·	·	·	·
Phoenicopsis		—			×	×	×	×	×	×	·	×	·	·	·	·	×
Eretmophyllum		—			·	·	·	×	·	×	·	·	·	·	·	·	·
Coniferales																	
Araucarites	—	—	—		·	·	×	×	·	·	×	·	·	×	×	×	×
Pagiophyllum	—	○	—		×	·	·	×	×	×	×	·	×	·	×	×	×
Brachyphyllum	—	○	—		·	·	·	×	×	×	×	·	×	·	×	×	×
Pinites, etc.	—	—	—	→	×	×	×	×	×	×	·	·	·	·	·	·	·
Podozamites	—	○	—		×	×	×	×	×	×	·	×	×	×	·	·	×
Caytoniales																	
Sagenopteris	—	—	—		·	×	·	×	×	×	·	×	×	×	×	·	×

during the whole period the earth's crust was relatively stable; evolution was more leisurely, and development continued in continental areas unaffected by the far-reaching climatic disturbances inseparable from crustal revolutions and the upheaval of mountain ranges.

As we pass in review the several groups we shall be better able to grasp the fact that the plant-world has assumed a more familiar guise; archaic forms, which cannot be accommodated within any living genus or family, are rapidly becoming replaced by plants that are essentially modern. We shall also find that on close examination some of the dominant types baffle us by an admixture of what appear to be contradictory characters: at first sight they seem to be closely akin to existing genera, but in those features which are the surest guides to affinity they are found to be remote from any member of the present plant-kingdom.

The Jurassic vegetation differed essentially from that of the present day in the almost complete absence of plants recognizable as members of the class that is now dominant; but, it must be added, in recent years evidence has been steadily growing in amount and in value which leads us to believe that the angiosperms had begun to play a part, though as yet a humble part, in the world's vegetation. In all probability descendants of Palaeozoic pteridosperms grew side by side with ferns, conifers, and other plants which differed in no fundamental features from living members of these classes. There were still living several kinds of plants which have long been extinct and cannot confidently be allocated to a definite systematic position; though these were far outnumbered by genera and species hardly distinguishable from plants that are now subtropical or tropical.

Could we visit the delta and estuary of the river which built up the beds of sediment that have lain for countless years in the Cleveland hills and in the cliffs of the Yorkshire coast, our first impression might be that we were in some tropical region. Conifers of familiar habit dominated the forest; river banks hidden under a tangled riot of luxuriant ferns; groups of horsetails (*Equisetum*) spreading a green mist over flat stretches of sand. We should be struck by the absence of broad-leaved trees festooned with rope-like climbers and their branches lifting up to the sunlight clusters of humbler flowering plants. We might well think that we had wandered into a tract of country where for some reason the ground was monopolized by ferns, conifers, and cycadophytes to the apparent exclusion of all flowering plants.

Equisetales

Beginning with the Equisetales certain species, or groups of closely related forms, were almost cosmopolitan and though some of them had thicker stems than any existing species, in external features they were practically identical: they are connecting links between the still larger Triassic species and the modern horsetails. One of the commonest species is *Equisetites columnaris*, usually represented by comparatively robust stems clasped at intervals by sheaths of fused leaves, and occasionally preserved where they grew with roots spreading through the sand in which they were eventually entombed. There still lingered a few examples of the genus *Neocalamites*, a type which was abundant and widely distributed in Keuper and Rhaetic floras and probably became extinct before the end of the Jurassic period. Similarly the equisetaceous *Phyllotheca* (fig. 74, *P.*), with its slender shoots and more spreading cup-like sheaths of partially coalescent leaves, still persisted in some regions as a less conspicuous member of the group.

Lycopods and Ferns

Jurassic lycopods, so far as is known, were exclusively small and herbaceous, indistinguishable in habit from some modern lycopods and selaginellas; and it has been shown that some of the Jurassic fossils included in the genus *Lycopodites* bore two kinds of spores and should therefore be transferred to the genus *Selaginellites*. The Rhaetic *Lycostrobus* was the last relic of the arborescent Palaeozoic stock, while on the other hand the herbaceous Mesozoic *Lycopodites* and *Selaginellites* are links in a chain which extends from the present to the Palaeozoic era. Passing to the ferns we find many genera which, having travelled over a considerable part of the earth's surface before the close of the Triassic period, extended their range and produced new forms in the course of the Jurassic period. The genus *Marattiopsis*[1] carried on the traditions of the Marattiaceae: the leaflets and the structure of the spore-capsules demonstrate a close affinity to some living members of the family which is now mainly tropical. The Osmundaceae, a family that had already become vigorous in the Rhaetic period, were represented by the

[1] Thomas, H. H. (13[2]).

world-wide *Todites Williamsoni*, with its long pinnae bearing closely set, short and broad leaflets, and by other types. In this connexion it may be pointed out that some of the most characteristic and common fern fronds found in Jurassic strata are almost always preserved in a sterile condition and cannot therefore be referred with any certainty to a family position. A good instance of this is a species, or, more correctly a group of species known as *Cladophlebis denticulata*. This form of frond is very similar to some of the *Cladophlebis* species common in Rhaetic floras (e.g. *C. Roesserti*, fig. 88); it is characterized by its more or less curved leaflets often with sharp and small teeth on the edge of the lamina near the pointed tip, a form of frond that is familiar in the South African *Todea barbara*, a member of the *Osmunda* family, and in many other living genera. Sporangia have been found on some fronds of *Cladophlebis denticulata*, and Prof. Halle[1] has described one type of spore-capsule on a *Cladophlebis* frond, under a distinctive generic name, *Cladotheca*, which he compares with those of the recent fern *Todea*. Evidence furnished both by fertile fronds and exceptionally well preserved petrified stems from Jurassic rocks in New Zealand[2] justifies the assertion that ferns very close to *Todea*, also a few nearer akin to *Osmunda*, were more widely distributed in the Jurassic world than are the living members of the family, even though the royal fern (*Osmunda regalis*) is now almost cosmopolitan.

Among characteristic Jurassic genera are *Laccopteris*, *Matonidium*, *Dictyophyllum*, *Clathropteris* and *Hausmannia*. The first of these, *Laccopteris*, has already been mentioned as a far-travelled Rhaetic fern: it became still more abundant in the Jurassic period. The genus *Matonidium*, characteristic of Jurassic and Cretaceous floras, resembles *Laccopteris* in the fan-shaped arrangement of the long arms of the frond bearing two ranks of narrow, linear leaflets and spreading in orderly series from the two equal and divergent branches at the summit of a long and slender stalk, as in the Malayan fern *Matonia pectinata* (fig. 94) and the Bornean species *Matonia Foxworthyi*.[3] *Dictyophyllum*, another legacy from the earlier Mesozoic vegetation, was one of the more conspicuous and handsome of Jurassic ferns: not unlike *Matonia* and *Laccopteris* in the form of the frond, it differs in the subdivision of its broader arms, each superficially resembling a large oak leaf, with lobed or deeply cut segments, and in the reticulate character of the finer venation. This

[1] Halle (11^2). [2] Sinnott (14). [3] Copeland (08).

FIG. 94. *Matonia pectinata*. Gunong Belumut, Johore, 3000 ft. above sea-level. (Photo. Mr R. E. Holttum.)

genus, though not identical in the size of the sporangia with any of the living ferns which come nearest to it in the shape and venation of the fronds, bears a sufficiently close resemblance to the Indian and Malayan *Dipteris* (fig. 95), both in the architectural plan of the large leaves and in the structure of the sporangia, to find an appropriate place in the *Dipteris-Matonia* alliance. The sporangia of the common Jurassic species, *Dictyophyllum rugosum*,[1] are not arranged in well-defined groups (sori) as in *Dipteris*, but are scattered over the leaf as in the tropical *Platycerium*, the elk's horn fern. *Hausmannia*, with its smaller fronds similar in venation to *Dictyophyllum*, apparently began its career in the Rhaetic period and became more widely spread in Jurassic floras and in the early part of the Cretaceous period. Among fossil ferns it is the nearest in sporangial characters as in the form of the leaf to *Dipteris*. *Dictyophyllum* and *Hausmannia* may be compared also with two other existing genera *Neocheiropteris* and *Cheiropleuria* (fig. 96), the former an ancient Chinese genus and the latter a fern which extends from Malaya to Formosa and New Guinea. *Cheiropleuria bicuspis* was formerly called *Acrostichum Vespertilio*, the specific name having reference to a resemblance of its fronds to the wings of a bat. Other ferns related to *Dictyophyllum*, the genera *Thaumatopteris*, *Camptopteris* and *Clathropteris* are more characteristic of Upper Triassic and Rhaetic than of Jurassic floras.

An enquiry into the past history of the Gleicheniaceae, another family of ferns which is now characteristic of the Tropics, reveals an additional example of the contrast between past and present geographical ranges. The genus *Gleichenia* (figs. 26, 95), using the name in a comprehensive sense, includes a large number of ferns, most of which have forked fronds endowed with a power of unlimited growth: the leaf-stalk may reach a length of over 20 ft., giving off as it grows pairs of divergent branches bearing pinnae with linear leaflets, usually 3–4 cm., and in some species 10 cm. in length. In some forms the leaflets are only a few millimetres long and approximately semicircular. *Gleichenia* ranges from Japan to the Straits of Magellan, reaching as far north as Mexico (fig. 110); it is found at a height of 10,000 ft. on the slopes of Ruwenzori in tropical Africa and at an altitude of 12,000 ft. on the mountains of New Guinea.[2] Appearing for the first time in Keuper floras as a type with undoubted relationship to living species, the genus continued through

[1] Thomas, H. H. (22). [2] Seward (22).

FIG. 95. *Dipteris conjugata* (the fern with broader fronds) and *Gleichenia glauca* on Western Hill, Penang, 2500 ft. above sea-level. (Phot. Mr R. E. Holttum.)

FIG. 96. *Cheiropleuria bicuspis*: Malay Peninsula, 4000 ft. above sea-level. (Phot. Mr R. E. Holttum.)

the Rhaetic and Jurassic periods and reached its maximum development in the Cretaceous vegetation: by the Tertiary period it had become comparatively rare in the northern hemisphere and is now unknown in North Africa, Europe, western Asia and practically the whole of the North American continent.

Another family of special interest from a distributional point of view is the Schizaeaceae. This family, including the genera *Lygodium*, *Schizaea*, *Aneimia* and *Mohria*, is now unrepresented in Europe; distinguishing features are the occurrence of sporangia singly and not in groups and the presence of a transverse ring of thickened cells (annulus) near the apex of the sporangium. One of the oldest examples of a schizaeaceous fern is the genus *Norimbergia*[1] discovered in Lower Jurassic (Liassic) rocks near Nürnberg. Most of the Jurassic examples of the family are included in the genus *Klukia*, characterized by fronds with small leaflets of the *Cladophlebis* form and sporangia with the apical annulus distinctive of the Schizaeaceae. In habit the leaves do not conform closely to those of any living member of the family; they illustrate, like those of many other extinct plants, a former range in design greater than is illustrated by surviving types of the same stock. In *Eboracia*, *Coniopteris*, and possibly *Stachypteris*,[2] we have widely spread genera belonging to the Dicksoniaceae[3] (usually included in the Cyatheaceae, but recently promoted to a family position), a family which is now characteristic of warmer countries and includes the largest tree-ferns; it is no longer represented in Europe. The genus *Eboracia* is recorded also from the Rhaetic plant-beds of Tongking. One of the commonest Jurassic species is *Coniopteris hymenophylloides* distinguished by its small, lobed leaflets and sporangia encircled by a complete and oblique annulus borne on reduced leaflets with little or no lamina. It is worthy of note that there is a close resemblance in the form of the frond and in the nature of the fertile segments between some widely distributed Jurassic Dicksoniaceae and the solitary *Thyrsopteris elegans* (fig. 97) which is now restricted to Juan Fernandez, the island—365 miles to the west of Valparaiso—made famous by *Robinson Crusoe*.[4] The genus *Eboracia* is distinguished by its larger leaflets and by the occurrence of an abnormally developed leaflet at the base and on the lower side of each of the lateral pinnae: it is an almost cosmopolitan Jurassic member of the Dicksoniaceae (see Table D, p. 343).

[1] Gothan (14). [2] Thomas, H. H. (12). [3] Bower (26). [4] Schmidt (28).

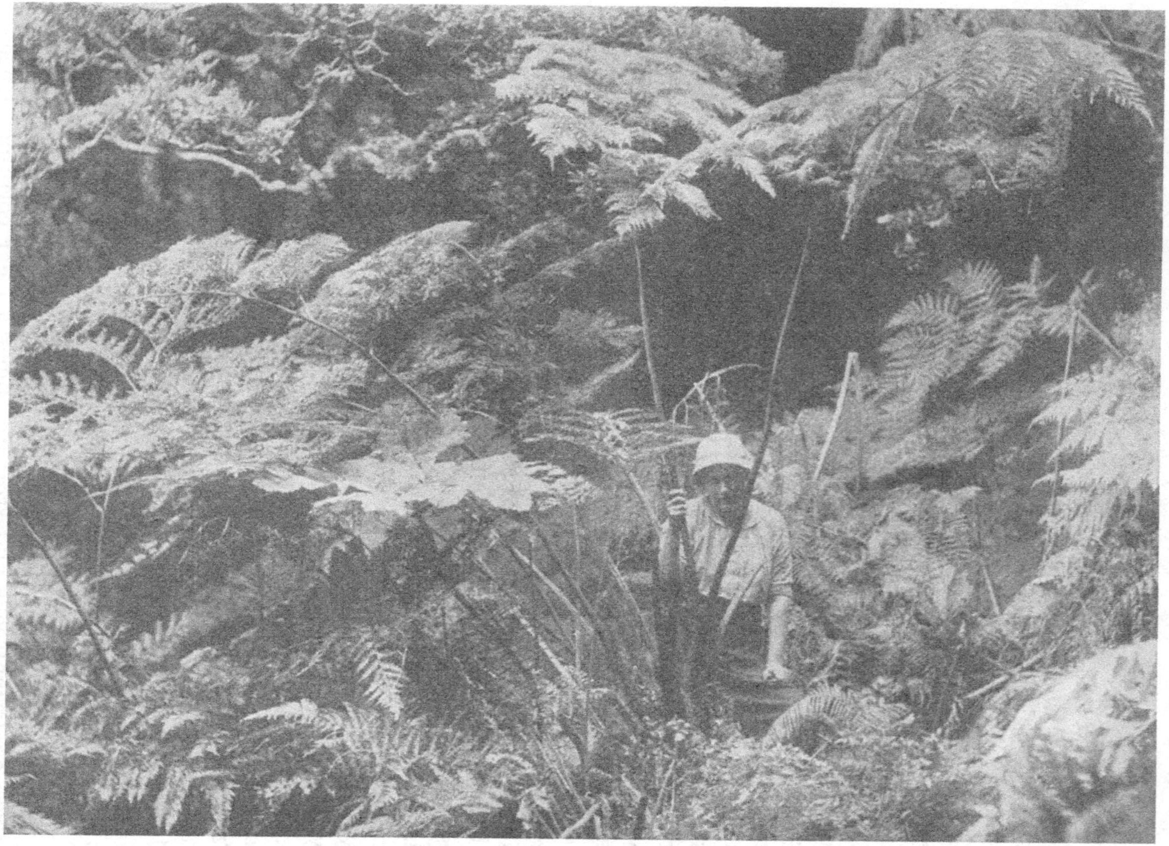

FIG. 97. *Thyrsopteris elegans*, in the upper part of the Villagen Valley, Juan Fernandez. (Photograph originally reproduced in the *National Geographical Magazine* (1928) and given to me by Dr C. Skottsberg.)

A Dominant Group: the Cycadophyta

It is often said that the Jurassic period was an age of cycads and this implies that the few living genera of the family, which are sparsely scattered over certain tropical regions and occur in some places both north and south of the equator, are the direct descendants of a group which in the Triassic, Rhaetic and Jurassic periods overspread the world. If resemblance, amounting almost to identity, in leaf-form and in the structure and habit of stems could be accepted as proof of near relationship, then the above statement would be beyond reproach; but the remarkable fact is that the reproductive organs of the great majority of such Mesozoic plants as are usually called cycads are entirely different from those of the living genera. The nature and degree of the relationship of the extinct genera to the true cycads raises a problem which has not been solved: we do not know whence the Mesozoic type came; nor by what steps, if any, their progeny led to the cycads of the present day. As it is difficult to give up a term that has been long in use, we may employ the designation "cycads" for an extinct group separated by essential features from the existing Cycadaceae provided we remember that its use is based on a supposition of affinity which rests on insufficient grounds. It is preferable to adopt the less misleading name cycadophyte. Fronds varying widely in size and bearing leaflets (pinnae) differing considerably in shape from one another, but most of them reminding us of the foliage of living cycads, are among the commonest Jurassic fossils. Many of these cycadean fronds occur among the débris of Keuper and Rhaetic floras, but they are still more abundant and represented by a greater variety of leaf-form in Jurassic rocks. Genera such as *Zamites*, *Ctenis*, *Pseudoctenis*, *Nilssonia* and *Pterophyllum* have already been mentioned. Others, more characteristic of Jurassic than of earlier Mesozoic floras, are *Otozamites*, *Dictyozamites* and *Ptilophyllum*. A feature shared by most of the fronds is the comparatively thick lamina of the leaflets; some of them give the impression of plants able to withstand scarcity of water and exposure to intense light.

It is also significant that in many of the fossil leaves the structure of the epidermal cells and stomata shows a clear correspondence with that characteristic of the true cycads.[1] Fronds included in *Otozamites* bear two rows of leaflets attached to the upper face of

[1] Thomas and Bancroft (13); Thomas, H. H. (30); Kräusel (21²).

the axis, varying in shape from short and broad, or almost orbicular, to long linear leaflets with approximately parallel or slightly spreading veins and a more or less lobed, or eared, base. These and other leaves nearly always occur as separate fossils and not attached to or directly associated with stems. In *Dictyozamites*[1] the leaflets are similar in form to those of most species of *Otozamites*, but the veins form a fairly regular network. *Dictyozamites* which is recorded from Jurassic floras of Bornholm, England, Patagonia, India and Japan, affords an interesting example of discontinuous distribution. *Ptilophyllum* is one of the most widely distributed of all Jurassic genera: its fronds, like a double comb with regular linear segments, are not always easy to distinguish from those of *Otozamites* and *Zamites*, particularly if we are unable to make microscopical preparations of the superficial cell-layer or the cuticle.

Convincing evidence has been given of the cycadean nature of some fronds known as *Taeniopteris*, a genus which was always regarded as a fern, in shape similar to the simple leaves of a *Scolopendrium vulgare* and the tropical *Oleandra*. *Taeniopteris* leaves may be compared also with the separate leaflets of the South African cycad *Stangeria*, a plant originally described as a fern because of the form and venation of its fronds. It is now known that a species of *Taeniopteris*, *T. vittata*, a characteristic Jurassic leaf, was borne on a stem of one of the extinct cycadean plants, *Williamsoniella*, a genus assigned with the great majority of its allies to the Bennettitales, a subdivision of the class Cycadophyta.

Williamsoniella: Williamsonia

Williamsoniella[2] is characterized by a slender, regularly forked stem similar to that of the Rhaetic genus *Wielandiella*, bearing scattered leaves to which the name *Taeniopteris* had long been applied before their true nature was discovered. In the forks of the stem were small fertile shoots bearing at the end of a short stalk a cluster of reproductive organs which may conveniently be termed flowers. Each flower when fully expanded may be compared with a fairly large buttercup (*Ranunculus*), the coloured petals being replaced by a number of thick, oval appendages forming a regular series, each member attached by a narrow base and broadening upwards to a bluntly rounded apex: when young these were folded inwards against the short axis of the flower; on maturity they spread

[1] Seward (03[2]); Halle (13). [2] Thomas, H. H. (15).

outwards and the pollen, contained in sacs borne on their sloping inner faces, was discharged. These pollen-bearing scales (microsporophylls) are the male organs and represent the stamens of an ordinary flower. The flowers were unprotected by any covering of sterile leaves. The axis of the flower, similar in shape to the central column of a *Ranunculus*, was covered, except at the tip, with small, crowded appendages, their truncate ends forming a pattern of hexagonal areas arranged in groups of five or six as rosettes having in the centre of each, in place of a hexagon, a smaller circular tube of microscopical dimensions projecting a short distance beyond the general level and leading to the female organ at the summit of a slender stalk attached to the flower-axis. One may compare each female organ (ovuliferous scale) with the ovule of a yew tree (*Taxus*), which consists of a mass of tissue enclosed in a thin envelope, or integument, prolonged above the ovule as a tube through which the pollen enters and effects fertilization of the egg-cell within the ovule destined to become a seed. In *Williamsoniella* the ovules were not borne singly as in *Taxus*, but there were many of them in each flower; and in place of the little membrane which envelops the base of the ovule and later grows into the red fleshy cup of the yew seed there are five or six separate appendages (interseminal scales) which are clustered round each ovule of *Williamsoniella*: it is their expanded tips which form the hexagonal pattern on the flower-axis. One may also compare the ovules of *Williamsoniella* with the carpels (that is the green, swollen bodies containing the ovules and later the seeds) crowded on the column of a *Ranunculus*;[1] but there is this essential difference: the ovules of *Ranunculus* and of all flowering plants (angiosperms) are hidden inside a closed case, in the ovary or cavity of the carpel, while in *Williamsoniella*, as in all gymnosperms, the ovules are naked. Comparisons such as these cannot be carried far, though they may serve to give a general idea of the bisexual flowers of this extinct and, as it seems to us, aberrant member of the Cycadophyta.

In the Bennettitales as a whole the flowers, particularly the female organs, conform to those of *Williamsoniella*: small ovuliferous scales are protected and surrounded by clusters of sterile interseminal scales which do not actually enclose the ovules. The male organs of Bennettitalean flowers, though in essentials similar, are

[1] A more familiar type of carpel is the pod of a pea enclosing the ovules which become the seeds (peas).

less uniform in plan than the female. In modern cycads male and female organs never occur on the same axis: except in the genus *Cycas* the ovules are produced in pairs on the greatly modified scale-like leaves of a cone, and the pollen-bearing scales are also borne on cones.

It is hardly possible to believe that the flowers of *Williamsoniella* and the cones of existing cycads are the reproductive shoots of members of one set of gymnosperms: they are poles apart. Moreover the stem of *Williamsoniella*, forking into spreading branches and bearing comparatively small leaves, differs very widely from the modern cycadean type. On the other hand it would seem from their relative scarcity—which may be rather apparent than actual—that *Wielandiella* and *Williamsoniella* are exceptional, or perhaps primitive, among the Bennettitales, in most of which the stem was tall and columnar or thick and tuberous as in living cycads, and bore a crown of spreading leaves at the summit (cf. fig. 30).

We will now pass to *Williamsonia*, a more typical genus of the Bennettitales, first recorded from Upper Triassic and Rhaetic rocks, but more widely spread in the Jurassic period. The stems agree closely with those of several living cycads: they were covered with contiguous bases of old leaves and bore relatively large palm-like fronds at the apex; but the flowers were totally different. We are not certain whether the flowers were unisexual or bisexual: such evidence as there is seems to point to the separation of the sexes. A *Williamsonia* flower[1] is enclosed by overlapping, narrow scale-leaves comparable with the protecting leaves on the flower-heads of the globe artichoke (*Cynara*). On the thick, pear-shaped axis of the flower there were innumerable interstitial scales grouped, as in *Williamsoniella*, around the fertile scales and forming a protective covering over the ovules. The male organs consist of a ring of reduced and transformed leaves united below into a collar, varying in depth, and bearing pollen-sacs. In some species each of the male appendages gives off two lateral rows of slender arms to which the pollen-sacs are attached; in others the male organs are smaller and unbranched and of a pattern nearer to that in the still smaller stamens of *Wielandiella* and *Williamsoniella*. It would seem that in the older members of the group the male organs were simpler and smaller and later became branched and in appearance resembled on a reduced scale fertile fronds of some ferns. *Williamsonia* bore

[1] For photographs see Nathorst (11^3); and for other references, Seward (17).

its flowers at the ends of branches projecting several inches from the thick stem. Fronds and flowers are usually preserved apart, but specimens obtained from several localities—in England, Italy, Mexico[1] and India—show that *Williamsonia* flowers were associated with stems having foliage of the *Zamites*, *Otozamites* and *Ptilophyllum* type. A well-preserved *Williamsonia*, *W. scotica*, has been described from Jurassic rocks in Scotland[2] showing very clearly the structure of the interseminal and seed-bearing scales and of the enveloping leaves (bracts). More recently a very similar type has been found in the Rajmahal series of India and named by Prof. Sahni *Williamsonia Sewardi*: this "flower" was almost certainly borne on a stem having the characters of *Bucklandia* and closely resembling those of some existing cycads: the fronds were of the form known as *Ptilophyllum*. Several species of *Williamsonia* have been described from Yorkshire, Sardinia, Mexico and other regions. The genus is characteristic of Jurassic floras, particularly in Europe, India and Mexico: it persisted as a comparatively rare plant into the Cretaceous period and began, so far as we know, in the latter part of the Triassic period.

Cycadeoidea

There is another and still more prominent genus, *Cycadeoidea*, which reached its maximum development in the early part of the Cretaceous period and was a conspicuous plant in some regions of the world in the latter part of the Cretaceous period. Several hundred beautifully preserved specimens have been collected from Upper Jurassic and Lower Cretaceous rocks in North America, specially from the Black Hills, a spur of the Rocky Mountains (fig. 103, XX, B). The American stems and flowers are fully described and illustrated by Dr Wieland[3] of Yale University to whom palaeobotanists are greatly indebted. Some excellent material has been described also from Upper Jurassic beds of southern England and from Lower Cretaceous localities in northern Italy and Galicia. Externally and in several anatomical features a *Cycadeoidea* stem recalls those of many existing cycads. One of the largest examples (1 metre 18 cm. in height) from an Upper Jurassic quarry in the Isle of Portland[4] may be seen in the Fossil Plants Gallery of the British Museum. The thick stem, with a conical bud at the summit, is encased in an armour of leaf-bases which were covered with chaffy

[1] Wieland (14[2]). [2] Seward (12). [3] Wieland (06), (16). [4] Seward (97).

scales as in modern cycads: the fluffy mass of the scales became saturated with the petrifying water more readily than the harder and less porous stumps of the fronds (leaf-bases), with the result that the scales are preserved as a prominent network enclosing hollowed out areas (meshes) produced by the partial decay of generations of leaf-stalks before petrification was complete. This stem, unlike the great majority of similar fossils, affords no indication of the presence of any flowers or fertile branches.

A feature characteristic of *Cycadeoidea* is the occurrence on a single stem of numerous flowers borne at the ends of short lateral branches which project hardly at all beyond the general level of the armour of persistent leaf-bases. In some examples several hundred flowers occur on one stem and all of about the same age: it is possible that the plant reached a certain stage of development, when the necessary store of food had been prepared, and then burst into blossom and died, a procedure illustrated by the so-called century plant (*Agave americana*). Each flower is terminal on a short stalk prolonged apically into a broad tapering cone or in some species forming a rounded dome. The great majority of flowers are bisexual. At the base of the conical column is a collar formed of the concrescent bases of branched pollen-bearing appendages, which at maturity bend outward as a fringed circle of leaf-like stamens. The higher portion of the column is covered with appendages, most of which are sterile interseminal scales, while others, the ovuliferous scales, are more slender stalks each bearing a terminal ovule enclosed in an integument prolonged just above the level of the surrounding interstitial scales as a tubular micropyle for the reception of the pollen. There is reason to believe that the two sets of reproductive organs were not ripe at the same time, the pollen being shed before the full development of the ovules, a feature familiar in many flowers in which cross-pollination is favoured by the separation in time of the ripening of the stamens and the pistil. Each flower is protected, as in *Williamsonia*, by a series of flat overlapping scales. The portion of a splendid stem, *Cycadeoidea* (*Raumeria*) *Reichenbachiana*,[1] in the Dresden Museum, reproduced in fig. 98, shows several flowers bearing both male and female organs interspersed among the bases of old fronds. The Dresden specimen is a piece of a stem half a metre in diameter and was probably at least a metre high when complete: it was discovered in Cretaceous beds of the

[1] Schuster (11); Kräusel (25^2).

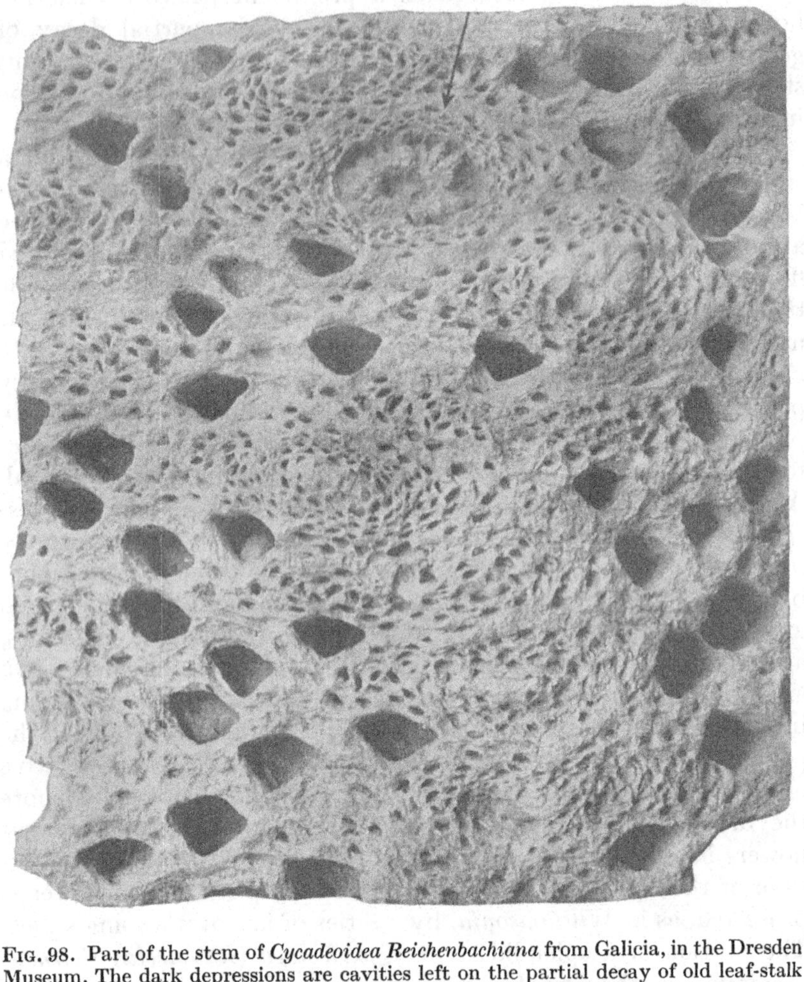

FIG. 98. Part of the stem of *Cycadeoidea Reichenbachiana* from Galicia, in the Dresden Museum. The dark depressions are cavities left on the partial decay of old leaf-stalk bases before petrification: the arrow (upper edge of photograph) points to one of several fertile shoots ("flowers") surrounded by a spiral series of small scale-leaves (bracts). About ½ nat. size. (The photograph, given to me by Dr Wieland, is reproduced by permission of Dr Wanderen of the Geological and Mineralogical Museum, Dresden.) For a photograph of the whole specimen see Schuster (11).

Galician Carpathian Mountains more than 150 years ago. The larger and roughly triangular depressions show the position of old leaf-bases which were destroyed before petrification occurred, and the numerous, smaller depressions mark the position of the much smaller leaves (bracts) associated with the fertile branches; the large oval area, to which an arrow points, includes 16 pollen-bearing leaves (staminate appendages) borne near the tip of a bisexual flower. On the surface of the half-stem of an unusually large American species, *Cycadeoidea Dartoni* (fig. 99) there are between five and six hundred flowers all produced during one blossoming period which may well have proved fatal. The arrows on the side of the stem point to some of the laterally borne fertile branches packed among the stumps of old fronds. In fig. 100 some of the apical portions of these flowering shoots are shown in cross-section: in the centre of each is the solid axis of the receptacle—the conical end of the flower-axis—and arranged in a circle at the periphery can be seen several small seeds marking the position of the fertile scales among the more numerous interseminal scales. This specimen affords a good example of what is known as a monocarpic plant: all the flowers are at the same stage of development, an indication that the stem had reached the critical stage in its life when it was petrified.

Cycadeoidea differs from existing cycads in the bisexual nature of the flowers, in the structure of both male and female organs and in the very much smaller seeds. The embryo practically filled the seed-cavity, whereas in modern cycads the embryo is embedded in a mass of reserve food. For a full account of this remarkable extinct group of plants reference should be made to the descriptive volumes by Dr Wieland.

The important point is that in the Jurassic period and the early stages of the Cretaceous period the Bennettitales played a prominent part in the vegetation of the world. The wide range in form and size of the fronds is evidence of the existence of many genera and species. The stems, particularly in the late Jurassic and early Cretaceous examples, bear a striking resemblance to those of living cycads: but those of a few older types, such as *Wielandiella* and *Williamsoniella*, have an entirely different habit. It is the nature of the reproductive shoots which presents the greatest puzzle: can we connect them with the corresponding organs in the cycads of the present day, with the flowers of angiosperms, or with any other

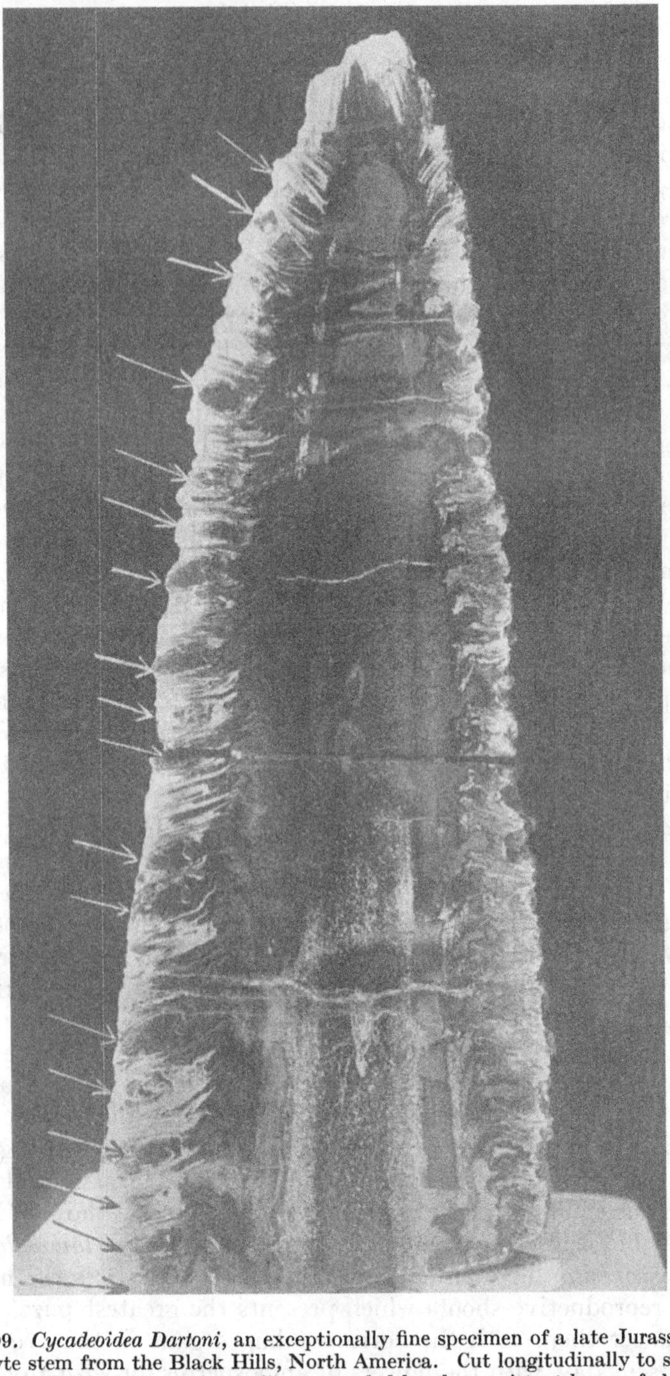

FIG. 99. *Cycadeoidea Dartoni*, an exceptionally fine specimen of a late Jurassic cycadophyte stem from the Black Hills, North America. Cut longitudinally to show the numerous fertile shoots ("flowers") surrounded by the persistent bases of old foliage fronds. The arrows point to the flowers (darker colour) some of which are seen in transverse view in fig. 100. $\frac{1}{5}$ nat. size. (Photograph supplied by Dr Wieland.)

plants? It has been suggested that resemblances between the ovules of *Cycadeoidea* and those of the Gnetales—an existing group of gymnosperms remotely connected with other members of that class —may be evidence of affinity, but the features in common may not be more than instances of parallel development. Attention has also been called to certain apparent resemblances between bennettita-

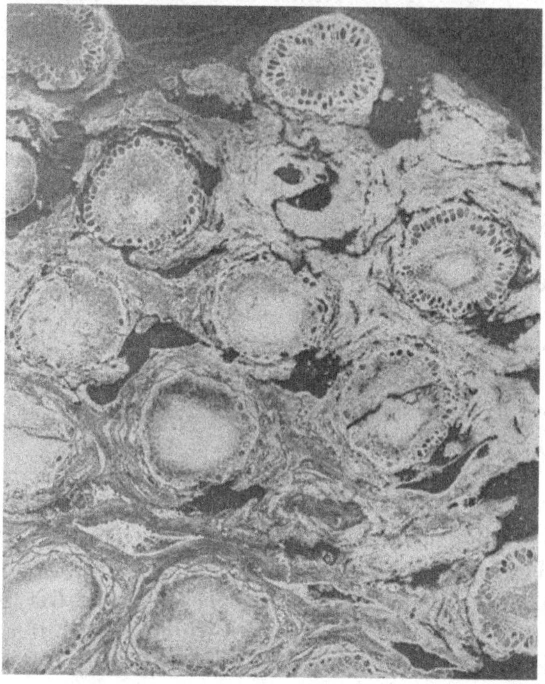

FIG. 100. *Cycadeoidea Dartoni*. Surface-view of a small portion of the stem: the fertile shoots are seen in transverse section; on each of them are several small seeds (white) arranged round the cylindrical axes. (Photograph supplied by Dr Wieland.)

lean flowers and those of the flowering plants,[1] though it is acknowledged that the similarity is rather superficial than real. Superficially there is a striking similarity between the covering of interseminal scales and ovules on the axis of a *Cycadeoidea* flower and the massed flowers in an inflorescence of some existing aroids and in such inflorescences as those of the tropical *Cecropia*[2], a member of the fig and mulberry family; but in these and other flowering plants the ovules are enclosed in a carpel.

[1] Scott (24). [2] Bailey, I. W. (22).

The living cycads are members of a class—the Cycadophyta—another section of which reached its zenith in the early days of the Cretaceous period and then rapidly dwindled, and, so far as we know, left no direct descendants. There are hardly any satisfactory records in Jurassic or in other rocks[1] of cycadean remains affording definite evidence of the existence of cycads with reproductive organs like those of the present representatives of the Cycadales. We know nothing of genuine connecting links between the Bennettitales and recent cycads. All that can be said is that the cycads, as we know them, may represent a branch of some ancient stock from which in the course of the Mesozoic era was evolved another set of plants, *Cycadeoidea* and its allies, destined to attain the rank of a ruling dynasty at a stage in geological history immediately preceding that at which the flowering plants assumed, with startling suddenness, an ascendancy that is still maintained. It would appear that the Bennettitales left no direct offspring; that their relationship to the plants represented by a few traces of what appear to be cycadean fronds in the later Palaeozoic strata is unknown; and that the surviving cycads came from ancestors distantly related to the Bennettitales whose history has still to be written.

Ginkgoales: Conifers

As in the Rhaetic vegetation so also in that of the Jurassic period the Ginkgoales flourished over the greater part of the earth. *Baiera*, though still conspicuous as a Jurassic genus, had previously reached its culminating point in the Rhaetic period, while *Ginkgoites* seems to have been even more abundant and widely distributed in Jurassic floras. It is unfortunate that, despite the great number of leaves, we know very little about the reproductive shoots; but the discovery of a few specimens of male organs, especially in Jurassic rocks of Siberia, furnishes some evidence confirmatory of the close relationship, suggested by leaf-form and structure, of the later Mesozoic *Ginkgoites* with the sole living relic, *Ginkgo biloba*. The two genera *Czekanowskia* and *Phoenicopsis*, previously mentioned as Rhaetic plants, reached their greatest development in Jurassic floras. Another less abundant representative of the Ginkgoales is the genus *Eretmophyllum*[2] distinguished by its simple paddle-like

[1] For a brief account of some Triassic cycadophytes which may be more closely related to living cycads, see Chapter XIII, p. 318.

[2] Thomas, H. H. (13).

leaves, 10–12 cm. in length, agreeing in the structure of the stomata and in the presence of secretory tracts between the veins with the leaves of *Ginkgo*.

A notable feature of Jurassic vegetation is the abundance of conifers: they appear to have played a more important part in Jurassic than in Keuper and Rhaetic floras. To this statement a word of caution should be added: we estimate the abundance or scarcity of a group on a basis which may be misleading. Jurassic conifers have a more modern look than some earlier plants which are perhaps equally entitled to be included in the group. As we follow the history of a set of plants backwards through the Mesozoic era we find a gradual falling off in the familiar types; we get farther away from our modern standards, and we are apt to forget that extinct forms, having but a remote likeness to living genera, may be ancestors whose strangeness obscures their affinity to distantly related members of one and the same line of evolution. The genus *Araucarites* became more widely spread in the course of the Jurassic period, particularly species reminding us vividly in their foliage and seed-bearing cones of existing species such as *Araucaria excelsa, A. Cookii* (fig. 102), and others included in the Eutacta section of the genus, in contrast to *Araucaria imbricata* (the monkey puzzle) and a few other species forming the section Colymbea characterized by their broader and flatter leaves and by distinctive features of the seed-scales. Petrified wood of conifers is exceedingly common, and much of this exhibits clearly marked araucarian affinities. There were conifers bearing needles like those of pines and other examples of the Abietineae, a family which, especially in the northern hemisphere, has now the greatest number of genera and species. It is significant that although several of the older conifers, including a few Palaeozoic and early Mesozoic genera, such as *Voltzia,* furnish indications of an abietineous connexion, it is the Cretaceous and Tertiary rocks which afford the first proof of the success of a family which is now the most vigorous of its class.

The genus *Sequoiites,* closely allied to the living sequoias of California, is recorded from a few Jurassic localities: it became much more prominent in the course of the Cretaceous period and was almost cosmopolitan in the early Tertiary vegetation. It would be interesting to know why the genus *Sequoia,* which is now confined to a narrow strip of country in western America where its stems grow taller than those of any other living conifer, should have long ago disappeared from the rest of the world where it was recently, in a geological sense, ubiquitous.

Our knowledge of Mesozoic conifers is lamentably incomplete:[1] there are many genera and species represented by pieces of sterile foliage-shoots, some bearing long and narrow leaves in two ranks as in the yew, the redwood tree (*Sequoia sempervirens*) and several others; some with crowded and more or less sickle-shaped leaves like those of *Araucaria excelsa* and *Cryptomeria*, and others with still smaller leaves closely pressed against the branches as broad triangular scales like those of the Tasmanian *Athrotaxis*. It is seldom that the fossil twigs bear cones well enough preserved to be used as tests of affinity. In the absence of cones or other reproductive organs we turn to such anatomical characters as are furnished by petrified wood and to the information supplied by a microscopical examination of the mummified surface-layers (cuticles) of leaves. The examination of fossil wood has shown that with araucarian conifers there were others having stems agreeing structurally with members of the cypress family, with living species of the podocarp family that is now characteristic of tropical forests, with abietineous and other genera. Dr Florin of Stockholm has in preparation a monograph of fossil conifers based on the characters of the surface-layers of leaves which will lay the foundations of a more satisfactory record of the family-histories of the class than it has so far been possible to obtain.

Reference should be made to a few common Jurassic genera represented as a rule by sterile foliage-shoots. A widely spread genus is *Pagiophyllum* with *Pagiophyllum Williamsoni* as a common Jurassic species: this is often described as araucarian, because of a resemblance to the leafy branches of *Araucaria excelsa*, though its actual nature is not absolutely certain. Another characteristic species, *Brachyphyllum expansum*, or *Thuites expansus*, is one of several forms of foliage-shoot characterized by small triangular leaves closely appressed to the branch and varying between a spiral and a verticillate arrangement: when the small leaves are in regular alternating pairs the specimens bear a close resemblance to twigs of *Cupressus* (cypresses) and *Thuja* (e.g. the *Arbor Vitae*), but proof of affinity is generally lacking. Other forms of branches having spirally attached, narrow leaves spread out in one plane are fairly abundant in Jurassic fossils, but it is seldom possible accurately

[1] See Sahni (28[2]) for a revision of Jurassic conifers from India.

to determine their relationship to existing conifers: shoots such as these are often included in *Elatocladus*,[1] one of several form-genera, that is generic designations which have reference to external characters insufficient to serve as criteria of relationship to living genera.

The gymnosperm *Podozamites* of uncertain affinity, already well represented in Rhaetic floras, was still more plentiful in Jurassic forests. It is undoubtedly true that in the course of the Jurassic period conifers of a modern type increased numerically and in variety and played a less subordinate part as trees than in earlier ages. Research into the structure of the abundant stems, often preserved as pieces of drift-timber in Jurassic rocks, has shown that many of the anatomical types illustrated by existing genera had already been evolved; but not infrequently the admixture in a single stem of characters now found in distinct genera adds to the difficulty of strict correlation between extinct and living conifers.

Possible Representatives of the Pteridospermeae: Jurassic Angiosperms

The genus *Thinnfeldia*, previously described as a characteristic element in Triassic and Rhaetic floras, persisted in reduced numbers into the Jurassic period: with it were other genera such as *Pachypteris*, *Dichopteris*,[2] *Lomatopteris*, and others all characterized by comparatively thick leaflets with obscure venation and by the absence of any reproductive organs on the foliage-shoots. It may be that in these plants we have a few survivors of the Palaeozoic pteridosperms which were shrubs and trees bearing branches with rows of small leaves like the leaflets of ferns: the branches are often forked as in the southern hemisphere thinnfeldias (*Dicroidium*), and in the Permian fronds of similar habit described in a previous chapter under the generic name *Callipteris*.

A single and imperfectly preserved specimen obtained many years ago from Stonesfield in England has a special interest for students of evolution who search for traces of flowering plants among the remains of Jurassic vegetation. The fossil[3] is an impression of an oval leaf a few centimetres long resembling in form and in the arrangement of the main ribs the leaf of a dicotyledon

[1] For an account of this and other conifers and references to original sources see Seward (19).
[2] Grandori (13).
[3] For a drawing of the leaf see Seward (04), Pl. XI, figs. 5 and 6.

such as a poplar (*Populus*); unfortunately the imperfect state of preservation precludes any comparison of the finer venation or of the structure of the surface-cells with the corresponding features of modern foliage. In itself this solitary specimen hardly proves the existence of an angiospermous tree, though in view of a recent discovery by Dr T. M. Harris of what appears on more substantial evidence to be the leaf of a dicotyledon in the Rhaetic plant-beds of Greenland, the Stonesfield specimen acquires an enhanced importance. It is probable that the almost complete absence of fossil angiospermous leaves in Jurassic and older Mesozoic rocks is due, not to the lack of flowering plants in the world, but to their failure to be preserved as fossils because they occupied a tract of country remote from localities where the conditions were favourable for fossilization. Some French fossils described by Prof. Lignier as *Propalmophyllum*[1] bear a striking resemblance to pieces of a palm leaf, though they cannot be accepted as proof of the existence of palms in a Jurassic flora.

Reference has already been made to the genus *Sagenopteris* (fig. 85, *Ss.*):[2] this Upper Triassic and Rhaetic plant had an almost world-wide distribution in the Jurassic period. In the restoration of a Rhaetic estuary shown in fig. 85, *Sagenopteris* (*Ss.*) is represented as the foliage of a plant bearing reproductive shoots of the type described by Dr Thomas under the names *Caytonia* and *Gristhorpia*, some forming female catkins and others male catkins. The generic name *Gristhorpia*, after the famous locality Gristhorpe on the Yorkshire coast, is given to fertile stalks a few centimetres long bearing two rows of berry-like fruits 2–5 mm. in diameter. One of these fruits is shown in fig. 101: within it are several closely packed seeds: a small lip projecting from the wall of the fruit near its attachment to the pedicel is described as a stigma, that is the knob-like or expanded summit of the pistil in an ordinary flower which receives the pollen. *Caytonia* differs from *Gristhorpia* in the less prominent stigma and in having fewer seeds. In both genera the seeds are enclosed and not

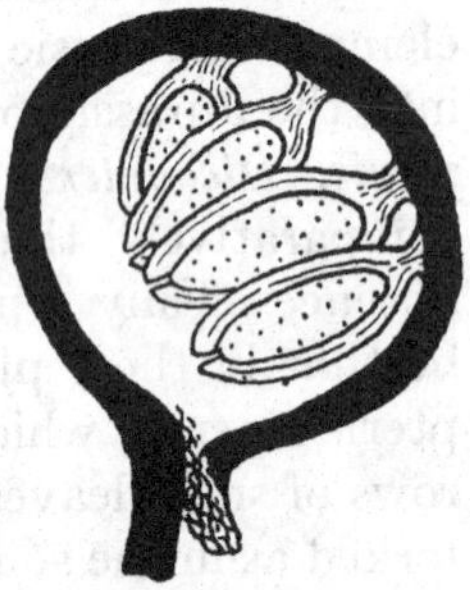

FIG. 101. Diagrammatic sketch of a longitudinal section of the Jurassic *Caytonia Sewardi* showing the arrangement of the seeds within a carpel-like case, also the stigma projecting from the wall close to the stalk. ×6. (After Dr Hamshaw Thomas.)

[1] Lignier (07), (07[2]). [2] Page 309.

exposed: the enclosing organ is compared with the carpel, or seed-vessel, of a flowering plant. It has been pointed out that the angiospermous carpel represents a single, modified leaf which has become folded over the ovules which it bears; whereas in the Jurassic fossils the whole fertile stalk is regarded by Dr Thomas as equivalent to a single leaf bearing several carpels. There are undoubtedly, as one would expect, differences between the Caytoniales and the present flowering plants: the important fact is the occurrence of seeds within a closed case provided with a pollen-catching stigma, a feature not found in any other Jurassic plants. Whether or not the term carpel should be used for the seed-containing vessels of the extinct plants is merely an academic point and of secondary importance. The male organs[1] of *Gristhorpia* and *Caytonia*, described under a separate name, *Antholithus* (literally a stone flower), because they have not been found in actual union with the female specimens, consist of an axis bearing several short lateral branches to which are attached clusters of stamens containing winged pollen-grains. It is noteworthy that pollen of this type was formerly described from Lower Jurassic beds in Sweden and assigned to a conifer: a similar form of pollen-grain was recorded also from older rocks in Antarctica.

Examples of Caytoniales have been found in Rhaetic beds in eastern Greenland and in Jurassic rocks in Sardinia.[2] The grounds on which *Sagenopteris* is believed to be the foliage of caytonialean plants are: its constant association with the reproductive organs and the resemblance in microscopical structure of the epidermal layer of the leaves to that of the corresponding layer of the seed-bearing stalks of *Caytonia* and *Gristhorpia*.

There are fragments in Jurassic as in other strata which are too small or incomplete to be assigned to the parent-plant or to any precise position in the vegetable kingdom: some of these scraps may belong to plants which, had they been preserved intact, might fill many a gap in the plant-record. An example of a fossil of this kind is a comparatively small oval leaf-like structure described by Prof. Halle[3] from Cloughton Wyke (Yorkshire) as *Cloughtonia rugosa*: what it is we cannot tell. As Halle suggests, it resembles a large petal; possibly it is part of the floral apparatus of a gymnosperm; or it may have been a coloured appendage from the fertile shoot of some primitive flowering plant. Reference may also be made to

[1] Thomas, H. H. (25). [2] Edwards (29^2). [3] Halle (11^3).

some seed-like bodies recently described from a Jurassic flora in Tiang-Chan (fig. 103, XVI, T) under the name *Problematospermum elongatum*:[1] the published photographs suggest the possibility that they may be fruits, but their precise nature is uncertain.

It has been established that as far back as the Rhaetic and Jurassic periods there were plants which bore seeds in closed cases or ovaries: a character that is one of the essential attributes of the angiosperms. The occurrence of several kinds of dicotyledonous wood, showing no sign of primitive characters, in Lower Cretaceous rocks[2] strongly supports the view that the trees were far removed in time from the earliest representatives of the present dominant class. Prof. Kräusel has recently described some imperfectly preserved wood from Jurassic rocks in Germany under the name *Suevioxylon zonatum*,[3] distinguished by the occurrence of large tubes (vessels), a character unknown in living gymnosperms except the Gnetales, and a characteristic feature of angiosperms. We shall return to the problem of the antiquity of the angiosperms in the next chapter.

A Jurassic Flora from the borders of Antarctica

In illustration of the wide geographical range of Jurassic plants and of the impressive similarity between floras of approximately the same age separated by many thousands of miles and preserved in regions which are now characterized by sharply contrasted climatic conditions, a comparison is made in tabular form (Table E) of a flora from Graham Land with floras from North America, Europe and India. Of the 61 different kinds of plant-remains described by Prof. Halle[4] from Hope Bay in Graham Land in latitude 63° 15′ S., 25 species are spoken of by him as ferns, though probably a few may be pteridosperms; 19 are classed as Cycadophyta, 13 as conifers, 2 as gymnosperms of uncertain relationship, and 1 is described as of unknown affinity. In some respects, notably in the absence of Ginkgoales, of *Podozamites*, and of some common Jurassic cycadean leaves, the Graham Land vegetation differs from that in most other parts of the world. It is worthy of notice that similar indications of regional peculiarities in Jurassic vegetation are given by floras in other parts of the world: in India, for example, the Ginkgoales

[1] Turutanova-Ketova (30).
[2] Stopes (12).
[3] Kräusel (28).
[4] Halle (13[2]).

TABLE E. *The Geographical Distribution of Jurassic plants from Graham Land*

Note. + = identical species; – = closely allied species.
The species marked with an asterisk are peculiar to Graham Land, though often hardly distinguishable from European or other types

	North America (fig. 103, XX, Ov and O)	England and other parts of Europe (fig. 103, VIII)	INDIA: Rajmahal series (fig. 103, XIII, R)	INDIA: Madras coast (fig. 103, XIII, M)	INDIA: Cutch and Jabalpur series (fig. 103, XIII, C and J)	
**Equisetites approximatus*	.	–	–	.	.	—
Sagenopteris paucifolia	+	+	.	.	.	Probably identical with an Australian species
Dictyophyllum, sp.	.	.	.	.	.	A fragment too small for identification. The genus is characteristic of Rhaetic-Jurassic floras in many parts of the world
Todites Williamsoni	+	+	.	.	.	Recorded also from Eastern Asia and elsewhere
Cladophlebis denticulata	+	+	.	.	.	One of the commonest forms of Jurassic ferns
**Cladophlebis oblonga*	.	–	.	.	.	Very similar to a Polish species
**Cladophlebis antarctica*	.	.	.	.	.	Resembles *Cladophlebis denticulata*
Cladophlebis (*Coniopteris* ?) *arguta*	.	+	.	.	.	Probably a member of the Cyatheaceae, a family that is now mainly though not exclusively tropical
Cladophlebis (*Eboracia* ?) *lobifolia*	–	+	.	.	+	This species occurs also in China (fig. 103, XII, D): probably cyatheaceous
Cladophlebis (*Klukia*?) *exilis*	.	+	.	.	.	Recorded also from the Caucasus (fig. 103, XI, Cs), Japan and Madagascar, a member of the Schizaeaceae
Coniopteris hymenophylloides	+	+	.	.	.	A cyatheaceous fern very common in the Jurassic period
Coniopteris, cf. *nephrocarpa*	.	+	.	.	.	Similar to *Coniopteris hymenophylloides*
Coniopteris (?) *lobata*	.	+	+	.	.	—
Sphenopteris (*Ruffordia*?) *Goepperti*	+	+	.	.	.	This species occurs also in Lower Cretaceous rocks in the Far East, Europe, and North America
**Sphenopteris Nordenskjöldii*	.	.	.	.	.	—
Sphenopteris Nauckhoffiana	.	.	.	.	.	A Lower Cretaceous Greenland species
Sphenopteris Fittoni	.	+	.	.	.	A Lower Cretaceous species recorded also from South Africa and New Zealand
**Sphenopteris antarctica*	.	–	.	.	.	Closely resembles a Jurassic species from Portugal
Sphenopteris Leckenbyi	.	+	.	.	.	—
**Sphenopteris Anderssonii*	.	.	.	.	.	Not unlike *Coniopteris hymenophylloides*
**Sphenopteris pecten*	.	.	.	.	.	—
**Scleropteris crassa*	.	.	.	.	.	Similar to species from Jurassic rocks in France
**Scleropteris furcata*	.	.	.	.	.	(as above)

Table E. *The Geographical Distribution of Jurassic plants from Graham Land* (cont.)

Note. + = identical species; – = closely allied species.
The species marked with an asterisk are peculiar to Graham Land, though often hardly distinguishable from European or other types.

	North America (fig. 103, XX, Ov and O)	England and other parts of Europe (fig. 103, VIII)	INDIA: Rajmahal series (fig. 103, XIII, R)	INDIA: Madras coast (fig. 103, XIII, M)	INDIA: Cutch and Jabalpur series (fig. 103, XIII, C and J)	
Pachypteris dalmatica	.	+	.	.	.	This genus is probably allied to *Thinnfeldia* and may not be a true fern: it occurs in Lower Cretaceous rocks in the island of Lesina off the Dalmatian coast
**Thinnfeldia constricta*	.	.	–	.	.	Compare Indian and Australian species
**Nilssonia taeniopteroides*	–	–	.	.	.	One of the most characteristic and conspicuous species in the Hope Bay flora, nearest to a Californian species and one described from Kamenka (fig. 103, XI, K)
**Pseudoctenis ensiformis*	.	–	–	.	.	—
Pseudoctenis, cf. *P. Medlicottiana*	–	.	+	.	.	—
**Zamites pusillus*	.	.	.	.	.	The species referred by Halle to *Zamites* may well be generically identical with *Ptilophyllum*
**Zamites Anderssonii*	.	.	.	.	.	This and other fronds from Hope Bay are hardly distinguishable from Lower Cretaceous *Ptilophyllum* species described from Greenland
**Zamites antarcticus*	.	.	.	.	.	Resembles a Pennsylvanian Triassic species
**Zamites pachyphyllus*	.	.	.	.	.	—
**Otozamites linearis*	.	–	.	.	.	—
Otozamites latior	.	+	.	.	.	Occurs also in Triassic rocks of eastern North America
Otozamites Hislopi	.	.	.	+	+	—
Otozamites abbreviatus	.	.	+	+	.	—
Ptilophyllum (?*Williamsonia*) *pectinoides*	.	+	.	.	.	—
**Williamsonia pusilla*	.	–	.	.	.	A small *Williamsonia* flower resembling a species from Bornholm
Araucarites cutchensis	.	–	.	+	+	Cone-scales of the *Araucaria excelsa* (Norfolk Island pine) type
Pagiophyllum, cf. *P. crassifolium*	.	+	.	.	.	A Wealden, European species
Pagiophyllum, cf. *P. Heerianum*	.	+	–	.	.	—
**Pagiophyllum Feistmanteli*	.	.	.	+	.	Recorded from India under another name
Sphenolepidium ? *oregonense*	+	.	.	.	.	—
**Elatocladus heterophylla*	.	.	.	.	.	A species characterized by dimorphic foliage shoots; cf. the Rhaetic *Stachyotaxus*
Elatocladus conferta	.	.	+	+	.	Very similar to an Australian species
Elatocladus jabalpurensis	.	.	.	+	+	—

were barely represented and by none of the common species. The Graham Land fossils are preserved near the place where the plants grew, in a land that has now an Antarctic climate and on which members of the Swedish South-Polar Expedition collected only one flowering plant,[1] a grass, *Deschampsia* (*Aira*) *antarctica*, in addition to mosses and lichens. As Halle points out, and the illustrations fully confirm his statement, the plants show no sign of stunted growth or any of the features which we now associate with the vegetation of polar regions. If the present position of Graham Land in relation to the South Pole is the same as it was in the middle of the Jurassic period the problem of climate is raised in an acute form. On the borders of Antarctica, in a glaciated country practically barren of vegetation, we find a comparatively rich flora strikingly similar in its general composition, and in no way inferior in the development of the fern fronds, or of the leaves and foliage-shoots of gymnosperms, to the flora furnished by the Jurassic rocks of Yorkshire. This astonishing fact may well incline us favourably to consider the possibility of drifting continents and to hazard the opinion that Graham Land was not always where it is to-day.

[1] Skottsberg (12).

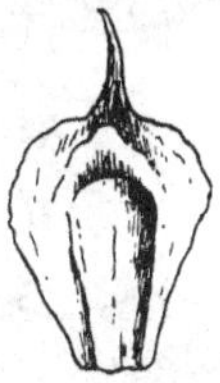

FIG. 102. Seed-scale from a cone of *Araucaria Cookii*; slightly reduced.

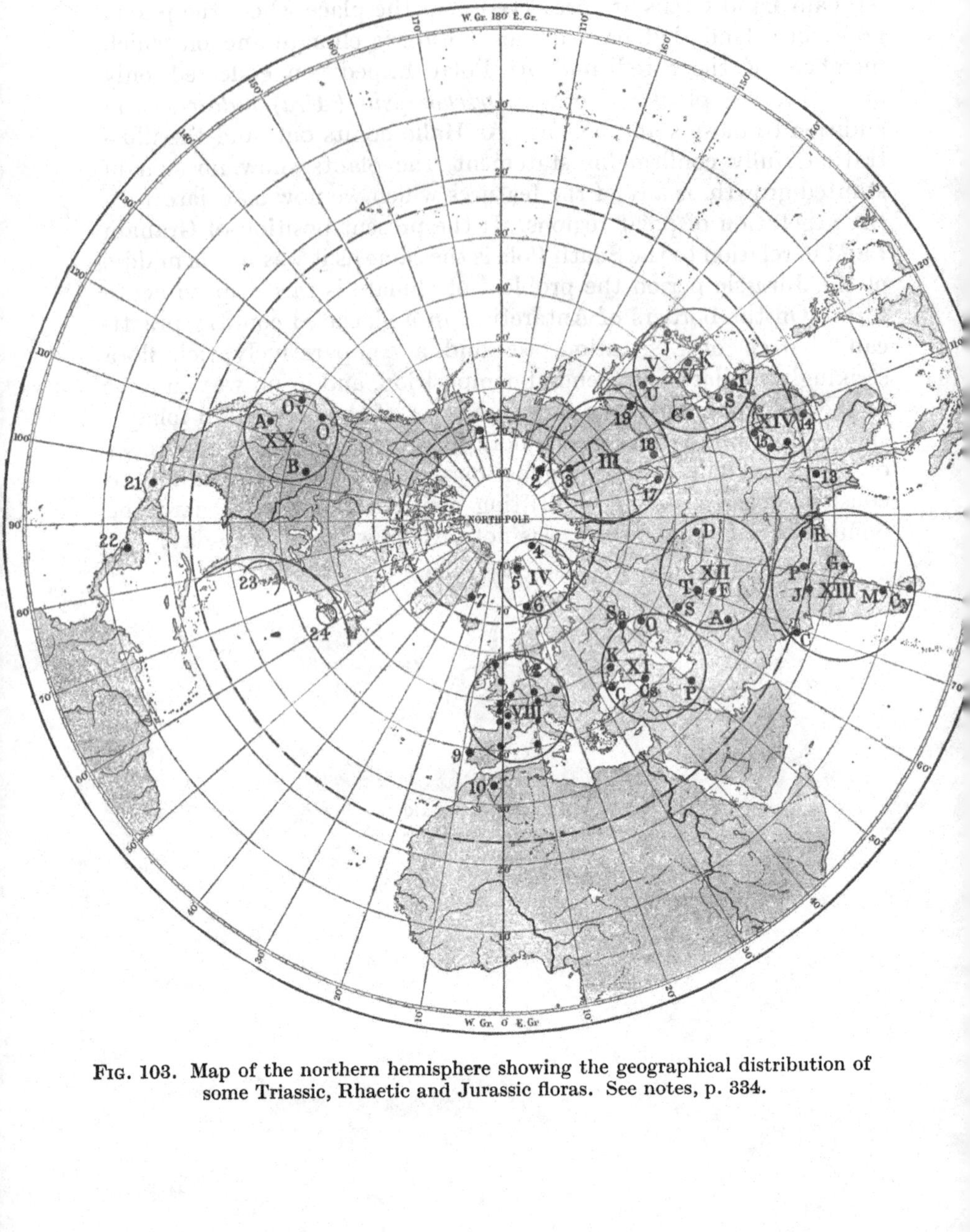

FIG. 103. Map of the northern hemisphere showing the geographical distribution of some Triassic, Rhaetic and Jurassic floras. See notes, p. 334.

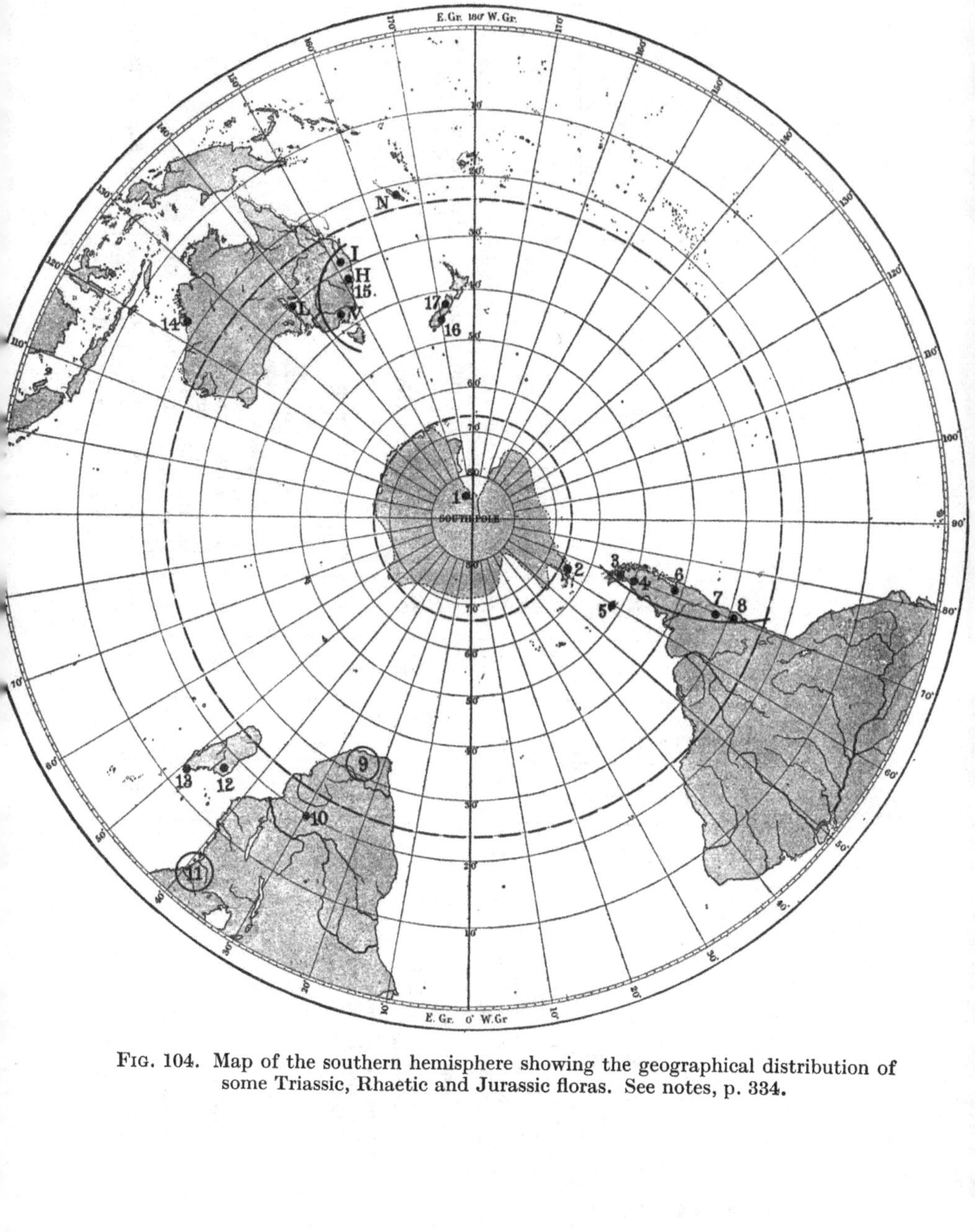

FIG. 104. Map of the southern hemisphere showing the geographical distribution of some Triassic, Rhaetic and Jurassic floras. See notes, p. 334.

Number on Map (fig. 103)	Localities	Trias: horizon not specified	Trias. L = Lower (Bunter or Muschelkalk). U = Keuper	Rhaetic	Jurassic	References	
	ARCTIC REGIONS						
1	Alaska	.	.	.	J	Knowlton (14), (16[2])	
2	New Siberian Islands	T	.	.	J	Nathorst (07)	
3 (III)	Northern Siberia (Lena River)	.	.	.	J	Heer (78)	
4 (IV)	Franz Josef Land	.	.	R	J	Newton and Teall (97), (98); Nathorst (99[2]); Solms-Laubach (04)	
5	Spitsbergen	.	.	.	J	Nathorst (97), (97[2]), (13); Gothan (10), (11); Walton (27[2])	Some of the beds are probably Lower Cretaceous
6	Lofoten Islands (Andö)	.	.	.	J	Johannson (20)	
7	East Greenland	.	.	R	J	Hartz (96); Harris (26)	
	EUROPE WITH NORTH AFRICA AND ASIA (excluding India, etc.)						
VIII.	British Isles	T	L	R	J	T. Wills (10); Vernon (10); R. Seward (04); J. (Scotland), Seward (11); Stopes (07); (England), Seward (00), (04); Thomas, H. H. (12–15[2]), (22), (25), (30); Fox-Strangways (92); Fox-Strangways and Barrow (15); Black (29)	
	Sweden	.	.	R	J	Nathorst (78), (78–86); Antevs (19); Chow (24); Johannson (22); Möller and Halle (13)	
	Bornholm	.	.	R	J	Möller (02), (03); Bartholin (92), (94), (10)	
	France	T	L	R	J	T. Fliche (10); Schimper and Mougeot (44); R. and J. Zeiller (11[3]); J. Saporta (73–91); Lignier (95), (07), (13)	See also Saporta (73–91)

		T	L	R	J		
	Spain	.	.	.	J	Zeiller (02^{2})	
	Germany	T	L	R	J	T. Blanckenhorn (86); Hörich (10), (12); Frentzen (20), (22), (26); Schlüter and Schmidt (27); Lilienstein (28); Kräusel (28); R. Schenk (67); Salfeld (07); J. Gothan (14); Salfeld (07), (09)	The plants described by Schenk (67) include several Jurassic species. See Gothan (14)
	Central European States	.	U	.	J	Krasser (08), (09), (09^{2}), (19); Kräusel (21), (21^{2})	
	Switzerland	.	U	.	J	T. Leuthardt (03), (04); J. Heer (76)	
	Poland	.	.	R	J	Raciborski (92), (94)	
	Sardinia	.	.	.	J	Krasser (13), (20); Edwards (29^{2})	
	Italy	.	.	.	J	Zigno (56–85); Grandori (13)	
9	Portugal	.	.	R	J	Saporta (94)	
10	North Africa	.	.	.	J	Edwards (26^{3})	
XI.	C. The Crimea	.	.	.	J	Kryshtofovich (13)	
	K. Kamenka	.	.	R–	J	Thomas, H. H. (11)	
	Cs. Caucasus	.	.	.	J	Seward (07^{3}); Kryshtofovich (26^{2})	
	O. Orenburg	.	.	R–	J	Kryshtofovich (12)	
	P. Persia	.	U–	R	.	Schenk (87); Krasser (91)	
	Sa. Samara region, South Russia	.	.	.	J	Prinada (28)	
XII	S. Turkestan (Syr-Darja)	.	.	R–	J	Seward (07^{3})	
	T. Turkestan	.	.	.	J	Turatanova-Ketova (29)	
	A. Afghanistan	.	.	R–	J	Seward (12^{2})	
	F. Ferghana (Turkestan)	.	.	.	J	Seward (07^{3})	
	D. Chinese Dzungaria	.	.	.	J	Seward (11^{2})	
XIII.	INDIA						
	C. Cutch	.	.	.	J	Feistmantel (76)	For a general account of the Mesozoic floras of India, see Sahni (22); for descriptions of plants, see also Seward and Sahni (20) and Sahni (28^{2}). Matley (21) suggests a Lower Cretaceous age for the Jabalpur beds
	J. Jabalpur	.	.	.	J	Feistmantel (77^{3})	
	P. Parsora	.	U–	R	.	Sahni (22)	
	R. Rajmahal	.	.	R–	J	Feistmantel (77)	
	G. Godaveri	.	.	.	J	Feistmantel (77^{2}), (82), (85)	
	M. Madras	.	.	.	J	Feistmantel (79)	
	Cy. Ceylon	.	.	.	J	Seward and Holttum (22)	
13	Burma (Shan States)	.	.	.	J	Sahni (28^{2})	
14 (XIV)	Tongking (Indo-China)	.	.	R	.	Zeiller (03); Pelourde (13)	
15	China, Kia-ling River; Szechwan, etc.	T	.	R	J	Krasser (00); Halle (27^{2}); Zeiller (00)	
XVI.	C. Manchuria	.	.	.	J	Krasser (05)	
	S. Shantung	.	.	R	J	Yokoyama (06); Yabe and Ōishi (28)	Plants from several localities in China are described by Yokoyama

Number on Map (fig. 103)	Localities	Trias: horizon not specified	Trias. L = Lower (Bunter or Muschelkalk). U = Keuper	Rhaetic	Jurassic	References	
	T. Tiang-Chan	.	.	R	.	Zeiller (00); Turutanova-Ketova (29)	
	U. Ussuriland, etc.	.	.	.	J	Kryshtofovich (10)	
	V. Vladivostock	.	L	.	.	Kryshtofovich (23)	
	J. Japan	.	T	R	J	Yokoyama (89), (05); Ŏishi (30); Yabe and Toyama (28)	Some of the Japanese beds are Lower Cretaceous in age
	K. Korea	.	.	R	J	Yabe (22); Kawasaki (25), (26); Ogura (27); Kon'no (28)	
17 (III)	Irkutsk	.	.	.	J	Heer (77); Seward and Thomas (11)	
18	Transbaikalia	.	.	.	J	Krasser (05); Kryshtofovich (15[3])	
19	Amurland	.	.	.	J	Heer (77), (78), (82); Seward (12[3]); Novopokrovskij (12); Kryshtofovich (15[2])	
XX.	Western North America						
	A. Arizona	T	.	.	.	Ward (00), (05)	
	B. Black Hills	.	.	.	J	Ward (99), (05); Wieland (16)	Most of the Black Hill plants are Lower Cretaceous in age
	O. Oregon	.	.	.	J	Ward (05)	
	Ov. Oroville (California)	.	.	.	J	Ward (00)	
21	Mexico	T	.	R	J	Wieland (14[2]), (16), (26)	Wieland states that the ten to twelve thousand feet of Mesozoic strata are about equally divided between Jurassic and Cretaceous, any Trias that may occur not being as yet separated or defined
22	Honduras	.	.	R	.	Newberry (88)	
23	Virginia and North Carolina	.	U	.	.	Fontaine (83); Berry (12)	
24	New Brunswick and Prince Edward Island	T	.	.	.	Holden, R. (13)	

	SOUTHERN HEMISPHERE (fig. 104)						
1	Antarctica	T	.	.	.	Seward (14); Walton (23[2])	
2	Graham Land	.	.	.	J	Halle (13[2])	
3	Patagonia	.	.	R	J	Halle (13); Berry (24[4]); Gothan (25)	Also Lower Cretaceous plants
4	Santa Cruz	.	.	R	.	Wieland (29); Gothan (25)	For illustrations of South American Mesozoic plants see Kurtz (21)
5	Falkland Islands	T	.	.	.	Halle (11)	
6	The Argentine	.	.	.	J	Kurtz (01)	
7	Chile. Cacheuta	.	.	R	.	Steinmann (21)	Szajnocha (88)
8	La Ternara	.	.	R	.	Solms-Laubach (99); Steinmann (21)	
9	South Africa	T	U	.	.	du Toit (26), (27[2]); Kräusel and Range (28)	For records of Jurassic plants in East Africa see Gothan (27[4])
10	Southern Rhodesia	T	.	.	.	Seward and Holttum (21); Walton (26[2])	
11	Kenya and Tanganyika	T	.	.	.	Seward (22[3]); Gregory (26)	*Equisetites arenaceus* was identified from specimens collected by Mr E. Parson in the coastal region of Kenya Colony
12	Madagascar	T	.	.	.	Zeiller (11)	
13	Madagascar (Nossi-Bé)	.	.	.	J	Zeiller (00[2])	
14	Australia. Derby	.	.	R–	J	Arber, E. A. N. (10[2]); Antevs (13); Walkom (21[2])	
	L. South Australia	.	U–	–R	.	Chapman and Cookson (26)	
15	V. Victoria	.	.	.	J	Seward (04[2]); Chapman (09)	
	H. Hawkesbury and other localities in New South Wales	T	.	.	J	Walkom (19[2]), (21[3])	
	I. Ipswich and other localities in Queensland	.	U–	–R	J	Walkom (15), (17), (17[2]), (19), (24), (28[4])	
16	New Zealand (Otago)	.	.	R	J	Arber, E. A. N. (17); Sinnott (14)	
17	New Zealand (Canterbury)	.	.	R	.	Arber, E. A. N. (17)	
	N. New Caledonia	T	.	.	.	Crié (89)	

Postscript. Germany: Lower Trias, Mägdefrau (30).

CHAPTER XV

THE CRETACEOUS PERIOD

Hearing you sing, O trees,
Hearing you murmur, "There are older seas,
That beat on vaster sands,
Where the wise snail-fish move their pearly towers
To carven rocks and sculptured promont'ries",
Hearing you whisper, "Lands
Where blaze the unimaginable flowers". *J. E. Flecker*

THE practice of applying to this chapter of geological history the term Cretaceous has been responsible for a common error of regarding the Chalk[1] as the one kind of rock that is characteristic of the period. Chalk is a deposit formed on the floor of a clear but not a very deep sea; it is only one of many different kinds of material composing strata included in the Cretaceous system.

Cretaceous Rocks

In most of the English districts where Cretaceous rocks occur they are separated from those of the Jurassic period by an unconformity, which implies a time interval and affords evidence of an uplift and of exposure of the elevated land to denudation before the deposition of the Cretaceous beds was rendered possible by a subsequent submergence. In the south of England the Upper Jurassic beds pass gradually into a series of strata at the base of the Cretaceous system. The lowest Cretaceous rocks are represented in the Weald district[2] of Kent, Surrey and Sussex, in the Isle of Wight, and in the Isle of Purbeck. In the Weald the beds form an oval dome bounded on the west and south by chalk downs and reaching the sea between Eastbourne and Folkestone where the strata pass below the English Channel to reappear in northern France. At the beginning of the Cretaceous period a delta was being built up by rivers from the north in a large lake extending considerably beyond the present limits of the Weald. In the cliffs at Ecclesbourne on the east side of Hastings and in many other places the delta-sediments contain samples of the vegetation which flourished on the river banks and

[1] Creta = Chalk.

[2] Some geologists, though only a few, prefer to regard the Wealden series as the final stage of the Jurassic period. See p. 340.

on the shores of the Wealden lake. Fig. 105 shows one of many coniferous stems carried by flooded waters from some neighbouring forest. One of the most interesting localities where Wealden fossils have been found is at Bernissart (between Mons and Tournai) in Belgium: in the old river-mud at the bottom of a gorge cut through Palaeozoic rocks were discovered the complete skeletons of more than twenty iguanodons which form an impressive group in the Brussels Museum, and with them fragments of some of the plants on which they browsed.[1]

The plants from the Lower Cretaceous beds in England, Belgium, France, Portugal and Germany, and from rocks of the same age in North America and elsewhere, demonstrate the persistence into the earlier part of the period of an association of genera differing only in minor features from that which is typical of the preceding Jurassic age. The Wealden lake was separated by a barrier of land from an open sea to the north, records of which occur in Lincolnshire and Yorkshire. As the land in southern England gradually sank the lake was invaded by sea, and the incursion is recorded in the marine beds overlying the freshwater Wealden strata. These marine beds, in which drifted plants are abundant, constitute the Lower Greensand or Aptian series. A further incursion of the sea is recorded in the Gault which is seen in the cliffs at Folkestone: as the sea became deeper conditions were established favourable to the accumulation of calcareous shells of the innumerable marine creatures of which the chalk consists. The cliffs of Dover and Flamborough and the rocks to which Salisbury Plain owes its characteristic scenery are a few examples of the upraised floor of the Cretaceous sea. This spreading of the waters at the end of the first half of the Cretaceous period has been described as one of the greatest marine transgressions in the course of geological history; the Mediterranean Sea spread to the heart of Africa and for the first time since the Palaeozoic era the Sahara Desert was submerged. The map reproduced in fig. 106 is an attempt to give a general impression of the distribution of land and water at the time of the Cretaceous flood.[2] The effect on the vegetation is difficult to estimate: it may be that this marine transgression—the flooding of the land—was one of the factors directly influencing the evolution of the more modern type of vegetation. The rapid development of the flowering plants during the Cretaceous period was the cause of the

[1] Seward (00^2).

[2] See p. 403.

Fig. 105. A petrified stem of a Wealden conifer (20 ft. long) draped with bladderwrack (*Fucus*) on the foreshore near Brook, Isle of Wight. (Geol. Surv. Phot. 1810.)

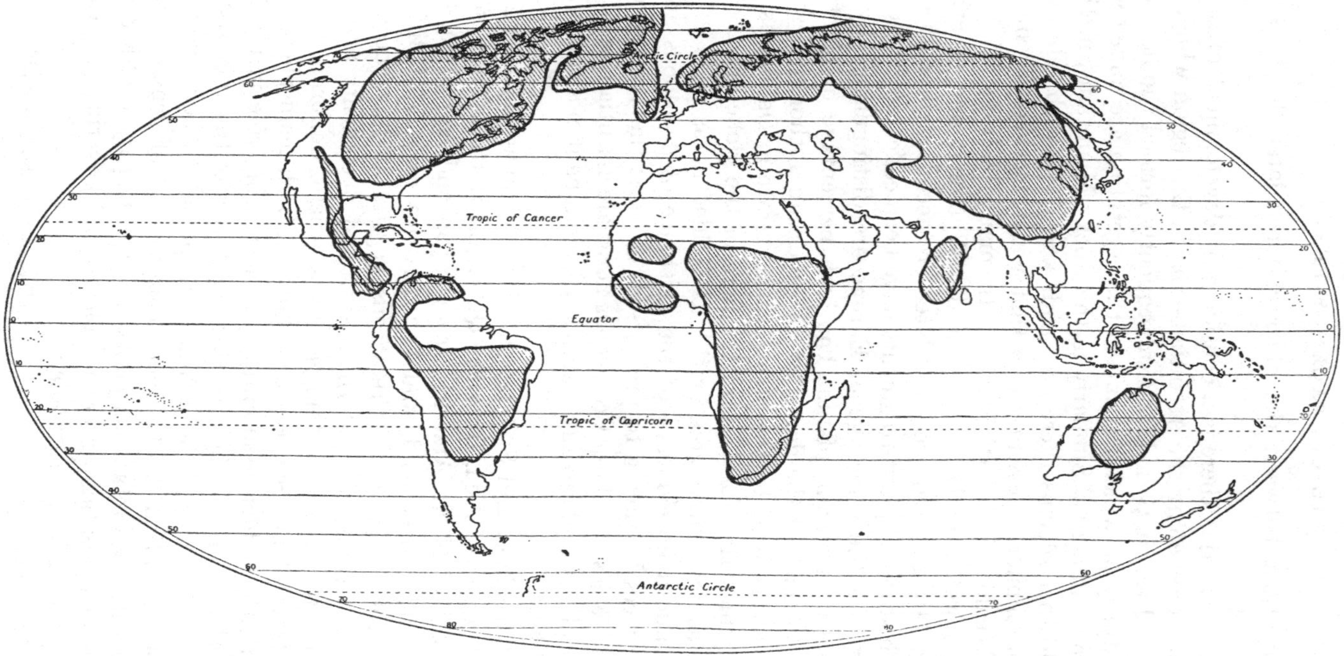

Fig. 106. Map of the world at the time of the great marine transgression which occurred about the middle of the Cretaceous period. The shaded areas are land. (Modified from a restoration given by Arldt.)

marked contrast between the older and the later floras of this age.

In North America there is a considerable development of Lower Cretaceous rocks along the Atlantic border in the region which was land in the Jurassic period, and subsequently formed the floor of a lake or estuary on which remains of the vegetation are mixed with river-borne sediments. Similarly in the Rocky Mountains, in Montana and other regions there are sands and clays with relics of a vegetation similar in essentials to that which clothed the margin of the Wealden lake in Europe. In many districts of North America, notably in Dakota, Colorado, Wyoming and elsewhere (Map, fig. 110) there are series of later Cretaceous sediments from which rich stores of plant-fragments have been collected: these and other floras of the same age in Europe are distinguished by the abundance of flowering plants. The high percentage of plants of this class in some of the floras led an American author to describe the stage represented by the plant-bearing strata in Dakota as "an epoch in which the character of the flora of the globe has become modified as by a new creation".[1] The older floras such as the Wealden afford no evidence of the existence of deciduous angiospermous trees belonging to present-day genera: the vegetation is Jurassic in facies. The later Cretaceous floras are essentially modern: flowering plants had already assumed the dominant position which they still hold.

From Greenland and Spitsbergen, from Sakhalin Island off the north-east coast of Asia, from the Queen Charlotte Islands off the Pacific coast of North America, from Japan and Argentina, from New Zealand, Queensland, and South Africa records of different stages in the evolution of Cretaceous vegetation have been discovered. The comparison of these scattered documents reveals a world-wide dispersal of certain genera, a general uniformity in the main features of Lower Cretaceous floras in both hemispheres, and an apparently sudden assumption of the leading rôle in the later phase of the period by vigorous representatives of the angiosperms, the class which marks the culminating point of plant evolution. A notable exception to the general prevalence of genial climatic conditions in the Cretaceous period is furnished by the discovery of glacial beds containing erratic blocks and covering an area of 40,000 square miles in Australia.[2]

[1] Lesquereux (91).

[2] Woolnough and David (26).

The classification given in the following table may serve as a guide to the general account of Cretaceous vegetation: it includes only the subdivisions with which we are more especially concerned.

Upper Cretaceous	Danian. Not represented in Britain: rocks of this age occur in Denmark, etc.	
	Senonian = Upper Chalk	
	Turonian = Middle Chalk	
	Cenomanian = Lower Chalk	
Lower Cretaceous	Albian	Upper Greensand
		Gault
	Aptian = Lower Greensand	
	Neocomian (marine) and Wealden (freshwater)	

The Dawn of a New Era

The change from the ancient to the modern plant-world revealed in the succession of Cretaceous floras appears to have been sudden: the impression of suddenness in the dawn of a new era, the era of flowering plants, is no doubt in part the result of imperfect knowledge; the documentary evidence is scanty; there are only a few pages, or parts of pages, taken at random and often at long intervals from a narrative which will never be accessible in its entirety. We are still searching for a solution to the problem of the origin of the flowering plants: there is reason to believe that the modern forms began their extraordinarily successful career in high northern regions whence they rapidly colonized the world.

Towards the close of the Cretaceous period in North America there was an upheaval of the crust into mountain ranges: this so-called Laramide revolution,[1] which continued into the Tertiary era, was no doubt one of the more effective factors responsible for the change in the organic world which is clearly shown by a comparison of Cretaceous and Tertiary fossils. The Deccan traps of India, which cover an enormous area, bear striking witness to volcanic activity in the latter part of the Cretaceous period on a scale comparable to that illustrated by the later outpouring of lava in north-western Europe at the beginning of the Tertiary era.

In the course of the Cretaceous period the vegetation of the world was transformed. As in the spring there is the annual "miracle of an earth reclad", so, it may be said, in the earlier part of the Cretaceous period there was added to the world's carpet of vegetation a new design and a new pattern. The records furnished by the oldest strata in North America and Europe give little indication of an

[1] Pirsson and Schuchert (20).

approaching change: Jurassic genera and some species continued to flourish over wide areas: ferns, cycadophytes, conifers, and some members of the Ginkgoales still occupied prominent positions in the plant-world. Jurassic floras were followed with no break in continuity by floras distinguished only by a few new introductions. Indications of the insertion of a new element into the vegetation are given by a few specimens obtained from North American beds[1] near the base of the Cretaceous sedimentary series and by others from Portugal and elsewhere. Passing to the Arctic regions we find a Greenland flora, intimately linked by several ferns and gymnosperms with the earliest Cretaceous floras of western Europe and eastern America, characterized by the occurrence of dicotyledonous leaves hardly distinguishable from the foliage of existing trees. This is the first appearance of broad-leaved forest trees in a vegetation which in its general features was essentially similar to that of the latter part of the Jurassic period; we have reached the first steps of a steeply ascending stairway leading to the plant-world as we know it to-day. It is the Lower Cretaceous rocks of Greenland[2] which have furnished the most striking demonstration of the modification of the earlier Mesozoic aspect of the vegetation of the world by, as it seems to us, the sudden evolution of familiar representatives of a class destined to take its place as the most vigorous and efficient section of the vegetable kingdom. During the earlier stages of the Cretaceous period flowering plants became conspicuous members in an association of ferns, conifers, and other gymnosperms which, with comparatively slight modification, had maintained a general uniformity in composition from the latter part of the Triassic to the end of the Jurassic period and, in North America and Europe, persisted into the initial stages of the Cretaceous period. The contrast between the earlier and later Cretaceous floras may be described as a contrast between ancient and modern: the plant-world of the latter half of the period differed in no outstanding features from that of our own time.

A botanist transported to an early Cretaceous forest in northern Europe would be impressed by the abundance of species and by the presence of some genera which could not be identified with living plants; he would not feel that he was in a wholly unfamiliar world, but in a region tenanted by plants which had established themselves many degrees of latitude farther north than they should be. As we

[1] Fontaine (89); Berry (11). [2] Seward (22^2), (25), (26).

shall see later, the most difficult and interesting questions raised by researches into the composition of the Arctic and north temperate floras are connected with the recurrent problem of climatic change.

An Early Cretaceous (Wealden) Flora

Let us now take a rapid survey of the vegetation which left abundant samples in the sediments of the Wealden lake which was bounded on the south by the rugged mountains of Brittany. The trees preserved as drifted logs, such as those exposed at low tide on the coast of the Isle of Wight (fig. 105), and as foliage-shoots and cones, were mainly conifers, some of them pines with clusters of long needles and long pendulous cones recalling the Himalayan *Pinus excelsa* that is often cultivated in our parks and gardens, with other members of the Abietineae, a family which had attained a position comparable with its present prominence in the northern hemisphere. There were araucarian trees, most of them like their Jurassic forbears, closely akin to the section of the genus represented by the Norfolk Island pine (*Araucaria excelsa*); members of a family that by the end of the Tertiary period had migrated across the equator. There were also trees with branches covered with a regular series of small and flat triangular leaves arranged in the regular pattern familiar in cypresses and other genera of the Cupressineae, together with others less closely linked with existing conifers. No leaves of *Ginkgoites* have so far been found in the Wealden beds of England; though we have evidence of the persistence of trees practically identical with *Ginkgo biloba* in scraps of foliage from French and German localities: the Jurassic members of the class, such as *Baiera*, *Czekanowskia*, and *Phoenicopsis*, played a very subordinate part in the Cretaceous vegetation.

A striking feature of the Wealden flora as of the Jurassic floras is the abundance of Cycadophyta: *Williamsonia* and *Cycadeoidea* were still common, the Jurassic *Williamsoniella* was not yet extinct,[1] and the variety of fronds shows that this section of gymnosperms continued to hold its own. Some of the cycadean stems bear a close resemblance to those of *Microcycas* (fig. 30) and other living genera; some were of a different type and had numerous flowers embedded in the covering armour of leaf-bases (figs. 98–100). Fronds such as *Otozamites*, *Zamites* with *Zamiophyllum*, *Ptilophyllum*, and *Nilssonia*

[1] Edwards (21²).

are widely spread Lower Cretaceous forms; and with them occur others bearing long and narrow leaflets superficially similar to those of the commonest cycad in the Tropics of the present day, the genus *Cycas*: these early Cretaceous fronds, referred to as species of *Pseudocycas*, are distinguished by the presence of a median pair of veins enclosing a groove on the lower surface of the leaflets.

Ferns continued to flourish, many of them differing but little from their Jurassic precursors: *Matonidium* was still a widely distributed genus and it is interesting to find that the close relationship to the Malayan fern *Matonia* shown by the fronds has been confirmed by the discovery, in Wealden beds of Belgium, of stems constructed on a plan that is now peculiar to the living genus. The allied *Laccopteris* is another member of the same family common to Jurassic and the earlier Cretaceous floras. *Hausmannia*,[1] recorded from several European localities, from the Pacific coast of Siberia and elsewhere, was still vigorous; and in this respect it affords a contrast to *Dictyophyllum* and some allied Jurassic and Rhaetic genera which are not quite so closely related as *Hausmannia* to the Malayan and Indian *Dipteris* and its allies. Among the Schizaeaceae *Ruffordia*[2] was one of the most abundant and characteristic plants in the older Cretaceous floras; and the same may be said of another member of the family recognizable by a different form of frond, *Cladophlebis Browniana*, a fern strongly reminiscent of the Jurassic *Klukia*. One of the most interesting ferns characteristic of Wealden and other Lower Cretaceous floras is the genus *Tempskya*; its family position is not certain, but the structure of its stem—and that is practically the only criterion available—suggests close affinity to some existing members of the Schizaeaceae. *Tempskya* is exceptional among ferns in the plan of its stem: the surface was covered with matted roots such as one sees in tree-ferns, but in place of a single stem bearing leaf-stalks there were several slender stems which, by repeated branching and the continuous development of roots, produced an axis which became thicker with age and may be described as a compound or false stem. The stem broadened upwards to a blunt apex and was covered with a tangle of fibrous roots: a transverse section[3] through it shows some stems with tubular conducting tissue embedded in a mass of roots and leaf-bases, an unusual type of construction recalling that in the Palaeozoic *Clepsydropsis* (fig. 59). Another Lower Cretaceous example of the Schizaeaceae is

[1] Richter (06). [2] Halle (21). [3] Seward (24).

Schizaeopsis expansa described by Berry[1] from the Patuxent formation of Virginia.

The Gleicheniaceae reached their maximum development both in variety and in geographical distribution in the early part of the Cretaceous period. No undoubted example of cyatheaceous fern fronds has so far been found in the Wealden flora of Western Europe, but petrified stems assigned to the genus *Protopteris* agree in anatomical characters with those of tree-ferns belonging to the Cyatheaceae which are now an attractive feature of southern floras. One of the most characteristic Wealden ferns is *Onychiopsis psilotoides*, so called because of the close resemblance of the fronds, both fertile and sterile, to those of some living species of *Onychium* especially the Japanese species *Onychium japonicum*. Though no well-preserved sporangia have been discovered, it is very probable that *Onychiopsis* is a representative of the Polypodiaceae, the family that now includes by far the greatest number of genera and species and is the most widely spread of all sections of the fern group, but has left remarkably few traces in Mesozoic floras.

The past history of the Polypodiaceae affords an illustration of one of the most striking lessons derived from a study of ancient plants, namely that wide distribution at the present day is often an indication of comparative youth in the history of a family or genus; while on the other hand many genera now confined to a few regions are the direct descendants of Mesozoic plants that were cosmopolitan. It has been argued that wide geographical range of a group of species is evidence of high antiquity, that area of occupation is an index of age;[2] but the records of the rocks show very clearly that genera now confined within narrow boundaries were often preceded by cosmopolitan ancestors. No osmundaceous ferns have been recognized in the older European Cretaceous floras, but it is probable that some of the common, sterile fronds such as *Cladophlebis Albertsii*, similar in habit to its larger, widely distributed *Cladophlebis denticulata* of Jurassic floras, may be the foliage of a member of the Osmundaceae, a family which we know from fossils discovered in many parts of the world, e.g. New Zealand, South Africa, and Egypt, to have been abundantly represented. Among the more characteristic plants in the older Cretaceous floras, and therefore the more valuable indices of geological age, is *Weichselia reticulata*,[3] a species distinguished by its large fronds divided into radially spreading arms

[1] Berry (11[2]). [2] Willis, B. (22). [3] Florin (19).

bearing two series of long pinnae having closely set, thick leaflets with a strong median rib and a regular network of smaller veins reminding us of the Palaeozoic *Lonchopteris* (fig. 63, N). *Weichselia* previously known from many localities has recently been recorded with a few other Cretaceous plants from Syria.[1] The genus has been placed by some authors in the Marattiaceae; but its affinity has not been definitely established.

The genus *Sagenopteris*, though long passed its prime, was still in existence. Among a few genera which, through lack of reproductive organs or of petrified material, cannot be assigned to a definite place in a scheme of classification, reference may be made to pieces of fairly robust stems bearing stout recurved spines and immediately above them what appear to be leaves of circular form or perhaps flattened leaf-like branches: this plant, *Sewardia armata*[2] (originally described as *Withamia armata*), may be a conifer, though it differs widely from any known form; it is recorded from England and France. No mention has been made of the Equisetales or Lycopodiales as the early Cretaceous examples of these groups seem to agree very closely with modern species: the *Equisetites* are smaller than their Jurassic ancestors, and an English Wealden *Selaginellites*,[3] producing two kinds of spore, is obviously very nearly related to existing Selaginellas.

Cretaceous Charophyta and Algae

The occurrence of small oval fruits (oogonia) and stem-fragments of *Chara* affords evidence of the occurrence in fresh or brackish water of representatives of the modern stoneworts—plants included in a group between the Algae and Bryophyta—which were links in a long evolutionary chain stretching far back into the Palaeozoic era. Many examples of algae have been described from Cretaceous rocks; some of them much too obscure and fragmentary to throw light on the evolution of the group, others demonstrating the occurrence in clear seas of reef-building calcareous species similar to living forms of the Siphoneae or allied to the more robust *Lithothamnium* (figs. 25, 114) and other lime-secreting members of the Red Algae.[4] The various fossil species are interesting mainly because they throw light on the past history and development of living types: they do

[1] Edwards (29).
[2] Seward (95); Zeiller (00[3]).
[3] Seward (13).
[4] Pia in Hirmer (27).

not reveal any striking peculiarities in the algal life of Cretaceous seas as compared with that of modern seas. For an account of the fossil genera reference should be made to the illustrated summary given by Dr Pia[1] in the section on Algae in Dr Hirmer's text-book.

An Arctic, Cretaceous Flora

The vegetation of the first part of the Cretaceous period, as revealed by the plants preserved in the Wealden lake, was mainly composed of ferns and gymnosperms most of which are either generically identical with existing plants or closely allied to them. An important point is that no specimens of any undoubted flowering plants have been found in the oldest European sedimentary deposits, nor have any been discovered in the equivalent strata of North America, disregarding a few impressions of leaves, which may be the incomplete remains of dicotyledons, recorded from the Potomac river area. We now pass a thousand miles farther north and glance at the earliest known vegetation of western Greenland. It must be pointed out that the geological age of these Arctic plant-beds has not been determined with a precision equal to that which we can apply to the strata in more intensively studied districts. Some of the Greenland rocks are closely linked by several common species with the Wealden series of Europe, and we can at least say that they are very near to the base of the Cretaceous system. The beds form part of a continuous series of sedimentary rocks deposited in the estuary of a river or arm of the Cretaceous sea: the lowest correspond with the Wealden series of America and Europe, and they are succeeded without any break by sands and clays agreeing closely in the fossil plants with North American and European strata higher in the Cretaceous system. It is not possible at present accurately to correlate the Greenland rocks with the various stages of the period as defined by American and European geologists: the significant fact is that the lowest Cretaceous flora of Greenland is clearly distinguished from flora of similar geological age in temperate regions by the presence of very modern-looking flowering plants.

The reconstruction (fig. 107) of a Cretaceous scene is based on data collected from the rocks of Disko Island, Upernivik Island, and the neighbouring mainland 300 miles north of the Arctic circle:[2] in the picture are included plants which were probably not all

[1] Pia in Hirmer (27); Yabe and Toyama (28); Romanes (16); Pfender (26); Umbgrove (27).

[2] For a brief account of the Greenland localities see Seward (22[2]), (26).

FIG. 107. A scene on the west coast of Greenland in the earlier part of the Cretaceous period. (Drawn by Mr Edward Vulliamy.) *At. Artocarpus*; *Cn. Cinnamomum*; *Cp. Cladophlebis*; *Db. Dalbergites*; *Gk. Ginkgoites*; *Gl. Gleichenites*; *H. Hausmannia*; *Lc. Laccopteris*; *Ls. Laurus*; *M. Magnolia*; *Mp. Menispermites*; *O. Onychiopsis*; *Pc. Pseudoctenis*; *Pt. Platanus*; *Q. Quercus* (oak).

actually contemporaneous, though the vegetation as a whole may be described as characteristic of the earlier phases of the Cretaceous plant-world. We are looking across a broad bay with its northern shores bounded by a range of Pre-Cambrian mountains. In the left-hand upper corner is one of several plane trees (*Pt.*), leaves of which occur in the rocks on the shores of a sea littered with icebergs, side by side with typical Wealden ferns and other plants: farther to the right is a branch of an oak tree (*Q.*); and in the right-hand upper corner boughs of other kinds of plane trees (*Platanus*). Below the highest plane there is a branch of an *Artocarpus*[1] (*At.*) with large, deeply dissected leaves and flowering shoots essentially similar to those of the bread-fruit tree (*Artocarpus incisa*) which is now exclusively tropical. On the left side, below the plane tree, a branch of a *Ginkgoites* (*Gk.*) is shown at closer range; this and the tree farther away and to the right are practically identical with the sacred maidenhair of the Far East, which is now unknown in a wild state. Beyond the Ginkgo are several conifers, recalling in habit cypresses, *Araucaria*, *Sequoia*, and trees such as pines and firs. One of the most abundant conifers in the early Cretaceous flora of Greenland is represented in some of the plant-beds by innumerable leaves lying in masses on the surface of the rock as dead needles of firs and pines now litter the ground in a modern forest. These fossil leaves, rarely exceeding a centimetre or two in length, are exactly like the parallel-sided and short needles of familiar fir trees, but an examination of their microscopical characters has shown that they can be much more closely matched with the needles of the Umbrella Pine of Japan (*Sciadopitys*),[2] an isolated conifer which is relatively remote from the Abietineae. The family Sciadopitineae may have reached its maximum in the early part of the Cretaceous period; it is now represented by a single survivor far distant from Greenland. The Greenland forests contained more than one representative of the Sequoia family which in the Cretaceous and Tertiary period had outposts in many parts of the Arctic regions. On the left, behind a row of flowering plants, a sloping bank is clothed with Gleichenias and other ferns and, farther to the right, is seen a partially uncoiled bud rising from the fork of one of the common species of *Gleichenia* (*Gl.*). Other ferns include species of *Laccopteris* (*Lc.*), conspicuous by their branched and spreading fronds, and in close association with them is a plant of *Hausmannia* (*H.*)

[1] Nathorst (90). [2] Florin (22).

recognizable by its smaller, wedge-shaped leaves on a slender stem. There are also ferns with fronds identical in form with those of *Onychiopsis* (*O.*), one of the genera characteristic of the European Wealden flora; and another bearing handsome fronds (*Cladophlebis frigida*)[1] similar in habit to those of *Cladophlebis denticulata* (*Cp.*). To the left is a solitary example of the Cycadophyta, a group which was represented by several members in the Greenland vegetation—*Otozamites*, *Pseudocycas*, *Pseudoctenis* (*Pc.*) and *Ptilophyllum*.

Several flowering plants are conspicuous in the foreground: some of these flourished in the earlier stage of the Cretaceous period in the Greenland area; others appeared rather later. It is probably true to say that in no other part of the world have familiar types of angiosperms been discovered in rocks as old as those of Greenland. The twig with large rounded leaves belongs to a tree believed to be a member of the Menispermaceae (*Mp.*), a family that is now tropical and warm temperate in its distribution: a flowering branch of a cinnamon is recognized by its oval leaves with three prominent veins and near it is a piece of a laurel bush. The genus *Cinnamomum* (*Cn.*) is now characteristic of the Tropics and the laurel-like leaves found in the same beds closely resemble the tropical and subtropical genus *Ocotea* (*Ls.*), which, with *Cinnamomum*, belongs to the laurel family. A species of *Ocotea*, known as the camphor tree, is a dominant element in some of the forests on the lower slopes of Mount Kenya in tropical Africa. Near the cinnamon is a plant of *Dalbergia* (*Db.*) distinguished by oval leaflets with a small piece cut out of the apex: this genus, now living in tropical countries (see Map, fig. 110), is a member of the Leguminosae. It should be pointed out that the leaflets of this plant were formerly described as leaves of the tulip tree (*Liriodendron*), a type which has not been proved to exist in the Cretaceous vegetation of Greenland,[2] though it flourished in the Tertiary forests of Iceland. In the right-hand corner is a branch of a *Magnolia* (*M.*) with flower and large leaves. It must be acknowledged that no fossil flowers, with the exception of pieces of the inflorescence of *Artocarpus* have been found in the Greenland rocks; but the numerous leaves are well enough preserved to serve as a satisfactory basis for the identification of the flowering

[1] Seward (26): more recently recorded also from Ussuriland by Kryshtofovich (26).

[2] Seward (25). *Liriodendron* is said by Ihering (28) to occur in Upper Cretaceous beds in Patagonia: no illustrations have been published. For a map showing the present distribution of *Liriodendron*, see Fernald (29).

plants included in the restoration. For additional information on the composition of the Greenland Cretaceous vegetation reference should be made to Table F: see also page 409.

It needs no additional words to point the contrast between Greenland as it was in the earlier part of the Cretaceous period and Greenland as it is to-day.[1] To-day there is an arctic flora with prostrate willows and the dwarf birch—the stunted representatives of forest trees—growing in company with numerous herbaceous plants which in temperate countries we find for the most part among the rocks and on the slopes of mountains far above sea-level, a flora which bears eloquent witness to the rigours of a polar climate. In the Cretaceous period there were forests of conifers and flowering trees and an abundance of ferns belonging to families that are now mainly tropical; it was a vegetation which in the size of leaf, as in the breadth of the rings of growth seen in petrified logs of wood, gives no sign of a physical environment such as we now find on the edge of an arctic land most of which lies under a storm-lashed mantle of ice. It must, however, be remembered that the wood may have been drifted by currents before it became petrified: we cannot be sure where the trees grew.

Other Cretaceous Plants

In the short sketch of the Wealden flora of England and western Europe and of the corresponding and slightly younger flora of Greenland mention is made of most of the more characteristic genera. We will now supplement this incomplete description by a reference to additional plants, several of which are not represented in the Cretaceous collections so far considered. Some imperfectly preserved stems described from Cretaceous rocks of Germany as *Nathorstiana*[2] are worth mentioning because of their superficial resemblance, except in their greater length, to the tuberous stems of the living genus *Isoetes* (quillworts): the fossil stems reach a length of 12 cm.; they are swollen in the middle and have spirals or whorls of scars which were left on the fall of needle-like leaves. This obscure type may be related to the Triassic genus *Pleuromeia* (fig. 84) and possibly to *Isoetes*: our knowledge of *Nathorstiana* is too small to admit of any definite statement; it is one of many relics in which it is easy to see missing links in an evolutionary series, though usually impossible to support fancy by fact.

[1] Seward (29^2).

[2] Richter (09).

Ferns

The ferns *Matonidium*, *Hausmannia*, *Gleichenites* and *Ruffordia* reached their maximum development in the earlier part of the Cretaceous period: the genus *Onychiopsis* is another type characteristic of the same phase of the fern-world. It is preferable to employ the less-committal name *Sphenopteris* for fronds agreeing in form with English and other examples of *Onychiopsis psilotoides*, but which afford no evidence of the nature of the spore-bearing pinnae; and for this reason the generic designation *Sphenopteris* (*Onychiopsis?*) is often used for this widely distributed type of Lower Cretaceous frond. The geographical range of this and other fossil ferns is shown in Table F. Reference may be made to a few other Lower Cretaceous ferns which are less well known: some pieces of fronds bearing leaflets not unlike those of *Matonidium* are recorded from North America under the generic name *Knowltoniella*[1] and compared with the Bornean fern *Matonia sarmentosa* though on grounds that are by no means convincing. *Knowltoniella* is recorded also from Lower Cretaceous rocks in Ussuriland.[2] Similarly a French Wealden genus, represented by a single species, *Feronia Sewardi*,[3] may be another extinct member of the Matonia family. Allusion has already been made to *Tempskya* with its peculiarly constructed axis made up of a number of branches given off by the repeated subdivision of an originally single stem. It is significant that the great majority of early Cretaceous ferns belong to families represented only by a few species north of the equator, or now restricted to the southern hemisphere: the few examples referred to *Onychiopsis*, *Adiantites* and other genera implying relationship with the Polypodiaceae, the family that is now dominant in the northern hemisphere, are for the most part too imperfectly known to be of much value as trustworthy records.

The genus *Taeniopteris* was represented in the older Mesozoic floras by many species and by a few species in the latest Palaeozoic floras: in the Cretaceous vegetation it was comparatively rare and most of the examples differ but little in the size and venation of the leaves from the Wealden species *Taeniopteris Beyrichii* or from living species of the polypodiaceous genus *Oleandra*. The difficulty is to determine the systematic position of *Taeniopteris* fronds: some of the Palaeozoic leaves of this form may have been the foliage of

1 Berry (11). 2 Kryshtofovich (26). 3 Carpentier (27).

pteridosperms; some, such as the Jurassic *Taeniopteris vittata*, undoubtedly belong to cycadophytes and are not ferns, while others may be the leaflets and not complete fronds of Marattiaceous species. The generic name is used merely for a certain form of simple leaf resembling the small fronds of the hart's tongue (*Scolopendrium vulgare*) or those of *Oleandra*: its employment does not necessarily imply close relationship with ferns.

Ginkgoales: *Cycadophyta*

The Ginkgoales reached their culminating point before the end of the Jurassic period. *Baiera* and *Czekanowskia* persisted into the Cretaceous period, though fewer in number and more restricted in geographical range; a new genus, *Feildenia*[1]—founded on slightly sickle-shaped leaves not unlike short blades of grass—which may be a member of the same class, made its appearance and continued to flourish as a member of Tertiary floras far within the Arctic Circle. The most vigorous representative of the Cretaceous Ginkgoales was the genus *Ginkgoites* (which might well be called *Ginkgo*), a type that merits the title of the most conservative of all living trees.

The continued existence in the earlier Cretaceous floras of several genera of Cycadophyta has already been mentioned in the account of the Arctic vegetation in which they were well represented. In addition to fronds of *Ptilophyllum* from Greenland, the Rocky Mountains, and Japan, species of *Otozamites*, e.g. *Otozamites Klipsteinii* from the Wealden beds of England and western Europe, fronds of *Pterophyllum*, and large fronds from Greenland of a type very close to a form first recorded from Upper Jurassic rocks in Scotland as *Pseudoctenis latipennis*, demonstrate the continued success of cycadean plants in the opening stages of the Cretaceous period. Cycadophyte fronds are represented also by a few examples of the genus *Nilssonia*: this form of leaf resembles *Taeniopteris* but is distinguished by more crowded, usually simple and not forked veins, also by the structure of the surface-layer and some other characters. *Nilssonia* was much commoner and represented by many more species in Jurassic floras; it was a diminishing genus in the Cretaceous period, though it seems to have survived into the Tertiary period[2]. An outstanding feature of the older Cretaceous vegetation is the extraordinary abundance of large stems of *Cyca-*

[1] Nathorst (97). [2] Berry (29).

deoidea, particularly in the Lower Cretaceous bed of the Black Hills (fig. 110, B), in beds of corresponding age in the Potomac River district, in northern Italy, England,[1] other parts of Europe and in Japan.[2] Hundreds of almost perfectly preserved stems, bearing numerous flowers and differing in no important respect from those furnished by the upper series of Jurassic plant-bearing beds, have been found in American localities where presumably the conditions were exceptionally favourable for the growth of these extinct plants.[3] It has already been said that they agree as closely in their habit of growth and in the form of stem and leaf with living cycads as they differ from them in their reproductive apparatus.[4]

Conifers

In addition to the genera of conifers already mentioned attention should be called to a few species of uncertain position that are characteristic of Lower Cretaceous floras. One of these is *Elatides curvifolius*, a type of foliage-shoot abundant in Spitsbergen[5] in rocks which are probably Lower Cretaceous in age: this species may be araucarian, but we are without proof of its affinity. Other species which have not been assigned to a definite position are mentioned in Table F. Reference may be made here to some well preserved remains obtained from lignites of Upper Cretaceous age in Staten Island (New Jersey) off the western coast of North America.[6] Among these are dwarf-shoots (that is very small branches bearing a cluster of leaves) referred to the genus *Prepinus*, similar, except in the larger number of needles, to those of living pines; also other leafy twigs still more nearly allied to shoots of modern species. In existing pines each dwarf shoot has a pair of needles, as in the scotch fir, three, or five needles, and in one Californian pine a single needle. Prof. Jeffrey has drawn attention to the occurrence in several of the New Jersey conifers of anatomical characters which he regards as indicative of araucarian affinity, though these are by no means always associated with a habit of foliage-shoot such as we now connect with members of the Araucarineae. A less familiar type described from Staten Island and many other Upper Cretaceous localities is illustrated by *Androvettia*, characterized by flattened

[1] For an account of one of the largest known English cycadeoidean "flowers" see Stopes (18).
[2] Endo (25).
[3] Wieland (06), (16), (19), (21).
[4] See page 359; also figs. 98–100.
[5] Nathorst (97).
[6] Hollick and Jeffrey (09).

branches reminding us of the leaf-like shoots (phylloclades) of the Tasmanian and New Zealand conifer *Phyllocladus*. Frequent allusion is made in accounts of the later Cretaceous floras to a conifer known as *Widdringtonites Reichii*[1] distinguished by its small spiky leaves attached in regular alternating pairs and closely resembling in habit branches of the African *Widdringtonia* and other members of the family Callitrineae, which now occurs in Africa, Australia, New Caledonia and in a few other southern regions. A more appropriate generic name is *Callitrites*[2] because it is impossible to determine the exact relationship of the Cretaceous examples to any one living genus. It is, however, clear that among the Cretaceous conifers of North America and Europe were trees closely related to members of the Callitrineae, a family which persisted in the northern hemisphere well into the Tertiary period and now has its northern limit in the Atlas Mountains of northern Africa. An older Cretaceous conifer which is of special interest as evidence of the vagaries in distribution demonstrated by a comparison of the past with the present, has been described by Prof. Halle[3] from Lower Cretaceous beds in Patagonia as *Athrotaxites Ungeri*, a species first recorded from Upper Jurassic beds in Germany: the foliage-shoots and cones closely resemble those of species of the genus *Athrotaxis* which is now confined to Tasmania.

Considering the conifers as a whole the facts which stand out most clearly are: the much greater variety in genera and species in the Cretaceous than in the present northern forests, and the wide distribution in north temperate and Arctic regions of types which have long been strangers to the northern hemisphere.

Flowering Plants

The occurrence of masses of fallen leaves of different kinds of *Platanus* (plane tree of England; sycamore in North America) in the oldest Cretaceous rocks in Greenland has already been emphasized: the genus *Platanus*,[4] now commonly planted in Europe as a hardy and clean shade tree in places some hundreds of miles north of its home in southern Europe, the Balkan Peninsula, Greece and elsewhere, lives in a wild state also in Mexico and farther north in the eastern and western states of North America. In Crete there is said to be an evergreen variety of the oriental plane. The group

[1] Seward (19); Berry (14); Hollick (06).
[2] Seward (19).
[3] Halle (13).
[4] Henry and Flood (20).

of plane trees ("sycamores" as they are called in America and Scotland) shown in fig. 108 might serve as a reconstruction of part of a Greenland Cretaceous forest. This forest-tree exhibits no features which stamp it as a primitive type or as one of the earliest members in an evolutionary series: it is clearly derived from much more ancient ancestors, though it is one of the oldest dicotyledons so far discovered which bears obvious marks of generic and almost specific identity with existing broad-leaved trees.

This and other facts of similar import illustrate the lamentably incomplete story furnished by scraps of vegetation which happen to have become embedded in water-borne sand and mud. The earlier links in the chain of successive products of evolution are missing: many no doubt were not preserved or have not been discovered, while others may be familiar to us as fossils which it has not been possible to identify because they are the leaves of trees that have long ceased to exist, and in their imperfect state cannot be brought within the compass of our standards based on a knowledge of living plants. *Platanus* is one of the oldest of broad-leaved trees; with other genera it has been found not only in Arctic Cretaceous rocks but in many parts of the temperate zone north of its present range.

Before dealing with the flowering plants characteristic of Upper Cretaceous floras, a brief reference must be made to some leaves obtained from beds in the lowest series of the Potomac formation in Maryland and Virginia: these beds may be correlated with the oldest Cretaceous series in Greenland. The Potomac leaves are assigned to the genera *Rogersia*, *Proteaephyllum*, and *Ficophyllum*,[1] and have been described as archaic dicotyledons: their affinity is admittedly problematical and all that can be said is that while they may be torn fragments of the foliage of some extinct angiosperms, we cannot accept them as evidence that is above suspicion. Ascending the Cretaceous system to the rocks known in England as the Lower Greensand, and more widely as the Aptian stage, we find definite proof of the existence in one of the early Cretaceous floras of dicotyledonous trees: five genera have been founded on petrified wood exhibiting anatomical characters of an elaborate type and showing no primitive traits. One of the genera described under the name *Woburnia*,[2] from Woburn in Bedfordshire, is compared with the wood of members of the family Dipterocarpaceae well represented

[1] Fontaine (89); Berry (11). [2] Stopes (12), (15).

FIG. 108. A grove of plane trees (*Platanus occidentalis*) in the State Park, Parke County, western Indiana. (Photograph, taken by Mr F. M. Hohenberger, obtained from the Department of Conservation, Division of Forestry, Indianapolis.)

to-day in the tropical vegetation of Africa and Asia. These stems, as Dr Marie Stopes says, afford no better clue to the ancestry of the flowering plants than we find in the stems of living genera. In other words, these oldest known samples of petrified wood confirm the conclusion, based on the still older leaves of the Arctic plane trees, that the oldest dicotyledons which clearly reveal their kinship with flowering plants are old only in a geological sense and astonishingly modern in their anatomical features. They confirm our belief in an antiquity of angiosperms antedating by many millions of years, probably by several geological periods, the first appearance of recognizable pioneers of the present ruling dynasty in the modern world. Specimens of stems, which may also be of Lower Cretaceous age, are recorded from Madagascar[1] and are believed to show points of contact with trees of the laurel family.

Various Cretaceous Floras. The Potomac Floras of Eastern North America

The boundary between the Lower and Upper series of the Cretaceous system is not always clearly defined—no boundaries are universally applicable—but in general it is possible to adopt a classification which has in part a physical foundation based on an unconformity in the succession of strata, and in part coincides with a striking transformation in the forest floras of most regions of the world. In North America, particularly in the United States, there are many areas in which plant-bearing beds are well developed and their botanical records have been carefully studied. A few examples supplemented by the notes on Table F will suffice to illustrate the solid basis underlying a separation of the vegetation of the period into an earlier and a later phase. In the coastal plain, east of the Appalachian Mountains, which extends along the Atlantic seaboard of North America from Massachusetts to Florida, there is a thick series of freshwater sands and clays of Cretaceous age resting on a foundation of an ancient eroded surface of crystalline rocks which was submerged at the close of the Jurassic period: these delta-deposits are known as the Potomac formation[2] from their occurrence in the Potomac River district. The basal beds are called the Patuxent series after a river in Maryland: above these in ascending order are the Arundel and Patapsco series. It is interesting to find a close resemblance both in the plant-remains of the Potomac strata and

[1] Fliche (05).

[2] Berry (11) with references to earlier literature.

in the nature of the rocks between the American delta-deposits and those on the west coast of Greenland: it is possible that these two now widely separated land-masses may have been much closer in the Cretaceous period. The flora of the Patuxent series, like that of the corresponding Wealden beds of England, is Jurassic in character: ferns and gymnosperms predominate and the only fossils which it seems possible to compare with angiosperms are the leaves referred to on a previous page (p. 398). The Arundel flora is very similar to the richer Patuxent flora. From the younger Patapsco series (named after a Maryland river) about thirty kinds of flowering plants have been obtained. It is only in the upper part of the Lower Cretaceous Potomac formation that the Jurassic type of vegetation is partially transformed by an admixture of angiosperms.

Other American Floras

During the first half of the Cretaceous period most of the North American continent was above sea-level: the eastern edge was under water, but the freshwater sedimentary rocks of the Kootanie and Montana series, much farther to the west, afford evidence of the existence of lakes in which sediments containing Lower Cretaceous plants were deposited. As in the lower series of the Potomac group of rocks, so also in the Kootanie and Montana beds angiosperms are practically unrepresented. From several localities in New Jersey within the coastal plain collections have been made which reveal a much more fundamental change in the vegetation: the plant-beds were deposited later than those of the Patapsco series and are Upper Cretaceous in age. The oldest set of plants obtained from the New Jersey strata is known as the Raritan flora with which may be correlated the flora of the Amboy clays:[1] the vegetation was composed of about 70 per cent. of angiosperms associated with a very much reduced number of Lower Cretaceous plants. The boundary between the Upper and Lower divisions of the Cretaceous system in eastern America is drawn above the Patapsco series and below the Raritan series.

Among other later Cretaceous floras in North America are the Dakota flora,[2] obtained from plant-beds in Dakota, Nebraska and neighbouring states, which includes about 90 per cent. of angiosperms; the Woodbine (Tennessee) and Cheyenne (Kansas) floras,[3]

[1] Newberry (95). [2] Lesquereux (91).
[3] Berry (16), (19), (22), (22[2]).

and the rich Tuscaloosa flora[1] (Carolina and Alabama), all of which are distinguished by the marked predominance of flowering plants. One of the youngest Cretaceous floras, known as the Vermejo series[2], has been described from beds in Colorado and New Mexico: this consists of a few ferns and conifers with abundant leaves of flowering plants including a palm, a species of *Artocarpus* (bread-fruit tree), many species referred to the genus *Ficus* (figs), species of walnut (*Juglans*), oaks, and many other living genera.

Cretaceous Floras in other parts of the World

Turning to other parts of the world we find abundant confirmation of the replacement in the course of the Cretaceous period of the older, Jurassic aspect of the vegetation by one which is essentially modern. It would necessitate a disproportionate treatment of the numerous Cretaceous floras were we to consider them in detail: most of the richer floras are included in Table F to which reference should be made for further information. A word may, however, be added on two other regions which have furnished rich collections of Cretaceous plants: we will take first the plant-beds of Sakhalin Island (fig. 110, S) which have been partially described in a preliminary account by Dr Kryshtofovich.[3] Plants from this island were recorded several years ago by the great Swiss Palaeobotanist Oswald Heer as Tertiary in age: most of them have now been shown to be Cretaceous. The oldest beds are particularly rich in ferns, especially *Gleichenia*, and the flora agrees very closely with the Lower Cretaceous flora of western Greenland. In the middle Sakhalin plant-beds ferns are still abundant, but with them are several dicotyledons; and in the uppermost series angiosperms form an even larger proportion of the flora. The luxuriance of the early Cretaceous fern-vegetation is a noteworthy feature of the Greenland and Sakhalin floras: this suggests conditions in high northern latitudes exceptionally favourable to the vigorous development of associations of *Gleichenia* and its companions. There have recently been recorded from the Suchan coal basin in the Maritime province of Siberia (Ussuriland),[4] (fig. 110, XIII), several Lower Cretaceous species, including *Ruffordia Goepperti*, *Onychiopsis*, with a few others of Jurassic facies: and from the upper strata in the same series the leaf of a flowering plant, described as *Aralia lucifera*, which agrees

[1] Berry (19).
[2] Lee and Knowlton (17).
[3] Kryshtofovich (18).
[4] Kryshtofovich (29), (29^3).

closely with specimens from the early Cretaceous beds of Portugal, Greenland, and North America.

The second region chosen for special mention is Bohemia, where numerous collections have been made from freshwater sediments known as the Perucer series.[1] The oldest beds have afforded several species of ferns, gymnosperms, and dicotyledons identical with Greenland and Sakhalin Island plants: their age is estimated by Prof. Velenovský as Cenomanian, that is at the base of the Upper Cretaceous; in many respects the earliest flora of Bohemia agrees closely with the later phase of the Lower Cretaceous vegetation.

We cannot as a rule draw a clearly defined boundary line between one stage and another in places where there is a continuous series of Cretaceous plant-beds: the important point is that before the end of the first half of the period there had been a large influx of flowering plants which substantially changed the character of the vegetation. Moreover there is striking proof of the rapid migration of identical genera and species over wide areas in the northern hemisphere, in all probability from a starting-point in the Arctic regions. It is believed that the transformation of the Cretaceous vegetation began earlier in the Arctic regions than elsewhere; and thence the host of recently evolved plants, vigorous and endowed with the qualities which make for success, spread along divergent routes into the temperate zone. In this connexion it is permissible to draw attention to the probable co-operation of physical factors in the creation of a plant-world of modern aspect. During the Cenomanian stage of the Cretaceous period (fig. 106) there was a remarkably widespread marine transgression; vast regions which had previously been dry land were flooded by an invading sea. May we not see in this sinking and flooding a possible influence on the course of evolution in the organic world, an almost world-wide interference with the physical environment which had its repercussion in the altered trend of plant development?

Cretaceous Flowering Plants

We will now attempt to convey a general idea of some of the more widely distributed flowering plants in the later Cretaceous floras without entering into controversial questions regarding the precise correlation and geological position of the several plant-

[1] Frič, A. and E. Bayer (01) with references; Velenovský and Viniklař (26–27).

bearing strata. An approximate correlation is given on pages 409–411. It is seldom that any trustworthy records of reproductive organs are available: by far the larger proportion of fossil angiosperms consist of impressions of leaves and their interpretation is fraught with no little difficulty. One of the most widely spread of the earlier Cretaceous dicotyledons is the genus *Platanus*: in the size and variety of the leaves, and in their wide distribution over temperate and Arctic regions of the old and new worlds, the Cretaceous plane trees surpass the surviving members of the family. Fruiting specimens precisely similar to the fertile branches of modern species have been found in Bohemia, Germany, and elsewhere. A type of leaf known as *Credneria* (fig. 109)[1] is one of the more characteristic early Cretaceous genera: this is most probably the foliage of a *Platanus* or of some allied tree. The genus *Magnolia*, or *Magnoliaephyllum* as it should perhaps be called to avoid the danger of implying more than the material justifies, ranged from Greenland through Europe and many parts of North America: the present northern limit of the family is shown on the map, fig. 110.

Another common dicotyledon is *Cinnamomum*, a genus which includes the cinnamon and the true camphor trees: its leaves may easily be confused with those of other plants having a similar venation, but as we know on good evidence that *Cinnamomum* existed in Europe in the Tertiary period, we may confidently assert that many of the Cretaceous specimens are correctly assigned to that genus: in this instance also the generic name *Cinnamomoides*[2] is often more appropriate than *Cinnamomum*. Other genera include *Menispermites*, a name implying close relationship to living species of the family Menispermaceae; *Artocarpus* (the bread-fruit tree), together with oaks, walnuts (*Juglans*), willows (*Salix*), *Myrica* (a genus including the sweet gale or bog myrtle of our fenland), figs (*Ficus*), and many other trees. The genus *Dalbergia*, already mentioned as a Greenland dicotyledon, is recorded from several Cretaceous localities in America and Europe; and *Bauhinia*, another member of the family Leguminosae, has been found in Upper Cretaceous beds in New Jersey and Alabama. *Sterculia*, an additional illustration of the former existence in the temperate zone of plants that are now tropical, is recorded from Bohemia, North America, and from other countries. Many references occur in accounts of later Cretaceous floras to species of *Eucalyptus*, the tree

[1] Engelhardt (85[2]); Krasser (96).

[2] Seward (25).

that is now dominant in the forests of Australia. Despite the fact that many fossil leaves and reproductive organs have been erroneously referred to *Eucalyptus*, there is fairly good evidence of its presence in the Cretaceous vegetation of the northern hemisphere.

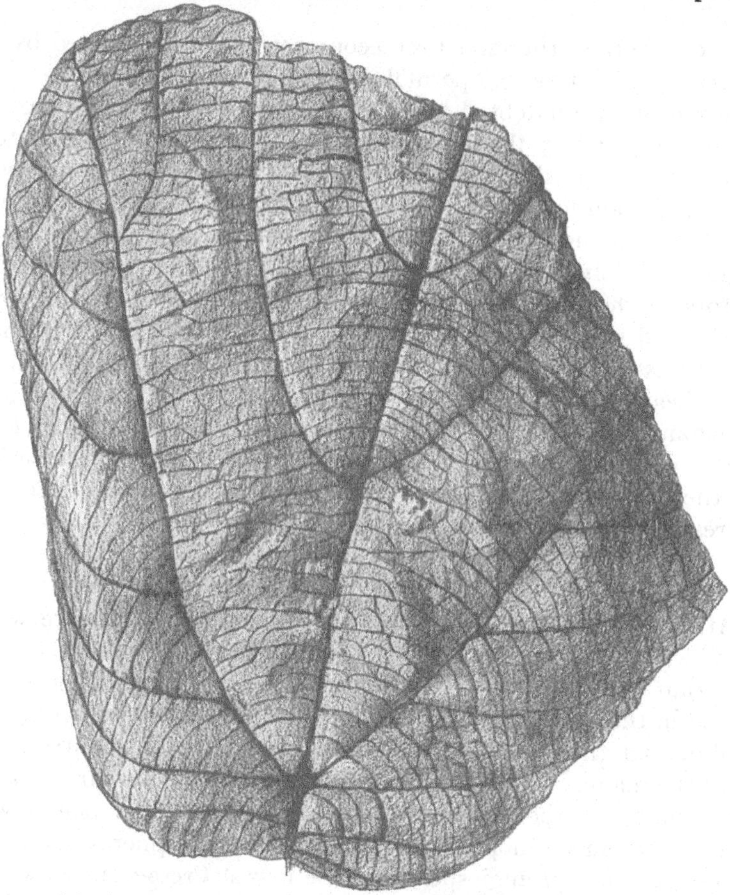

FIG. 109. A leaf of *Credneria* from a Cretaceous sandstone, Germany; partially rolled as a leaf that has dried in the sun. (M.S.)

These are only a few of the Cretaceous dicotyledons selected from a long list; they are chosen as examples of identification based on evidence that is for the most part satisfactory. Our knowledge of the early history of monocotyledons is very meagre and of little value as a trustworthy guide to the course of evolution of this group of flowering plants. It has been clearly established that palms flourished in the latter part of the Cretaceous period hundreds of miles beyond the northern limit of their present range: leaves, seeds, and some petrified stems are recorded from New Jersey, New Mexico and elsewhere in North America, also from a few European localities.

One feature of the later Cretaceous vegetation illustrated by the composition of the angiosperm floras, pointed out by Mr E. W. Berry to whom we are indebted for much of our knowledge of the Cretaceous and Tertiary floras of America, is the commingling of temperate and tropical genera, an admixture which he compares with that in the rain-forests of southern Chile, southern Japan, Australia, and New Zealand. In the course of the latter part of the Cretaceous period successive waves of flowering plants, influenced by controlling factors in the inorganic world, spread progressively farther from Arctic and north temperate regions. Some of the wanderers have left traces of their temporary resting-places in Europe only in the plant-bearing rocks of the Tertiary period: others, better able to withstand the lowering of temperature in the latter part of the Tertiary period—a change in climate which became more serious as the Quaternary ice-sheets and glaciers advanced—are still represented in the present vegetation.

Cretaceous Floras of the Southern Hemisphere

Attention has been focussed on the northern hemisphere which provides by far the greater number of valuable sources for the historian of Cretaceous vegetation. Most of the localities are mentioned in the notes on the map, fig. 110. It is, however, interesting to find that the information available for southern countries confirms the main conclusions drawn from a comparative survey of the floras north of the equator: the earliest Cretaceous plants were many of them cosmopolitan, and in both hemispheres there is a general absence of angiosperms in the lowest Cretaceous rocks. In India[1] plants have been found in the uppermost part of the Umia stage in Cutch, referred on evidence derived from marine animals to a Lower Cretaceous age, which demonstrate a persistence of the Jurassic type of vegetation and an apparent lack of recognizable flowering plants. Remains of angiosperms are recorded from Rajputana in the Balmir series which should be classed as Cretaceous and not Jurassic as formerly supposed. The Lameta and other fresh-

[1] Sahni (26); Wadia (19).

water beds of central India lying immediately below and associated with the Deccan basaltic lavas of late Cretaceous age, have afforded dicotyledonous leaves, "fruits" of *Chara* and reproductive organs of the water-fern *Azolla*. Reference has been made to the occurrence of angiospermous wood in what are believed to be Lower Cretaceous rocks in Madagascar. One of the richest collections of southern hemisphere plants comes from the Uitenhage series of Pondoland in South-East Africa.[1] The flora is undoubtedly equivalent in geological age and in its general facies to Wealden floras of Europe and North America; it consists of ferns and gymnosperms, some of which appear to be specifically identical with northern types; and there are no angiosperms. Remains of a Lower Cretaceous vegetation including European Wealden species have been found at more than one locality in Peru[2] and Patagonia,[3] and from the latter district a younger Cretaceous flora has been described in which angiosperms are represented. It has already been pointed out that in the Jurassic flora of Graham Land are included a few Wealden types which are believed to be specifically identical with European species: the occurrence of these plants at a slightly lower geological horizon than that of which they are characteristic in the northern hemisphere supports the possibility of an Antarctic origin.

Floras preserved in strata of the Burrum, Styx, and Maryborough series of Queensland[4] afford additional evidence of the general uniformity in composition of the earlier Cretaceous vegetation. *Hausmannia* and other cosmopolitan Lower Cretaceous genera have been discovered at Plutoville in the Cape York district. The discovery in central Australia of Cretaceous beds containing numerous and large erratic blocks is of special interest as evidence of climatic conditions very different from those characteristic of the period in other parts of the world. Petrified stems of conifers obtained from the glacial series show well-marked and relatively narrow rings of growth.[5] A few Lower Cretaceous plants including well-preserved stems of *Osmundites*[6] are recorded from Auckland in New Zealand, and specimens of *Araucarites* and *Taeniopteris* have been discovered in beds of Upper Cretaceous age at Kaipara Harbour.[7]

[1] Seward (03). [2] Zeiller (14); Neumann (07). [3] Halle (13).
[4] Walkom (19). [5] Walkom (29).
[6] Kidston and Gwynne-Vaughan (07). See also Arber, E. A. N. (17).
[7] Edwards (26).

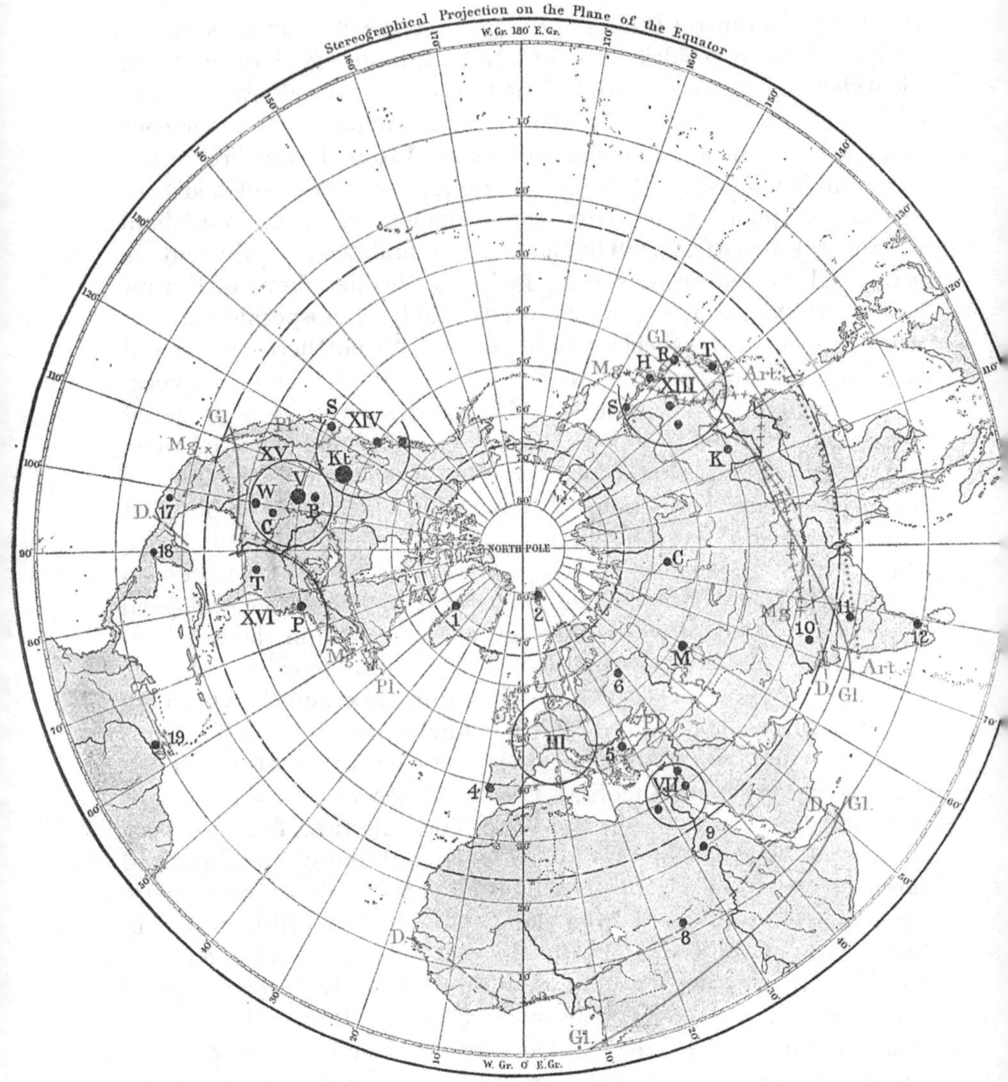

FIG. 110. Map of the northern hemisphere showing the geographical distribution of some Cretaceous floras; also, in red, the present northern limits of distribution of certain Angiosperms. See notes, pp. 409–411.

NOTES ON MAPS (FIGURES 110, 122)

Selected Cretaceous localities are shown on the map, fig. 110. In the map of the southern hemisphere, fig. 122, both Cretaceous and Tertiary localities are given; the former as • and the latter as ⊗. The approximate northern boundaries at the present day of the fern-family Gleicheniaceae (Gl.), the flowering plants *Magnolia* (Mg.), *Platanus* (Pl.), *Dalbergia* (D.), and *Artocarpus* (Art.) illustrate the striking contrasts between the past and the present geographical range of these plants.

Map Numbers and Letters NORTHERN HEMISPHERE (fig. 110)		References	Cretaceous Lower	Cretaceous Upper	Remarks
1.	Greenland	Seward (25), (25^2), (26). The bibliographies at the end of these papers should be consulted for other accounts of the Greenland floras, notably those by Heer and Nathorst. See also Drygalski (97); White and Schuchert (98); Florin (22); Nathorst (07^2), (11), and Walton (27^2)	×	×	The Greenland beds form a continuous, conformable series including both Lower and Upper Cretaceous floras: the precise boundaries have not been determined
2.	Spitsbergen	Nathorst (97), (97^2)	×	.	
III.	Europe	For a general account and correlation of the Cretaceous floras see the volumes by Berry and Stopes mentioned in the notes on this table			
	England	Seward (94), (95), (13^2); Stopes (12), (13), (15), (18)	×	.	In the paper of 1913 Dr Stopes describes several types of Lower Cretaceous petrified stems of dicotyledons
	France	Carpentier (27) with bibliography, (29^3)	×	.	
	Belgium	Seward (00^2); Bommer (10); Fraipont (21)	×	.	An account of a rich flora of which a few examples have been described is promised by Dr Bommer
	Holland	Kräusel (22^2) with bibliography; Kräusel and Jongmans (23)	.	×	
	Germany	In addition to the sources cited in the general accounts of Cretaceous floras, see Lipps (23); Gothan (28^2); Richter (06), (09) with bibliography	×	×	The rich Upper Cretaceous flora partially described by Debey and Ettingshausen (59) needs revision: see Kräusel (22^2)
	Bornholm	Bartholin (10)	×	.	
	Sweden	Möller and Halle (13)	×	.	
	Italy	Cappelini and Solms-Laubach (91); Fucini (28)	×	.	Prof. Fucini submitted to me photographs of his specimens and I suggested that some of them resembled *Baieropsis*: subsequently when I saw the actual specimens I felt sure that the supposed leaves are merely inorganic structures and not the remains of plants
	Poland	Nowak (07)	.	.	
	Czechoslovakia and adjacent territories	For descriptions of the rich Bohemian flora see Frič and Bayer (01) for references, also Velenovský and Viniklař (26–27); Krasser (96), (06); Kerner (95) who describes a small collection from the Island of Lesina	.	×	These floras mostly of Middle Cretaceous age (Cenomanian) include many dicotyledons

Map Numbers and Letters Northern Hemisphere (fig. 110)	References	Cretaceous		Remarks
		Lower	Upper	
4. Portugal	Saporta (94)	×	×	
5. Bulgaria	Zeiller (05)	.	×	
6. Russia	See references in Berry (11) and Stopes (13)	×	.	
VII. 8, 9. N. Africa, Syria and Transjordania	Kräusel and Stromer (24) with bibliography; Hirmer (25); Seward (07[4]); Edwards (26[4]), (29)	×	.	Some of the petrified wood in the Libyan desert is probably Cretaceous, though much of it is Tertiary. The specimen described as *Clathropteris* [Seward (07[4])] may, as Kryshtofovich believes, be incorrectly named
M. Siberia	Tamir River. Kryshtofovich (14[2]), (26). Mugojar Hills. Kidston and Gwynne-Vaughan (11) who describe a specimen of the Lower Cretaceous fern *Tempskya* from rocks said to be Tertiary	×	×	
C. Siberia	Chulyna River. Kryshtofovich (29)	.	×	This flora has not been described
XIII. S. Sakhalin Island	Kryshtofovich (18), (18[3]), (29), (29[3])	×	×	
XIII. Amurland: Ussuriland	Halle (21). Wealden plants from Manchuria	×	.	
XIII. Japan				
XIII. H. Hokkaido Province	Stopes and Fujii (10) with bibliography; Endô (25	.	×	
XIII. R. Rickuchu Province	Yabe (13)	×	.	
XIII. T. Tosa and other provinces	Nathorst (90[2]); Yabe (27); Yabe and Toyama (28); Yokoyama (94)	×	.	
K. China	Kalgan (North-west Peking). Barbour (24), (29)	×	.	
India 10, 11, 12	Balmir beds (Rajputana); Lameta beds, below the Deccan volcanic series; Trichinopoly (Madras). Sahni (22)	.	.	Cretaceous floras are poorly represented in India
North America XIV. Pacific Coast and part of Montana and Alberta	Ward (05). References to earlier papers by Dawson and other authors; Penhallow (02). See also Berry (16)	×	×	The Kootanie flora (Wealden) is preserved in beds occupying a large area in Alberta and Montana
S = Shasta; Kt = Kootanie flora	Shasta (California) formation. Ward (05)	.	.	

XV. Dakota, Nebraska, part of Montana, Wyoming, Kansas, Oklahoma, part of Texas. B = Black Hills; C = Cheyenne; V = Vermejo; W = Woodbine	Dakota flora—Lesquereux (91). Black Hills "Cycad" beds (B) Vermejo flora (V) of the Raton Mesa region—Lee and Knowlton (17) with bibliography; Knowlton (22); Animas flora of the San Juan basin—Knowlton (24). See also Cockerell (16); Berry (29). Cheyenne flora (C) (Kansas), Berry (22). Woodbine flora (W) (Texas), Berry (22^2)	×	× ×	A very rich flora (Cenomanian) For reference to other floras in this region see Knowlton (24). The name Laramie used to be employed for a flora in this region which was believed to be Upper Cretaceous in age: see Lee and Knowlton for a revision of the age of the Laramie plant-beds
XVI. Atlantic Coastal Plain. T and P	The Island floras: Long Island, Martha's Vineyard, etc. New Jersey floras: Amboy clays, etc. Raritan and other important floras. Potomac formation (P). References to the voluminous literature will be found in the following books and papers: Berry (14), (15), (16); Newberry (95); Hollick and Jeffrey (09); Hollick (06). For an account of the Tuscaloosa flora (T) and others in the Gulf region see Berry (19), (25)	×	×	
17. Mexico	Nathorst (93); Steinmann (Algae) (99)	×	×	Steinmann describes some well-preserved Calcareous Algae: the few plants described by Nathorst may be of Lower Cretaceous age
18. Central America: Guatemala	Stephenson and Berry (29)	.	×	
SOUTHERN HEMISPHERE (fig. 122)				
1. Madagascar	Fliche (05)	×	.	A description of wood of one of the older Cretaceous dicotyledons
2. Africa				
East Africa	Potonié (02)	.	×	Fossil coniferous wood
3. South Africa	Uitenhage flora. Seward (03)	×	.	
South America, etc.				
4 (and B). Graham Land: Snow Hill	Gothan (08)	×	.	Petrified wood
V. Argentina: Patagonia	Halle (13) with bibliography. Berry (24^4), (24^5); Kurtz (99)	×	×	
6. Peru	Zeiller (14); Salfeld (09^2); Neumann (07)	×	.	
19. (N. Hemisphere, fig. 110), Venezuela	Schlagintweit (19)	×	.	
7. New Zealand	Arber, E. A. N. (17); Edwards (26); Stopes (14^2), (16)	×	×	
Australia				
8. Central Australia	Woolnough and David (26); Walkom (29)	×	.	Plants associated with glacial deposits
9. Queensland	Walkom (18), (19), (19^3); Burrum flora; Styx flora	×	.	
10. York Peninsula	Walkom (28^5). Plutoville flora	×	.	
N. New Caledonia	Crié (89)	.	.	

NOTES ON THE LIST OF CRETACEOUS FLORAS (pp. 409–411)

The literature on Cretaceous floras, particularly those of North America, is much too voluminous to be cited at length. References are given to papers and books containing bibliographies and many of the earlier accounts are therefore not included in the foregoing list. For a general account of the floras see Berry (11), (16); Stopes (13), (15); also Knowlton (19) for a list of North American species. An interesting summary of plant-life in Cretaceous times is given by Knowlton (27) in his *Plants of the Past*.

The Table of Distribution of Cretaceous plants is deliberately very incomplete: a few genera are selected mainly to illustrate the wide range of some of the commoner plants.

Ferns. Under *Laccopteris* is included the genus *Nathorstia* for reasons given elsewhere—Seward (25), (26): Prof. Halle, whose opinion I greatly value, does not agree with me. Some well known Cretaceous ferns are omitted either because of their restricted range or because we know very little of their affinity to existing ferns.

Gymnosperms. It is noteworthy that the cycadean genus *Williamsoniella* has been recorded from the English Wealden beds: Edwards (21[2]).

Many species of conifers are omitted on the ground that it is usually impossible accurately to determine the systematic position of sterile foliage shoots. Many specimens are described by authors as species of the genus *Widdringtonites*: some afford satisfactory evidence of relationship to the Callitrineae; but only a few. The genus *Athrotaxites* described by Halle from Patagonia is one of the few examples of fossil conifers which affords good evidence of affinity to the Tasmanian genus *Athrotaxis*: the species from the Potomac formation named *Athrotaxopsis* may be similarly related, but proof is lacking.

Angiosperms. A distribution table compiled from published lists of genera and species would be misleading, and for this reason no attempt has been made to show the range in space of most of the flowering plants, named almost entirely from leaves, recorded from Cretaceous localities. It is certain that before the end of the Cretaceous period most of the families of dicotyledons were represented: this will be apparent to students who consult the papers by Mr Berry on American floras which are valuable sources well worthy the attention of systematic botanists. My point is that we shall not be in a position adequately to deal with the subject of geographical distribution until more is known of the microscopical structure of the leaf-cuticle and a more critical study, including comparisons with existing plants, has been made of European and other floras, several of which were described a good many years ago.

My primary aim in giving this and other tables of distribution is to stimulate students to devote themselves to a neglected and promising field of work. Botanists frequently speak of the untrustworthy data published by their palaeobotanical colleagues and yet on occasion they quote fossil species without question. It is freely admitted that the identification of leaves is often impossible and undoubtedly names of existing genera are frequently given to fossils on insufficient evidence. Nevertheless the available material is abundant and is well worth investigation at the hands of systematic botanists.

TABLE F. *The Geological and Geographical Distribution of some Cretaceous plants*

See Maps, figs. 110, 122, pp. 408, 479

	Jurassic	Cretaceous Lower	Cretaceous Upper	Tertiary	Greenland. 1	Spitsbergen. 2	Europe. III, 4–6	N. Africa: Syria. VII, 8, 9	Siberia: Sakhalin I. M, C, S	Japan: China. XIII, K	N. America. XIV	N. America. XV	N. America. XVI	S. America. 6, 19	S. America. V	S. Africa, etc. 1–3	New Zealand. 7	Australia. 8–10
Filicales																		
Gleichenites	—	—○—	—	—	×	×	×	·	×	·	×	×	×	·	×	·	×	×
Osmundites	—	—	—	—	·	·	·	×	·	·	×	·	·	·	·	×	×	·
Laccopteris } *Nathorstia*	—	—			×	·	×	·	×	·	·	·	·	×	×	·	·	×
Matonidium	—	—			·	·	×	·	·	·	×	×	·	·	·	·	·	·
Hausmannia	—	—			×	·	×	·	×	·	×	·	·	·	·	·	·	×
Ruffordia	—	—○—			·	·	×	·	×	×	·	·	×	×	×	·	·	×
Tempskya		—			·	·	×	·	×	·	×	·	×	·	·	·	·	·
Sphenopteris (*Onychiopsis*)		—			×	×	×	·	×	×	×	×	×	·	×	×	·	·
Cladophlebis Browniana		—			·	×	×	·	×	×	×	·	·	×	×	×	·	·
Plantae incertae sedis																		
Weichselia		—			·	·	×	×	×	×	·	×	·	×	·	·	·	·
Thinnfeldia	—	—			·	×	×	·	·	·	·	·	×	·	·	·	·	·
Ginkgoales																		
Ginkgoites	—	—	—	—	×	×	×	·	×	·	×	·	·	·	·	·	·	×
Baiera	—	—			×	×	×	·	×	·	×	·	×	·	×	·	·	·
Cycadophyta																		
Cycadeoidea	—	—			·	·	×	·	·	×	×	×	×	·	·	·	·	·
Williamsonia	—○—	—			×	·	·	·	·	·	·	×	×	·	·	·	·	·
Ptilophyllum	—	—			×	·	×	·	·	×	×	·	·	·	·	·	·	×
Otozamites	—	—			×	·	×	·	×	·	·	·	·	×	×	·	·	×
Zamites } *Zamiophyllum*	—	—			·	·	×	×	·	×	×	×	·	·	×	×	·	·
Pseudocycas		—	—		×	·	×	·	×	·	·	·	·	·	·	·	·	·
Pseudoctenis	—○—	—			·	×	×	·	×	·	×	·	·	·	·	·	·	·
Nilssonia	—	—	—	—	·	·	×	·	×	×	×	×	×	·	·	×	·	×

Table F. *The Geological and Geographical distribution of some Cretaceous plants* (cont.)

See Maps, figs. 110, 122, pp. 408, 479

	Jurassic	Cretaceous: Lower	Cretaceous: Upper	Tertiary		Greenland. 1	Spitsbergen. 2	Europe. III, 4–6	N. Africa: Syria. VII, 8, 9	Siberia: Sakhalin I. M, C, S	Japan: China. XIII, K	N. America. XIV	N. America. XV	N. America. XVI	S. America. 6, 19	S. America. V	S. Africa, etc. 1–3	N. Zealand. 7	Australia. 8–10
Coniferales																			
Araucarites *Araucarioxylon* }	—	—	—	—		×	×	×	×	×	×	·	×	×	×	×	×	×	×
Dammarites		—	—			×	·	×	·	×	·	·	×	×	·	·	·	·	·
Sequoiites	—	—	—	—		×	×	×	·	×	·	×	×	×	·	·	·	·	·
Moriconia			—			×	·	×	·	·	·	·	×	×	·	·	·	·	·
Widdringtonites		–	—			×	·	×	·	·	·	·	·	×	·	·	·	·	·
Athrotaxites	–	—	–			·	·	·	·	·	·	·	·	·	·	×	·	·	·
Pinites	—	—	—	—		×	×	×	·	·	·	×	×	×	·	·	·	·	·
Podozamites	—○—	—	—			·	·	×	·	×	×	×	×	×	×	·	·	·	×
Angiospermae																			
Sagenopteris (affinity uncertain)	—	—				·	·	×	·	×	×	×	·	×	·	·	·	·	·
Magnoliaephyllum *Magnolia* }		—/—	—	—	→	×	·	×	·	·	·	×	×	×	·	·	·	·	·
Cinnamomoides		—	—	—		×	·	×	·	×	·	×	×	×	·	×	·	·	·
Platanus *Platanophyllum* }		—	—	—	→	×	·	×	·	×	·	×	×	×	·	×	·	·	·
Credneria		—	—			·	·	×	·	×	·	·	·	×	·	·	·	·	·
Artocarpus		—	—	—	→	×	·	·	·	·	·	·	×	·	·	·	·	·	·
Menispermites		–	—	—		×	·	×	·	×	·	×	×	×	·	·	·	·	·
Dalbergites		—	—	—		×	·	×	·	·	·	×	×	×	·	·	·	·	·
Palmae	–	—	—	—	→	·	·	×	·	·	·	×	×	×	·	·	·	·	·

CHAPTER XVI

THE CAINOZOIC ERA
THE TERTIARY PERIOD

The lowness of the present state
That sets the past in this relief. *Tennyson*

THE Cretaceous period, which is taken as the upper limit of the Mesozoic era, was followed in many parts of the world by an elevation of the land: as the upraised areas became exposed to denudation the products of erosion were deposited in estuaries, lakes, and deltas. These sediments are the lowest members of the Eocene series, the opening stage of the Tertiary period. The change in surface features due to the displacement of the Cretaceous sea by land and shallow-water basins in England, in many European districts north of the Alps, and in parts of North America, is marked by a discordance in the succession of strata. In some other regions Mesozoic and Tertiary rocks form a continuous series; but it is not only because of the frequent occurrence of a clearly defined dividing line between the strata of the two eras that they are recognized as separate chapters of geological history; it is largely because of the difference in the animal population of the two seas. Southern Europe, across the yet unborn chains of the Alps and Carpathians as far as northern India, remained under the waters of the Tethys Sea. At a later date the sea-floor was slowly raised in the process of mountain-building: this we know from the occurrence in Tibet at an altitude of 20,000 ft. and in the Alps at a height of 10,000 ft. of limestones full of the disc-like shells of *Nummulites*, a genus of Foraminifera (a group of simple marine animals) still found in the Pacific Ocean, and one of the most widely distributed organisms in Eocene seas.

Tertiary Rocks

Tertiary sediments though not infrequently as hard and compact as those of earlier periods are often indistinguishable from modern sands and gravels. Resting on the Eocene series in ascending order are the Oligocene, Miocene, and Pliocene series. These names were chosen because of the gradually increasing percentage of species of living marine molluscs represented in the fossil content of the several sediments. In the Eocene beds (ἠώς = dawn, καινός = new, recent) the percentage of recent species was estimated at about 5 per cent.: in the Miocene beds the percentage, though higher, was found to be below that of extinct species: in the Pliocene strata the living types outnumber the extinct. At a later date the name Oligocene was proposed for a series of beds intermediate between the Eocene and Miocene.

TERTIARY		
	PLIOCENE	Cromerian series: Forest bed and freshwater beds with plants
		"Crag" desposits
		Teglian } Holland
		Reuverian } Holland
		"Crag" deposits
	MIOCENE	Not represented in England
	OLIGOCENE	Hamstead beds; chiefly freshwater
		Bembridge beds; chiefly freshwater
		Osborne and Headon beds
	EOCENE	Bagshot beds
		Bracklesham beds: Alum Bay and Bournemouth plant-beds
		London clay
		Oldhaven and Blackheath beds
		Woolwich and Reading beds
		Thanet sands

The Eocene beds of England occur in two separate basins; the London basin and the Hampshire basin. These two sets of Tertiary rocks are no doubt parts of a once connected series which was involved in crustal folding, the more exposed portions being subsequently removed by denudation. In the London basin the oldest beds are the Thanet sands exposed in the cliffs of Ramsgate: overlying these is a set of deposits known as the Woolwich and Reading beds, and above them are the Oldhaven and Blackheath beds which have yielded samples of the contemporary vegetation. Next in ascending order is the London clay, a thick mass of sea-mud underlying London and exposed in the cliffs of Sheppey in east Kent. Many seeds and fruits have been found in the London clay; among others, fruits closely allied to those of the stemless palm *Nipa*[1] which is to-day a common plant in estuaries and deltas in the Tropics. The uppermost Eocene beds are represented by the marine Bagshot sands of the Surrey heaths and commons. In the Hampshire basin the strata, though similar to those in the London area, present some distinctive features: the sands in the cliffs of Alum Bay in the Isle of Wight and at Bournemouth have furnished many

[1] Seward (11^{3}); Chandler (25).

leaves and other remains which, from their resemblance to species now living in more southern countries, are usually regarded as evidence of a sub-tropical climate. The basalts of the Giants' Causeway in northern Ireland, of Staffa, Mull, and other islands off the coast of Scotland are detached pieces of enormous sheets of lava which were poured out through cracks and fissures in an over-strained crust in the early days of the Tertiary period. The occurrence of basaltic lavas and beds of ash in Disko Island (fig. 1) and the adjacent mainland of west Greenland, in the Scoresby Sound district on the east side of Greenland, in Iceland, the Faroe Islands, and in other northern regions justifies the description of the lavas and beds of ash as examples of the "most stupendous succession of volcanic phenomena"[1] in the whole course of geological history. In the cliffs of Mull there are occasional bands of ancient soil and lake deposits interbedded with the lava-flows, and some of these have furnished waifs and strays from forests which had become established, in quiet intervals, on the volcanic rocks bordering a lake.

Oligocene deposits are found in the Isle of Wight, in Dorset, and at Bovey Tracey near Newton Abbot in Devonshire. A rich flora has been described from deposits known as the Bembridge beds in the Isle of Wight.[2] Plant-bearing beds occur in the Paris basin, and in Switzerland Tertiary strata play a conspicuous part in the hills around the lakes between the Alps and the Jura. Oligocene sands on the Baltic coast near Königsberg are one of the chief sources of amber,[3] the resinous secretion which exuded from wounded pines and other trees and as it hardened enclosed fragments of flowers and leaves which had been caught on the sticky surface.

The stage of geological history represented by Miocene strata in Europe and North America is unrecorded in England where elevation of the land precluded the accumulation of detritus under conditions favourable to its permanent preservation. In North America the Oligocene conditions continued with little change into the Miocene: in the Yellowstone National Park[4] a succession of forest trees embedded in ash (fig. 21) furnishes impressive evidence of repeated volcanic eruptions which foreshadowed the much more widespread and stupendous crustal dislocations which brought into being the Alps, the Himalayas, and other great mountain ranges.

[1] Geikie (97). [2] Reid, E. M. and Chandler (26) with many references to literature.
[3] Conwentz (86). [4] Knowlton (99).

Miocene lakes occupied depressions in the older rocks of North America and Europe, and from the sands and mud of their floors large collections of plants have been obtained. Over the German plain were large lakes and freshwater lagoons bordered by marshes in which flourished a species of *Taxodium* closely allied to the bald

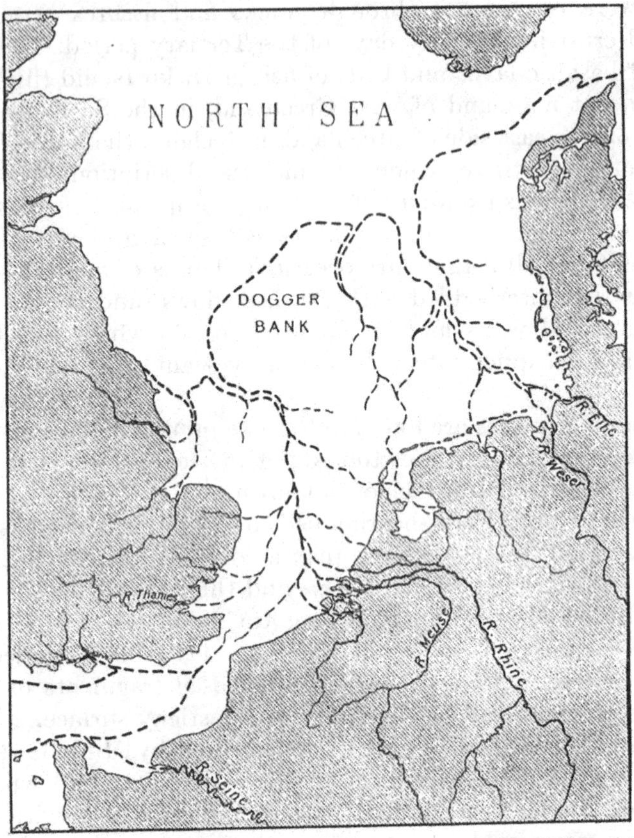

FIG. 111. Map showing the approximate coast-line over the North Sea at the period of the oldest submerged forest (late Quaternary, see p. 510). (From C. Reid's *Submerged Forests*. Camb. Univ. Press.)

(swamp) cypress of North America (fig. 115). Reference has previously been made to the preservation of Miocene plants in the volcanic beds of Florissant in Colorado. An exceptionally rich flora has been described from the Miocene beds on the borders of Lake Constance in Switzerland. These and other floras from temperate regions, which consist largely of genera that are now characteristic of more southern countries, raise problems relating to plant dispersal and fluctuations of climate.

In England the Pliocene series consists mainly of marine shelly sands known as "crag" which are best developed in Norfolk and Suffolk. By far the most important, botanically, of these East Anglian Pliocene strata are those known as Cromerian from their occurrence in the cliffs at Cromer. These beds are part of the old

FIG. 112. A view near Happisburgh on the Norfolk coast showing in the foreground drift-wood from the Cromer forest bed exposed at low tide, and cliffs composed mainly of glacial deposits. (After Dr Duckworth (11), from a photograph taken by Mr J. J. Lister.)

delta of the Rhine which formerly reached as far as the coast of Norfolk (fig. 111). The Cromerian beds include an estuarine deposit containing drifted trees and other plant remains known as the Cromer forest bed[1] (fig. 112): the vegetation seems to have differed very little from that in the same neighbourhood to-day. The spruce fir (*Picea excelsa*) and the water chestnut (*Trapa natans*) (fig. 120) are examples of a very few species that are no longer members of the British flora. Above the Cromerian series is a band of clay known as the Arctic plant-bed[2] because of the occurrence of an Arctic

[1] Reid, C. (99).

[2] This bed is usually regarded as the lowest member of the Quaternary system.

willow (*Salix polaris*) which is now a common plant in Spitsbergen, and the dwarf birch (*Betula nana*), a widely distributed prostrate shrub in northern countries. It is clear from a comparison of the plants of the forest bed with those of the Arctic bed which lies above it, that during the interval represented by the layers between the two plant-bearing horizons there must have been a considerable lowering of temperature. The Arctic plant-bed is usually regarded as the base of the Quaternary period: the botanical evidence which it affords prepares us for the more intense Arctic conditions which are the outstanding feature of the earlier stage of the Quaternary period. Records from English localities enable us to follow the history of our vegetation, or at least to recover a few scraps of the history, from the pre-glacial Arctic plant-beds upwards to the forest bed at the foot of the Cromer cliff, through the post-glacial neolithic period to the time of the Roman occupation.

In the following account of Tertiary floras attention is practically confined to flowering plants, conifers and ferns;[1] but before describing these higher forms it is worth while briefly to consider some of the records of algae and fungi furnished by Tertiary strata.

Tertiary Algae

Comparatively little attention has been paid in the preceding chapters to the lower plants—algae and fungi—on the ground that our knowledge of the geological history of these groups is hardly sufficient to justify more than a brief reference in a general account of ancient floras. Among the numerous fossils cited in palaeobotanical literature as examples of Tertiary algae some are described as species of existing genera; others are regarded as extinct genera: with the exception of the calcareous forms, few have any value as trustworthy records.[2] Though examples of Blue-Green Algae have been found in Tertiary rocks, they do not throw any light on the ancestry of existing members of this extremely ancient group. Reference may be made to a recent description of several well preserved specimens very similar to living species found in beds of oil-shale which were deposited in an Eocene lake in Colorado[3] and

[1] For a good, concise account of Tertiary floras, see Reid, E. M. and Chandler (29).

[2] For references to literature and for illustrations see Pia in Hirmer (27); Pia (20), (26); Lemoine (17); Morellet, L. and J. (13), (22). For a general account of living algae see Oltmanns (04–05) and, for freshwater algae, West and Fritsch (27).

[3] Bradley (29).

adjacent regions of North America. The most satisfactory and interesting fossils are those which owe their preservation as recognizable algae to the fact that the plant-body was impregnated with lime, or rendered permanent by the flinty nature of the cell-walls as in the siliceous coverings of the freshwater and marine diatoms. Calcareous Algae discovered in Tertiary strata include several genera belonging to the group Siphonales, that is, plants which consist of narrow tubes—a central axis giving off scattered or whorled lateral branches—or of irregularly interwoven tubes forming a more or less compact mass. The plant-body does not consist of cells each of which is bounded by a cell-wall, but of units which may be described as compound cells not sub-divided by complete septa. Reference has been made in earlier chapters to the family of siphoneous algae known as the Dasycladaceae: these algae, most of which live in tropical or sub-tropical seas, are characterized by a central axis formed of a relatively broad tube, which may be a few millimetres in diameter, either wholly calcified or made up of calcareous segments separated by very narrow zones in which the lime is lacking: from the main axis are given off numerous and more delicate branches which are seldom preserved but have left traces in small perforations marking their attachment to the central tube (fig. 113, C). Members of this family, as we have seen, were abundant in the Tethys Sea in the early part of the Mesozoic era: their remains played an important part in building up an enormous thickness of calcareous ooze. This calcareous material, as the result of stupendous upheavals in the Tertiary period—the so-called "Alpine storm"—which produced the Alps and other mountain-ranges, was incorporated in the jumble of rocks of the Tyrolese and Himalayan dolomites and limestones.

The Dasycladaceae continued to flourish abundantly in the later Mesozoic and in Tertiary seas: genera now living in the Mediterranean, Adriatic and tropical seas were probably represented in the Tertiary period by a greater variety of forms; it is certain that they ranged much farther north than the existing species. It has been pointed out that several of the extinct species differed from the living in having the branches scattered and not restricted to regular whorls. From the Eocene beds of the Paris basin fourteen different genera of Dasycladaceae have been described. The genus *Cymopolia* (similar to fig. 113, B), now living in the Gulf of Mexico and on the shores of the Canary Islands, is a small alga with a repeatedly forked body, the slender branches composed of series of calcareous cylindrical segments or joints separated by narrow bands of soft material and bearing numerous branchlets, some sterile, others supporting spore-cases: this genus lived in the Eocene lake of northern France.[1] In another genus, *Dactylopora* (fig. 113, C), the

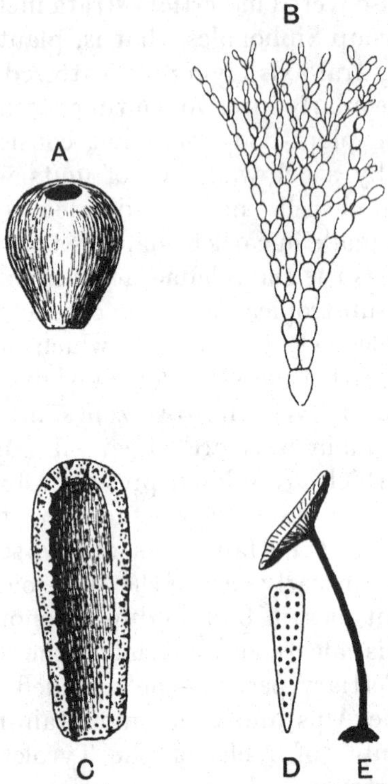

FIG. 113. Calcareous Algae from the Eocene Paris basin (A–D). A. *Ovulites*. A simple segment of an alga considerably enlarged (cf. B). B. *Ovulites margaritula*. Reconstruction after Munier-Chalmas. Slightly enlarged. C. *Dactylopora cylindracea*. × 2. (After Morellet.) D. *Acicularia Michelini*. × 12. Spicule from one of the compartments of the "umbrella" showing cavities originally occupied by the spores. (After Morellet.) E. *Acetabularia mediterranea*; a recent specimen. Nat. size.

central axis is a tube, reaching a length of 2 cm., open at one end, closed and rounded at the other, freely perforated by fine canals indicating the possession of branchlets: this genus is recorded also from northern Africa where it is one of many marine organisms proving the existence of an Eocene sea on the site of the present

[1] Morellet, L. and J. (13), (22).

Libyan desert. Among other Eocene genera is *Neomeris* which now occurs in the warmer seas.[1] A genus of a different type, though included in the same family, is *Acicularia*, which was originally described as a Tertiary alga and subsequently found living. *Acicularia* (fig. 113, D) is very closely allied to the better known *Acetabularia* (fig. 113, E) of the Gulf of Naples and other warm seas: a slender stalk supports a terminal horizontal disc about 1 cm. in diameter almost completely divided into long and narrow wedge-shaped compartments containing reproductive cells, which are embedded in a gelatinous substance that becomes converted into a solid matrix of lime. *Acicularia*, recorded from the Eocene beds of the Paris basin, from Miocene strata in the Crimea, Italy, and elsewhere, is usually represented in a fossil condition by very small spicules of carbonate of lime a few millimetres long with cavities (fig. 113, D) that were originally occupied by minute spore-capsules: the spicules are the calcified contents of the compartments of the umbrella-like caps which fell apart as the alga began to decay.

Mention may be made of the extinct Cretaceous *Triploporella*,[2] intermediate between *Acetabularia* and *Dasycladus* (fig. 83), which has been recorded also from Lower Tertiary beds in northern India. These are a few of the Dasycladaceae which have been found in Tertiary beds in France and in many other parts of the world.

Another type of Siphoneous alga is illustrated by the family Codiaceae,[3] one genus of which, *Codium*, is represented by British species, though most of the members are characteristic of tropical and sub-tropical seas. *Codium* is not a calcareous alga; it consists of soft, cylindrical, forked branches reaching a length of several centimetres and is hardly likely to be preserved as a fossil. The greater part of the plant-body of the closely related genus *Halimeda* is calcareous and made up of branches composed of broad and relatively short, flat segments strung together on a slender axis of interwoven tubes. The alga is either prostrate or erect; it is abundant as an important constituent of tropical coral reefs; it occurs on the reefs of Florida and was common in the earlier stages of the Tertiary period in European seas.

Another and much smaller genus of the Codiaceae is *Penicillus*—the merman's shaving brush: the body of this alga, which is almost entirely tropical in range, consists of fairly thick stalks a few centi-

[1] Church (95).

[2] Walton (25^2).

[3] Gepp, A. and E. S. (11); for fossils see Glück (12).

metres high bearing a dense mop of branches made of numerous, very slender calcareous joints. Several years ago some small fossils, 2–6 mm. long, resembling minute eggs, were found in the Paris basin beds and referred to the animal kingdom as species of *Ovulites* (fig. 113, A, B). It was subsequently demonstrated that these egg-like bodies, made of a thin shell with an aperture at each end of the long axis, are the detached segments of an Eocene alga closely allied to living species of *Penicillus* or *Cymopolia*.

Another family of Calcareous Algae included in the Rhodophyceae (red algae) is represented by the genera *Lithothamnium* (fig. 114), *Archaeolithothamnium*, *Lithophyllum* and *Goniolithon* (fig. 25), all of which are still living and were widely spread in Tertiary seas: these differ widely from the Siphonales in being constructed of masses of small cells and not of undivided long and narrow tubes. *Lithothamnium*[1] is particularly abundant in Arctic seas; it occurs on coral reefs and in temperate waters: the oldest known examples are from the highest stage of the Cretaceous system, but it was much more plentiful in the Tertiary period. The other two genera have not been recorded before the Tertiary period. The photograph reproduced in fig. 114 shows the floor of a rock-pool on the coast of the Isle of Man covered with a pavement made of the hard chalky bodies of living *Lithothamnias*; and in the left-hand corner is another calcareous alga (*Corallina*) of the same family though of very different habit. This modern picture would serve as a reconstruction of a scene on the beach of a Cretaceous or Tertiary sea.

Attention has been called by an authority[2] on these genera to the remarkable uniformity in the structure of the plant-body revealed by a comparison of the extinct and living species, an architectural plan which seems to be proof against environmental influences. It may also be pointed out that in the Palaeozoic genus *Solenopora* (fig. 41, B) the construction of the calcareous skeleton is very similar to that in *Lithothamnium* and its allies.

Another and even more striking instance of persistence through the ages is afforded by the geological record of the Charophyta (fig. 136)—relatively complex plants which are often included in the Algae though they differ in many respects from the more typical members of the class. Fragments of the calcareous stems and the calcified parts of the "fruit" (oogonia) are common in many

[1] Foslie (29); Lemoine (29); Howe (18), (19), (22) for photographs of fossil species.
[2] Lemoine (17).

FIG. 114. Photograph of a pool at low water on the coast of the Isle of Man, taken through 6 ins. of water, showing the floor covered with *Lithothamnium*; in the left-hand corner a cluster of another calcareous alga (*Corallina*), and a limpet shell in the right-hand upper corner. (Photograph by Dr Margery Knight.)

Tertiary strata, from the Eocene to the Pliocene. In a book[1] dealing with living and fossil species of the genus *Chara* (stoneworts) the authors speak of the highly specialized form of "fruit" having remained "practically unchanged" from the Carboniferous period to the present day.

The siliceous cases of diatoms are present in countless millions in some Tertiary localities, notably in Miocene rocks in the Richmond district of Virginia, in beds of the same age at Bilin in Bohemia, and in several other parts of the world. No undoubted examples of these algae have been found in rocks older than the early part of the Jurassic period: the family became cosmopolitan during the Tertiary period. Readers interested in diatoms should consult the comprehensive list of living and extinct species drawn up by Prof. Frenguelli[2] and published in Dr Hirmer's *Handbuch der Paläobotanik.*

Fungi

It is not often that fossil fungi can be used as evidence of climatic conditions; in this connexion attention may be called to a few records of a particular type of fungus from Cretaceous and Tertiary rocks. On examining microscopically fragments of carbonized leaves collected from a peaty layer in the Cretaceous sands exposed on the face of a cliff in Disko Island, off the west coast of Greenland (lat. 70° N.), some well preserved fruit-bodies were discovered having the form of flattened discs composed of radially arranged rows of small cells, bearing a striking resemblance to the fructifications of certain living fungi included in the family Microthyriaceae. Precisely similar fossils were found on leaves of a *Sequoia* in the Tertiary beds of Ellesmere Land[3] (lat. 77° N.) and in the late Tertiary plant-beds of Frankfurt on the Main.[4] More recently, good illustrations have been published of the same type of disc-like fructifications and of spores from the Eocene plant-beds of Mull.[5] The living fungi which agree most closely with these fossils are mainly tropical in distribution and are characteristic of regions with a high rainfall.

Among the numerous examples of fungi described and illustrated in accounts of Palaeozoic, Mesozoic, and Tertiary floras there are very few which make any appeal to the student of evolution. Dark spots on the surface of fossil leaves bearing a superficial resemblance

1 Groves and Bullock-Webster (24).
2 Hirmer (27).
3 Nathorst (15^2).
4 Engelhardt and Kinkelin (08).
5 Edwards (22).

to the spore-producing fructifications of existing fungi are fairly common, and in some instances they are well enough preserved to be compared with modern genera; but the great majority have no botanical value. One thing is clear: from the Devonian period onwards and even from a more remote age there were parasitic and saprophytic fungi—fungi thriving on living hosts or deriving food from dead tissues—which so far as we can tell differed in no essential respects from living representatives of the class. We can safely say that bacteria and many other fungi are entitled to be included among the most ancient members of the plant kingdom. It may confidently be assumed that through the ages these plants occupied the position which they now hold as scavengers, transforming the complex bodies of organisms into the raw material essential for the nutrition of green plants. There is reason to believe that from very early times there have been two kinds of association between the higher plants and fungi: fungi preying upon their hosts and others beneficial to the hosts in which they lived.

General features of the Vegetation of the Tertiary Period

We pass now to the history of the Tertiary vegetation as revealed by a study of the higher plants, more especially the flowering plants. By the end of the Cretaceous period the vegetation of the northern hemisphere had become essentially modern: the old order had passed and an apparently new and exceptionally vigorous class of plants had proved its efficiency as a successful competitor in the struggle for position and as a colonizer of large spaces. It is seldom possible to follow step by step the passage from the Mesozoic to the Cainozoic era. Geographical changes, including elevation of the sea-floor and in some regions the uplifting of mountain-ranges, precluded the formation of a continuous and conformable series of sedimentary rocks suitable for the preservation of botanical records. There is usually a break in the succession as we pass beyond the highest Cretaceous strata: marine beds are suddenly succeeded by deposits formed in estuaries and lagoons, and the eroded surface of the older rocks at their junction with the younger affords proof of a long interval of time—one of many gaps in the history of the world. On the other hand, although there are differences between the later Cretaceous and the early Tertiary floras, they are differences in degree rather than in the general composition of the vegetation. One of the more striking contrasts between the Cainozoic vegetation as a whole and the older vegetation of successive stages of the Cretaceous system is the much greater uniformity and stability of the plant-world throughout the Tertiary period. In the course of the Cretaceous period the aspect of the vegetation was fundamentally changed: ferns and gymnosperms, which had long occupied a dominant place, became subordinate groups in a company consisting mainly of flowering plants. No such revolutionary shifting of the balance of power is revealed by the history of Tertiary floras. What, then, are the characteristic features of the vegetation with which we are now concerned, and what do we learn from a review of the floras as they follow one another through the interval—reckoned in millions of years—which separated the end of the Cretaceous period, or more correctly the latter part of that period, from the later stages of the Tertiary period?

From first to last Tertiary floras agree very closely with those in various regions of the modern world: the present domination of the flowering plants, firmly established before the end of the Cretaceous period, still continues. There is, however, one clearly defined difference between the earlier Tertiary floras and those of the present day: among the hundreds of fossil plants described from the older Tertiary rocks there are few examples of herbaceous genera: trees are greatly in excess of the smaller flowering plants such as now clothe our meadows and moors or live far above the ground on the boughs of forest trees. This preponderance of trees is no doubt more apparent than actual: our knowledge of the floras is based very largely on leaves and twigs carried by rivers and preserved in the sediments of lakes and deltas, and it is natural that the larger and stronger fragments of trees should be found in more abundance than scraps of the more delicate herbaceous plants. In some instances where layers of peat have been searched for fossils a good harvest of fruits and seeds has been gathered. The proportion of one class of plants in relation to another depends in part at least on the conditions of deposition. But it is true to say that in the earlier part of the age of flowering plants herbaceous forms were relatively few: their subordinate position in the vegetation finds a closer parallel to-day in tropical than in temperate regions; yet, speaking generally, and making allowance for the misleading information furnished by the comparatively incomplete collections of fossils, the relatively greater number of arborescent as compared with herbaceous plants is characteristic of the earlier floras in which angiosperms were predominant. Palaeobotanical evidence supports the view that herbaceous flowering plants as a class are the more recent products of evolution.

A fact of considerable interest which emerges from a comparative study of the successive phases of the Tertiary plant-world is the contrast presented by the earlier floras, which have left samples in the Eocene, Oligocene, and Miocene stages, to those of the Pliocene stage. A similar contrast is noticed when we compare tropical or sub-tropical forests with those in the northern part of the temperate zone, in North America, or in central and northern Europe. There are two extremes: in the first half of the Tertiary period the vegetation of that part of the northern hemisphere which is now southern England was of the type we find in warmer countries, in the Far East, the Malay Peninsula, in the southern part of North America and regions still farther south. The other extreme is illustrated by the relics of temperate plants which, as the Tertiary period passed into the Quaternary period, became associated with Arctic species. Between these two sharply contrasted types there are intermediate floras; floras that are mainly tropical or sub-tropical, including among a preponderance of existing genera a few extinct forms, which serve as connecting links between clearly differentiated living genera.

As we watch the shifting scenes, which make up the long drama dimly presented by the scattered relics of the Tertiary plant-world, we seem to see a steady advance from one type of vegetation to another. The earlier vegetation foreshadows the eastern and southern forests of the present day, though differing from them in its occurrence in places that have now temperate floras, and in the inclusion of plant-forms that seem to us archaic. The later vegetation, which is less tropical and has fewer plants that cannot be closely matched with living species, shades into one that agrees in essentials with the vegetation of the regions where the fossils were found. The record is one of climatic change and of migrating hosts impelled by the stimulus of altering climates to wander farther from the north towards more genial lands.

Fossil Plants as Evidence of Climate

Many attempts have been made to estimate the fluctuation in temperature at a given locality by comparing fossil plants with their nearest of kin in the present world: leaves believed to have come from a tree, such as a cinnamon, a bread-fruit tree, or some other genus, which is now tropical in its distribution, are discovered in Lower Tertiary rocks in the south of England or in other temperate regions: the inference is that because existing species are now confined to warmer regions therefore the Tertiary species required much the same environment. A conclusion such as this seems to be almost obvious when we find not one or two species but a large group of plants the majority of which have their nearest relatives in countries separated from the home of the Tertiary flora by several degrees of latitude. We shall refer later to the use of fossil plants as thermometers of the past, but meanwhile it is worth while to keep before us a few considerations as we pass in review some of the better known examples of Tertiary floras. In the first place, the majority of Tertiary plants are not specifically identical with living forms. It is well known that many closely related species live under very different conditions: several genera and families that are now characteristic of the Tropics include members which flourish in temperate regions. We cannot escape from the conclusion that in North America north of latitude 45° N. and in a corresponding position in Europe the evidence of Tertiary floras is very definitely in favour of a more genial climate than at present; but it is not safe to assume that the difference is as great as we should be disposed to believe from a superficial comparison of the past with the present. It is impossible accurately to determine the precise relative susceptibility to external factors of extinct and living plants; moreover, we cannot assert with confidence that in the earlier days of its career a genus or a species responded to climatic influences exactly as it does today. In the course of thousands or millions of years plants may well have become modified in their constitution, that is in their ability to withstand a climate that would now be intolerable: changes may have occurred which are not expressed in external characters. Plants, like races of men, may have become less hardy and less resistant to changing conditions. All that is urged at the moment is that we cannot hope to estimate within narrow limits the range of temperatures indicated by groups of species which include many extinct types and are far removed in time from plants that are still with us.

There is another point of some importance: our reconstructions of Tertiary and other floras are often based on scraps of twigs, fallen leaves, seeds and fruits embedded in the sediments of an estuary; they are the remains of a vegetation which grew on low ground and represent only a particular section of the whole plant population. In some places the conditions of preservation were such as provided records of plants from upland stations; but it is the strand floras and their associates which bulk largest in the herbaria of the rocks. We know only in part, though what we know is full of interest and pregnant with problems.

A very pertinent question often raised by botanists and laymen alike when they look at collections of fossil leaves is the possibility of accurate interpretation of the records: is it possible to distinguish one kind of flowering plant, conifer, or fern from another if the material consists for the most part of leaves and sterile twigs, and reproductive organs, which are the surest guides to affinity, are lacking? The answer is that in many instances the documents are illegible and many which have been treated as legible should have been discarded: it is equally true that by careful attention to details and after as thorough a comparison as possible of fossil and living leaves, identification is possible. It is notorious that many of the published lists of Tertiary plants are misleading and untrustworthy through insufficient care in comparing the dead with the living, through failure to preserve a due sense of proportion under the stimulus of the passion for the search. Several palaeobotanists have shown that form and venation-characters can be employed with satisfactory results provided the material is well enough preserved. References to a few sources where fuller information may be found are given in the footnote.[1] In recent years[2] successful attempts have been made to employ the character of the cuticle, that is the superficial layer which can sometimes be detached from a fossil and, after treatment with certain chemical reagents, examined under the microscope. Occasionally, though the cuticle may be preserved, the carbonized film of a leaf cannot be separated from the rock, and recourse must then be had to another method of treatment by which the mineral matter attached to the leaf can be removed and a portion of the specimen prepared for microscopical examination. The pattern of the superficial layer of cells with the minute stomatal apertures is a valuable aid to identification. Another line of attack, applicable when the fossils are mainly fruits

[1] Laurent (07); Menzel in Potonié (21).
[2] Bandulska (23), (24), (26), (28).

and seeds, as in a layer of peaty substance, is to wash the material and separate the vegetable remains from the matrix in which they are embedded, and then by a laborious process of comparison endeavour to assign the fossil fruits and seeds to their systematic positions.[1] The task of interpreting the records of Tertiary plants, though difficult, is by no means hopeless. There are many collections in this country as well as in continental museums which have never been adequately described, and it is regrettable that students should be discouraged from a line of work, unquestionably rich in promise, by the repeated assertions of sceptics who have not experienced the thrill of reconstructing the vegetation of other days.

It is not my intention to give a full description of Tertiary floras; this would be far too ambitious an aim: it is rather to give an outline sketch of the plant-world of the northern continent in the course of the Tertiary period. At the same time, in order to further this object, certain genera are discussed that were widely spread in the northern hemisphere, though they were not represented in the standard floras selected as examples. Frequent reference will be made to facts having a direct bearing upon the plant-geography of former epochs: it is impossible to form a true conception of the meaning and significance of the present distribution of genera and species if attention is confined to the present.

Eocene Floras

It is customary to draw a distinction between the oldest rocks, that is, the first, or Eocene, stage of the Tertiary period, and those immediately following them. Thus, in many geological text-books this first stage is divided into Palaeocene and Eocene.[2] In this general account the whole of the oldest group of floras is spoken of as Eocene since there is little difference, in the broader features, between the vegetation of the two subdivisions.

In the southern states of North America bordering the Gulf of Mexico there is a considerable thickness of old delta-deposits formed at different times in the course of the Tertiary period when the shore of the sea lay much farther north: it was gradually silted up in the interval between the Eocene and the end of the Tertiary period. The richest flora preserved in the lower members of this

[1] Reid, C. and E. M. (15). For an account of methods employed in the investigation of peat and other deposits containing fruits and seeds see Reid, E. M. (07–09).

[2] Haug (20).

series of beds is known as the Wilcox flora[1] from the county of that name in Alabama: another flora, preserved in the same region, slightly younger but also Eocene in age, is the Claiborne flora[2] to which occasional reference will be made. Reference will be made also to the Raton flora[3] which is probably the oldest of the Eocene floras in America: this has been described from plant-bearing rocks in the Raton Mesa area of Colorado and New Mexico, a plateau to the east of the Rocky Mountains (within the circle C, fig. 121). The genera mentioned in the following account are merely a few examples selected from a much longer list.

The Wilcox flora, so far at least as the available data indicate, was made up of over 90 per cent. of flowering plants (angiosperms), the great majority of which are dicotyledons. Ferns are poorly represented and some of the specimens are too imperfect to be identified with any degree of certainty. In order to form an idea of the more characteristic members of the fern group in the early Tertiary floras we must extend our survey to other districts.

Ferns

An almost cosmopolitan genus in floras of this age is *Lygodium*, a member of the family Schizaeaceae which played a prominent part in the Mesozoic vegetation from the early stages of the Jurassic period upwards: *Lygodium* is now mainly tropical and includes the best known examples of climbing ferns: there is a species still living in temperate North America. The Tertiary species *Lygodium Kaulfussii* occurs in America, many parts of Europe and in Manchuria (fig. 121, 8); it is an interesting example of a fern, hardly distinguishable from a living species, which has now the most northerly range of any member of the family. There are no representatives of the Gleicheniaceae in the Wilcox or other floras of the Gulf region: the genus *Gleichenia* or, as the fossils are usually called, *Gleichenites*, was a conspicuous plant in the Cretaceous floras of Greenland and other parts of the northern hemisphere. It would seem that by the beginning of the Tertiary period the family had begun to desert its former northern territory for the southern Tropics where it is now abundantly represented: in the New World it still reaches as far north as the state of Louisiana. Among the few European Glei-

[1] Berry (16[2]) with references, (17).
[2] Berry (14), (24).
[3] Lee and Knowlton (17).

chenias recorded from early Tertiary beds (Eocene) in Europe are species from southern England and from Ireland.[1] Among the comparatively few ferns that are known from the older Tertiary floras several have been assigned, often on slender evidence, to the family Polypodiaceae which is by far the largest in the modern fern-world; it includes well over a hundred genera whereas the next largest family has only nine genera.

We know very little of the early history of the Polypodiaceae. The number of well-preserved examples in Mesozoic floras is insignificant, and such evidence as the rocks afford supports the view that not infrequently plants that are more or less cosmopolitan are comparatively late products of evolution. One member of the family that was widely spread in the Tertiary vegetation is the genus *Acrostichum*. The Claiborne beds (these and the Wilcox are from the area D on the map, fig. 121) furnished some pieces of fronds of an *Acrostichum* resembling the foliage of a living species, *Acrostichum aureum*, which is now a common fern in the mud-flats of tropical estuaries. The genus was widely spread over the northern hemisphere in the Old and the New World in the Tertiary period. Reference should be made to Table G, page 471, for information on the past and present range of this and other Tertiary genera. Another is *Onoclea*, now represented by the sensitive fern *Onoclea sensibilis*, which has a discontinuous range; it occurs in the north-eastern part of North America and in north-eastern Asia. Fossils which differ in no obvious feature from the living species are recorded from Lower Tertiary beds in northern Ireland and the Island of Mull[2] off the west coast of Scotland as well as in Greenland and in North America. We may picture *Onoclea* as migrating along divergent routes from the Arctic regions, permanently establishing itself in the west and east but not retaining a foothold in Europe. This is one of very many instances of plants which had a continuous and wide geographical range in the Tertiary period—several of them also in the latter part of the Cretaceous period—but in the course of time suffered partial extinction, as the climatic conditions changed and barriers of sea and mountain ranges prevented free migration. It is by following the history of genera and species into the past that light is thrown on the causes responsible for the present geographical distribution. The genus *Osmunda*, on the other hand, affords an

[1] Gardner (85^2), (86).
[2] Seward and Holttum (24); Gardner (85^2).

illustration of a member of one of the most ancient families (Osmundaceae) which still occupies a wide territory: species flourished over a considerable part of Europe in the Tertiary period and the existence of the genus in Spitsbergen has been definitely proved. The royal fern (*Osmunda regalis*) now lives in Scandinavia, throughout Europe, in many parts of North America, in Brazil, in tropical and South Africa, in China, Japan and many other regions: *Osmunda* would seem to be endowed with powers of endurance and adaptability which are by no means generally found in plants which trace their ancestry back well into the Palaeozoic period. The Marattiaceae, another of the more ancient fern families, have a history similar to that of the Osmundaceae: in the later Triassic and in Jurassic and Rhaetic floras the family was almost world-wide; few traces have been found in Cretaceous rocks. Among the rare examples from Tertiary localities are one from a North American flora and another from southern England.[1] Here we have an additional illustration of a family, now characteristic of tropical floras, which reached its maximum development north of the equator long before the dawn of the Tertiary period.

Conifers

We will now pass in review some of the conifers: the collections obtained from the Gulf region of North America afford little information, since coniferous trees would be found in the upland forests. The elements of the Wilcox flora are for the most part indicative of swamps and lowlands, and conifers are therefore poorly represented; but we learn from other sources that members of this group had a range in space very different from that at the present day. One of the most frequently recorded conifers in Tertiary floras is the genus *Glyptostrobus*, a tree closely related to *Taxodium*, the bald cypress of eastern America (fig. 115) and often confused with it. *Glyptostrobus* is now confined to China: in the Tertiary period it ranged over the northern hemisphere and was still living in Europe in the Pliocene stage. In the history of *Sequoia*, we have a similar example of a genus cosmopolitan in the Tertiary period and now confined within the limits of a narrow strip of country bordering the Pacific coast in California and Oregon. *Glyptostrobus* found refuge in the Far East, which is a rich storehouse of survivals[2] from a Tertiary vegetation that stretched from North America across

[1] Gardner and Ettingshausen (82). Gray, Asa (89).

Europe to the Pacific coast: *Sequoia*, with its two species, the red-wood and the big trees, is now one of the most impressive links with the past in the plant-world, a genus surpassing all others in the majesty of its towering columns and hardly exceeded by any other giants of the forest in the number of rings—which rings are years—that can be counted in its stems. Living in splendid isolation it carries us back through a length of years in comparison with which the three thousand years and more of the older individual trees is an insignificant fraction of the span of the family history. Abundant remains of a *Sequoia* hardly distinguishable from the Californian redwood (*Sequoia sempervirens*) have been found as far north as Ellesmere Land[1] (E, figs. 121, 137), in Alaska (A, fig. 121), in many widely separated localities in other parts of the United States, in the Island of Mull, at Bovey Tracey in Devonshire, in Chile, Japan, Manchuria, and elsewhere.[2] Specimens from the late Pliocene beds of the Rhone Valley and the valley of the Main in Germany[3] show that the genus was still represented in Europe in the latest stage of the Tertiary period. The inference is that it was finally destroyed by the fateful advance of the Arctic ice in the early part of the Quaternary period. Seeds from California were introduced into England in 1833: the height which many of the trees have already attained is by no means suggestive of decadence. We can trace a slowly rising curve in the progress of a genus or family and follow its steady, or it may be sudden, descent; but the nature of the controlling factors remains an enigma.

Another characteristic Tertiary conifer is *Taxodium*: it formerly existed in Spitsbergen, the Mackenzie River district of Canada, and in many other parts of the North American continent; it was a common tree in the forests of central Europe which furnished the material of which the Miocene Brown Coal[4] is formed, and it is recorded from Tertiary beds in Japan. There are now three species, the two best known being the bald cypress, *Taxodium distichum*, which grows in the Dismal swamp of Carolina and Virginia, in Louisiana (fig. 115) and other Southern States of North America; and *Taxodium mucronatum*, a famous example of which in the church-yard of Santa Marie de Tule in Mexico, over 60 yards in girth and about 120 ft. high, is said to be 5000 years old. The genus reached a world-wide distribution in the earlier part of the Tertiary period;

[1] Nathorst (15[2]).

[2] Chandler (22).

[3] Engelhardt and Kinkelin (08).

[4] Kräusel (28[2]) with references.

FIG. 115. The bald cypress, *Taxodium distichum*, at Grand Lake, Louisiana. (Photograph from the United States Department of Agriculture; Forest Service.)

it probably had its home in the Arctic regions whence it spread over the northern continents. It persisted in Europe into the Pliocene stage and now occupies a comparatively small area, from Mexico through the southern states of North America to Delaware.

The identification and classification of conifers are exceptionally difficult: their fossil remains are generally pieces of foliage-shoots, and it is comparatively seldom that recognizable cones or other reproductive organs occur in organic connexion with the branches. Petrified wood is common, but anatomical features, though they are valuable guides within relatively wide limits,[1] do not often enable us confidently to assign specimens to any one living genus. A good illustration of the difficulty of defining precisely systematic position is afforded by foliage-shoots having the external character of the Cypress family—small overlapping scale-like leaves arranged in regular, alternating pairs as in familiar cypress trees, the Arbor Vitae (*Thuja*), and *Libocedrus*[2] (incense cedar). Some Tertiary fossils have been described as species of *Libocedrus* and many of them, though by no means all, are no doubt correctly named. *Libocedrus* was certainly widely distributed in early Tertiary forests: it has not been recognized in the Wilcox flora, though it is recorded from floras in the north-western area (B, fig. 121) and from many parts of Europe where it continued to exist into the Pliocene stage. Its present distribution is discontinuous: the Pacific regions of North America, eastern Asia, New Zealand, and New Caledonia[3] the home of many trees of ancient lineage. The Cypress family as a whole had a much more extended range in Europe in the earlier stages of the Tertiary period than it has now. Araucarian conifers, as we have seen, were cosmopolitan in Jurassic floras. Members of the family still lived within the Arctic circle in the early part of the Cretaceous period. Evidence of the existence of *Araucaria* in the Tertiary vegetation is far from plentiful, at least in a form which carries conviction; but we know that the genus, or possibly a closely related conifer, lived in what is now the south of England and in other parts of Europe in the Eocene stage several thousands of miles north of the present home of the family in Brazil, Chile and Australasia. We know also that in the Tertiary period araucarian trees were established in Seymour Island[4] on the edge of Antarctica.

Additional examples of contrasts in distribution between Ter-

[1] Seward (19).
[2] Florin (30).
[3] Compton and others (22).
[4] Dusén (99).

tiary and existing conifers are supplied by *Callitris*, *Cryptomeria*, *Sciadopitys* and *Podocarpus*. It is preferable to employ the term *Callitrites* rather than *Callitris*, which is the name of a genus now confined to the Atlas Mountains of northern Africa, Spain and Malta, Australia and New Caledonia, because many of the fossils referred by authors to *Callitris* are more like twigs of some other members, such as the African *Widdringtonia* and the Australian *Actinostrobus*, of this small and widely scattered family which affords a striking example of discontinuous range. The family is now unrepresented in the forests of North and South America and Europe. In the early part of the Tertiary period *Callitrites*, using this name as a family designation, was abundant in North America and Europe and continued to flourish in the latter continent during part at least of the Pliocene stage.

Conifers afford many examples of contrasts in distribution between the Tertiary period and the present time. The genus *Cryptomeria*, often cultivated in our parks and gardens and wild only in Japan where it is familiar to travellers as the tree of the Memorial Avenue at Nikko, in the early part of the Tertiary period was a forest tree of northern Ireland.[1] Similarly the umbrella pine of Japan (*Sciadopitys*), distinguished by its tufts of long, double needles grooved on both the upper and the lower surface, was represented in the Cretaceous flora of Greenland and in the Tertiary vegetation of Europe as late as the Pliocene stage by trees with a similar type of leaf. Finally, as we should expect from the present geographical range of the genera in the north temperate zone, pines and their allies were widespread members of Tertiary forests as far north as Greenland and Alaska. Pines have now a very wide distribution: they are the most numerous conifers in the northern hemisphere and occur in Formosa, Japan, South Africa and other regions far south of the equator. Other members of the pine family (the Abietineae) such as firs and larches were also well represented in the Tertiary vegetation, but in the earlier part of the period they were eclipsed by genera such as *Taxodium*, *Glyptostrobus*, *Sequoia*, *Callitris*, *Podocarpus* and others which have long ceased to exist in Europe. The Tertiary records show very clearly that the vegetation of the northern hemisphere included a much greater variety of conifers than is seen in the forests of the present age: they show also that many of the genera which are now confined to restricted areas

[1] Gardner (86).

either in the Far East, North America, or in other widely separated regions, were common trees in a belt of forest stretching without interruption across the temperate regions to the north of the Tethys Sea. It would seem probable that many of these genera had their original home in the Arctic regions.

Ginkgoales: Cycadophyta

The genus *Ginkgo* carried on the line of the Ginkgoales which we have followed from the latter part of the Palaeozoic era: it was perhaps the sole relic of this remarkably persistent stock in the Tertiary vegetation. It is possible that some leaves discovered in the Tertiary beds of Grinnell Land (E, fig. 121), originally described under the generic name *Torellia* and subsequently transferred to *Feildenia*,[1] may belong to an extinct member of the same alliance. *Ginkgo*, though rather less widely spread in Tertiary than in Cretaceous floras, was still represented in the Arctic regions, in Spitsbergen and elsewhere, in Alaska, in England, and in other parts of Europe where it continued to live as late as the Pliocene stage. Fig. 116 shows a leaf of *Ginkgo* from the sediments of a lake which lay in a depression on the lava-covered ground of the vast Eocene basaltic plateau, a remnant of which now forms the cliffs of the Island of Mull.

Little need be said of the cycads in an account of Tertiary floras. A few specimens of cycadean fronds have been found in the older Tertiary rocks: pieces of leaves from the Wilcox formation are referred to the genus *Zamia* which still grows in Florida; other examples have been described from a few localities, in Europe and elsewhere. It is, however, clear that before the beginning of the Tertiary era the Cycadophyta had fallen from the dominant position they occupied in the early stages of the Cretaceous period: Tertiary floras give no indication of the survival of the Bennettitales. Such examples of the Cycadophyta as are recorded lead us to conclude that the cycads as we know them to-day—though by no means conspicuous features of the Tertiary vegetation—were sparsely scattered over the northern hemisphere.

Flowering Plants

We now pass to the flowering plants, selecting a few for brief description. Our knowledge of the monocotyledons is based mainly

[1] Nathorst (97).

on leaves that are often fragmentary and hardly identifiable, on petrified stems and many large specimens of palm leaves. The fossil material does not substantially help us to settle the question of the relative antiquity of monocotyledons and dicotyledons. Palms are represented by fruits which remind us of the tropical *Nipa*, a "stemless" palm that is now a characteristic plant in the tidal

FIG. 116. A leaf of *Ginkgo*, from the Tertiary plant-bed of the Island of Mull. Nat. size. (British Museum collection.)

estuaries of tropical rivers flowing into the Indian Ocean. *Nipa* is one of several palms the fruits of which occur in the old delta-mud—the London clay—which lies below London and comes to the surface in the Island of Sheppey. We can picture the Eocene rivers carrying floating fruits of *Nipa*[1] as they are now borne on the waters of the Ganges. This genus seems to have been rare in North America; it flourished in Europe through southern Russia and is recorded

[1] For map of distribution see Berry (16^2).

from Tertiary beds in Egypt and Borneo. There is no evidence of its survival in the northern hemisphere later than the Eocene or possibly the Oligocene stage. Leaves of another Wilcox palm agree closely with those of the living *Sabal*, the Palmetto of Florida, the West Indies and Mexico. This genus grew also in Europe, England, France, Germany and southern Russia in the earlier part of the Tertiary period: it continued to live in the Rhone Valley as late as the Pliocene stage. It has been pointed out that a comparison of the Tertiary leaves included in the genus *Sabalites*[1] reveals slight differences such as are now shown by the various forms of the genus *Sabal* in its present home in America. Some impressive examples of palm leaves of another type from the Lower Tertiary beds of Hampshire are displayed in the Fossil Plant Gallery of the British Museum. Other forms of leaf similar to those of the date palm (*Phoenix*) and allied living genera have been found at several European localities, in northern Italy and elsewhere. The palms tell much the same story as the conifers: a relatively continuous distribution throughout the north temperate zone in marked contrast to their present discontinuous range in warmer regions of the world. In the genus *Typha* (or *Typhacites*) we have an example of a different kind: *Typha* (the reed mace or bulrush) is now a common swamp-plant in tropical and temperate countries, the same species being equally at home in situations with very different climates. Specimens indistinguishable from a northern species have been described from an early Tertiary swamp in Nigeria[2] (fig. 121, Ng). The genus *Stratiotes* (the water-soldier or water-aloe of the British flora), now represented by a single species in Europe and Siberia, was plentiful in the lakes and pools of the Tertiary period; and it is interesting to find that a comparative examination of the several fossil forms illustrates gradual changes in the character of the seeds as the genus is followed from the Eocene stage to the present day.[3] *Stratiotes* has not been found in North America. This must suffice for the monocotyledons.

Of the numerous dicotyledons a few of the more characteristic and more trustworthy examples are selected: some, but not many, of the earlier Tertiary genera are extinct and many of the genera are now mainly tropical; others are still common in temperate floras, but the fossil forms from older Tertiary floras resemble most closely the species which now live in the more southern parts of the

[1] Fritel (10); Kryshtofovich (27[2]). [2] Seward (24[2]). [3] Chandler (23).

distribution-areas of the several genera. It should be noted that the early Tertiary vegetation in the northern hemisphere was less uniform in character than in many of the earlier periods: the Arctic Tertiary floras, probably Eocene in age, and those from the more northern parts of the temperate zone were much richer in plants that are now characteristic of temperate countries and relatively poor in representatives of genera which we associate with subtropical and tropical countries. It has recently been pointed out that this temperate facies was a distinguishing feature of Tertiary floras in Siberia and other parts of Asia in places considerably south of the contemporary American and European floras.[1] The dotted line stretching across Asia in fig. 121 shows roughly the southern boundary of the eastern floras which indicate climatic conditions less genial than those in corresponding latitudes in Europe and America. It has been suggested that the contrast between east and west affords evidence of a shifting of the earth's axis: this is a view which has long been a favourite explanation of climatic vagaries, but it receives no support from astronomers and on other grounds it cannot be regarded as adequate.

The genus *Magnolia* (family Magnoliaceae), already well established in the Arctic and temperate vegetation of the Cretaceous period (fig. 107), retained its hold on the northern hemisphere even to the latter part of the Tertiary period: it has now a discontinuous range in Asia and North America extending in America through the south-eastern United States as far north as Ontario (fig. 110). Species are recorded from the Wilcox flora, the Raton flora (C, fig. 121), and other North American areas, from the Arctic regions, England and many parts of Europe. By the Pliocene stage its range was restricted within much narrower limits, but it still lived in France and in a few other European districts. The occurrence of magnolias in the Arctic regions inspired a minor poet, attracted by the warm Eocene seas, to write with poetic license,

> "When, genial, a lost Alaska grew
> Broad-blossomed trees, and the Magnolia stole
> Warm-scented to the Pole."

Though not included in the lists of Wilcox plants the genus *Liriodendron* (tulip tree), another member of the Magnolia family, is a striking example of a tree—with two living species, one confined

[1] Kryshtofovich (29²).

to two small areas in China, the other, almost identical with the eastern type, to a large area in North America—which in the early days of the Tertiary period lived in that part of the northern continent that is now Iceland,[1] and, at a later stage (Miocene) in Switzerland: it persisted into the Pliocene stage as is shown by fossils from France, Holland and the Altai Mountains (5, fig. 121). The discovery of the genus in late Cretaceous rocks in Patagonia[2] raises an interesting problem, whether it began its career in the far south or in the north: it may well have started life in the temperate or Arctic zone of the northern hemisphere. *Liriodendron* is one of a large company of which Europe was deprived at the time of the Quaternary Ice Age. *Cinnamomum*, a member of the laurel family, familiar as the source of camphor and cinnamon, and now essentially tropical, is recorded from the Wilcox and many other Tertiary floras in North America and Europe, and from the Far East. Fossil leaves are referred to this genus in the great majority of instances because of the presence of three main ribs, a mid-rib and two curved laterals united by more slender veins: it is not as a rule possible to decide whether leaves having venation characters like those of a cinnamon tree are the foliage of that genus. Several other dicotyledons have a similar venation: to mention only one, the genus *Viburnum* (Caprifoliaceae) includes some Chinese species, e.g. *Viburnum Davidii*, with the cinnamon type of leaf.[3] We know that many of the Tertiary European plants are still living in China, or at least their nearest allies are frequently natives of the Far East, and it is highly probable that of the numerous fossil leaves described as species of *Cinnamomum* some at least may be examples of *Viburnum* or other genera. But allowing for mistaken or unconvincing citations of Tertiary species, there is proof of the occurrence of the genus in Europe: flowers which are undoubtedly those of a cinnamon were found in a piece of Baltic amber from the Oligocene beds in Prussia,[4] and it has been shown on the evidence of microscopical leaf-characters that the genus was a member of the rich Eocene flora of southern England.[5] There is no doubt of its wide distribution in North America and Europe in the earlier Tertiary floras: it is recorded also from a Pliocene flora in Caucasia.

The Sacred Lotus, once a favourite model for Egyptian artists and craftsmen, but no longer living in the Nile, is a species of the

[1] Heer (68). [2] Ihering (28). [3] Seward (25).
[4] Conwentz (86). [5] Bandulska (28).

genus *Nelumbium* (a member of the water lily family) which ranges from Japan to the Caspian, and as far south as north-east Australia. It occurs in the eastern states of North America. During the Eocene stage *Nelumbium*[1] grew as far north as Canada: it has been found in Oligocene beds in Hungary and elsewhere, and in the Miocene lake deposits of Switzerland. Fruits from the Wilcox beds demonstrate the former extension into North America of the genus *Trapa* (the water chestnut, a member of the family to which the willow herbs and fire weed belong); a floating water plant easily recognized by the hard woody portion of its fruit provided with two to four prominent pointed horns (fig. 120). It now lives in central and southern Europe, in Africa, India, but not in America: in the early part of the Tertiary period it had a much wider range and this was maintained into the Quaternary period.[2]

We turn next to a family which has been the cause of much discussion and difference of opinion, the Proteaceae. This is now confined to the southern hemisphere where its discontinuous distribution is strongly suggestive of antiquity: the genus *Protea*, with its wealth of species, some with magnificent flower-heads a foot in diameter, and *Leucodendron*, including the famous silver tree on the slopes of Table Mountain, are among the more striking elements in the varied and rich vegetation of the Cape Province of South Africa. Some genera play an equally conspicuous rôle in the Australian bush; other members of the family are peculiar to South America. The question is, whether the numerous species recorded from Tertiary rocks in the northern hemisphere as examples of proteaceous genera are trustworthy as evidence of the former occupation by this southern family of a wide territory north of the equator? It is certain that many of the fossil leaves belong to other families of dicotyledons: this is now generally admitted, but there remains a goodly number of specimens on which it is difficult to pronounce a definite opinion. The subject is too controversial to discuss in a general sketch: it must suffice to express the opinion that the family was represented in the northern hemisphere during the Tertiary period. A few fossils from Pliocene beds in Europe[3] have been assigned to the Proteaceae, and though the identifications are regarded

[1] Berry (17).

[2] A map showing the distribution of *Trapa*, recent and fossil species, is given by H. Gams (27).

[3] Reid, C. and E. M. (15).

by some authors with suspicion,[1] their validity has not been disproved. Similarly there has been much criticism, often perfectly sound, of the numerous leaves and fruits described as species of *Eucalyptus*,[2] the gum trees of Australia, which belong to the Myrtaceae or myrtle family. Several of the fossils have been shown to be cone-scales of a conifer allied to the kauri pine of New Zealand (*Agathis* or *Dammara*); but none the less there appears to be good evidence of the occurrence of the genus in Tertiary Europe.

The genus *Terminalia* (family Combretaceae), now characteristic of tropical strand-floras, is included in the list of Wilcox plants:[3] it is recorded also from Europe. The ability of the fruits of *Terminalia* to float long distances on sea-water without injury is a feature which may have contributed to wide dispersal. Many genera of the large family Leguminosae, which are now tropical and subtropical, were common elements in the northern hemisphere vegetation. *Dalbergia* (or preferably *Dalbergites*),[4] already mentioned as an Arctic Cretaceous tree, is known to have existed in the western area and in some European districts: the genus is now widely spread in the southern Tropics. Another Leguminous genus in the Wilcox and several other Tertiary floras is *Cassia*, represented to-day by many species in tropical and warm temperate countries, including the plant from which senna is obtained. It had an extended distribution in the northern hemisphere. The sweet gum of North America, *Liquidambar*[5] (Hamamelidaceae), a genus ranging from Connecticut to Florida, and occurring in the Mediterranean region of Europe, in Asia Minor and Formosa, was a common Tertiary type and flourished in Arctic lands. *Platanus* (Platanaceae), which was one of the commonest and most widely spread genera in the Cretaceous vegetation, continued to colonize fresh ground in the Tertiary period. Leaves exceeding the average size of the foliage of planes cultivated in temperate Europe occur in the early Tertiary beds of Spitsbergen (fig. 117) and unusually good specimens have been obtained from the plant-beds exposed on the cliffs of the Island of Mull (fig. 118). *Platanus* was a common genus of the northern hemisphere during the first half of the Tertiary period: it is recorded also from Patagonia and was a characteristic tree in the Miocene forests of the Far East. In the Pliocene stage it lived in

[1] Laurent and Marty (23).
[2] Deane (96), (00).
[3] For reconstruction of a fossil species see Berry (26).
[4] Seward (26).
[5] Laurent (19).

FIG. 117. Part of a leaf of *Platanus* from the Tertiary coal-bearing rocks of Spitsbergen. ⅓ nat. size. (Photograph of a specimen in the Stockholm Museum supplied by Dr Florin.)

FIG. 118. *Platanus* leaf from the Tertiary beds of the Island of Mull, Scotland. ⅔ nat. size. (British Museum Collection.)

North America and France, a considerable distance beyond its present limits, and west of its present range in Europe (fig. 110). *Platanus* is a good example of a tree which, though hardy and very often used as a shade-plant in the streets of northern Europe, is no longer native in Europe north of the Mediterranean province. There are several North American species and the oriental plane is a characteristic feature of the Adriatic coast and of parts of Asia Minor: why the genus should have failed to survive in its former, much more extended, territory is one of many similar problems presented to the student of ancient floras.

Salix (the willow) affords an example of a widely distributed Tertiary genus which is now most abundant in the temperate zone; though like pines and several other trees it has many representatives in the southern hemisphere. Very much the same type of distribution is illustrated by *Myrica*, familiar in England as the sweet gale or bog-myrtle and a common genus in North America: it was cosmopolitan in the northern hemisphere in Tertiary floras. *Myrica* grows also on tropical mountains and still lives in the northern regions which it occupied in the Cretaceous period. The genus *Acer* (maple and the English sycamore) was a vigorous tree in the Tertiary period: its past and present distribution has recently been plotted on maps included in a very useful series now being published.[1]

The three genera, *Betula* (birch), *Alnus* (alder) and *Quercus* (oak) are among the more abundant members of Tertiary floras, particularly of those in the colder parts of the temperate zone and in regions farther to the south in Siberia and Japan. Oaks are by no means peculiar to the north temperate zone; there are several in the Tropics; the dwarf birch (*Betula nana*) is a common prostrate shrub in the Arctic circumpolar vegetation and on high ground farther south. The Arctic birches of Tertiary floras in Grinnell Land and Spitsbergen were far taller and bore much larger leaves than their stunted successors in more southern Arctic countries to-day.

Mention may here be made of a common, extinct genus, *Dryophyllum*, one of a group of flowering plants characteristic of early Tertiary floras: it is represented by large leaves closely resembling in form and venation those of various species of *Quercus*, *Castanea* (chestnut), and other genera of the same family (Fagaceae), though apparently not identical with the foliage of any living tree. *Dryo-*

[1] Pax (26–27). See also Fernald (28) for criticism of Pax's map.

phyllum occurs in the early Tertiary floras of America;[1] in the Wilcox flora and in other American floras farther north, in Europe, and as far east as Manchuria. It has been pointed out by a French author[2] that examples of the genus from the Puy-de-Dôme district bear a strong likeness to the leaves of some living oaks. It is not improbable that this extinct genus may be a generalized type foreshadowing a subsequent splitting up into the various generic forms which we know to-day: it does not appear to have persisted as a well-defined genus later than the Oligocene stage. In contrast to the more temperate genera, such as *Quercus*, *Betula* and *Salix*, the tropical *Artocarpus* (bread-fruit tree), which has already been mentioned as a remarkable example of a tree, that is now essentially tropical, having lived in the Cretaceous period as far north as Greenland, was a member of the Wilcox flora. It is represented also by some particularly good specimens in the early Eocene Raton flora: and there is evidence of its occurrence in the Miocene vegetation of Switzerland.[3] *Ficus* is one of several genera frequently cited in Cretaceous and in Tertiary floras: identification of the fossil leaves with recent forms usually rests solely on form and venation. On the evidence usually presented, it is difficult to believe that figs and banyan trees were as ubiquitous and numerous as the published lists imply. It is, however, clear that this tropical genus was widely spread in the early Tertiary vegetation of the northern continents.[4]

The two genera, *Zizyphus*[5] and *Paliurus*,[6] both members of the Rhamnaceae, the family which contains the Buckthorns (*Rhamnus*), are recorded from the Wilcox area, from other American localities, and from several European floras: both genera still exist in southern Europe and are more abundant in the Far East, Africa, and other southern regions: they do not occur in the present vegetation of North America. *Zizyphus Lotus*, a north African prickly shrub with sweet fruits, is said to be the lotus of the lotus-eaters. There are many Tertiary records of the Juglandaceae (the walnut family): in the European area the genus *Juglans* (walnut),[7] which grows wild in Greece, is the only member of the family west of the Caucasus. In North America there are also the hickories (*Carya*) and in central America the genus *Oreomunnea*: a fourth genus, *Engelhardtia*,[8]

[1] Berry (24). See also Marty (07).
[2] Laurent (12).
[3] Heer (59).
[4] Laurent (99).
[5] Berry (25).
[6] Langeron (02[2]).
[7] Kryshtofovich (15); Berry (26[2]).
[8] Berry (24).

distinguished by its nuts attached to a conspicuous three-pronged wing, occurs in the Indo-Malay province and China. There is another genus, the Caucasian *Pterocarya*, resembling the walnut in the separation of the soft pith into transverse discs, which are easily seen on cutting a twig down the middle. *Juglans* appears to have been a cosmopolitan Tertiary genus: *Engelhardtia* was a member of the Wilcox flora, and in the Oligocene stage it grew as a forest tree on the borders of an estuary extending over part of the present Isle of Wight (see Reconstruction, fig. 119, *Ea.*). The walnut family as a whole ranged very much farther afield in the Tertiary period. It may be that in the extinct, early Tertiary genus *Dewalquea*,[1] characteristic of the older Tertiary floras, we have one of the earlier stages in the evolution of the various types included in the Juglandaceae. Finally, *Diospyros* (family Ebenaceae), the Persimmon of North America, mainly a tropical genus, had a wide distribution in European Tertiary floras and lingered on into the Pliocene forests.

In the choice of genera selected for brief notice attention has been concentrated, at least in the flowering plants, on such as have been recognized in the plant-beds of the Wilcox formation. It has been shown that very many Eocene genera continued to occupy areas in the succeeding Oligocene and Miocene stages much farther north than their present boundaries, and not a few did so even as late as the Pliocene stage. A distinctive feature of the majority of the Eocene floras is the association of genera that are no longer in existence—archaic forms which may be regarded as generalized types—with other genera which still live, either in temperate or tropical countries or in both.

Oligocene and Miocene Floras

An Oligocene River-estuary in the Isle of Wight

We may now glance at an Oligocene flora which has recently been described from the Isle of Wight: it agrees in several respects with the older Wilcox flora and others of similar age, but it differs in having many more herbaceous plants associated with the trees. The plant-beds from which the material was obtained occur on the north-west coast of the island near Cowes, and consist of clays, marls, and a bed of limestone: with the plants are bones of mammals,

[1] Johnson and Gilmore (21).

also remains of turtles, crocodiles and other animals. The majority of the plants are represented by fruits and seeds, some by leaves and other fragments. We may picture the vegetation growing on mud-flats and on ground overlooking the estuary of a river. The scene represented in fig. 119 is based on data given in the description of the flora by Mrs Reid and Miss Chandler.[1] A group of palms (*Sabal*), seen at the foot of a tree to the left of the middle of the picture, illustrates the presence of an American element: immediately behind the cluster of fan-shaped palm leaves is the trunk of a *Catalpa* (*Ca.*) with spreading branches bearing large, oval leaves: this genus belongs to the tropical family Bignoniaceae; it is now native in North America and eastern Asia, and is often cultivated in English gardens. To the left are hornbeams (*Carpinus*), beeches, and oaks. In front of the oak is a group of horsetails (*Equisetum*, *Eq.*). In the upper part of the left-hand side is a branch of a cinnamon tree (*Cn.*) bearing flowers and the three-ribbed type of leaf characteristic of the genus. Behind, and hanging obliquely down into the picture, is a long fruiting spike of *Engelhardtia* (*Ea.*), with a branch of a beech tree (*Fg.*) across it: near to this is an upright branch of a *Zizyphus* (*Z.*) shrub with leaves similar in the number of prominent veins to those of the cinnamon. In the left-hand corner, growing in the foreground, is a plant of *Incarvillea* (*I.*) with a large flower: this, like *Catalpa*, is a member of the family Bignoniaceae; next is a poppy (*P.*) and farther to the right a *Melissa* (*Ml.*), a genus of the Labiatae which includes the balm plant. The shrub with large leaves and pendulous flowers is a *Clematis* (*Cs.*): still farther to the right is *Abelia* (*Ab.*), a genus of the honeysuckle (*Lonicera*) family, and at the edge of the bank an *Acanthus* (*Ac.*). Floating on the water are, *Aldrovanda* (*Al.*), an aquatic insect-catching plant allied to the sundews (*Drosera*); farther to the right the large leaves of *Brasenia* (*Ba.*), a genus of the water lily family that is now unknown in Europe. The two plants in the right-hand corner are *Stratiotes* (water aloe, *St.*) and beyond them, against the farther bank, some plants of *Sparganium* (burweed, *Sg.*). There also grew in the water a species of *Azolla*, too small to be shown in the drawing, one of the water-ferns (Hydropterideae), a genus no longer native in Europe though in recent years naturalized in England and elsewhere: the fossil species is said to combine features that are now met with in two separate sections of *Azolla*, one in California, South America,

[1] Reid and Chandler (26).

FIG. 119. Reconstruction of an Oligocene landscape in the Isle of Wight. (Drawn by Mr Edward Vulliamy.) *Ab. Abelia*; *Ac. Acanthus*; *Al. Aldrovanda*; *Am. Acrostichum*; *Ba. Brasenia*; *Ca. Catalpa*; *Cn. Cinnamomum*; *Cs. Clematis*; *Ea. Engelhardtia*; *Eq. Equisetum*; *F. Ficus*; *Fg. Fagus*; *I. Incarvillea*; *Ml. Melissa*; *P. Papaver*; *Sg. Sparganium*; *St. Stratiotes*; *Z. Zizyphus*.

Australia and New Zealand, the other in Japan, Ceylon, Australia and Africa. Two clumps of the swamp fern *Acrostichum* (*Am.*) with long, pinnate fronds are conspicuous beyond the smaller water plants. Araucarias, cypresses and pines are seen on the ground rising from the water's edge, and hanging into view from the upper corner is a branch of a fig tree (*F.*).

The representative plants chosen for illustration show that the vegetation contained an admixture of elements now living in temperate Europe with others which are tropical or subtropical in range. In this respect the Oligocene flora of the Isle of Wight conforms to the usual character of European and American floras of the same or approximately the same age. If, however, we look more closely at the plants which belong to genera now represented in temperate Europe we find that the relationship of the fossil to existing species adds considerably to the interest of the Oligocene flora. The *Incarvillea* is most nearly akin to species now growing in western China and Tibet: the poppy is compared with a species now native in Spain and Morocco; *Abelia,* a genus extinct in Europe, is represented by a form recalling species living in the Old World, the Himalayas, and China. The clematis is nearest to a Chinese and North American species; the *Melissa* closely resembles a species still living both in Europe and the East. *Acanthus* resembles most closely a form from eastern Asia and Australia.

These are only a few of the plants composing the Isle of Wight flora: 32 per cent. of the genera are still European: 68 per cent. no longer live in Europe, but all occur in eastern Asia or North America, though not exclusively; 47 per cent. no longer exist in Europe and occur only in eastern Asia and America; 89 per cent. of the genera are represented by species allied to plants now living in eastern Asia and America; 21 per cent. of the non-European genera are now found in Africa and India, or in India alone. The most striking fact is the close resemblance of a large number of the plants to species that are now members of far eastern and American floras.

Other Floras: the Wanderings of Tertiary Plants

Reference should be made also to the Oligocene flora of Samland[1] on the Baltic shores reconstructed from the wonderfully preserved flowers and other pieces of trees, shrubs and herbs sealed in the amber that oozed as resinous trickles from the trees of a pine

[1] Conwentz (86).

forest, and pieces of wood with tissues rendered permanent by the resinous preservative.

We have seen that the analysis of Eocene and Oligocene floras in America and Europe reveals a dispersal of genera over the northern hemisphere much wider than at present. Moreover many of the flowering plants which were widely spread in Europe in the Cretaceous period tell a similar story, namely the occurrence of many genera in the circumpolar region, also in northern and central Europe, which are now characteristic of warmer countries, the Mediterranean area, China and elsewhere. The inference is that many flowering plants had their original home in the far north: in the course of the latter part of the Cretaceous period they spread rapidly along divergent routes into the Old and New Worlds. This migration continued into the Tertiary period even as late as the Pliocene stage when, as we shall see later, several of the genera which are no longer alive in Europe still occupied stations in the Rhone Valley, in the Netherlands and elsewhere. At a time, not long ago in a geological sense, when man had made his entrance on to the stage, the European floras were much richer in genera and species than they are now. The general lowering of temperature over the northern hemisphere and the consequent spread of ice-sheets and glaciers as far as the Thames and Severn valleys in England and far over the German plain acted as a driving force compelling a southern migration to more genial homes. On the American continent and in Asia there were no impassable barriers across the path of the wandering hosts. Wind, birds, other animals, and flowing water acted as agents of dispersal and by slow degrees the plants were able to reach refuges beyond the danger-zone which moved progressively farther south until the climax of the Ice Age. In Europe some of the hardier genera survived in the land north of the Mediterranean, but a large number of plants, cut off by the rising Alps and the west-to-east barrier of the sea, failed to gain access to regions where the climatic conditions were endurable: their existence as elements in the vegetation of Europe was ended. It must be remembered that in the earliest stage of the Tertiary period the Alpine and Himalayan ranges had not yet emerged from the ancient Tethys Sea; this barrier became operative in the latter part of the period as the crust, responding to pressure from the south, became involved in stupendous foldings; but the waters of the Tethys formed an effective obstacle to travellers from the north ages before the dawn of the Cainozoic era. To this subject we shall return in the fuller account of the Pliocene floras.

There are many other floras illustrative of the early Tertiary vegetation: the exceptionally rich flora of Florissant in Colorado reconstructed from scraps of vegetation buried in volcanic ash spread over the floor of a Miocene lake in the valley where the small town of Florissant now stands;[1] and another, still richer flora from Miocene strata at Oeningen on an arm of Lake Constance in Switzerland.[2]

Considered broadly, the older Tertiary floras of North America and Europe point to the conclusion that from the beginning of the period up to the early days of the Pliocene stage the climate was warmer than it is at present. Some years ago Prof. Nathorst[3] of Stockholm described a Tertiary flora from Japan which, in contrast to floras of similar age in the western world, he believed to indicate a temperate climate. More recently Dr Florin[4] of Stockholm made a thorough examination of some of the Japanese floras and arrived at the same conclusion. The question is discussed by Prof. Kryshtofovich in a still more recent paper:[5] he states that the early Tertiary Lozva flora in the Ural Mountains (fig. 121, 1) includes no tropical genera, only temperate plants such as now occur in eastern Asia and in the eastern states of North America; it shows hardly any features common to the contemporary floras of western Europe and the Ukraine (fig. 121, *b*), no palms, no cinnamons, or other tropical or subtropical plants which were common elements in the European vegetation. He states that none of the Siberian floras indicate tropical or subtropical conditions, and in the same category he includes several Japanese floras; in fact all the localities within the large area bounded by the dotted arc on the map, fig. 121. Through the territory east of the Black Sea, across Siberia to the Pacific Ocean, the Tertiary floras were characterized by the presence of such genera as *Fagus* (beech), *Ulmus*, *Alnus*, *Betula*, hazel (*Corylus*), poplar (*Populus*), *Juglans*, and other plants. On the other hand, a collection of early Tertiary plants obtained from Turkestan (fig. 121, K) is made up of types which recur in the lists compiled from southern Russia, central and western Europe and from many North American localities. A flora of similar composition is recorded from the southernmost part of Japan (fig. 121, Mg). During the earlier

[1] Knowlton (16); Cockerell (06), (08[2]), (08[3]).
[2] Heer (55–59).
[3] Nathorst (83).
[4] Florin (20).
[5] Kryshtofovich (29[2]).

stages of the Tertiary period the region including Greenland, Iceland, Spitsbergen, Alaska, and the greater part of Japan was the home of a "monotonous summer-green forest" in which the catkin-bearing hazels, poplars, and other familiar trees played a dominant part. This interesting conclusion raises in an acute form the question of cause and effect: why should the climatic conditions in this eastern province have been markedly different from those in the same degrees of latitude in other parts of the world? Reference was made to this problem in a previous page and it was suggested that whatever the solution may be, it is hardly likely to be found by taking liberties with the North Pole.

Pliocene Floras

The greater number and the more critically studied of the floras comprised within the uppermost series of the Tertiary system are in western Europe. In comparison with the abundance of material furnished by Eocene and other early Tertiary floras, the information on the later developmental stages of the plant-world in America is relatively scanty. A small collection of plants from the Gulf region (fig. 121, D) known as the Citronelle flora[1] shows that the vegetation was very similar to that of the same region to-day. A Pliocene flora from Texas[2] is described as coastal in facies and similar in its components to strand-floras in the present-day tropics. At another locality in the Kootenay Valley[3] (British Columbia) a small flora, probably Pliocene, has been obtained, which in its main features agrees with the vegetation of the New World, though it demonstrates the occurrence of a few genera, *Ficus*, *Menispermum*, *Platanus*, which subsequently spread farther south. Dr Florin[4] has described a relatively rich Pliocene flora from Japan which, according to the standard of comparison usually employed in measuring climate, indicates a lower mean temperature than is at present registered in the island. A noteworthy feature is that this Pliocene forest was composed of trees such as one would expect to find in the less genial districts of Japan. Though for the most part the Japanese species are either specifically identical with or very closely related to plants now living in the country, the commonest tree was an American beech (*Fagus ferruginea*), and with it grew the North American bald cypress (*Taxodium distichum*). An important fact

[1] Matson and Berry (16).
[2] Matson and Berry (16^2).
[3] Hollick (27).
[4] Florin (20).

is that the meagre collections from North America and the large one from the south island of Japan tell the same story, namely a close resemblance to the present vegetation of these two regions: a difference is that the British Columbian flora indicates a climate such as now prevails several hundred miles farther south, while the flora from Japan affords evidence of colder conditions in Pliocene days.

It has already been pointed out that the Tertiary vegetation of Japan was in its general features such as would be regarded as characteristic of the country at the present day. From a botanical point of view Japan and other parts of eastern Asia may be described as exceptionally conservative: the exotic elements in the flora are few and afford additional illustrations of the present linkage between the vegetation of the Far East and that of eastern North America.[1] When we look at the Pliocene floras we at once notice a striking contrast to the present European floras. The Pliocene vegetation of western Europe, though including several genera that are still represented in the district from which the fossils were collected, was distinguished by the occurrence of a large number of elements that no longer form part of the European vegetation, but are often specifically identical with species now living in China and Japan or on the Himalayan range.[2] In western Europe the Pliocene sources are much more informative: in the Rhone Valley,[3] in the Pyrenees,[4] in central France, in the Netherlands, and in England rich collections of fruits, seeds and leaves have been obtained: some are from deposits at the base of the Pliocene series, others from the middle, and some from horizons which bring us to the closing scenes of the Tertiary period.

It is convenient at this stage to note the more interesting conclusions drawn from a comparative study of the several phases of the Pliocene vegetation as represented in Europe. The Pliocene records in northern and central Europe bridge the interval between the rich Miocene floras and the existing vegetation, for it was during the Pliocene stage that the European vegetation was transformed.

If we compare a Miocene and an early Pliocene flora there is little

[1] Berry (27).

[2] Reid, C. and E. M. (15); Reid, E. M. (20). For an interesting account of the present vegetation of some parts of China see Wilson, E. H. (13).

[3] Depape (22), (24), (28).

[4] Reid, E. M. and P. Marty (20), (23).

difference between them. The vegetation which it is possible to reconstruct from the remains buried in the old Swiss Miocene lake, or in the Miocene brown coal of central Europe[1] and other localities was composed of a mixed company of plants. Of these many are now characteristic of warmer countries; many occupy discontinuous areas in the northern hemisphere; some having their present homes in the mountain gorges of China, or in the Himalayas. Others are now indigenous in southern Europe, northern Africa and the Canary Islands, and a smaller number of species are characteristic of temperate districts. The same constituents are to be found in the older Pliocene floras. In these are exotic elements whose ancestry can be traced far back into the early Tertiary or even to the Cretaceous period, and side by side with these are indigenous elements composing actual species still native to Europe.

As we ascend to the later phases of the Pliocene stage, the number of exotic types diminishes, while plants characteristic of temperate countries and species which are still native in Europe play a progressively more important part. Thus, in contrast to the varied elements which formed the earlier Pliocene floras, the later Pliocene floras possessed only a poorer vegetation of the type which we now associate with temperate Europe.

By far the greater number of Pliocene species are alive to-day; those of the older Pliocene floras usually in regions remote from the places where the fossils are found, while in the more recent Pliocene floras the majority of the species still exist as familiar elements in our present European floras.[2]

As the upper limit of the Pliocene stage is approached the changes in climate become more marked. We learn from Quaternary floras that these changes foreshadow the transformation of northern Europe into a vast Arctic land with the lower ground and many of the hills buried under a mantle of ice, which spread through the valleys as tentacles feeling their way over the land at the southern boundary of the ice-sheets and glaciers. Beyond this zone the ground was bare of vegetation and strewn with the gravel and sand carried by torrents from melting ice. Farther south there was a forest-belt composed of trees which are still familiar features in our north temperate forests.

[1] Kräusel (28[2]). [2] Reid, C. and E. M. (08).

French Pliocene Floras

In order more clearly to present the results obtained from an analysis of lists of Pliocene plants we will take a closer view of a few selected floras, beginning with the oldest. During the earlier part of the Pliocene stage the volcanoes of central France, which have long been extinct, were centres of activity. As on the slopes of Vesuvius and Etna to-day, there were forests and associations of shrubs and herbaceous plants on the lava-covered hills and the fertile valleys. From time to time showers of ash and occasional lava-streams overwhelmed the vegetation, destroying much and yet preserving fragmentary relics from which it is possible partially to reconstruct the vegetation. The remains identified from the Mount Dore area are comparatively few; they include trees that are still well represented in Europe: pines, firs, oaks, chestnut (*Castanea*), and planes, with *Torreya*, a conifer that is no longer European, allied to the yew (*Taxus*), the persistent *Ginkgo*, and several other genera. A richer flora has been described from the series of Pliocene beds which form part of the old delta of the Rhone[1] when its mouth lay near the present site of the city of Lyons. In the earlier part of the Pliocene stage vegetation on the shores of the enlarged Gulf of Lions included fan palms closely related to the American Palmetto (*Sabal*), the evergreen oak (*Quercus Ilex*), the Chinese conifer, *Glyptostrobus*, species of *Myrica* and other swamp-loving flowering plants. On the banks of the river were alders (*Alnus*), plane trees (*Platanus*), *Liquidambar* and many other trees. Ferns were represented by *Woodwardia*, a Tertiary genus which has not yet deserted the Mediterranean region, the royal fern (*Osmunda regalis*) and others. Among other legacies from the earlier floras were *Zelkova*, a member of the elm (*Ulmus*) family, now living in Caucasia; *Pterocarya*, which flourishes on the Caucasus range and farther east; at higher altitudes, a *Torreya* which is now native in Japan, and a *Sequoia*. These and many other Pliocene plants from southern France made up a flora which, if we compare it with its modern counterpart in the same region, gives the impression of a heterogeneous association consisting in part of species now alien to western Europe and others that are familiar elements in the existing vegetation, such as *Polypodium vulgare*, the male fern (*Dryopteris Filix-mas*), the scotch fir (*Pinus silvestris*), the European larch (*Larix europaea*), the

[1] Depape (28).

silver fir (*Abies pectinata*), with alders, birches, hazels, oaks, elms, hawthorn (*Crataegus*), ivy (*Hedera*), and cherry laurel (*Prunus Laurocerasus*). These and several others, though members of modern European floras, are some of them characteristic of the southern part of the continent while others extend far into the colder parts of the temperate zone. Members of the Rhone flora still characteristic of the Mediterranean province include the fern *Woodwardia*, a survival from the earlier stages of the Tertiary period and now reaching as far as the Indo-Malay region, central America, and the Canary Islands; *Callitris*, one of the conifers of the North African Atlas Mountains, the southern European cypress (*Cupressus sempervirens*), the dwarf palm (*Chamaerops humilis*), the sole survivor of a group of Tertiary palms in Europe, the horse chestnut (*Aesculus*), which has now a discontinuous distribution in the Balkan Peninsula, the laurel (*Laurus nobilis*), *Zizyphus* and *Paliurus*, two widely spread genera in the earlier floras, *Phillyrea*, a shrub closely allied to the olive; also *Arbutus* (strawberry tree) which is the most conspicuous and striking representative of a small company of Mediterranean plants long established in south-western Ireland.[1] There were several plants that are now native in China, Japan, and other parts of the Orient: *Glyptostrobus*, *Ginkgo*, *Cinnamomum*, the Japanese *Paulownia*, a genus of the Foxglove family (Scrophulariaceae) which occasionally produces clusters of handsome purple flowers in the gardens of southern England. North American types are represented by *Taxodium*, *Sequoia*, a *Libocedrus*, the Floridan *Sabal*, *Magnolia* and others. Still another alliance is illustrated by the fern *Adiantum reniforme* (maidenhair), a holly (*Ilex*), and a few other dicotyledons which are characteristic elements in the present vegetation of the Canary Islands.

Another early Pliocene flora, described from fruits and seeds found at Pont-de-Gail in the Commune of St Clément (Cantal)[2], contains a large proportion of exotic types, 64 per cent. of which are most closely related to Chinese and North American species.

Pliocene Plants from the Old Delta of the Rhine, etc.

A still richer flora has been found in the ancient delta of the Rhine and Meuse on the Dutch-Prussian border.[3] In the earlier part of the Pliocene stage the North Sea was a gulf of the Atlantic Ocean land-locked on the south; and the British Isles formed the western

[1] Seward (11[3]). [2] Reid and Marty (20), (23). [3] Reid, C. and E. M. (15).

peninsula of Europe. The Rhine was building up a delta (fig. 111), one edge of which reached as far as what is now the Norfolk coast, and the other lay across the area which to-day forms the low-lying country of the Netherlands and the adjoining district in Germany. Three large collections of plants have been made from the delta: the oldest from Reuver—the Reuverian flora; another of rather later date from a neighbouring locality—the Teglian flora; and a third, still younger, from the plant-bearing beds at Cromer and other parts of the East Anglian coast—the Cromerian flora. The Reuver flora, 300 species of which have been identified, furnishes a striking illustration of the large proportion of species in a relatively early Pliocene flora identical with or next of kin to plants now living in western China, Japan, the Himalayas, and Tibet. Some of the Reuver plants belong to genera which, though still represented in Europe and the Caucasus, are most closely related to Chinese or Himalayan species of these genera. The alliance between the Reuverian and the Chinese and Himalayan floras has been explained by the supposition that they radiated from the Far East just as they spread from the northern regions as the result of a cooling climate. Both regions formed "reserves" for the origin of temperate species and centres from which they were distributed.

The representatives of far eastern trees and shrubs include *Pterocarya*, *Zelkova*,[1] and a species of *Clematis* believed to be identical with one that is now native in China; a *Magnolia*,[1] *Cinnamomum*, a jasmine, and other flowering plants. North American types are represented by *Liriodendron*, *Carya* (hickory), *Sequoia*, and some other genera: there are also several plants which are still widespread members of the European vegetation. These and other identifications are based on fruits and seeds, and it is satisfactory to find that a critical examination of a collection of leaves by French authors substantially confirms the main conclusions reached by the English botanists. Of the Reuver plants selected for comparison with existing species, 88 per cent. are exotic or extinct; 54 per cent. are Chinese and American.

Passing to the middle of the Pliocene stage, we may intercalate in this brief account of the delta-vegetation a reference to a flora discovered at Castle Eden[2] on the east coast of England several miles north of the Cromer area: in this flora the percentage of exotics

[1] For other records of these genera see Stojanoff and Stefanoff (29).
[2] Reid, E. M. (20[2]).

has fallen to 64 per cent. as compared with 88 per cent. in the older flora of Reuver and with 94 per cent. in the still earlier flora of Pont-de-Gail:[1] the Chinese and North American types are reduced to 31 per cent. The Teglian flora, a successor to the Reuverian flora in the eastern part of the delta, contains a much larger proportion of herbaceous plants many of which are aquatic: this may be due in part to a different environment. A more significant difference is the much smaller representation of exotics and a closer approach to the present European type of vegetation: exotic and extinct species make up 40 per cent.; of Chinese and North American species there are only 16 per cent.

Finally, in the Main Valley near Frankfurt[2] a large number of seeds, fruits, and pieces of foliage-shoots were discovered some years ago in the course of excavating foundations for new buildings: these are samples of an Upper Pliocene flora comparable in age with the Cromerian flora of England. The German collection includes, in addition to several European species, many interesting examples of far eastern and North American plants: the conifer *Libocedrus*, now a Californian, Oriental, and New Zealand genus, a Japanese species of *Torreya*, a *Sciadopitys* (umbrella pine of Japan), and both leaves and seeds of *Ginkgo*; also *Sequoia*, an American pine (*Pinus Strobus*, the Weymouth pine), and the Japanese conifer *Keteleeria*.

In the Cromerian flora only 5 per cent. of exotic or extinct species are recorded and the Chinese-North American element has almost disappeared. It is therefore abundantly clear that the forests which had long been established across the northern hemisphere in the earlier part of the Pliocene stage were composed of an overwhelming number of plants which to-day are either extinct or for the most part still living in the Far East or in North America, but not in Europe. East and west were united by a middle section into one transverse forest belt. Conditions quickly changed, the factors controlling plant-life operated with varying intensity, and a gradual modification of the plant population was shortly accomplished. In the earlier part of the Tertiary period the Tethys Sea continued to form a barrier between Europe and Africa: in the Miocene stage began the great upheaval which culminated in the Pliocene stage, and a barrier of mountain ranges, stretching from the Alps to the eastern limit of the Himalayas, was thrown across Europe and Asia. The

[1] Reid and Marty (20); Reid, E. M. (23).
[2] Engelhardt and Kinkelin (08).

effect of these two transverse barriers—the Alps and the attenuated relic of the Tethys Sea—on the migrating host of European plants driven south by the advance of ice-sheets and glaciers has already been discussed: the connecting links in the Tertiary forests between the western and eastern vegetation were broken, and the European flora was shorn of its glory and reduced to its present inferiority. During the last few decades, thanks to travellers in the Far East, there has been a steady stream of introduced trees, shrubs and herbs from the highlands and valleys of China and Tibet; the descendants of the human race which inhabited Europe before the end of the Tertiary period are now restoring to the gardens of the impoverished middle region of their old domain a few of the innumerable treasures which were lost during the Quaternary Ice Age.

Tertiary Floras in the Southern Hemisphere

Cainozoic floras are poorly represented in regions south of the equator in comparison with the large number described from localities in the northern hemisphere: moreover, some of the few published descriptions of southern floras need revision on modern lines—notably those by Ettingshausen[1] of Australian and New Zealand collections. Several small collections from South America show that the difference between the Tertiary and the present vegetation in that part of the world was much less than it was in northern continents: there is no reason to suppose that the climate of localities where Tertiary plants have been found in South America differed to any appreciable extent from the climate in the same place to-day. The fossil plants of the southern hemisphere as a whole do not afford evidence of climatic changes during the Cainozoic era comparable in degree with those indicated by the floras from northern lands. The most interesting examples of contrasts between the present and the past are the records from Seymour Island and Snow Hill[2] (fig. 122) and from Kerguelen Island[3] (fig. 122). More information on Tertiary floras of the southern hemisphere is needed, not only in relation to climatic changes but to plant-migration: we should like to know whether the geographical distribution of genera favours the view to which Darwin referred when he wrote to his friend Hooker—"there must have existed a Tertiary Antarctic continent from which various forms radiated to the southern extremities of our present continents"; or whether most of

[1] Ettingshausen (87), (88). [2] Dusén (99); Gothan (08). [3] Edwards (21^3).

the genera are immigrants from centres north of the equator. No reference is made in the map to Tertiary localities in South Africa: fossil wood and other remains have been found at several places, but they have not yet been described.

FIG. 120. Fruit of *Trapa natans*. *Ca.* nat. size.

NOTES ON THE MAPS AND TABLE G

In the last column of the table references are given to papers which should be consulted for descriptions of floras and genera; many of the authors mentioned give bibliographies, including important sources of information omitted from the selected list in this table.

For the position in the Tertiary period of the floras from which the geographical distribution of genera has been compiled reference should be made to the following notes on floras, but it must be remembered that the precise geological age of several Tertiary floras is by no means definitely settled. The intention is to give a general idea of the geographical distribution of some of the genera in the course of the Tertiary period, and to provide a guide to readers who wish to make a comparative study of Tertiary floras.

Regions on Map (p. 478)		Notes on the floras	Geological age	References
ARCTIC with Iceland	M.	*Mackenzie River*. Tertiary plant-beds are spread over a wide area in British Columbia, Alberta, Saskatchewan and in the western United States	Eocene	Heer (82); Penhallow (08); Berry (26^4)
	E.	*Ellesmere Land*. Beds of shale rich in plant remains and coal. Numerous well-preserved twigs of *Sequoia Langsdorfii*; also pieces of stems more than 1 metre in diameter	? Miocene	Nathorst (15^2)
	,,	*Grinnell Land*		Heer (78); Feilden and Rance (78)
	G.	*Greenland*. A thick series of estuarine sediments with numerous plants associated with basaltic lava-flows, especially well developed in the cliffs and ravines of Atanekerdluk (70° N. lat.) on the west coast, also on Hare Island and Disko Island	Eocene	Heer (82); White and Schuchert (98); Nathorst (11); Walton (27^2); see also Seward (25), (26)
		A few Tertiary plants, e.g. leaves of *Ginkgo*, have been obtained from Sabine Island on the east coast: in the Copenhagen Museum (fig. 137, 3)	? Eocene	Seward (26)
	S.	*Spitsbergen and King Charles Land*. Well-preserved leaves of *Ginkgo*, *Platanus*, etc.; also fossil wood	Eocene	Schroeter (80); Gothan (07); Nathorst (11), (19)
	N.	*New Siberian Islands*	? Eocene	Schmalhausen and v. Toll (90)
	Tc.	*Tas-takh Lake* (Yakutia)	Eocene	Kryshtofovich (29^2)
	I.	*Iceland*. Plant-beds associated with basaltic lava	Eocene	Gardner (85)
TEMPERATE ZONE NORTH AMERICA	A.	*Alaska*. Plants collected from several localities	Eocene	
	B.	*Canada and the Western States* as far south as Nevada and Colorado	Eocene to Pliocene	Knowlton (93), (94), (04); Berry (26^4)
		Canada. Vancouver Island, Queen Charlotte Islands, etc. The Fort Union flora, not yet fully described, is from beds covering a wide area in Canada, Wyoming, the Dakotas, and in Montana east of the Rocky	Eocene	Dawson (95); Penhallow (02), (03), (08); Newberry (98); Knowlton (93), (27)
		Mountains. Another large flora (Puget flora) has been obtained from the Puget Sound district.	Eocene	Berry (26^4)

Tertiary Floras in the Northern Hemisphere (cont.)

Regions on Map	Notes on the floras	Geological age	References
TEMPERATE ZONE North America (*cont.*)	St Eugene Silts flora: Kootenay Valley, British Columbia: one of the few late Tertiary floras in North America	? Pliocene	Hollick (27). For other references to Pliocene floras in the Western States see Chaney (25)
	The United States. Latah flora (Washington): the richest Miocene flora next to that from Florissant (Colorado) in North America.	Miocene	Berry (29[3]); Knowlton (26)
	Green River flora. Plant-beds deposited in a large lake in western Wyoming, Colorado, and Utah.	Eocene	Knowlton (23); Berry (25[2]), (30); Cockerell (25)
	Mascall flora: from strata scattered over an extensive territory in Washington, Oregon, Idaho, Nevada, California, etc.	Miocene	Hollick (29); Chaney (25[2]); Berry (27[2])
		Pliocene	Dorf (30)
	John Day flora (John Day basin, Oregon)	Eocene	Knowlton (02), (23); Chaney (25), (27)
	Bridge Creek flora (Oregon), with many species closely related to plants now living in the Redwoods forest	Oligocene	Chaney (24), (25)
	Yellowstone Park: an impressive example of the preservation of forests in a mass of volcanic ash and other material 2000 ft. thick (fig. 21)	Miocene	Knowlton (99)
	Payette flora: plant-beds in the basaltic region of the Snake River, Idaho	Miocene	Knowlton (98); Chaney (22
C.	This area includes the rich plant-bearing beds of the Raton flora in the Raton Mesa region of Colorado and New Mexico;	Eocene	Lee and Knowlton (17)
	the exceptionally rich Florissant (Colorado) flora preserved in volcanic ash;	Miocene	Cockerell (08), (08[2]), (08[3]); Knowlton (16); Bather (09) with references
	and the flora of the Denver Basin	Eocene	Knowlton (22)
D.	The area bounded by the thick black line shows approximately the extension of the sea at the time of maximum transgression in the early days of the Tertiary period. The richest and oldest floras are: the Wilcox, Claiborne, and Jackson.	Eocene	Berry (14), (16[2]), (22), (22[2]), (24)
	Other and younger floras are: the Alum Bluff flora (Florida), one of the very few of this age in North America,	? Oligocene	Berry (16[3])
	and the Citronelle flora of Texas, Louisiana, Alabama, and Florida	Pliocene	Matson and Berry (16), (16[2])
E.	A small coastal flora has been described from Calvert County, Virginia and other regions.	Miocene	Berry (16[4])
	From the more northerly locality plants have been obtained from the Brandon lignites	? Miocene	Berry (19[2]); Jeffrey and Chrysler (06)
Europe	Within the circle are shown many, though by no means all, districts from which Tertiary floras have been described, from Mull and northern Ireland in the north-west to Greece in the south-east		

	Great Britain and Ireland. Ireland: plants from lacustric beds associated with the basaltic lavas of Austria and from County Tyrone, etc.	Eocene	Baily (61), (69), (80); Gardner (85[2]); Johnson and Gilmore (21), (21[2]), (22)
	Scotland: Plants from sediments deposited in the intervals between volcanic activity on the Island of Mull	Eocene	Seward and Holttum (24); Edwards (22)
a.	Within *a* are the plant-beds of England, Belgium, northern France, Holland, and part of Germany. England: the delta deposits of the Island of Sheppey (Kent) have yielded a rich flora which has not been thoroughly investigated	Eocene	Ettingshausen (79): see also Gardner (86)
	The older Tertiary floras of Hampshire and the Isle of Wight (Alum Bay, Bembridge, etc.); also the Bovey Tracey flora (Devonshire) from the deposits of a small lake in the Dartmoor granite district	Eocene and Oligocene	Gardner and Ettingshausen (82); Gardner (86); Chandler (25); Bandaluska (23–28); Reid, C. and E. M. (10); Heer (62)
	Later Tertiary floras from Cromer (Norfolk) and Castle Eden (Durham)	Pliocene	Reid, E. M. (20) Reid, E. M. (20[2])
	Belgium. Floras from Gelinden, Hainaut and other districts	Eocene, etc.	Saporta and Marion (78); Marty (07); Gilkinet (22[2]), (25); Seward and Arber (03); Langeron (07)
	Holland. A small collection of woods from Limburg; also large collections of fruits and seeds from the old delta of the Rhine slightly older than those obtained from Cromer in England	Miocene Pliocene	Kräusel and Schönfeld (24); Reid, C. and E. M. (15); Laurent and Marty (23)
	France. Plants have been described from several localities in northern France, from the Paris Basin and other districts, including the well-known flora of Sézanne, represented by numerous remains preserved in calcareous material deposited from the spray of waterfalls in a mountainous region	Eocene Eocene Eocene	Depape (25); Dehay and Depape (29) Watelet (66) Saporta (68); Langeron (99–02)
c.	Among other French floras are: Meximeeux; Menat (Puy-de-Dôme), plants preserved in a small basin; Célas (Gard, southern France), a flora consisting of plants bordering a lake and some water-plants; Chartres; Aix-en-Provence; also later floras from the Rhone Valley (*c* on the map), from the neighbourhood of Biarritz; Pont de Gail (Cantal), Finisterre	Pliocene Oligocene Oligocene Miocene Pliocene Miocene and Pliocene	Saporta and Marion (76). For a more general account of Tertiary floras see Depape (28); Laurent (19[2]) Laurent (99), (12); Saporta (62–73); Depape (24). For other references see Reid and Chandler (26); Depape (22) with references (28); Chauvet and Welsch (16); Welsch (15); Reid, E. M. (20), (23), (30)
	Balearic Islands. A small flora from Majorca	Miocene	Depape and Fallot (28)
	One of the richest of all Tertiary floras is from Oeningen and other localities in the valley between the Jura and the Alps: the plants are exceptionally well preserved (a large collection in the Zürich Museum)	Miocene and Oligocene	Heer (76)
d.	*Germany.* Numerous plants from the amber beds on the Baltic coast (*d*), including flowers beautifully	? Oligocene	Goeppert and Menge (83); Conwentz (86)

Tertiary Floras in the Northern Hemisphere (cont.)

Regions on Map		Notes on the floras	Geological age	References
EUROPE (*cont.*)		preserved in amber. Large collections of plants have been described from many localities where the brown coal is worked: references to the literature will be found in the papers cited	Miocene	Potonié (08); Gothan and Zimmermann (19); Gothan (24); Kräusel (19), (20), (20[2]), (20[3]) with bibliography, (28[2]); Kubart (24)
		A large flora from the Main Valley (Frankfurt)	Pliocene	Engelhardt and Kinkelin (08); Engelhardt (11)
		Austria. The more important floras, which need critical revision, are: Häring (Tyrol); Sotzka (Styria); Sagor (Carinthia)	Eocene	Ettingshausen (53), (58), (77); Unger (51)
		Czechoslovakia. The Bilin flora; a flora from Laun	Miocene	Velenovský (81); Ettingshausen (69); Engelhardt (85)
		Jugoslavia. Floras from Bosnia; Radoboj (Croatia); and from Mount Promina	Oligocene Eocene	Engelhardt (12); Unger (69); Ettingshausen (54)
		Hungary. Collections described from a few districts	Oligocene	Tuzson (14); Staub (87)
		Italy		Meschinelli and Squinabol (92)
		Poland. A collection of seeds and fruits		Zablocki (28)
		Bulgaria. A flora from Kurilo showing a mixture of plants now living in southern Europe and others which are extinct in Europe	Pliocene	Stojanoff and Stefanoff (29)
		Greece. A comparatively small flora from the Island of Euboea	Miocene	Unger (67)
	b.	*Ukraine.* Older and later Tertiary floras from southern Russia and the Crimea, etc.	Eocene, etc.	Schmalhausen (84); Palibin (06), (22); Kryshtofovich (12), (14), (27), (27[2])
NORTH AFRICA	L.	Tertiary plants are recorded from the Libyan desert, where most of the fossils are from Cretaceous strata, including petrified wood and Calcareous Algae	? Eocene	Kräusel and Stromer (24); Blanckenhorn (21)
SIBERIA	1.	Lozva River (Ural Mountains)	Oligocene or Eocene	Kryshtofovich (29)
	2.	Aral Sea and Turgai Province	Oligocene—Miocene	Florin (22[2]) with bibliography; Endo (28); Yabe and Endo (27); Palibin (06[2]); Kryshtofovich and Palibin (15)
	3.	Irtysh River (Tara)	Miocene	Palibin (04), (04[2]), (06), (06[3]); Schmalhausen (87)
	4.	Tomsk	Oligocene—Miocene	Florin (20[2]), (22[2]) (China)
	5.	Altai Mountains	Miocene	
	6.	Baikal Lake	Miocene	
	7.	Kalgan (Mongolia) and Chikli Province (China)	Oligocene	
	8.	The Mukden district (Manchuria)	Oligocene	
	9.	Korea		
	10.	Fushun (Manchuria); Sikhota-Alin Mountains; Amur	? Oligocene	

	13.	Sakhalin Island. Many of the plant-beds which were formerly believed to be Tertiary have recently been shown to be Cretaceous: there are also Tertiary strata		Kryshtofovich (18)
	14.	Commodore Islands (Kamchatka)	Oligocene	Kryshtofovich (29[2])
	15.	Anadyr	Eocene	Kryshtofovich (29[2])
	K.	A flora from Turkestan differing in its more tropical aspect from the floras north of the broken line	Oligocene	Kryshtofovich (29[2])
JAPAN	Mg.	The Mogi flora from localities near Nagasuki; many of the species are now confined to the Far East	Pliocene	Florin (19[2]), (20)
	11.	Honshu (Hondo)	Eocene—Pliocene	Kryshtofovich (29[2])
	12.	Hokkaido (Yezo)	Eocene—Miocene	Kryshtofovich (18[2])
INDIA AND ASSAM	Sd.	*Sind.* Marine strata with Calcareous Algae	Eocene	Walton (25[2])
	As.	*Assam.* A few leaf impressions	? Older Tertiary	Seward (12[4]); Sahni (22)
TROPICAL COUNTRIES				
MEXICO	Mx.	The plant remains from the regions in the northern tropics both of the new and old worlds are too few to throw much light on the Tertiary vegetation, but it is clear that there was a much greater resemblance between the present and the Tertiary floras than in the temperate zone	Miocene	Berry (23)
CENTRAL AMERICA AND NORTHERN SOUTH AMERICA				
	Cr.	*Costa Rica.* A small flora	? Miocene	Berry (18), (21), (21[2]), (21[3]), (24[2]), (29[4])
	Cl.	*The Canal Zone*	? Miocene	
	Cb.	*Colombia*	? Miocene	
	V.	*Venezuela*	? Miocene	
WEST INDIES	WI.	Plants from Dominica and Trinidad; also from the Island of Antigua where petrified palms and other flowering plants are abundant	Miocene, etc.	Berry (24[3]); Felix (83); Stenzel (04)
WEST AFRICA	Ng.	A swamp flora from beds 120 miles north-west of Calabar	Eocene	Seward (24[2])
BURMA	Br.	Some fossil wood (*Dipterocarpoxylon*) from Upper Burma; lignite and amber beds.	Eocene	Holden, R. (16); Stuart (25)
	Sm.	A few specimens described from the Mepale River near the Siam frontier	? Pliocene	Edwards (23)

Notes on *Tertiary floras* in the Southern Hemisphere

Map References (fig. 122)	Geological Age	References to original sources
A. Kerguelen Island	Tertiary	Seward (19), p. 185: coniferous wood (*Araucarioxylon*); Edwards (21^3), *Araucarioxylon* and *Cupressinoxylon*. An important point is the origin of the wood; whether the logs had drifted from a more northern source or came from trees which grew on the spot. Crié (89)
B. Seymour Island and Snow Hill	Tertiary	Gothan (08), gymnospermous and angiospermous wood
SOUTH AMERICA		
C. Magellan Straits and neighbouring districts	? Oligocene	Dusén (99), *Araucaria*, cf. *A. imbricata*; *Fagus*, *Nothofagus*, and other genera based on leaves. Gilkinet (09)
D. i. Patagonia	Tertiary	Kräusel (24^2). Fossil wood from rocks over a large area. Seward (19), p. 186
ii. The Argentine	Oligocene and Miocene	Berry (28^2). Several localities shown on a sketch-map: among other plants a species of the cycad *Zamia* is recorded
iii. Chile	? Miocene or Oligocene	Engelhardt (91), leaves, mostly dicotyledons
E. Paraguay	? Tertiary	Schuster (11^2)
F. Bolivia	Tertiary	Engelhardt (94), a small collection of little value; Berry (17^2)
G. Peru and Ecuador	Eocene and Miocene	Berry (19^3), (26^3), (29^2). Descriptions of palm seeds
	Tertiary	Engelhardt (95)
MALAYA		
H. Malay Archipelago	—	Kräusel (25) gives lists of plants from several islands and a summary of palaeobotanical data
i. Sumatra	Miocene	Kräusel (22). Dicotyledonous wood and a petrified palm stem. Kräusel (29^2). Descriptions of leaves
ii. Borneo	Tertiary	Kräusel (23). Fruit of *Nipadites* (the "stemless" palm)
	Miocene	Rulten (20). Records of the Calcareous Alga *Halimeda* from Borneo and some other islands
iii. Java	Quaternary	Schuster (11^4). Leaves from the *Pithecanthropus* bed
	Tertiary	Kräusel (26). Description of wood as *Dipterocarpoxylon*
	Miocene	Raciborski (09)
Other islands (Labuan and Bangka)	Tertiary	Kräusel (25); Crié (89)
I. AUSTRALIA		
Tasmania	Tertiary	Ettingshausen (88); Arber, E. A. N. (04); Seward (19), p. 211, a splendid specimen of an opalized stem of coniferous wood now in the British Museum (*Mesembrioxylon Hookeri*)
Gippsland	? Miocene	Chapman (18), wood referred to *Eucalyptus*
New South Wales	Tertiary	Ettingshausen (88), see also Süssmilch (22), p. 202, and Deane (96), (00)
Queensland	Tertiary	Cainozoic plants not uncommon; not yet described
K. NEW ZEALAND	Tertiary	Ettingshausen (87). A large collection which should be re-described. Crié (89)

TABLE G. *Geographical distribution of some Tertiary Plants*

Classes, Families and Genera	Regions shown on the Maps, figs. 121, 122: North America A	B	C	D	Arctic	Europe	Siberia: Japan	Genus extinct	Extinct in Europe	Living in Europe	Recorded from Cretaceous rocks	Notes on Present Distribution	References
FILICALES Ferns POLYPODIACEAE *Acrostichum*	.	×	×	×	.	×	×	.	×	.	.	By far the largest and most widely distributed family of ferns in both temperate and tropical countries. *Acrostichum* is characteristic of tropical estuaries and mangrove swamps in the Old and New Worlds	Berry (24); Reid and Chandler (26)
Onoclea	.	×	.	.	×	×	.	.	×	.	×	*Onoclea sensibilis* (sensitive fern) extends from Florida to Newfoundland and occurs in East China and in Japan	
SCHIZAEACEAE *Lygodium*	.	×	.	×		×	×	.	×	.		*Lygodium* ranges from North America to South Africa, South China, Queensland, New Zealand. A species (*L. palmatum*) ranges from Florida to near New York	Florin (20)
GLEICHENIACEAE *Gleichenia*	.	.	.	.	.	×	.	.	×	.	×		For distribution, See Seward (22)
OSMUNDACEAE *Osmunda*	×	×	.	.	×	×	×	.	.	×	×		
MARATTIACEAE HYDROPTERIDEAE *Azolla*	.	×	.	.	.	×	.	.	×	.	×	*Azolla* is now well established as an alien in England and in many parts of Europe: it is a widely distributed tropical and subtropical genus	Reid and Chandler (26). For reference to a Cretaceous species, see p. 407 (Chap. xv)
Salvinia	.	×	.	×	.	×	×	.	.	×	×	A warm temperate and tropical genus; unrepresented in North America	Fritel (10); Florin (19[2]); Chandler (25); Yabe and Endo (27); Kirchheimer (30), (30[2])

Table G. *Geographical distribution of some Tertiary Plants* (cont.)

Classes, Families and Genera	Regions shown on the Maps, figs. 121, 122: North America A	B	C	D	Arctic	Europe	Siberia: Japan	Genus extinct	Extinct in Europe	Living in Europe	Recorded from Cretaceous rocks	Notes on Present Distribution	References
GYMNOSPERMAE													
Conifers													
Sequoia	×	×	×	×	×	×	×	.	×	.	×	The genus is confined to a narrow strip along the Pacific Coast of North America (South West Oregon and California) about 450 miles long. *Sequoia gigantea* ("big tree") has a more restricted range than *S. sempervirens* (redwood)	Berry (23²); Chandler (22)
Taxodium	×	×	.	.	×	×	×	.	×	.	.	There are two species in eastern North America, the commonest (bald or swamp cypress, *Taxodium distichum*) from South Delaware to North Florida; and a Mexican species	Berry (23²)
Araucaria	.	.	.	.	.	×	.	.	×	.	×	Confined to the southern hemisphere, South America and Australasia; absent from Africa	Seward and Ford (06)
Libocedrus	.	×	.	.	×	×	.	.	×	.	? ×	This genus, in habit similar to *Thuja*, has a continuous distribution: California (incense cedar), Chile, Japan, Australia, New Zealand, etc.	Johnson and Gilmore (22) found cones and twigs in Co. Tyrone, Ireland. Florin (30)
Cryptomeria	.	.	.	.	.	×	.	.	×	.	.	In habit resembles *Araucaria excelsa*. It is confined to South China and Japan (Japanese cedar) and forms the impressive avenue at Nikko	Gardner (86)
Glyptostrobus	×	×	.	×	? ×	×	×	.	×	.	×	Related to *Taxodium*; confined to China	Florin (20). Many of the records of this genus are untrustworthy. Kräusel (19); Depape (22)

Callitris (*Callitrites*)	.	.	.	.	.	×	.	.	×	.	×	*Callitris* is confined to Australia and New Caledonia. Other members of the family live in tropical and South Africa, Australia	Many of the fossils named *Callitris* may be more nearly allied to other members of the Callitrineae and should be named *Callitrites*. Seward (19)
Pinus	×	×	×	?	×	×	.	.	.	×	×	Reaches the tree-limit in the northern hemisphere; grows on the mountains of Formosa, Malaya, and other tropical countries	
Crossotolepis	.	.	.	.	.	×	.	×	.	.	.		Fliche (99)
Sciadopitys (*Sciadopitytes*)	.	.	.	.	.	×	.	.	×	.	×	The umbrella pine of Japan, an isolated and ancient type	Halle (15); Florin (22)
GINKGOALES													
Ginkgo	×	×	.	.	×	×	×	.	×	.	×	*Ginkgo* no longer exists as a wild tree; its last home was no doubt in China	
MONOCOTYLEDONS													
Palms	.	×	×	×	.	×	×	.	.	×	×	There are now about 130 genera of palms, fan palms and feather palms; they are almost entirely tropical. The one European genus *Chamaerops* (dwarf palm) occurs in the Mediterranean region: the American sabal (palmetto) is confined to the south-eastern part of the United States	Berry (14) gives a map; also Laurent (99) reproduces a map from Drude
Nipa (*Nipadites*)	.	.	×	×	.	×	.	.	×	.	.	This "stemless" palm is characteristic of tropical estuaries and mangrove swamps on the coasts of the Indian Ocean	Many seeds recorded: Kräusel and Stromer (24) give references. A few leaves [Chandler (25)]
Sabal (*Sabalites*)	.	.	×	×	.	×	.	.	×	.	.	Confined to southern North America, the West Indies, Mexico, etc.	
LILIACEAE													
Macclintockia	.	.	.	.	×	×	.	×	.	.	×		Seward (25). The systematic position of this extinct genus is uncertain: the leaves resemble those of *Smilax*
TYPHACEAE													
Typha (*Typhacites*)	.	×	×	.	.	×	.	.	.	×	×	The only genus of the family; temperate and tropical, especially common in the temperate zone (reed mace or bulrush)	A fossil species identical with European forms is recorded from an early Tertiary flora in Nigeria [fig. 121, Ng: Seward (24^2)]

Table G. *Geographical distribution of some Tertiary Plants* (cont.)

Classes, Families and Genera	Regions shown on the Maps, figs. 121, 122							Genus extinct	Extinct in Europe	Living in Europe	Recorded from Cretaceous rocks	Notes on Present Distribution	References
	North America				Arctic	Europe	Siberia: Japan						
	A	B	C	D									
MONOCOTYLEDONS (*cont.*) HYDROCHARITACEAE *Stratiotes*	.	.	.	.	.	×	.	.	.	×	.	The water soldier or water aloe lives in Europe and Siberia	Chandler (23) traces the gradual change in the seed characters from the Eocene to the present
DICOTYLEDONS MAGNOLIACEAE *Magnolia*	×	×	×	×	×	×	×	.	×	.	×	*Magnolia* has a discontinuous distribution—eastern North America, India, China, Japan, Mexico, etc.	Dandy and Good (29) distribution map of existing species: also Hutchinson (26)
Liriodendron	.	×	.	.	×	×	×	.	×	.	×	The tulip tree (two very closely related forms) is confined to North America, Nova Scotia to Florida, and China, where in Hupek it grows at an altitude of 6000 ft.	See Seward (25) for a discussion on some supposed Cretaceous Arctic species. There are good Tertiary specimens from Iceland in the Copenhagen Museum
LAURACEAE *Cinnamomum* (*Cinnamomoides*)	.	×	×	×	.	×	×	.	×	.	×	The cinnamon and camphor trees are now tropical: China, Japan, India, Ceylon, Australia	Seward (25)
NYMPHAEACEAE *Nelumbium*	.	×	.	×	.	×	×	.	×	.	.	One species ranges from Pennsylvania to Colombia: the Sacred Lotus, no longer found in the Nile, occurs from Japan to the Caspian Sea and North-East Australia	Berry (17) and Depape (24) give maps
MENISPERMACEAE *Menispermum* (*Menispermites*)	.	×	.	×	.	×	.	.	×	.	×	Mainly tropical: the genus *Menispermum* lives in temperate East Asia and Atlantic North America	Laurent (12)
ONAGRACEAE *Trapa*	×	×	.	×	.	×	×	.	.	×	.	A floating water plant (horn nut); Central and South Europe, China, Japan, South Asia, and Africa. The hard part of the fruit is easily recognized by the 2–4 horns or spines (fig. 120)	Berry (27[2]), Gams (27) distribution map

PROTEACEAE	·	·	×	×	·	×	·	·	×	·	×	A family now mainly Australian, tropical and South African (silver tree, etc.). Some genera also in South America	There has been much controversy over the European and North American fossils assigned to this family: there can be no doubt that it existed in the earlier part of the Tertiary period in the northern hemisphere
MYRTACEAE *Eucalyptus*	·	·	·	·	·	×	·	·	×	·	? ×	*Eucalyptus* is almost confined to Australia (gum trees)	Many of the Cretaceous and Tertiary fossils referred to this genus are valueless as evidence; but some records seem to afford proof of its former occurrence in the northern hemisphere
COMBRETACEAE *Terminalia*	·	·	×	×	·	×	·	·	×	·	·	A tropical genus	Berry (26) gives a restoration of an Eocene species
TILIACEAE *Tilia*	×	×	×	·	×	·	×	·	×	·	·	The family is mainly tropical: *Tilia* is a north temperate genus (lime tree or linden)	
LEGUMINOSAE *Dalbergia* (*Dalbergites*)	·	×	·	×	·	×	·	·	×	·	×	A common tropical genus which extends into the temperate zone in Africa and East Asia	Seward (25)
HAMAMELIDACEAE *Liquidambar*	×	×	×	×	×	×	×	·	×	·	×	*Liquidambar* (sweet gum) resembles the maple but its leaves are alternate, not opposite. Scattered distribution: one species in Asia Minor, one in eastern North America, one or more in China, etc.	Laurent (19)
PLATANACEAE *Platanus*	·	×	×	×	×	×	×	·	·	×	×	*Platanus* is the only member of the family: the Oriental plane ranges from the eastern Mediterranean to the Himalayas; the Occidental plane from Mexico to Canada and occurs on the Pacific Coast of North America	For recent species see Jaennicke (99); Henry and Flood (20). For fossil records see Berry (23[2]); Seward (26)

Table G. *Geographical distribution of some Tertiary Plants* (cont.)

Classes, Families and Genera	Regions shown on the Maps, figs. 121, 122							Genus extinct	Extinct in Europe	Living in Europe	Recorded from Cretaceous rocks	Notes on Present Distribution	References
	North America				Arctic	Europe	Siberia: Japan						
	A	B	C	D									
DICOTYLEDONS (*cont.*)													
SALICACEAE													
Salix	×	×	×	.	×	×	.	.	.	×	.	Willows are now circumpolar and extend across Europe south of the equator to the south temperate zone	
MYRICACEAE													
Myrica	×	×	×	×	×	×	×	.	.	×	×	*Myrica Gale* (sweet gale or bog myrtle) occurs in western and northern Europe and North America; other species live on the mountains of tropical Africa, Asia, South America and other countries	Berry (17) gives maps; see also Laurent (99)
BETULACEAE													
Betula	×	×	×	.	×	×	×	.	.	×	.	The dwarf birch (*Betula nana*) reaches to the northern tree-limit: the genus is widespread in Arctic and temperate lands	
Alnus	×	×	.	.	×	×	×	.	.	×	.	Alders have a wide distribution in the temperate zone of the northern hemisphere and some occur in South America	Berry (23[2]) gives map
FAGACEAE													
Quercus	×	×	×	.	×	×	×	.	.	×	×	Northern hemisphere and well represented in tropical countries	
Dryophyllum	×	×	×	×	.	×	×	×	.	.	×	Large leaves reaching a length of 30 cm. resembling those of *Castanea* (chestnut) and oaks, probably an ancestral form of *Castanea*, *Quercus* and allied genera	Marty (07) figures very large examples from Belgium. Berry (24); Laurent (12)

MORACEAE													
Artocarpus	.	×	×	×	.	×	.	.	×	.	×	The bread-fruit tree is essentially tropical, Indo-Malay and China	Lee and Knowlton (17)
Ficus	×	×	×	×	.	×	.	.	×	.	×	A large tropical genus, chiefly East Indian and Polynesian	The name *Ficus* has been used by authors of palaeobotanical papers much more freely than the evidence warrants
RHAMNACEAE													
Zizyphus	.	×	×	×	×	×	×	.	.	.	×	Mediterranean; Indo-Malay, Africa and other tropical countries	
Paliurus	.	×	.	×	.	×	.	.	.	×	.	South Europe to China and Japan	Langeron (02[2]) gives a distribution map
SAPINDACEAE													
Sapindus	.	×	×	×	×	×	.	.	×	.	.	Tropical and subtropical excluding Africa and Australia	
ACERACEAE													
Acer	×	×	×	.	×	×	×	.	.	×	.	North temperate; China, Japan, Java, etc.	For existing maples, see Pax (27)
JUGLANDACEAE													
Juglans	×	×	×	×	×	×	×	.	.	×	.	The walnut is now native in Europe only in Greece; *Juglans* is widely distributed in the northern hemisphere and occurs also in the Tropics	Map given by Berry (23[2]), (26[2]); Florin (20)
Engelhardtia	.	.	×	×	.	×	.	.	×	.	.	Indo-Malay and China: a closely allied genus *Oreomunnea* occurs in Central America	For maps see Berry (23[2]); Reid and Chandler (26); Ettingshausen (69). For restoration of *Engelhardtia*, see Berry (30)
Dewalquea	.	.	.	.	.	×	.	×		.	×	This extinct genus of unproved affinity has been described as an ancestral walnut	
NYSSACEAE													
Nyssa	×	.	.	×	×	×	.	.	×	.	.	Eastern Asia and North America	
ARALIACEAE													
EBENACEAE													
Diospyros	×	×	×	×	×	×	.	.	×	.	×	*Diospyros* (persimmon) is now mainly tropical and subtropical	Berry (23[2])

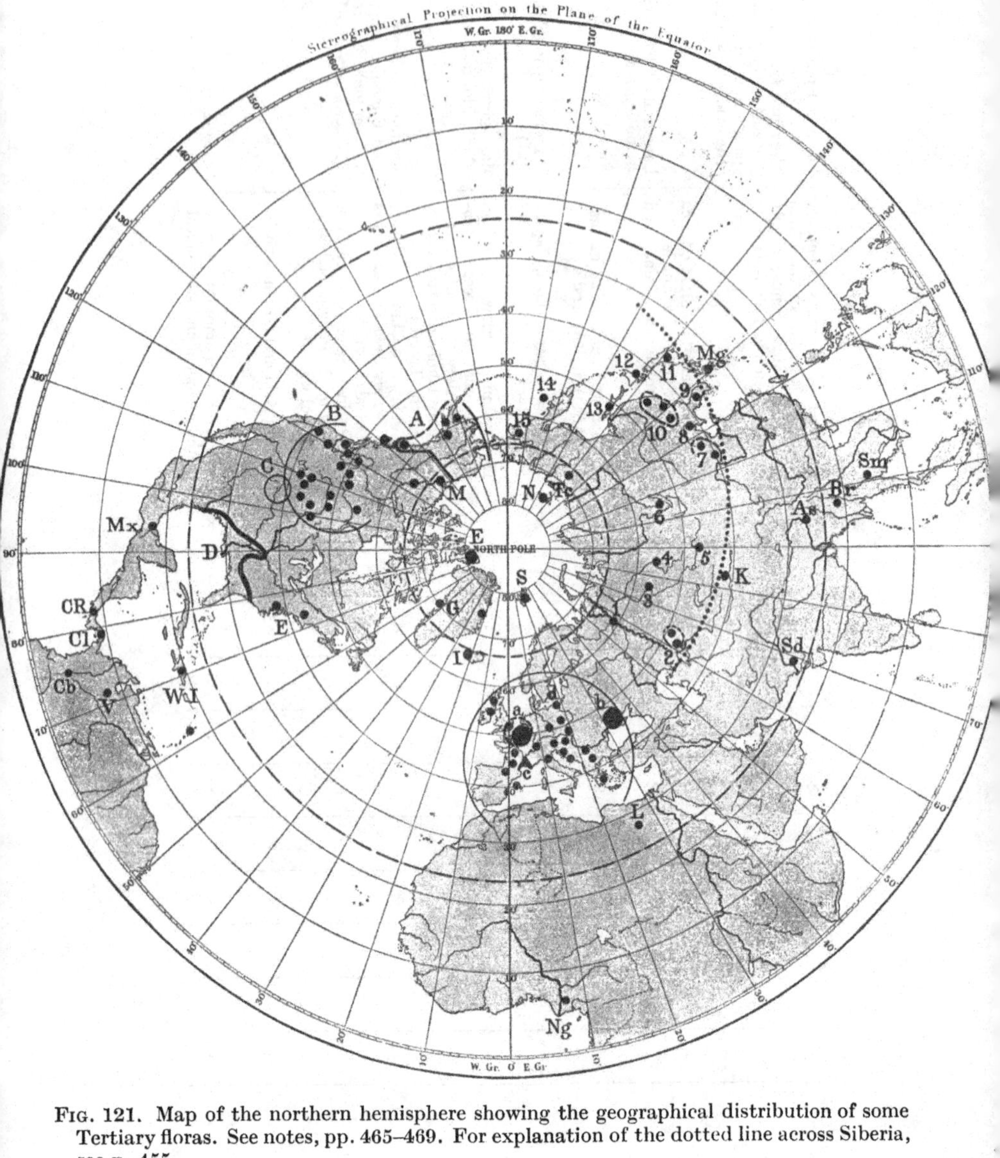

FIG. 121. Map of the northern hemisphere showing the geographical distribution of some Tertiary floras. See notes, pp. 465–469. For explanation of the dotted line across Siberia, see p. 455.

FIG. 122. Map of the southern hemisphere showing the geographical distribution of some Tertiary ⊗ and Cretaceous (•) floras. See notes, p. 470; also, for notes on Cretaceous floras, p. 411.

CHAPTER XVII

THE QUATERNARY PERIOD

Even the trifling irregularities were not caused by pickaxe, plough, or spade, but remained as the very finger touches of the last geological change. ***Thomas Hardy***

THE plant-world reconstructed from the records furnished by the later series of Tertiary rocks is in essentials the world as we now see it; but it is customary and on the whole convenient to draw a distinction between the penultimate and the last phase of organic evolution. No clearly defined break at this stage of geological history is revealed by the fossils or the rocks which contain them. Geologists do not invariably adopt the same boundary-line: for our present purpose the arguments for or against any particular classification need not be considered. The Quaternary, or Pleistocene, period may be defined as the last and much the shortest chapter of geological history; in it geology, archaeology, and human history overlap one another without any well-marked gap. The chapter opens with the records of the last Ice Age and the events immediately preceding it: it begins at the close of the Cromerian stage, and the first pages of the chronicle are written in the Arctic plant-bed and the boulder clay which lies above it. It may indeed be argued that we are still living in the Tertiary period: until a few years ago the Quaternary period was believed to embrace the whole of the time during which man has existed, and it was not unnatural to devote a separate chapter to the history of the organic and inorganic world in which the human race played a part. As Prof. J. L. Myres[1] says, the term Quaternary "signifying as it does those recent phases when man's presence can be demonstrated is a needless concession to self-esteem". Recent research has shown that man lived long before the end of the Pliocene stage; how long we do not know.

In the Cromer district and in other parts of East Anglia Mr Reid Moir has discovered implements which are generally accepted as bearing the impress of human craftsmanship. There is no doubt that the earliest representative of the human race wandered through Tertiary forests containing trees which we should now call exotics,

[1] Myres (24).

and enjoyed an English climate more genial than that endured by his sophisticated descendants. One reason for separating the Caino-zoic era into the Tertiary and Quaternary periods is the occurrence of an Ice Age, which reached its maximum development at a time not long after the colonization of the borders of the ancient delta of the Rhine by the vegetation which has left many traces in the Cromer forest-bed. Taking the Ice Age as the outstanding event in this the latest phase of geological history, it is useful to adopt a twofold division of the Quaternary period, including in the first the events immediately before and during the great glaciation,[1] when, through many thousands of years ice-sheets and glaciers devastated not less than 8,000,000 square miles of country in North America and Europe; and in the second stage the whole of the post-glacial records.

Post-glacial Deposits	Forest period Tundra and Steppe period: Loess, etc.
Glacial Deposits	The Glacial period with Interglacial stages Arctic plant-bed of Cromer Sands with Arctic shells

A. The Glacial Period

At the time of the Quaternary Ice Age the boundaries of continents and oceans were pretty much as they are to-day (fig. 123): the chief difference is the attachment to the mainland of small areas that are now islands. The darker patches indicate approximately the areas invaded by ice: if the scale were larger more land would be left ice-free in the extreme north of America and on the Gaspé Peninsula at the mouth of the St Lawrence River. In the large scale map (fig. 124) ice-sheets and glaciers are shown in black and the direction of flow by arrows. The dotted regions illustrate the wide distribution of loess[2] in northern Europe.

The cliffs on the Norfolk Coast (fig. 112) illustrate a succession of changes, geographical and climatic, from the time of the Cromer forest-bed, with its drifted stumps of trees and fragments of smaller plants washed by the waters of the greater Rhine into the collecting ground of its delta (fig. 111), to a bed containing remains of an Arctic flora and, higher in the series, the boulder clay. The forest-bed is assigned to the uppermost stage of the Tertiary period; the Arctic plant-bed, heralding the approach of the Ice Age, is selected

[1] Antevs (28).

[2] See p. 488 and fig. 126.

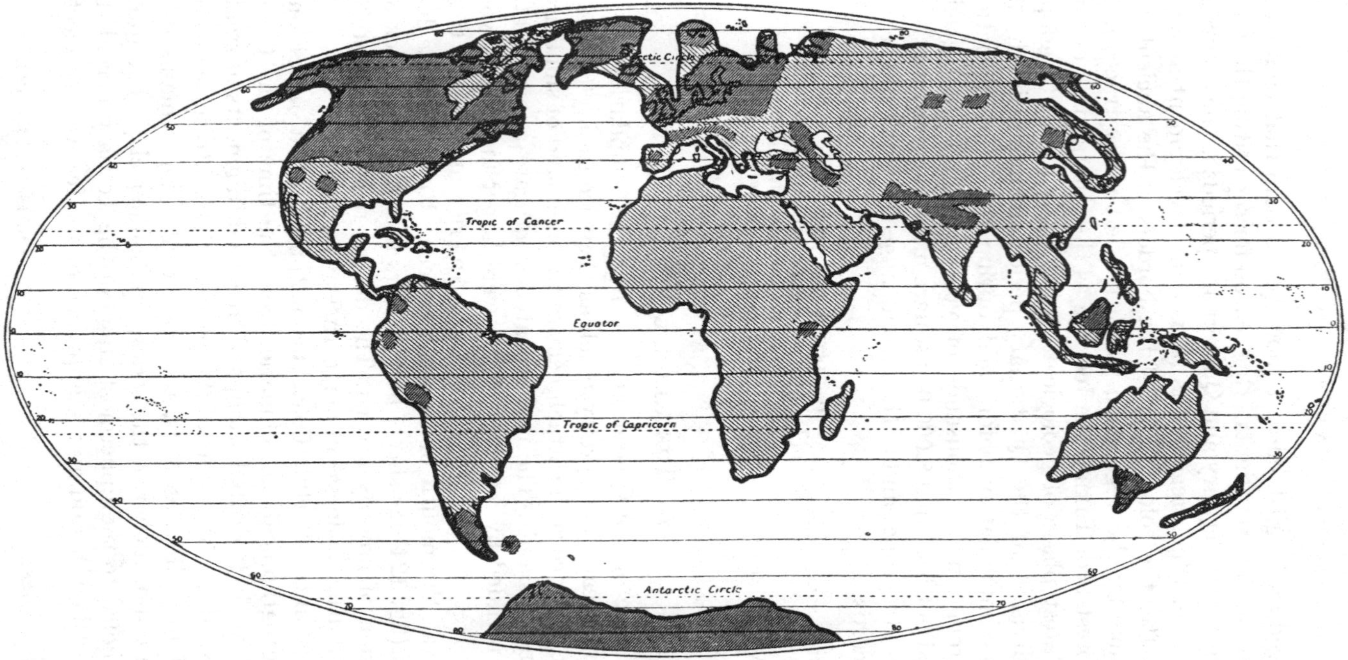

Fig. 123. Map of the world in the early part of the Quaternary period showing the maximum extension of the ice (darker areas) and the slight extension of continental edges to include portions that subsequently became detached as islands. (Based on a reconstruction by Arldt.)

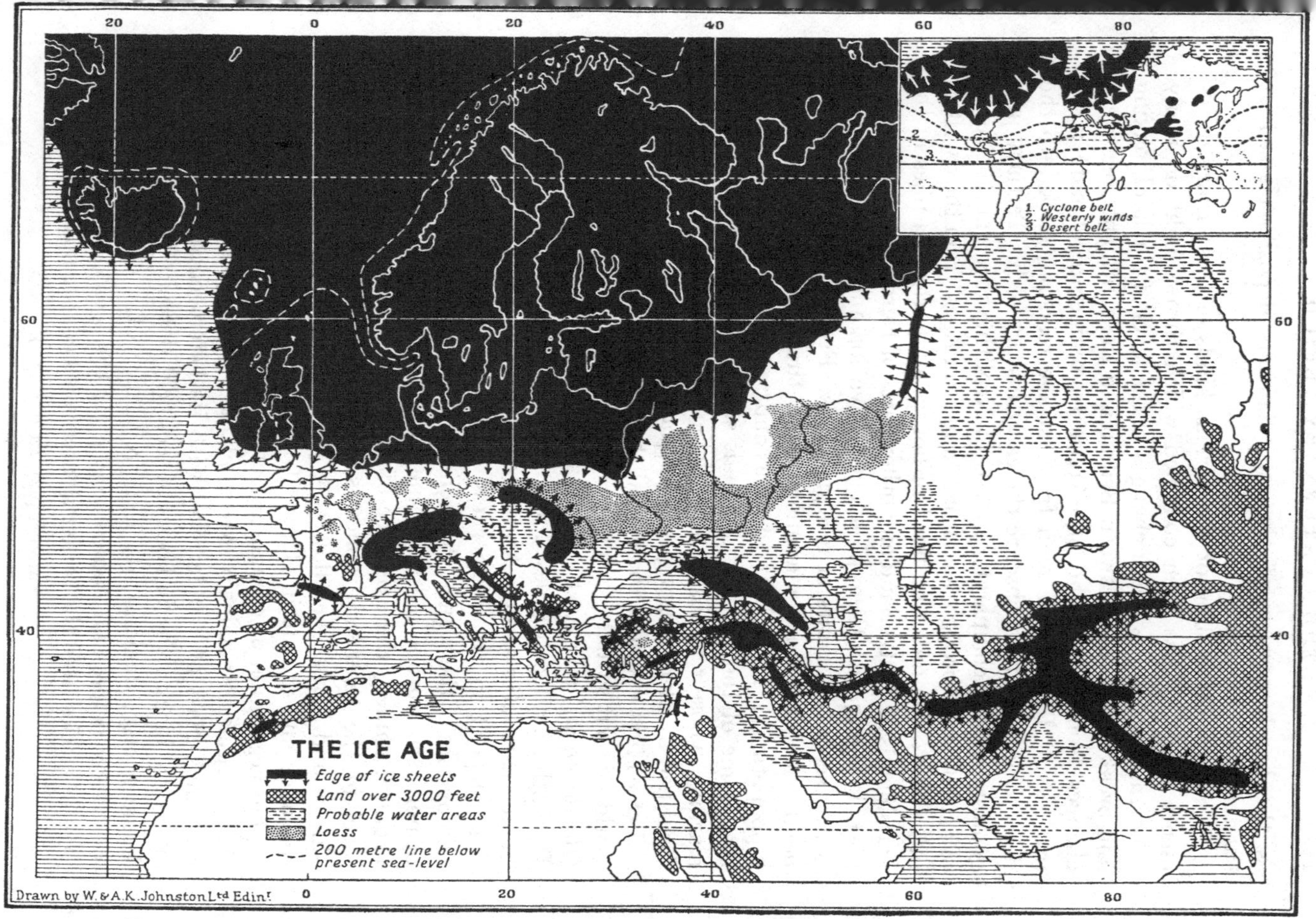

FIG. 124. Map of Europe during the Ice Age. (After Prof. Myres, from the *Cambridge Ancient History*, vol. I.)

as the basal member of the miscellaneous deposits classed as Quaternary.

From an examination of the various kinds of Quaternary documents a history has been compiled which throws light on many departments of knowledge: changes in the level of the land making and breaking contact between Britain and the continent; recurrent changes from Arctic and sub-Arctic to temperate climates; the progress of Palaeolithic man from an age reaching back into the deep shadows, which obscure the early history of his Tertiary progenitors, to the comparatively recent Neolithic race and the dawn of agriculture; through the Bronze and Iron Ages, from the dwellers in caves to the folk who lived in houses supported on piles in the marshlands of Switzerland and other parts of Europe; to the Roman hosts bringing with them plants of cultivation and, as unconsidered trifles, seeds and fruits of foreign weeds. From a botanical standpoint the Quaternary mammoth, with the bison, rhinoceros and hippopotamus, lions, bears, hyenas and reindeer are of less interest than some of the smaller mammals such as the lemming and the marmot, whose remains confirm the evidence for steppe conditions furnished by plants during certain phases of the Quaternary period.

Let us first consider the glacial stage in its relation to the plant-world. About seventy years ago some fragments of the dwarf birch (*Betula nana*)[1] and Arctic willows were found at Bovey Tracey in Devonshire, a locality best known as the site of an Oligocene lake. Some years later similar Arctic relics were discovered in a Quaternary deposit in the Holderness district of south-east Yorkshire. Discoveries of these and other relics of an Arctic flora in many parts of northern Europe tell the same story as the Arctic plant-bed of Cromer: namely the existence of plants, in places at or near sea-level in the temperate zones, specifically identical with stunted shrubs and herbaceous flowering plants now characteristic of the lowlands and mountain slopes of polar regions and of the Arctic-alpine vegetation of mountains several hundred miles south of the Arctic Circle. Remains of Arctic plants in beds immediately below the thick mass of boulder clay with its heterogeneous assortment of erratic blocks, and in layers of sediment deposited later than the boulder clay, demonstrate the prevalence of a cold climate immediately before and immediately after the Ice Age. Many of the Arctic species from the earlier Quaternary beds are now circumpolar

[1] Nathorst (92); Reid, C. (99).

in distribution; they live in Greenland, Spitsbergen, Ellesmere Land, and in other districts girdling the polar sea; some of them flourish on the Altai Mountains and the Himalayas as well as on the highlands of temperate North America and Europe.[1] Many years ago the English naturalist Edward Forbes[2] put forward an explanation of the scattered distribution of Arctic species on the mountains of temperate countries, separated from one another by lower ground occupied by an entirely different vegetation, and too far apart for an interchange of plants by ordinary methods of dispersal. He saw in the Arctic species of isolated hill-tops in temperate Europe relics of the glacial period; survivors of a flora which lived in the circumpolar zone before the change in climate compelled the northern plants to seek shelter within the temperate zone and, as the conditions became progressively arctic, to travel still farther afield, replacing the former occupants of the invaded territory which were themselves affected by the advancing ice. As time passed the lowlands and valleys of temperate regions put on an arctic dress. Eventually conditions began to improve: the ice-sheets shrank, and as the glaciers dwindled the Arctic immigrants were able to follow the retreating ice towards their original home: some re-entered the Arctic Circle; some found the conditions they needed by a vertical migration and gained a permanent resting-place on the higher slopes of mountains, where they have since remained as outliers of a circumpolar flora surrounded by reinstated temperate forms. The history of these wanderings directed by the advance and retreat of the ice is, however, not quite as simple as this brief outline suggests. There were many backward and forward movements both of the ice and the vegetation during the first half of the Quaternary period.

Pauses in the advance of ice-sheets and large glaciers are often marked by well-defined rows of the gravelly material which forms moraines, rubbish heaps tipped on the ground near the ice-front. Some of the best examples are found in Finland: the railway from Helsingfors to Viborg, near the Russian border, through the southern part of Finland[3] follows approximately the line of two moraine ridges deposited during a halt in the movement of the invading ice. In the Fenno-Scandian area clear evidence has been obtained of the up-and-down movements of the land which are a characteristic feature of the Quaternary period. Subsequent to the formation of

[1] Holm (22). [2] Hooker (62); Engler (04). [3] Sauramo (29).

the moraines and at a time coincident with the final withdrawal of the ice the land subsided, causing the Baltic basin to be occupied by an Arctic Sea which is spoken of as the Yoldia Sea because of the frequent occurrence of a fossil mollusc, *Yoldia arctica*, an inhabitant of the northern ocean. At a later stage elevation of the land converted the sea into a freshwater lake, the Ancylus Lake, so called from the freshwater limpet. These oscillations occurred in the post-glacial stage. Then ensued a sinking of the land—the date may be put at about 4000 B.C.—which caused a fresh incursion of the sea—the Littorina (periwinkle) Sea, warmer and salter than the Baltic. In the Yoldia stage the flora was Arctic, but in the latter part of it *Betula odorata* and the aspen (*Populus tremula*) were the dominant trees. In the Ancylus Lake stage the pine, mountain ash (*Pyrus aucuparia*) and bird cherry (*Prunus Padus*) made their appearance; and the pine replaced the birch as the characteristic tree and was followed by a mixed forest of elm, oak, hazel, hawthorn, lime and cornel (*Cornus*). The dwellers on the warm and moist shores, where the oak was successfully competing with the pine, were a people representing a stage of human development transitional between later Palaeolithic and the earlier Neolithic races. At the end of the Littorina stage the climate became gradually wetter and colder and the present conditions were established.

Interglacial Periods

By some geologists it is believed that researches in the Swiss Alps have shown that there were four different periods during which there was a general advance of the ice-sheets and glaciers and a lowering of the snow-line: each of these glacial stages is called after a locality, Günz, Mindel, Riss and Würm. The Mindel glaciation was the most extensive: the snow-line fell 4250 feet below its present level. There were three interglacial periods, the Günz-Mindel—that is the interval in which the ice retreated between the Günz and Mindel glaciation periods—the Mindel-Riss, and the Riss-Würm: the duration of the first of these is estimated at 60,000 years. The interglacial climate was at least as warm as it is now in the same region. Climatic fluctuations such as those indicated by interglacial periods could hardly be attributed to any change in the physical features of the earth's surface. Dr Simpson[1] considers that changes in solar radiation seem to be the only

[1] Simpson (29). See also a more recent paper, Simpson (30).

possible cause: he points out that the net result of decrease in solar radiation would be a lower mean temperature and less snow-fall, consequent on the decrease in circulation of the world's atmosphere and decreased cloud-production; the thickness of ice would decrease. A small increase in solar radiation would produce a reverse effect; a higher mean temperature, more precipitation, and a greater thickness of snow and ice—that is to say, an increase in solar radiation—would lead to increased glaciation. If, however, the increase in radiation continued, the annual melting might be as great as the annual snow-fall and the ice covering would disappear. Here, Dr Simpson adds, is a "possible clue to the meteorological conditions during the great Ice Age in the Pleistocene period".

Though there are not many records from the western hemisphere which are universally accepted as conclusive evidence of inter-glacial periods, reference may be made to some delta-deposits, known as the Toronto formation,[1] which rest on boulder clay and consist of clay and beds of sand about 25 ft. thick. Several trees have been identified, such as oaks, elms, maples, hickory and others. This Quaternary delta, preserved between two sheets of boulder clay, affords clear evidence of an interglacial interval when the land east of the Rocky Mountains was free from ice. In the northern hemisphere generally or in that part of it which was to a greater or less extent overridden by ice there were oscillations in temperature and in the advance and retreat of the glaciers, but exactly how many interglacial periods there were and how long each lasted are questions which have not yet been definitely answered. Contro-versy over the number and even the occurrence of interglacial periods in western Europe has been actively waged by geologists for many years. For our present purpose this is a relatively un-important matter. Reference is made later to plants found in some interglacial beds in Denmark. A well-known example of an inter-glacial flora was described nearly forty years ago from the neigh-bourhood of Innsbruck:[2] 70 per cent. of the plants which it was possible to recognize still live in the same or similar districts, and the flora as a whole most nearly resembles that on the mountains surrounding the Black Sea (Pontic flora). One species, *Rhododen-dron ponticum*, a plant of the Caucasus and southern Spain, is said to require a temperature higher than that of Innsbruck. This inter-glacial flora was probably contemporary with the steppe vegetation

[1] Coleman (26). [2] Wettstein (92).

of central Europe. When the southern part of middle Europe was under steppe conditions the Pontic flora had spread to the Alps and remained there as an island-oasis in the steppes of the lowlands.

Tundra and Steppe Conditions

Each serious advance of the ice was one phase in a regular cycle: as the ice slowly withdrew a belt of barren, treeless and frozen ground near the ends of the glaciers assumed the character of tundra—a region where bare rock and groups of the hardier Arctic plants share possession of the ground (fig. 125). The landscape on the edge of the ice in North America and Europe resembled that of the Eurasian belt north of the Arctic Circle, a land described by Marco Polo as "inaccessible because of its quagmire".[1] At a later stage as the tundra zone kept pace in its slow northward trend with the retreating front of the ice, the southern part of it admitted the entrance of fresh types of plants; the tundra shaded into boundless, rolling plains of steppe. The nature of the steppe phase of Quaternary history may be illustrated by a brief description of two regions, one in Turkestan and a much smaller one of a different character in East Anglia. A formation known as loess, from a German word meaning "loose" (pronounced lus, as bus) covers a large area in northern Europe, middle Asia and China (fig, 124): it is a friable clay containing a large percentage of sand and is believed to consist of masses of fine dust blown by the wind from glacial moraines and deserts. The wind-driven loess was the soil of post-glacial steppes: the loose material was invaded by vegetation and held by a network of roots as shifting sand dunes are often controlled by plants selected by man for that purpose. Characteristic features of the loess are the abundance of fine tubular spaces or capillaries left on the decay of roots and stems of the natural sand-binders, and the occurrence of skeletons of lemmings and other animals of the steppe. In the west of Europe the dry post-glacial stage is marked by "straggling loess hills" from Hungary to France, but farther east evidence of steppe conditions becomes more and more conspicuous. The photograph, reproduced as fig. 126 from one of Mr Rickmers' negatives, shows a river valley in the loess of middle Asia. Mr Rickmers,[2] in his description of the Duab, the land between the Oxus and Jaxartes, writes: "Whosoever has seen, smelt and breathed

[1] Haviland (26). See also Gates (28).
[2] Rickmers (13). See also Barbour (29), p. 64.

Fig. 125. Siberian tundra in summer. (After Mrs H. H. Brindley [Maud D. Haviland].)

Fig. 126. A river-valley in the loess, Turkestan. Mr Rickmers, from the *Duab of Turkestan*. (Camb. Univ. Press.)

loess, to him it is middle Asia, a fragrance, a colour, a sensation; to him it is the East which he hears a-calling". The loess landscape of the East recalls also a post-glacial scene[1] in the West where a vast expanse of steppe lay across the zone recently deserted by the retreating ice. The barren tundra lands of Russia and Siberia and the loess canyons and cliffs of Turkestan, through a fortunate persistence in those regions of certain climatic conditions, have been preserved as natural monuments of incidents of the post-glacial stage in western Europe. Remains of Neolithic man with the bones of mammoths, rhinoceros, and hyena have been found in the loess terraces on the banks of the Yenisei.

The second example is an East Anglian district known as Breckland which in a minor degree is reminiscent of the post-glacial steppe phase: the name Breckland has reference to common lands which were ploughed and broken up and were known as brecks. Breckland[2] lies in the north-west corner of Suffolk and the south-west corner of Norfolk on the eastern border of the Fenland: it covers about 400 square miles and is from 20 to 150 ft. above sea-level. This East Anglian sandy plain with sheets of bracken (*Pteridium aquilinum*), wide stretches of heath and grassland, occasional patches of swamp, with earthworks and implements of Palaeolithic and Neolithic men, bears the impress of antiquity and takes us back to a time when a similar landscape, though probably a different vegetation, was characteristic of a much larger region which had recently been overridden by ice. A significant feature is the occurrence of several flowering plants nearly all of which are peculiar to this part of England, though many are common enough on the continent and some range far to the east. Among other species which may be regarded as immigrants across the North Sea basin and as relics of the steppe phase are: *Silene Otites* (Spanish catchfly), *Silene conica* (common in central and southern Europe and central Asia), *Artemisia campestris* (allied to the wormwood, widely spread over Europe and temperate Asia), *Carex ericetorum* (a sedge of Arctic Europe and northern Asia), *Muscari racemosum* (grape hyacinth).

Large areas of country must have had the open, almost treeless aspect of Breckland during the earlier stages of the retreat of the ice; but during the waning of the glacial period there was a steady

[1] For a map of steppe areas in post-glacial time see Florav (27).
[2] Farrow (25).

improvement in conditions, dryness gave place to humidity and the inauguration of a forest zone.

"The vast tract of the parched and sandy waste
Now teems with countless rills and shady woods."

The present treelessness of Breckland is probably due only in part to the combination of dry, infertile soil combined with low rainfall, for the very abundant rabbits prevent tree regeneration over the whole area save in special enclosures. The geological, botanical and zoological records reveal recurrent changes as different regions came under the potent influence of the ice-invasion, and demonstrate the importance of the glacial epoch as a factor in the development of the present vegetation during the thousands of years of the earlier stage of the Quaternary period.

An English Landscape in the latter part of the Glacial Period

Fig. 127 represents a landscape at a time subsequent to the climax of arctic conditions in the Pennine Hills of England: the plants in the foreground are all species recorded from the Quaternary peaty layers in the gravel pit at Barnwell[1] (Cambridge). Glaciers from a gathering ground behind the hills fill the valleys: icebergs detached from the end of a glacier on the left, with a median moraine of stones collected from the edge of cliffs on the higher slopes, are stranded in the shallows of the lake. A river laden with scourings of the rocks below the glacier on the right meanders among the moraine hillocks on the ground between the crevassed margin of the glacier on the right and the rocky slope in the foreground. The angular blocks of the Millstone grit in the right-hand corner show no signs of the grinding and polishing of overriding ice: at the foot of the unglaciated bastion on the extreme right is a clump of *Saxifraga oppositifolia* (purple saxifrage, *Sx.*); next to this and slightly farther back the alpine poppy (*Papaver alpinum, Pv.*); below, a plant of *Dryas octopetala, D.*; to the left, *Primula scotia, Pr.*, *Carex arenaria* (sand carex, *Cx.*) with the taller *Polygonum viviparum, Pg.*, and beyond a shrubby *Vaccinium uliginosum* (bog *Vaccinium, V.*), closely allied to the cranberry and bilberry; farther to the left the plumed heads of *Eriophorum polystachyum* (a cotton grass, *E.*); in the water a tall plant of *Menyanthes trifoliata* (bog-bean, *M.*) and a pondweed (*Potamogeton gramineus, Pn.*); on the left-hand edge of the pool *Thalictrum alpinum, T.* (a near

[1] Marr and Gardner (16); Marr (20); Chandler (21).

FIG. 127. An English landscape in the latter part of the Quaternary Ice Age. (Drawn by Mr Edward Vulliamy.) *Bt. Betula*; *Cp. Carpinus*; *Cx. Carex*; *D. Dryas*; *Da. Draba*; *E. Eriophorum*; *M. Menyanthes*; *Pg. Polygonum*; *Pn. Potamogeton*; *Pr. Primula*; *Pt. Potentilla*; *Pv. Papaver*; *Sl. Salix*; *Sx. Saxifraga*; *T. Thalictrum*; *V. Vaccinium*.

relation of our meadow rue); to the left of it *Draba incana, Da.*; and farther to the left, *Potentilla anserina* (silver weed, *Pt.*). In the left-hand corner, *Salix reticulata, Sl.*, to the right of it a branch, with smaller foliage, of *Betula nana, Bt.*; and, coming into the picture above the *Salix*, a branch of *Carpinus Betulus* (hornbeam, *Cp.*).

FIG. 128. A view from Upernivik Island (lat. 70° N.), west Greenland. (Photo. A. C. Seward, from *A Summer in Greenland*, Camb. Univ. Press.)

The effect of the Glacial conditions on the Vegetation

A general account has already been given of the Quaternary Ice Age. The nature of the evidence attests the presence of ice-sheets and glaciers both in regions where the high ground is now permanently below the snow-line and in more mountainous countries where the glaciers are comparatively small. The two photographs reproduced in figs. 128, 129 help us to visualize scenes in the British Isles when they were as Greenland and Spitsbergen are now. Fig. 128 is a view from a valley on the western side of Upernivik Island (lat. 70° N.) off the Greenland coast looking across a sea littered with icebergs towards another island farther west. Fig. 129

shows the face of a glacier at the head of Cross Bay, Spitsbergen, with a few nunataks projecting through the ice.

In North America the first ice-sheet to reach its full development had its centre in the highlands on the western side of the continent: this is known as the Cordilleran ice-sheet: the second in point of time spread fan-like from a gathering-ground to the north-west of the Hudson Bay region, and finally the great Labrador ice-sheet swept over the eastern territory. What was the fate of the vegetation in the glaciated areas? The glaciers of Scotland and England increased in size and became sheets of ice, as the still larger ice-sheet from the Scandinavian Alps reached across the North Sea basin and impinged on the eastern border of England. Is it conceivable that any of the pre-glacial vegetation could survive? In the British

FIG. 129. The Lilie Hoot glacier, at the head of Cross Bay, Spitsbergen, with nunataks. (Photo. A. C. Seward.)

area by far the greater part of the land north of a line connecting Bristol and London was under ice, but not the whole. We can picture the western edge of the British ice-sheet beyond the present coast of Ireland as a floating cliff similar to the great Antarctic Barrier. Glaciers from the Scottish Highlands left records of their progress to the south in trails of boulders scattered over Wales and England. The direction of flow of rivers of ice from the English Lake District both towards the south and across the Pennine Chain to the east is clearly shown by the travelled blocks of shap (Westmorland, north-west England) granite, recognizable by the pink crystals of felspar, which have been found as far south as Robin Hood's Bay on the coast of Yorkshire.[1] From Scandinavia the mass of ice invaded the North Sea and left traces of its contact with the English coast in scattered erratics from Norwegian sources. Rocks from

[1] Kendall and Wroot (24); for a general account of the Ice Age see Wright (14); Antevs (28); Evans and Stubblefield (29).

Scandinavia, Scotland, Cornwall and Devon occur in the glacial gravels in the Oxford district,[1] which were spread far and wide during a partial submergence when Britain was an archipelago in the midst of a frozen sea. Some of the higher ridges and peaks even in the most intensely glaciated places remained as nunataks (fig. 129), islands of dark rock in a frozen billowy waste, as in the Pennine Chain and the Cleveland Hills farther east. An obvious method of approaching the problem, whether or not any plants of the pre-glacial vegetation were able to live through the Ice Age, is to go to a region which affords the closest parallel to north temperate America and Europe when approximately 8,000,000 square miles were in the grip of the Ice Age. The granitic headlands of Cape Farewell at the southern extremity of the large continental island of Greenland lie a few miles to the south of lat. 60° N., and the northern shores of this most northerly land in the world project into the polar sea a short distance beyond lat. 83° N. Nearly 1200 miles of the length of Greenland lie within the Arctic Circle. During the short summer there is a relatively narrow fringe of snow-free ground on the edge of the island, in some places more than a hundred miles broad; in others, particularly on the east coast and in some districts on the west, much narrower, and here and there replaced by stupendous cliffs of glacier ice overlooking an iceberg-littered sea. Almost the whole of Greenland is hidden under ice of enormous thickness, a vast shield-shaped mass sloping towards the heads of the fiords and with a gentler curve bending towards the northern coast, where there is a comparatively broad, bare strip in July, the one month of the year when the temperature rises above freezing-point. At varying distances from the edge of the ice-field a few mountain peaks protrude as nunataks.

Approximately 400 species of flowering plants, ferns and their allies are recorded from Greenland:[2] of these probably not more than four are peculiar (endemic) to this section of the Arctic regions. A considerable number of the Greenland plants are circumpolar in distribution and several are well-known elements in the Arctic-alpine flora of Great Britain and North America; some occur in the Alps, Pyrenees, Himalayas and other mountain ranges. The following examples are selected as a few of the more characteristic species. It is noteworthy that by no means all the Greenland plants are in the strict sense Arctic types; though many are confined, in the

[1] Sandford (29). [2] Ostenfeld (26).

warmer regions south of the Arctic Circle, to the higher slopes of mountains, not a few of them are common in the lowlands and valleys of more southern countries and occur as characteristic elements in temperate floras.

It is significant that in the ponds and small lakes in some of the boggy depressions near the western coast in the latitude of Disko Island (70° N.) and other districts there are several water plants, species of pondweed (*Potamogeton*) and others, which would be regarded as temperate forms. This association of temperate aquatic plants, in places where the water is warmed by continuous sunlight with typical Arctic species on the tundra and heath moors, is worthy of note because of similar mixtures described from Quaternary plant-beds in England and elsewhere. In the course of a Danish expedition to the north coast of Greenland in 1916–18,[1] seventy species of flowering plants were collected: these outposts of the Arctic vegetation include a *Ranunculus*; *Dryas integrifolia*, differing but little from *Dryas octopetala* (fig. 131) which is characteristic of the east coast and is a member of the Arctic-alpine flora in temperate Europe; *Silene acaulis* (moss campion); a *Taraxacum* (dandelion); two species of *Equisetum*, both of which have a wide range, horizontal and vertical, in the temperate zone; *Eriophorum* (cotton grass); *Polygonum viviparum*;[2] Saxifrages, including *Saxifraga oppositifolia*, and other flowering plants; also two ferns, a species of *Woodsia*, and *Cystopteris fragilis* (brittle fern) which has wandered as far south as the Island of South Georgia in lat. 55° S. This polar vegetation illustrates an amazing power of existing under circumstances which might fairly be described as intolerable. During nine or ten months the plants hibernate and the advanced state of development of the resting buds enables them to expand with almost explosive suddenness when growth is resumed. The discovery of a few flowering plants on nunataks slightly north of latitude 82° N. caused Prof. Ostenfeld to remark: "One must be extremely careful in asserting that an ice-bound country, that is a country under conditions answering to an Ice Age, is completely destitute of flowering plants". Ice-bound as Greenland is to-day there is abundant evidence, in the occurrence of moraines on ground that is now beyond the reach of glaciers, of a more intense glaciation during the early stage of the Quaternary period. The problem is:

[1] Ostenfeld (25).

[2] Plants of this species are shown in the Greenland photograph, fig. 139.

Was the pre-glacial vegetation or any part of it able to live in the same region through the whole of the Ice Age? Prof. Warming of Copenhagen staunchly maintained that the present plant population consists in large measure of species which have lived in Greenland since pre-glacial days: Prof. Nathorst of Stockholm was strongly opposed to the view that the plants could survive the great ordeal: his opinion was that the present flora colonized the country at the end of the Ice Age, travelling from more southern regions by way of a land-bridge of which Iceland formed a part.

The latest contribution to this subject is from Prof. Ostenfeld[1] of Copenhagen who thinks that about sixty species may have survived in sheltered stations when the Ice Age was at its maximum. He suggests that perhaps 13 per cent. of the present flora were introduced by the first Norse settlers who sailed from Iceland late in the tenth century with Eric the Red, while a large proportion had returned to Greenland after the maximum extension of the ice by way of the narrow channel bordering the north-east coast of the Arctic Archipelago, and others had travelled as wind-driven seeds to the east coast by way of the frozen sea. It may be assumed therefore that the majority of the Greenland plants were unequal to the strain; but it is almost certain that several members of the pre-glacial flora lived in their northern home through the Ice Age in Greenland and in other regions. One argument in favour of the view that the greater number of the present species arrived after the partial retreat of the ice is the almost complete absence of endemic forms; that is plants peculiar to a country or locality. In a comparatively modern flora endemic species are rare; in long established communities there has been more time for the evolution of types peculiar to the district.

We now turn to a region where the flora furnishes definite evidence of survival from the pre-glacial age. The following account is based on an exceptionally interesting paper by Prof. Fernald[2] of Harvard on the "Persistence of plants in unglaciated areas of Boreal America". On the bleak headlands of the Schickshock Mountains (often called on maps the Notre Dame Mountains), which form the backbone of the Gaspé Peninsula parallel to the curved north coast overlooking the Gulf of St Lawrence, there is a large group of plants occupying localities untouched by the Labrador

[1] Ostenfeld (26).

[2] Fernald (25). See also Fernald (29).

ice-sheet. Geologists have shown that the eastern ice-sheet did not completely override the peninsula,[1] and it is on the unglaciated areas that the plants we are now considering are found. One striking feature of this flora is its richness in endemic forms, not less than eighty: another very significant fact is that these plants of the St Lawrence region form an isolated outlier of a vegetation which is now broken up into widely-separated groups in the American Arctic Archipelago, the Altai Mountains, and in other parts of Siberia, while the largest group forms part of the flora of the Cordillera in western America. The Cordillera include the Rocky Mountains, the Cascade and the Sierra Nevada separated by 2000 miles or more from the Gaspé Peninsula.

Prof. Fernald gives many distribution-maps showing the remarkable discontinuous range of Gaspé species. A few examples chosen from a long list will serve as illustrations of the striking affinity of the eastern and western groups: one of the Gaspé ferns, *Polystichum mohrioides*, var. *scopulinum*, a northern variety of *P. mohrioides* which ranges as far south as the Falkland Islands and is closely related to the holly fern *Polystichum* (*Aspidium*) *Lonchitis*, of arctic and temperate countries. This fern is confined to about twelve stations, two in southern California, two others more than 500 miles farther north, another 300 miles still farther north; other stations on the Cascade Mountains and 300 miles to the east on the Rocky Mountains, also one station, 2000 miles farther east in the Gaspé Peninsula. Similarly a willow (*Salix brachycarpa*) occurs in British Columbia, Washington, and Oregon and then 1500 miles away in Gaspé county, Quebec. Two species of *Empetrum* (crowberry, etc.), which have their geographical centre in the St Lawrence region, are very closely related to a species, *Empetrum rubrum*, on the Andes, 4000 miles distant, Juan Fernandez, Chile, and Tierra del Fuego. These and many other Gaspé plants are more directly allied to Cordilleran than to circumpolar species and, as we have seen, representatives of these occur several thousand miles to the south on the borders of Antarctica and on the South American Andes. It is reasonable to assume that plants which occur both in the southern hemisphere and in disconnected areas in the northern hemisphere are relics of an ancient stock: migration over the world is a slow process. Many different lines of argument strongly support the contention that in the group of plants on the Gaspé Peninsula

[1] Fernald (30).

and neighbouring districts we have survivors of a relatively ancient flora, which in its wealth of endemics, in the discontinuous distribution of its elements, and in the enormous range of some species differs from the younger flora which now forms a girdle round the polar sea. In this connexion one may draw attention to the occurrence in the typical circumpolar flora of plants which vie with some of the Gaspé species in the extent of their present territory. The brittle fern, *Cystopteris fragilis*, lives in Greenland and other Arctic lands, in temperate Europe and in the southern part of South America, in South Georgia (lat. 54°–55° S.), in tropical Africa, and on the Himalayas. Another example is afforded by a species of *Draba* (allied to the British whitlow-grass) recorded from the north coast of Greenland and almost identical with a form described from the Magellan Straits. Possibly these and other far-flung members of the circumpolar flora may, like the Gaspé plants, be relics of a group which was able to retain possession of favoured stations during the Ice Age.

If it is true, as Prof. Fernald believes—and his conclusions are based on a substantial foundation of fact—that the Gaspé plants are now living in places which they occupied in the glacial period, what is the explanation of their present isolation? In the first place, one would expect to find representatives of this flora in the broad territory between the Gulf of St Lawrence and the Pacific mountain ranges. How may we account for the lack of intermediate stations? After the final retreat of the North American ice-sheets, plants which had hung on to life might be expected, when the way was open, to extend their range and colonize new ground as did many of the circumpolar plants. The failure to do this, Prof. Fernald thinks, may be attributed to lack of vigour in an ancient, senile group. Be this as it may; the assumption of decadence is in harmony with an opinion expressed in a previous chapter in reference to plants of ancient lineage being inferior in resisting power to more recently evolved and more vigorous races. It should, however, be remembered that the absence of certain plants may be due in part to the slow rate at which species are able to spread under natural conditions. The second point is the explanation of the presence of the Gaspé species in the Arctic Archipelago and on the Cordillera. It is generally agreed that the confluent glaciers and ice-sheets swept across the North American continent from west to east; the Cordilleran ice-sheet was the first to attain full development; as it gradually retreated other ice-sheets farther east slowly spread over the Hudson Bay region, and at a still later date the Labrador sheet reached its maximum. During the whole of this time a flora persisted in the Arctic Archipelago in places which were left comparatively ice-free: as the Cordilleran ice-sheet began to wane, plants migrated along the open road on the hill-ranges and travelled eastward as far as Newfoundland and the Gaspé Peninsula before the occupation of the country by the central and eastern ice-sheets. As the middle region was invaded by the Keewatin ice-sheets (so called from the Indian word ki-wé-tin, which means north or north wind) the wanderers from the west and north were overwhelmed during the complete glaciation of the area. On the other hand the more fortunate species, which had established themselves on the Gaspé highlands before the Labrador ice-sheet had reached an advanced stage of development, continued to occupy unglaciated stations and still remain as impressive relics of a pre-glacial vegetation. There are many other points of no less interest raised by Prof. Fernald, but enough has been said in illustration of his main thesis—that in the Gaspé Peninsula and in other parts of the eastern area there exists a group of plants which furnishes strong evidence of the ability to survive in unglaciated districts the effect of the Quaternary Ice Age.

Attention has been confined to the records of the earlier part of the Quaternary period furnished by the boulder clays, gravels and sands of the northern hemisphere: there are, however, clear proofs in the Andes, the Himalayas, also in Australia and New Zealand of a greater development of glaciers than at present; but the extension of the ice was relatively small. There was no true glacial period in the southern hemisphere.

B. The Post-Glacial Stage

Passing to the post-glacial stage, the most instructive records have been obtained from peat. Over a large area in northern and central Europe and farther east beds of peat, often reaching a depth of 10 ft. or more, cover wide stretches of moorland and occur in valleys and depressions where lack of free drainage has favoured the growth of marshy vegetation through many generations. The most convenient method of presenting the main facts relevant to the history of plant-life during the relatively short interval between the last retreat of the ice and the present day is to give a general account of results obtained in a few of the more thoroughly investigated localities. By the employment of Prof. de Geer's method,[1] of estimating the rate of recession of the ice by counting the coarse and fine layers of sediment deposited in summer and winter respectively by the water issuing from the ends of glaciers, it has been possible to state with some confidence that the ice finally left Sweden about 6500 B.C. The length of the post-glacial epoch, from the last year of the Ice Age to the present time, may be reckoned at about 8700 years. Europe has not yet recovered from the effects of the reign of cold: it may be that our views on climates of the past are influenced and perhaps in part vitiated by thinking of the present climate as normal. Short, in a geological sense, as the post-glacial stage of the Quaternary period was, it was long enough for the erosion of the Niagara gorge; long enough to include oscillations in the relative position of land and sea in the north of Europe through several hundreds of feet. We have already referred to the sequence of events subsequent to the passing of the ice-sheets: the next step is to trace the development of vegetation by the employment of data derived from various sources.

Plant-records from Peat

Peat is one of the best sources of information on the succession of trees during the time subsequent to the passing of the ice. Fig. 130 shows part of a pollen-diagram compiled by Dr Erdtman[2] from data obtained by a microscopical examination and a numerical analysis of samples of peat taken by means of a boring apparatus at Chat Moss, Lancashire. The numbers on the left give the depth in metres from which samples were taken; those on the right are the successive sample numbers: the frequency of the pollen found at different levels is expressed as percentages of the total tree pollen along the horizontal scale, from 0 to 100 per cent. or over. A frequency of more than 100 per cent. means that the pollen of *Corylus* (not counted as a forest tree) is more abundant than that of all the forest trees put together. The diagram (fig. 130) gives the proportion of trees represented in the peat, from a depth of 7·5 metres to a short distance above a depth of 5 metres. In the lower layers, which rest on glacial beds, the birch and *Corylus* are dominant; next in order the pine reaches a maximum and later the alder

[1] Geer, G. de (12); Fairchild (20).
[2] Erdtman (28). See also Erdtman (24); Woodhead (28), (29).

appears and gradually increases in abundance. References are given in footnote 1 (p. 506) to papers in which bibliographies will be found of the numerous contributions to post-glacial botany based mainly on pollen-statistics. It must be remembered that conclusions are based on statistics furnished by such pollen as is preserved; the apparent absence of some trees is due to the inability of the pollen to resist decay.

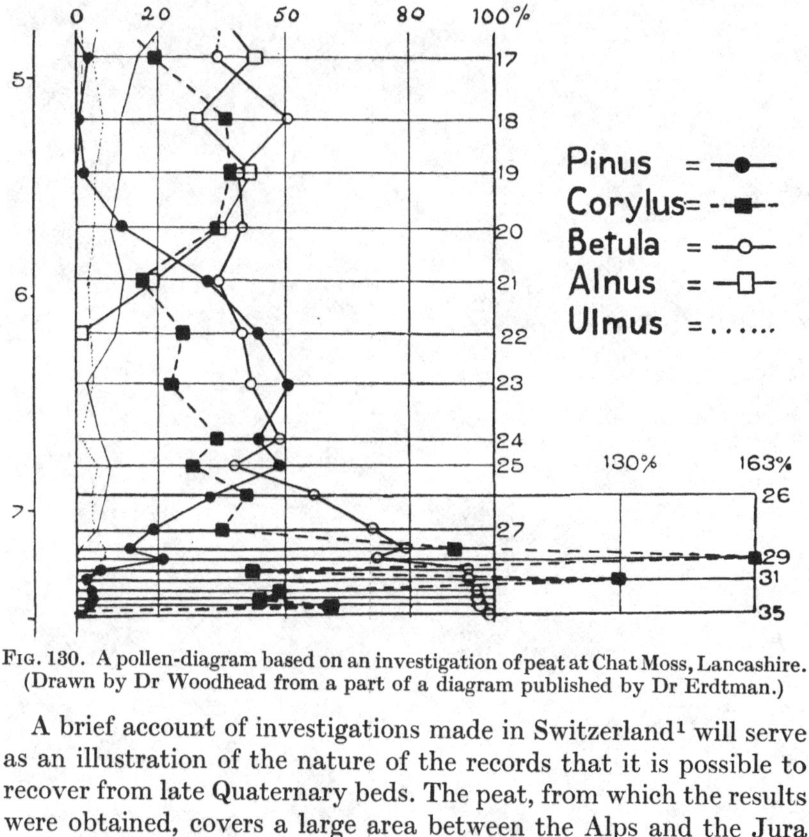

FIG. 130. A pollen-diagram based on an investigation of peat at Chat Moss, Lancashire. (Drawn by Dr Woodhead from a part of a diagram published by Dr Erdtman.)

A brief account of investigations made in Switzerland[1] will serve as an illustration of the nature of the records that it is possible to recover from late Quaternary beds. The peat, from which the results were obtained, covers a large area between the Alps and the Jura Mountains at a height of about 500 metres above sea-level. The peat began to be formed soon after the end of the Ice Age. At the lower limit of the peat evidence was discovered of the existence of *Dryas octopetala* (fig. 131), the dwarf birch (*Betula nana*), species of *Salix*, *Loiseleuria procumbens* (the trailing azalea, a circumpolar species),

[1] Keller (28).

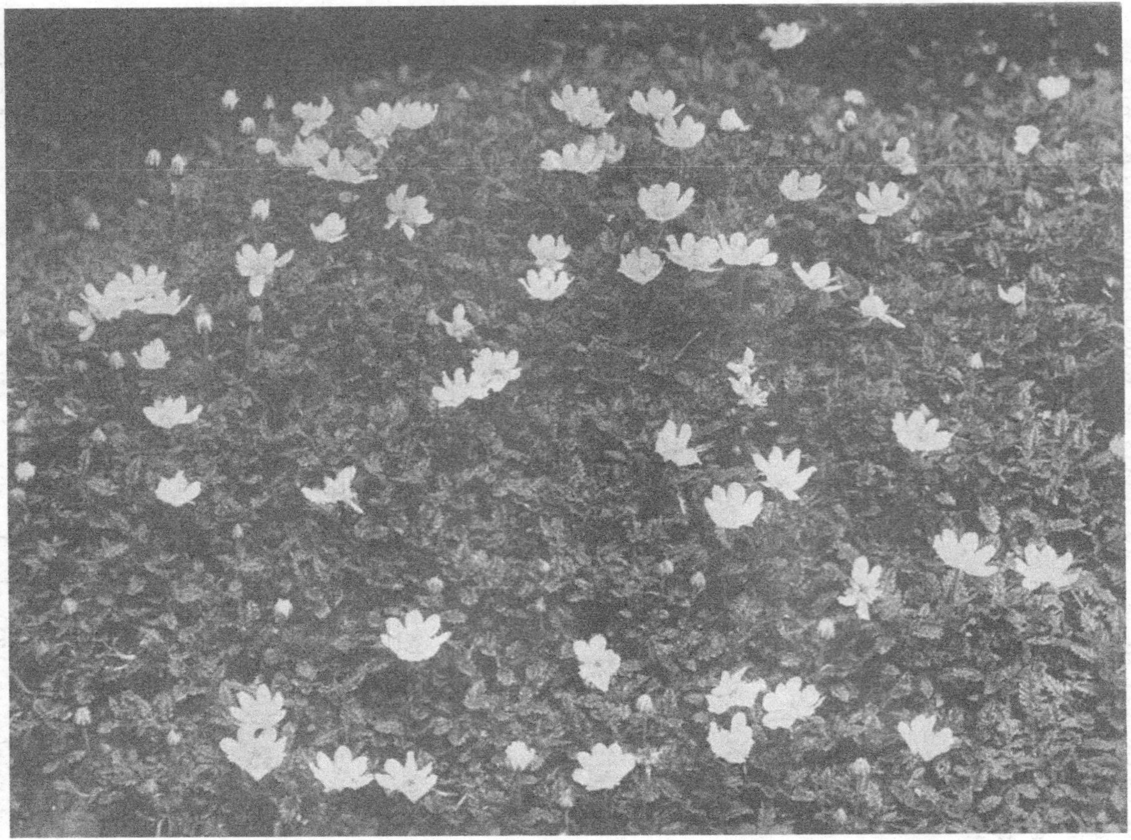

Fig. 131. ***Dryas octopetala***, **Ben Laoigh, Scotland, 2000 ft. above sea-level. (Photograph by Mr R. M. Adam.)**

Polygonum viviparum (another circumpolar plant which occurs also in the Alps and in many parts of the temperate zone) and other flowering plants. This lowest horizon is spoken of as the Dryas stage: it is characterized not only by the occurrence of many plants which are now Arctic but by remains of the banded lemming and other animals which live in the tundra or steppe regions of Russia and Siberia. Closely associated with the *Dryas* beds and immediately above them were found traces of several aquatic and marsh plants, *Potamogeton*, *Typha*, *Myriophyllum*, *Sphagnum* (bog moss), and other genera some of which, though usually regarded as temperate in range, are common in Greenland tarns in close proximity to localities occupied by *Dryas* and its associates. These two sets of plants enable us to visualize a treeless tract such as is seen to-day in Greenland and in other Arctic lands, a vegetation with a few prostrate willows profusely sprinkled with woolly catkins, clumps of white-flowered *Dryas* and many other flowering plants, with occasional lakes or tarns in marshy hollows tenanted by *Potamogeton* and other aquatics able to endure an Arctic climate tempered by the concentrated rays of the summer sun. At a higher level in the peat pollen of birch trees becomes predominant; the dwarf shrubs gave place to taller species as the climate improved. The birch marks a climax of forest development. The birch was followed by the pine; and it is interesting to notice frequent records of pollen of the scotch fir (*Pinus silvestris*) from peat beds in England and parts of France where this familiar tree no longer exists in a wild state. The pines finally ousted the birch and then came the hazel (*Corylus*), first establishing itself in the undergrowth of the pine woods, later monopolizing the ground. *Corylus* marks the maximum phase of the relatively warm continental climate; its present northern limit is approximately defined by an east to west line along latitude 60° N., though it extends farther north in Norway. The next stage which began in the hazel period, was the spread of mixed forests of oak, elm, lime with some examples of *Abies*[1] (silver fir) and *Picea* (spruce fir). Palaeolithic man[2] was a witness of the series of events up to the beginning of the mixed forests and was then succeeded by the earliest representatives of the Neolithic race. At a still higher level the peat-analyses enable us to follow the spread of beech (*Fagus*) woods at the expense of the mixed oak

[1] For distribution of *Abies* see Mattfeld (26).
[2] For plants of the Palaeolithic stage see Bonč-Osmolovskij (29).

forests. The dominance of the beech marks the transition to the present era.

The next illustration is taken from descriptions of excavations and pollen-analyses in the Pennine Hills of Yorkshire by Dr Woodhead[1] and his co-workers in the Huddersfield district. It is pointed out that not only are the most complete records of the post-glacial plant-world found in peat, but the "area of our country covered by moorland species is much greater than that of any other natural plant community; and as these seem to be the oldest members of

FIG. 132. Part of a Roman road at Blackstone Edge, on the Yorkshire moors, laid on peat. (Photograph by Dr Woodhead, from the *Journal of Ecology*, 1929.)

our flora, they call for special consideration". A section of the peat shows the following succession: (1) at the base sand resting on the Millstone grit (Carboniferous) and in it remains of old flint workings, tools used by late Palaeolithic man with charcoal of birch (*Betula*) and in the peaty débris of the floor pollen grains of birch, oak, alder (*Alnus*), hornbeam (*Carpinus*),[2] and other plants; (2) a layer of trees in the lowest bed of the peat containing birch and oak and

[1] Woodhead (29) with references to literature. For a more general account of post-glacial forests see Woodhead (28). For a full list of papers on pollen-statistics, etc. see Erdtman (24^2), (27), (30); Gams (27); Overbeck (28); Dokturovsky (29) and Gerasimov (30).

[2] For the present distribution of *Carpinus* see Christie (24).

History of the Vegetation of the Southern Pennines (Yorkshire)

Post Glacial	1000 A.D.				Cotton grass (*Eriophorum*) on the summit plateau, bracken on the higher treeless slopes. Open oak-birch forest, etc. Oak woods on moist soils; alder-willow swamps on the river-plains
		Moist and cold climate	Historical and Iron Age	Romano-British pottery; Roman road on Blackstone Edge (fig. 132)	Cotton grass dominant on wet, summit plateau. Grass-heath and degenerate birch-oak forest on the higher slopes
	1000 B.C.	Warm and dry			Cotton grass association invaded by heath plants
	2000 B.C.		Bronze Age	Bronze Age arrowheads	
	3000 B.C.				
	4000 B.C.	Warm and moist	Neolithic Age	Horn cases of *Bos primigenius*; Neolithic tools; arrowheads	Formation of cotton-grass peat and degeneration of forest on summit plateau
	5000 B.C.		Epipalaeolithic age (an intermediate period)	Late Tardenois tools. The name Tardenois (from La Fère-en-Tardenois, 23 miles west of Reims) is given to a special industry which produced very small flint implements: it is dated by some authorities late Palaeolithic, by others early Neolithic	Heath ground-flora invaded by cotton grass. Disappearance of pine
	6000 B.C.				Forest climax; woodland, and alder-willow swamp on river plains
		Warm and dry		Early Tardenois tools	Birch-heath forest with oak, alder, hazel
Late Glacial	8000 B.C.	Sub-Arctic	Palaeolithic	Aurignac tools (an upper Palaeolithic culture, from a cave 40 miles south-west of Toulouse)	Birch-heath forest with pine
	9000 B.C.	Arctic			Tundra with birch-heath scrub; marsh peat, with Arctic plants, forming in lowlands
Glacial	10,000 B.C.				Tundra with many common moorland species, relics of interglacial period, on nunataks and unglaciated areas
	11,000 B.C.	Glacial			

some Neolithic implements; (3) peat above the tree layer with horn cases of *Bos primigenius* (a Quaternary ox still preserved in a semi-wild state in Chillingham Park) from which were scraped peaty fragments yielding 146 pollen-grains of several trees; (4) Bronze Age arrow-heads; (5) Romano-British pottery. The oldest part of the peat was probably formed soon after the last retreat of the ice, but except for a few tools nothing is known of the interval between the end of the glacial period and the age of the latest representatives of Palaeolithic man. Plants of the tundra flora which immediately followed the receding glaciers probably continued to occupy some of the ice-free hills in this district during the Ice Age. Late Palaeolithic man was contemporary with the forests of birch, hazel, oak, pine and elm: as the climate became wetter and encouraged the development of peat the pine rapidly disappeared with the oaks and elms, and eventually the more persistent birches, alders and hazel succumbed. Before the Bronze Age the Pennine moorland had put on its present covering. It was formerly thought the paved Roman road (fig. 132) which stretches across the moors had been constructed before the growth of the peat, but it has now been proved that the Romans used the peat as a foundation. The destruction of the forests cannot be laid wholly to the charge of Roman soldiers; their disappearance was mainly the result of natural causes. The table shown on p. 507, based on one given by Dr Woodhead, serves as a summary of the records of post-glacial plant-life furnished by the Pennine peat.

Interglacial Periods

Reference was made on a previous page to interglacial periods: the number of such warm interludes in the long course of the Ice Age has no special significance to the historian concerned with the passing pageant of vegetation in its broader features. It is certain that the Ice Age was not an unrelieved age of desolation; not only were mountain-peaks and ridges of hills left as refuges on which a comparatively small though virile company of plants maintained a precarious foothold; there were many climatic fluctuations registered in the backward and forward movements of the glaciers and ice-sheets. There were also periods of comparatively long duration, in some regions one or two, in others three or four, when the rise in temperature rendered previously glaciated areas fit for the growth of a temperate vegetation: these are the interglacial periods. As an example of the sequence of events during an interglacial period a brief account is given of records furnished by a series of beds in Jutland and described by the Danish authors, Dr Jessen and Dr Milthers.[1] These records were obtained from the deposits of an interglacial lake, the site of which is a basin-shaped depression. A surface layer of stony earth rests on some fine sand, probably wind-blown; below this a bed containing remains of *Trapa*, *Acer* (maple), alder, birch, hornbeam, *Corylus*, pine, oak, lime, yew, willows, water-lilies and other plants. This prolific bed rests on Arctic clay with the dwarf birch and Arctic willows in its deeper layers; lower still is a glacial clay. From the data collected including analyses of the pollen a table was constructed which is here given in simplified form.

The History of Vegetation during the last Interglacial Period in Jutland and north-west Germany

(The oldest beds are at the foot of the table)

Clay with sub-Arctic flora	*Betula nana* (dwarf birch) heaths; poor aquatic flora	The last Scandinavian ice-sheet advancing
Peat with temperate flora	*Betula pubescens* (common birch), *Pinus silvestris*, *Picea excelsa* (spruce fir), *Betula nana*, etc.	Retreat of the ice-front
	Maximum development of deciduous trees, also records of *Brasenia* (waterlily family), *Trapa*, etc.	Temperate climate in Jutland
Clay and sand with a sub-Arctic flora	*Betula nana* heath; sub-Arctic swamp, *Picea excelsa*, juniper	Advance of the ice in Scandinavia: sub-Arctic conditions in Jutland
	Pinus silvestris, *Picea excelsa*, *Betula pubescens*, *Populus tremula* (aspen), *Betula nana*	Swampy forests and high moors
	Picea dominant; *Pinus silvestris* and *Betula* more abundant	Rising land; lowering of temperature
	Maximum development of the hornbeam (*Carpinus*): continuation of mixed oak forests	
Mud and peat with temperate flora	Mixed oak forests; *Alnus*, *Corylus*, *Quercus*, *Tilia*, *Carpinus*, etc.	Temperature relatively high
	Pinus silvestris and *Ulmus*; maximum development of aquatic vegetation. *Pinus silvestris* and *Betula pubescens* dominant	Cooler at first, becoming gradually milder
Clay with sub-Arctic and Arctic floras	*Betula nana*, Arctic willows, *Dryas*, Arctic mosses	Melting of the ice-sheet

[1] Jessen and Milthers (28).

The period of Submerged Forests, and later

Another type of evidence bearing on post-glacial conditions is presented by the submerged forests[1] which are a well-known feature on many parts of the English coast, also on the coasts of Wales and France. Beds of peat broken into patches by the scour of the tides containing occasional stumps of trees, oak, yew (*Taxus*), birch, pine, and others which lie well below low-tide level demonstrate a sinking of the land through sixty or seventy feet: their date is fixed by the discovery of many remains of Neolithic man. A good account of the submerged forests, one of which on the Cheshire coast[2] is seen in fig. 133, will be found in Mr Clement Reid's book cited in the footnote. Some interesting exposures of buried forests have been made in the course of digging the foundations of docks; a forest on the site of Tilbury Dock was found to lie below a bed containing Roman remains and is no doubt Neolithic in age.

Before the close of the Neolithic Age the vegetation of Britain had acquired its present composition: some of the commonest forest trees such as the scotch fir (*Pinus silvestris*) and species of herbaceous plants from Neolithic localities reveal changes in geographical distribution in the interval between prehistoric man and his modern descendants. Seeds and other remains obtained from Roman sites[3] are specifically identical with plants that are still native in Britain; and other remains obtained from refuse heaps belong to cultivated plants such as we should now find in similar circumstances.

The history of the plant-world in the latter part of the Quaternary period is a chronicle not only of plant-migrations and fluctuating climate, but also of human agency as a factor in plant-dispersal. Through the mists of the Neolithic Age we see the dawn of civilization, the cultivation of cereals and other food plants,[4] the domestication of animals, the manufacture of pottery, and the practice of a more refined technique in the fashioning of implements.[5] Men of this race roamed the forests which reached beyond the present limit of the English coast and were able to wander across the swamps

[1] Reid, C. (18). For an account of the Scottish peat-moors see Lewis (11) and Erdtman (28).

[2] Erdtman (28).

[3] Reid, C. (01), (02).

[4] For the history of agriculture see Werth (29). For information on the origin of the present flora of Britain see Stapf (14); Stomps (23); Matthews (23); Yapp (23).

[5] For a discussion on the subdivisions of the Quaternary period in relation to pre-history see Osborn and Reeds (22).

Fig. 133. Submerged forest seen at low water at Leasowe, on the Cheshire coast, England. (After Seward, *Fossil Plants*, vol. I, Camb. Univ. Press.)

connecting the southern part of eastern England with the Continent (fig. 111). Dr Rice Holmes[1] in his book on "Ancient Britain" suggests that one should "ascend the hill on which stands Dover Castle, and gaze upon Cape Grisnez. Let the waters beneath you disappear: across the chalk that once spanned the Channel like a bridge men walked from the white cliff that marks the horizon to where you stand. No arithmetical chronology can spur the imagination to flights like these". The rough-hewn massive stones of the Dolmens, good examples of which may be seen not far from Groningen in Holland, are themselves relics of a still earlier age when the Scandinavian ice scattered rocks from the Norwegian highlands over the north European plain and so provided material with which Neolithic man and probably his successors in the Bronze Age raised massive monuments to the dead. Implements and bones scattered in river-gravels or buried in moorland peat and the refuse littered in caves, are among the miscellaneous scraps of evidence which it is the business of the archaeologist to interpret. The geologist hands over the documents to the antiquarian and the historian, but retains as an essential part of his domain the study of the play of forces on the present surface of the earth. "The very finger-touches of the last geological change" no less than the catastrophic revolutions of earlier periods come within his purview: from a knowledge of the present he gains experience and inspiration which render him better fitted to decipher the story of the past.

It is often impossible to determine whether or not man has been responsible intentionally or by accident for the occurrence of plants in unexpected places; his share in contributing to the composition of our flora, though undoubtedly very small, is not entirely negligible.

The inferiority of the British flora in point of number of species to that of continental Europe, mainly due to the effect of the Ice Age, is in part a consequence of isolation: from time to time in the later stages of geological history, even in the Quaternary period, Britain was united to France, the Netherlands and Germany, but the boggy morass which stretched across the North Sea was by no means an ideal route for most kinds of plants (fig. 111). The western outlier of Europe we may be sure retained some of its pre-glacial vegetation, though the greater part was expelled or destroyed: the re-stocking of the plant-population with its Germanic, Atlantic,

[1] Holmes, T. R. (07).

Mediterranean and other elements is a subject beyond the scope of this history. Having followed the changing pattern of the carpet of vegetation through the last stage of geological history we pass from the fossils preserved in sands, clays and peat to the vegetation of the present age. The palaeobotanist endeavours to interpret the records of the past in the hope that they may be helpful to the student of existing floras: the living plant takes the place of the dead; as, in human history, the tablets of Assyria are superseded by the actors which now occupy the stage.

FIG. 134. Oak-leaf and acorns (*Quercus pedunculata*) in a Spandrel of a Stall in the Chapter House of Southwell Minster (*c.* 1290), England.

CHAPTER XVIII

CONCLUSION

It's a poor sort of memory that only works backwards. *The White Queen*

As we follow the procession of floras through the ages certain questions keep intruding themselves: what do the records of the rocks tell us of evolution, of the development of the plant-world; what is the significance of the sharply contrasted geographical range of groups, families, and genera shown by a comparison of extinct and living members of the vegetable kingdom; what light is thrown by the wanderings of plants over the earth's surface on the climates of the several geological periods? As we try to piece together the scraps of history disinterred from the refuse heaps in nature's workshop, accumulated in the course of hundreds of millions of years, what is it that makes the strongest appeal to our imagination?

Passing from age to age we gain an insight into the development of the modern world; we see land and water changing places; upheavals of the crust into alpine chains; the cumulative effect of the ceaseless action of denuding agents reducing mountains to hills, and hills to diminishing irregularities in a vast plain of erosion. We follow the operation of forces which are still imperceptibly but surely, and at times with startling manifestations of their stupendous power, altering the features of the earth's face. Man sees the world for a moment and, as it seems to his limited vision, in a relatively static state. As he searches through the archives of the rocks he realizes that the world and its teeming population of animals and plants have been ever changing; that the present insignificant changes of which he is witness are but the last finger touches of a hand that has been fashioning the earth's crust through countless ages.

Vast as seems to us the duration of the world it is as nothing in comparison with the age of the universe. The human period is but a fraction of geological time; and yet there is reason to believe that within this brief space the deep and tortuous gorge of the Zambesi below the Victoria Falls has been carved out of a basaltic plateau. Discoveries made by Sir Aurel Stein in innermost Asia furnish one among many proofs of the rapidity with which substantial changes have occurred within comparatively recent years: sites that were thickly populated are now arid desert. The vagaries of climate registered in the inequality of the rings in the stem of a *Sequoia*,[1] which had lived through three thousand years or more, afford an additional illustration of the importance of adjusting our standard and extending our purview beyond the confines of the immediate present.

The History of the Plant-World

It is clearly impossible in the concluding section of a general account of the floras of the past to discuss the thorny problem of evolution;[2] it must suffice to draw attention to a few of the more obvious trends of thought suggested by a consideration of the recorded facts.

Let us think of the vegetation which is spread over a large part of the earth's crust as a carpet which, though constructed throughout of the same material—the cells and tissues of plants—displays many different designs and colour schemes as we follow it from one region to another. On this rough analogy we may think of the prevailing colours as representing the larger groups of plants; the areas occupied by the various colours indicate the relative prominence of each group in the vegetation as a whole; individual genera or groups of closely allied plants being the threads of the fabric. Continuing the simile, we imagine spread out before us as we pass from one age to another so many samples of nature's workmanship; the changing pattern and the kaleidoscopic sorting of colours enable us to recognize some of the more striking transformations which have occurred since the first design was impressed upon the fabric. Through the mists, which become more opaque as we carry our retrospect to its utmost limits, we seem to see a land that is still unclothed; we see through the lens of imagination plants too small for definition under the most powerful microscope floating on the waters of a primeval sea—and yet endowed with miraculous potentialities—the germs of the plant-kingdom. The latter part of the Pre-Cambrian era is often spoken of as an Age of Algae, a title based on unsubstantial evidence which, as we give rein to our fancy,

[1] Douglass (19); Huntington (14).

[2] The latest contribution to this subject is a book by Prof. W. Zimmermann (30) of Tübingen which was received after practically the whole of the present volume had been written.

assumes a disproportionate importance. Such facts as there are lend support to the view that the oldest carpet of vegetation lay in patches on the edges of continents and islands having a design primitive in its simplicity and woven in threads that were formed in the sea.

It is not until we reach the beginning of the Devonian period that the solid ground is chequered with patches of green for the most part on swampy areas, the individual plants differing one from another in certain external and internal features, though their resemblances suggest a common line of ancestors stretching beyond our ken. These dissimilar and yet as we believe related land-plants are grouped together as members of an extinct class, the Psilophytales; and it is they that form the most conspicuous patterns in the oldest sample of carpet. The Psilophytales flourished for a comparatively short time and seem to have reached their culminating point before the latest phase of the Devonian period. What was their progeny? It may be that in the Psilophytalean epoch were laid the foundations of the great group of Lycopodiales, and that in the genus *Asteroxylon* we have a connecting link between some of the leafless Psilophytales, such as ***Rhynia*** and ***Hornea***, and the later Devonian *Protolepidodendron* with its near relatives *Lepidodendron* and *Sigillaria* which were among the giants of the Carboniferous forests. There are some grounds for believing that the extinct Palaeozoic genus *Cladoxylon* should be regarded as another offshoot of the Psilophytalean stock. It is conceivable that in *Psilophyton* and genera of similar habit, characterized by leafless or spinous branches, we have the pioneers of plants with large leaves such as *Eospermatopteris*, *Aneurophyton* and others (see fig. 45). The spreading fronds of *Aneurophyton* may be the derivatives of the much simpler, lateral branches of a plant like *Psilophyton*, branches which in the earliest stage of their evolution might be regarded as the bare ribs or scaffolding of a future frond, and by degrees developed small appendages which later assumed the form of flat leaflets. In this connexion it is noteworthy that in *Aneurophyton* there is little difference in structure between the branched lateral members, sparsely set with small leaflets, and the main axis: this may be an early stage in the development of a partially transformed branch-system into the well-defined fronds of the late Devonian and the Carboniferous pteridosperms. In the latter part of the Devonian period *Archaeopteris* and similar plants (fig. 48) bearing large fern-like fronds were conspicuous features of the undergrowth, and with them were lycopodialean trees rivalling in girth and height the Lepidodendra and Sigillarias of the Coal Age.

Another line of evolution is illustrated by *Callixylon*, a tree with a thick cylinder of wood (fig. 64, Q), its minute structure differing only in details from that of existing araucarian conifers. We know nothing of the ancestry of this and other Devonian trees of similar habit. The probability is that our knowledge of the plants of the Mid-Devonian period is based mainly on the remains of a fenland flora. There were no doubt larger and differently constructed plants on higher ground, and of such *Callixylon* and a few others of the more robust types may be examples. The Mid-Devonian genera *Bröggeria*, *Hyenia* and *Calamophyton* foreshadow the great class of articulate (jointed) plants of which *Calamites* and *Sphenophyllum* became the typical representatives in the succeeding period. Similarly *Psygmophyllum*, which was a member of one of the earlier Devonian floras and became more abundant in later Devonian and in Carboniferous floras, is one of the oldest known plants with large, simple leaves: we know only the leaves and nothing of the structure or relationships of the genus. The carpets of the first and middle stages of the Devonian period were fashioned on similar lines, the chief differences being an advance in the complexity of the design and the introduction of some new colours. On the other hand, during the latest phase of the period the archaic features almost disappeared. It is at this stage—the beginning of the Upper Devonian period—that the change in the vegetation was sufficiently definite and far-reaching to be spoken of as a transformation.

The Carboniferous period was distinguished by exceptionally vigorous development in the plant-world as measured by the evolution of new forms and the production of a large number of new genera and species. In the late Carboniferous and early Permian age we see Palaeozoic vegetation at its zenith: this may be called the age of pteridosperms and giant pteridophytes. It is here for the first time that we find more clearly foreshadowed several types of existing plants. Whether or not the Devonian and Carboniferous *Psygmophyllum* is related to the living *Ginkgo* we cannot tell, but there can be little doubt that in the genus *Saportaea* (fig. 67) we have a late Palaeozoic representative of the group Ginkgoales. So far no plants have been mentioned which could be included with existing pines, firs and other cone-bearing trees in the Coniferales.

Before the end of the Carboniferous period there were trees such as *Walchia*, *Dicranophyllum* and others, and in greater abundance the genus *Cordaites*, all of which exhibit signs of affinity, more or less close, to living conifers. *Cordaites* is placed in a group of its own—the Cordaitales—because, despite the close resemblance of its wood to that of *Araucaria*, the reproductive apparatus differs in important respects from that characteristic of living conifers. It is probable that *Cordaites* and its allies were evolved from some ancient stock which had direct connexion along another branch with the conifers. It was at the last stage of the Palaeozoic era—at the end of the first half of the Permian period—that there occurred a second transformation: the large patches of colour given by the luxuriance and variety of pteridosperms, the arborescent Lycopodiales and Calamites, the Cordaitales and other groups became not merely faded but reduced to inconspicuous vestiges or completely obliterated. The carpet was almost threadbare; spots of the older colours remained, but the design had been greatly simplified and many of the threads snapped. What was the cause of this drastic change?

Through the whole course of geological history, as we have endeavoured to show, cyclic disturbances in the earth's crust were powerful factors in directing the course of evolution in the organic world: it was the Hercynian and Appalachian revolutions which, by completely changing the physical conditions, gave a new impulse and a new direction to evolution. Reference has already been made to the complete or almost complete disappearance of many of the more prominent families during this critical stage in the earth's history, and attention was called to the difficulty of discovering connecting links in the chains joining the Palaeozoic to the Mesozoic world.

Passing through the relatively barren period, of which there is abundant evidence in the later Permian and the older Triassic rocks in the northern hemisphere, to the latter part of the Triassic period and to the Rhaetic stage, we see an almost new set of colours and patterns which are much more conspicuous than the old. Ferns, Cycadophyta, and the Ginkgoales take the place of pteridosperms, arborescent Lycopodiales and Equisetales: the few Palaeozoic fronds of cycadean habit probably belong to some of the earlier members of the cycadophytan alliance, the group which bulked so largely in Mesozoic floras. It is at least clear that a new directive influence was operative; the plant communities were radically changed and yet the persistence of not a few of the Palaeozoic groups has been well established. The pteridosperms, though greatly reduced in number, were by no means extinguished: one of them—*Thinnfeldia*[1] —was a common plant over a large part of the world in the earlier stages of the Mesozoic era. The ferns of the two eras are united by a few persistent types; but it is not until the Mesozoic era had well begun that the modern plan of construction became stereotyped. So far as we know ferns first appeared in the latter part of the Devonian period: they seem to have been comparatively late products of evolution, at least in the form which is now regarded as characteristic of the group. It was in the earlier part of the Mesozoic era that the Coniferales began to assert themselves as prominent constituents of the world's vegetation; and this is true also of the Cycadophyta and Ginkgoales. The Calamites and Sphenophyllums disappeared: the latter seem to have become extinct by the end of the Permian period. The Equisetalean group was henceforward represented by plants of less robust habit and of a type strikingly similar to the modern *Equisetum*. From the Carboniferous period to the present day some threads and bundles of threads can be traced without interruption: side by side with the Lepidodendra and Sigillarias in the forests of the Coal Age lived small herbaceous lycopods hardly distinguishable from those of our own time. Another example of persistence of type from the latter part of the Carboniferous period to the present day is afforded by members of the Hepaticae (liverworts) and some other groups.

A review of the Jurassic and Rhaetic vegetation reveals no evidence of any widespread change: in both periods the same colours provide the main features of the design; some occupy a subordinate position in the general scheme, others disappear and new patterns begin to take shape. The pteridosperms dwindled to a small company: the Pteridophyta, Cycadophyta and Coniferales were the dominant groups. The Upper Triassic, Rhaetic and Jurassic periods may be spoken of as an age of gymnosperms and ferns. Between the later Jurassic vegetation and the earlier vegetation of the Cretaceous period there is no very strongly marked distinction, minor rather than major modifications were made in the general design of the plant-cover.

[1] Evidence in support of the pteridosperm nature of *Thinnfeldia*, which was suggested several years ago (Seward (10)), and the allied genus *Lepidopteris* has been furnished by Dr T. M. Harris and Dr Hamshaw Thomas; the facts will be published later.

The Evolution of the Flowering Plants: an unsolved problem

At this point another transformation occurred, again with a suddenness that is not merely apparent because of the imperfection of our knowledge, but actual: the colour scheme was substantially altered by the introduction of a new decorative feature which rapidly became aggressively distinct and remains the most conspicuous character over the greater part of the world. This transformation is the result of the rise to pre-eminence of the flowering plants which appeared as a new creation. But, one may ask, are there no records of earlier members of the class? In the description of the Rhaetic-Jurassic vegetation reference was made to the Caytoniales, a small group of plants of which two genera have been described, *Caytonia* and *Gristhorpia,* characterized by enclosed seeds in contrast to the naked seeds of all other early Mesozoic and Palaeozoic seed-bearing plants. The seeds of the Caytoniales are borne in a closed case (fig. 101) corresponding in its protective function, its adaptation for catching the pollen and in its general relation to the seeds, with the carpel or fruit of the angiosperms. If we regard the occurrence of seeds in a closed vessel as the distinguishing attribute of the angiosperms, then it is reasonable to consider *Caytonia* and *Gristhorpia* as members of that class. Several of these seed-cases or fruits are borne on a single branched organ which is believed to be equivalent to a leaf, though in form unlike a foliage-leaf: it slightly resembles a fertile leaf (megasporophyll) of a *Cycas*, though in that genus the seeds are naked. In the flowering plants the carpel which contains the ovules (seeds) is equivalent to a leaf, whereas the seed-vessel in the Caytoniales corresponds not to the whole leaf but to a small part of it. Abnormalities occasionally seen on shoots of female trees of *Ginkgo biloba*[1] suggest early and unsuccessful efforts towards the enclosure of the naked seeds in a transformed leafy covering. Normally at the base of a *Ginkgo* seed, which is borne on a long stalk, is a rounded collar-like investment; but sometimes a seed occurs on the edge of a leaf which is partially folded over it like a loose cloak. The leaves known as *Sagenopteris*, and which are believed to be the foliage of *Caytonia* and *Gristhorpia*, are characterized by a simple network of veins and differ in this respect from the more elaborately veined leaves of living angiosperms. The point is that in one character, and that an important one, the

[1] Sakisaka (29).

Caytoniales are angiosperms: the seeds were enclosed in a receptacle, not left exposed on the surface of a scale; though in some features members of this extinct group so far discovered differ rather widely from any known flowering plants. Granting that we are not in a position definitely to assert that the extinct genera are undoubted angiosperms, there remains the fact that the female organs are much more angiospermous than gymnospermous. They may be representatives of a branch of some ancestral stock from which was evolved, possibly at a later date, the whole angiosperm alliance.

FIG. 135. A sacred relic of the past—trunk of a *Ginkgo* at Sendai, northern Japan, with stalactite-like air-roots, 'Titi', a feature characteristic of old trees. (Photograph supplied by Hofrat Prof. Hans Molisch.)

It is generally agreed that the early Cretaceous angiosperms, such as *Platanus*, *Magnolia* and other genera which seem to spring from the earth as Melchizedeks of the vegetable kingdom, with no apparent lineage, cannot be the oldest representatives of the class. We recognize Cretaceous leaves as those of dicotyledons by their practical identity with modern foliage; but whence did these plants come? If it is conceded that *Caytonia* and *Gristhorpia* may be admitted as archaic types into the hierarchy of flowering plants, they must be the descendants of more ancient ancestors.

A comparison of pteridosperms with angiosperms shows several features in common: the fronds of many pteridosperms, though on the whole fern-like, are constructed on a plan that is not uncommon in several large-leaved flowering plants. The general structural plan of some pteridosperm stems is not very different in its broader features from that in living angiosperms, both the normal type and the unusual form characteristic of certain tropical lianes. Moreover, when we compare the pollen-bearing organs of some pteridosperms, such as *Potoniea* (fig. 61), with the male flowers of a few existing angiosperms, such as *Populus*, certain resemblances become apparent. Similarly some pteridosperm seeds, such as *Lagenostoma* (the seed of *Lyginopteris*), and *Trigonocarpus* (borne on *Alethopteris* fronds, which are the foliage of a *Medullosa*) may be compared with the seeds and carpels of dicotyledons. The cup-like husk (cupule) of *Lagenostoma* is analogous to the protective wall of a carpel though it never, so far as we know, completely covered the seed. A resemblance in certain anatomical features between *Trigonocarpus* and the seeds of *Myrica Gale* was pointed out several years ago.[1] These resemblances, it is true, do not in themselves carry much weight as evidence of actual relationship: there are many and obvious differences between pteridosperms and such angiosperms as now exist. Allusion has already been made to a similarity in the structural plan of some pteridosperms and angiosperms: misled by the resemblance of a transverse section of a leaf-stalk of a pteridosperm (*Medullosa*), an American author described a Carboniferous fossil as a piece of a monocotyledonous stem. In the leaf-stalk of the extinct plant the scattered strands of conducting tissue naturally suggest comparison with the arrangement familiar in stems of Indian corn (*Zea Mays*) or in a sugar cane; but a closer comparison reveals differences in detail: the agreement is only apparent and cannot be regarded as indicative of relationship. It is unfortunate that this supposed Carboniferous angiosperm, described under the misleading name *Angiospermophyton americanum*,[2] has been quoted by more than one author as evidence of the existence of angiosperms in the forests of the Carboniferous period. In the latter part of the Palaeozoic era the pteridosperms occupied a place in the vegetation similar to that now filled by the angiosperms: is it possible that such resemblances as are revealed by a comparison of these two great classes may be more than mere coincidences or analogies in plants evolved

[1] Kershaw (09).
[2] Noé (23); Seward (23[2]).

along two independent lines? The evidence hardly justifies an affirmative answer; nor does it necessitate a direct negative. It is suggested that as our knowledge of the pteridosperms, both Palaeozoic and Mesozoic, increases, we may find that these two groups, which reached their maximum development in periods separated by many millions of years, may be on the same line of evolution; and that the high degree of specialization which is an astonishing feature of the pteridosperms foreshadowed the still higher type of plant-architecture characteristic of the angiosperms.[1]

By the middle of the Cretaceous period the plant-world had become almost completely modernized: few archaic types survived: the old order had passed and the new had become established.

Persistence, Progression, and Retrogression through the Ages

If we could see the floras of Europe and North America, which have come and gone since the beginning of the Devonian period, pass before us as a moving picture, we should notice groups of plants intruding themselves into prominence then rapidly fading into obscurity or vanishing; their place taken by new and more vigorous groups which in turn fell to a subordinate position as still newer types gained the ascendant. The plant-world seen in perspective through the long vistas of geological time may be graphically represented as a series of curves slowly rising to a peak and rapidly falling, each curve marking the rise to prominence and the subsequent decline of a class or group. There are only a few types which have remained relatively unaffected by the passage of time and changing circumstances, and the history of these can be represented by a straight line stretching from the Palaeozoic era to the present day (see fig. 136). Reference has been made to the Charophyta, Calcareous Algae, the Blue-Green Algae, the herbaceous lycopods, the Equisetum family, and to other groups of plants remarkable for their superior antiquity and conservatism. On the other hand, the Psilophytales held for a comparatively short time the position of a ruling dynasty, though it may be that in the living genera *Psilotum* and *Tmesipteris* we have two solitary units at one end of the long Psilophytalean chain, the intermediate links of which have not been discovered. The Carboniferous Lepidodendra with their towering

[1] The reader is referred to the following contributions to this difficult subject of the evolution and interrelationships of the angiosperms: Hutchinson (26); Bews (25); Campbell (29); Wieland (29²), (29³).

columnar stems and large cones, and some of them with reproductive organs entitled to be called seeds, afford an example of a degree of specialization and dimensional development far higher than are found in the distantly related members of the group which play a humble part in the present vegetation. From the Mesozoic age may be selected the Cycadophyta and Ginkgoales as instances of exceptionally successful lines of evolution. The extinct cycadophytes differ widely in the form and structure of the fertile shoots from the living cycads, so widely indeed as to make one hesitate to regard the extinct and the living as members of one group: the Ginkgoales, which trace their ancestry back to the latter part of the Permian period, rose to the full height of their vigour in the course of the Upper Triassic and the Rhaetic-Jurassic age. The solitary representative of the race, *Ginkgo biloba,* now survives through its well-merited veneration as a sacred relic. From the boughs of the venerable maidenhair tree at Sendai,[1] Japan, reproduced as fig. 135, there hang root-like branches which are credited with miraculous power: the superstition that still clings to this living fossil seems a fitting attribute to one of the oldest trees in the world.

In fig. 136 an attempt is made to convey a general idea of the geological history of some of the main divisions of the plant-kingdom. It is impossible from the available data to follow through the ages the various families of algae. The subdivisions selected are both represented by many fossil species which owe their preservation to the deposition of carbonate of lime in the living cell-walls. Algae assigned to the Dasycladaceae, a family of the Siphonales, are recorded from Ordovician and other Pre-Devonian rocks and species are still living in the warmer seas; members of the family were abundant in Triassic seas and again in the earlier stages of the Tertiary period. The height of the line in the diagram above the horizontal indicates the relative abundance of these algae at different periods. The upward slope of the second line in the first half of the Carboniferous period indicates the abundance of *Solenopora* and other genera in the calcareous marine beds of this age, and the continuation of the line through the Jurassic and later periods indicates the persistence of related genera such as *Lithothamnium* and others to the present day. In the Charophyta and Bryophyta we have examples of persistence from the Devonian and Carboniferous periods respectively to the present day.

[1] Molisch (27); Fujii (95).

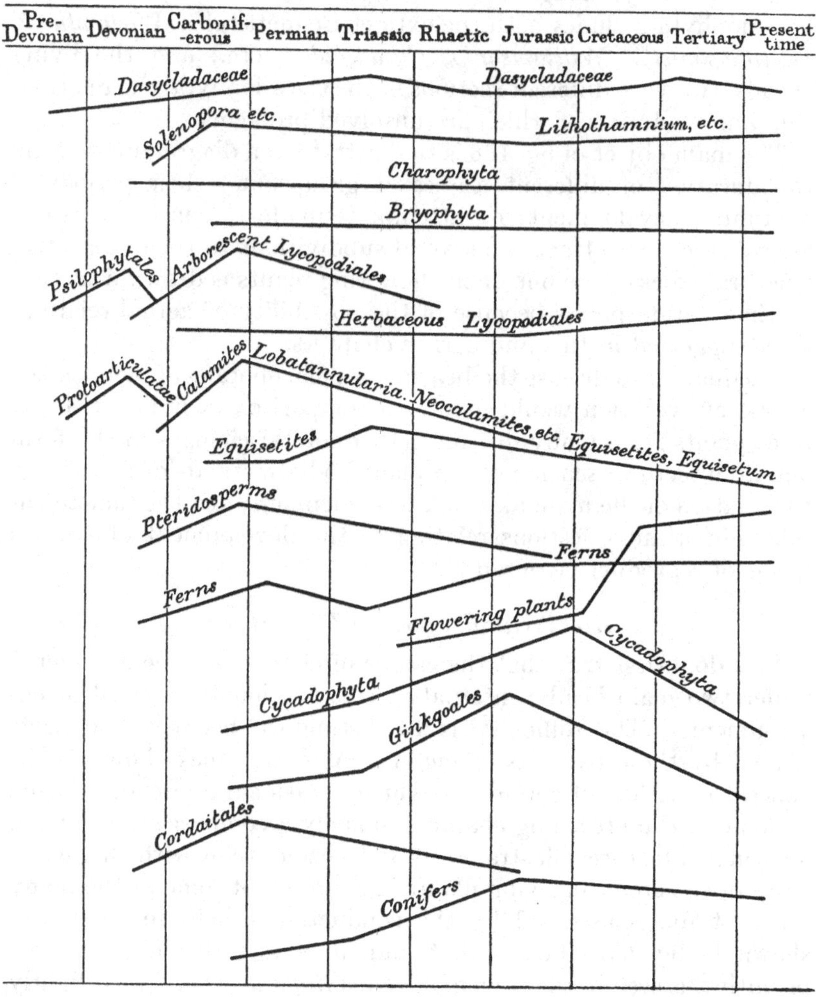

FIG. 136. Diagram summarizing impressions made by a review of evidence, gathered from the rocks, bearing on the past history of certain extinct and recent groups and families of plants.

The Psilophytales and arborescent Lycopodiales, e.g. *Cyclostigma*, *Lepidodendron* and *Sigillaria*, are two extinct groups, the latter of longer duration than the former. The other lines in the diagram are self-explanatory, though it should be pointed out that the line of the Cycadophyta includes both the extinct Bennettitales (*Wielandiella*, *Williamsoniella*, *Williamsonia*, *Cycadeoidea*, etc.) and the living cycads, two very different sections of the Cycadophyta, the relationship and evolution of which are unsolved problems.

The main object of fig. 136 is to illustrate in a diagrammatic form the duration of different classes or groups and their periods of maximum development: no attempt is made to construct genealogical trees or to trace the several subdivisions to their respective ancestral stocks. The line of the flowering plants is drawn near that of the pteridosperms because of the possibility of actual relationship, suggested in this and earlier chapters.

Adequately to discuss the bearing of palaeobotanical facts on the course of evolution would involve a comparison of living with extinct plants from many different standpoints: changes in the form and architectural scheme of the plant-body, methods of branching, the relation of the main axis to lateral members, and the fundamentally important questions relating to the development of various forms of reproductive organs.

Architectural Types in Plant-stems

It is doubtless true that the causes of change may be considered under two main heads: internal or constitutional; external or environmental. The difficulty is to distinguish between the parts played by these two sets of factors. Attention may, however, be called to one line of enquiry to which occasional reference is made in some of the preceding chapters, namely a comparison of certain anatomical features illustrated by Palaeozoic stems with characteristics of the stems of living plants. Examples of some of the many different forms assumed by the conducting tissues in stems are shown in fig. 64. The simplest and, it is safe to add, the most primitive, is seen in the Devonian genus *Rhynia* (A) and in a slightly more advanced form in stems of some of the older species of the Carboniferous *Lepidodendron* (F). In two other Devonian genera, *Asteroxylon* (B) and *Schizopodium* (C), the solid rod of xylem is deeply fluted, and in the latter stem the wood at the ends of some of the arms shows the regular radial seriation characteristic of secondary xylem. The same type, though without any secondary tissue, is found in some existing lycopods (E); also, in an extreme form, in the Permian fern *Asterochlaena* (L). In *Cladoxylon* (D) the structure is more elaborate as though a single fluted column had broken up into discrete units: the conducting tissue consists of a number of apparently separate strands, each band or loop being composed of a central core of primary wood surrounded by a secondary zone. It would seem that along one line of development the primitive, solid column became extended radially not by the simple method of adding to its diameter, but by the more economical method of adding to its circumference so as to form projecting ribs. This type, though still found in certain lycopods, was much commoner in Palaeozoic stems and is not confined to one class of plants: the arrangement of conducting tissue seen in some stems of *Cladoxylon* and in the pteridosperm *Medullosa* (N) is practically unknown in modern plants possessing a cambium, and therefore able to grow in thickness by the addition of secondary tissue to the original wood and phloem: the nearest approach to it is met with in some tropical, dicotyledonous lianes. There is no evidence in the Palaeozoic stems of any such correlation of a climbing habit with the separation of the conducting tissue into independent strands.

Another line of evolution is shown in some *Lepidodendra*: namely the conversion of a solid rod (F) into a hollow cylinder (G), the centre of the stem retaining scattered groups derived from the primitive column of solid wood. A later stage is seen in another type of *Lepidodendron* (H) in which the primary wood forms a hollow cylinder and the centre is entirely occupied by the soft, pith-like tissue.

The changes so far mentioned may in part at least be correlated with an increase in diameter of the plant-stem: as the stem grows in girth and bears a larger number of leaves a greater amount of conducting tissue is needed: the demand is met in the most economical way either by the conversion of a solid cylindrical rod into a fluted column, or by the substitution of a hollow and wider cylinder for the smaller central rod.[1] One feature of the latter method is the gradual disappearance of all primary wood from the axial region. Another illustration of the gradual decrease in amount of primary wood is afforded by a comparison of such plants as *Botrychioxylon* (M), a species of *Heterangium* (S), and *Stenomyelon* (T), in

[1] Bower (23), (30).

which the primary wood remains as an axial column with stems such as *Lyginopteris* (P), *Callixylon* (Q), and *Protopitys* (V) in which the primary wood, though much reduced, is still a prominent feature. Similarly in a species of *Pitys* (R) primary wood occurs as strands scattered through the whole of the pith: in another form of the same genus (U) it has disappeared from the centre and taken up a peripheral position.

We pass now to living plants in which the regular production of secondary wood is a normal feature: conifers, arborescent dicotyledons, and some cycads in which the secondary tissues form a single, continuous cylinder such as we are familiar with in a pine or an oak. In these plants the primary wood is an insignificant part of the whole; it is represented by small groups of conducting tissue produced in the juvenile stage which, as the stem grows in girth, are recognizable, though with some difficulty, as a narrow and interrupted strip on the inner margin of the cylinder of secondary wood. The modern type of stem differs from that of *Callixylon* (Q) chiefly in the lack of separate strands of primary wood: in the stems, say of an oak and a pine, the primary wood of the young twigs does not retain its individuality; it persists only as an inconspicuous, inner lining of the cylinder of subsequently formed secondary wood. The wood of an oak, as seen in a transverse section of the stem, appears to consist of a single cylinder enclosing a pith: on microscopical examination the primary wood can be detected as a series of points projecting slightly into the pith from the edge of the main mass.

These contrasts between extinct and modern types suggest that in the course of evolution there has been a gradual reduction in the quantity of primary wood in plants capable of producing secondary wood. In many Palaeozoic stems primary strands of xylem, and in some genera a solid axial column, persisted despite the development of a thick cylinder of secondary wood. Simplification led to the production of a stereotyped plan which presumably is more efficient than the more elaborate types which under many different forms were characteristic of Palaeozoic plants.

Another instance of simplification and greater efficiency is illustrated by the comparatively common occurrence in extinct plants of separate strands of conducting tissue each of which gradually surrounded itself with secondary xylem and phloem (e.g. *Cladoxylon* and *Medullosa* (fig. 64, D and N)). It is obvious that an arrangement such as this would not permit of more than a limited development of secondary tissue; a stage would soon be reached when mutual interference of several strands, each of them adding to its diameter, would be a limiting factor.

Another example of a complex anatomical plan is seen in some forms of *Medullosa* (N and O) in which secondary wood was added to the older wood both centrifugally and centripetally: in some existing cycads, which are comparable with *Medullosa*, occasional relics of the inner cylinder (centripetal) of secondary wood are found as abnormalities. The formation of secondary wood became confined to a direction away from the centre of the stem (centrifugal) and the centripetal wood gradually disappeared or persisted occasionally as a vestigial character.

A striking contrast between many ancient and modern plants is the frequent production of secondary tissue in extinct members of groups or classes, the existing representatives of which have only primary wood or at most mere vestiges of secondary wood. In *Equisetum* (fig. 64, I) the conducting tissue is entirely primary: in *Calamites* (K) nearly all the wood is secondary. In living ferns there is only one genus, *Botrychium*, in which any secondary tissue occurs, and that to a very slight extent: in *Botrychioxylon* (M) and a few other Palaeozoic ferns secondary wood was either a normal feature or commoner and more abundant than in living genera. Again in existing lycopods, all of which are herbaceous, there is no cylinder of secondary wood: on the other hand the conducting tissue of many *Lepidodendra* was largely secondary in origin.

This is too wide a subject to discuss at length: comparative anatomy is one of many branches of palaeobotanical research which throws light on the evolution of the plant-world; the more we study the structure of the widely divergent types of architecture met with among the relics of Palaeozoic vegetation, the more clearly we realize that evolution has not been a simple progression; it has been a process of trial and error, a series of experiments leading eventually to the selection of a few structural plans simpler and more efficient than many that were long ago superseded and, as it seems to us, discarded as nature's failures.

In the plants of our own day we have the outcome of an age-long series of experiments, the result of selection of certain designs which, like those of a good architect, owe their superior quality to simplicity combined with efficiency and the absence of features that are unessential.

The problem of evolution cannot be solved by a study of the plant-world in its present state; nor can we expect to discover a solution from the records of the rocks: the fragmentary relics of the past enable us with more confidence to make guesses at the truth; they supply facts which convince us that "all authority must go for nothing".

The interpretation of palaeobotanical documents—some astonishingly legible, others almost undecipherable—enhances our appreciation of the difficulties which Darwin encountered as he searched among the relics of the past for light on the problems of evolution. Since the publication of the *Origin of Species* the progress that has been made is rather in the accumulation of additional and more trustworthy data than in a closer approach to a solution. But none the less we seem to see more clearly that the history of the vegetable kingdom, as also the history of peoples, was marked by recurrent periods when certain races occupied dominating positions. It has been suggested that the more vigorous and widely spread plants of one age may be the descendants of an equally successful race of an earlier age. A possible instance of this has already been adumbrated in the tentatively expressed view of a possible relationship of the pteridosperms and angiosperms.

As the form of a pillar in stone suggests derivation from bundles of Papyrus stems, so the botanist confronted at the outset of his enquiry with a vegetation already far advanced, a vegetation which flourished in the Devonian period, may detect in its general features the impress and reflexion of more primitive ancestors, though the discovery of the ancestors may be beyond his power. The theoretically primitive type eludes our grasp; our faith postulates its existence but the type fails to materialize. In his address to the British Association in 1914 Prof. Bateson said: "But as we have got to recognize that there has been an evolution, that somehow or other the forms of life have arisen from fewer forms, we may as well see whether we are limited to the old view that evolutionary progress is from the simple to the complex, and whether after all it is conceivable that the process was the other way about".[1]

Though we may fail to discover on the tattered pages of the history book of the plant-world facts that enable us to construct genealogies which carry conviction, we learn enough about the ancestry of some survivals from a remote antiquity to quicken our

[1] Bateson (15).

interest in them, and to inspire an almost reverential admiration. The late Prof. Froude spoke of the sound of church bells as "that peculiar creation of mediaeval age which falls upon the ear like the echo of a vanished world": through the links with the past[1] that are still with us we obtain glimpses of other days. This is true not only of a select number of individual plants but, in various parts of the world, of the vegetation as a whole: Dr Holloway,[2] for example, speaks of the general appearance of the flora of Stewart Island, New Zealand, as suggestive of a past age when gymnosperms and pteridophytes were dominant rather than phanerogams.

Arctic Floras of the Past

Attention has been directed in the descriptive chapters to the geographical distribution of genera and families and to many impressive contrasts between the past and the present, not only in the size of the areas occupied by extinct as compared with living plants, but in the presence of luxuriant floras in regions which are now treeless and support a flora bearing the impress of Arctic conditions. We may briefly consider some aspects of these two nearly related subjects, the geographical distribution of plants, and the light thrown by fossil plants on the climates of past ages.

The accompanying map (fig. 137) has been prepared in illustration of the occurrence of floras of many geological periods within the Arctic zone. The age of the plant-beds is given in the appended list, also references to original sources. The approximate northern boundary of trees in the Arctic regions at the present day is indicated by the sinuous line.[3] Some of the collections of Arctic fossil plants are too small and the specimens too incomplete to afford satisfactory evidence of geological age. The most striking feature of the fossil floras is the marked difference in the size of leaves and stems between extinct and living plants: there is no indication in the remains of former vegetations of any dwarfing effect of extreme arctic conditions on the dimensions of leaves and stems. The plants as a whole are not inferior in size to those of the same type obtained from the temperate zone.

It is hoped that the map with the notes will be self-explanatory.

[1] Seward (11^3).
[2] Holloway (18).
[3] Based on a boundary line on one of Messrs Stanfords' maps (London).

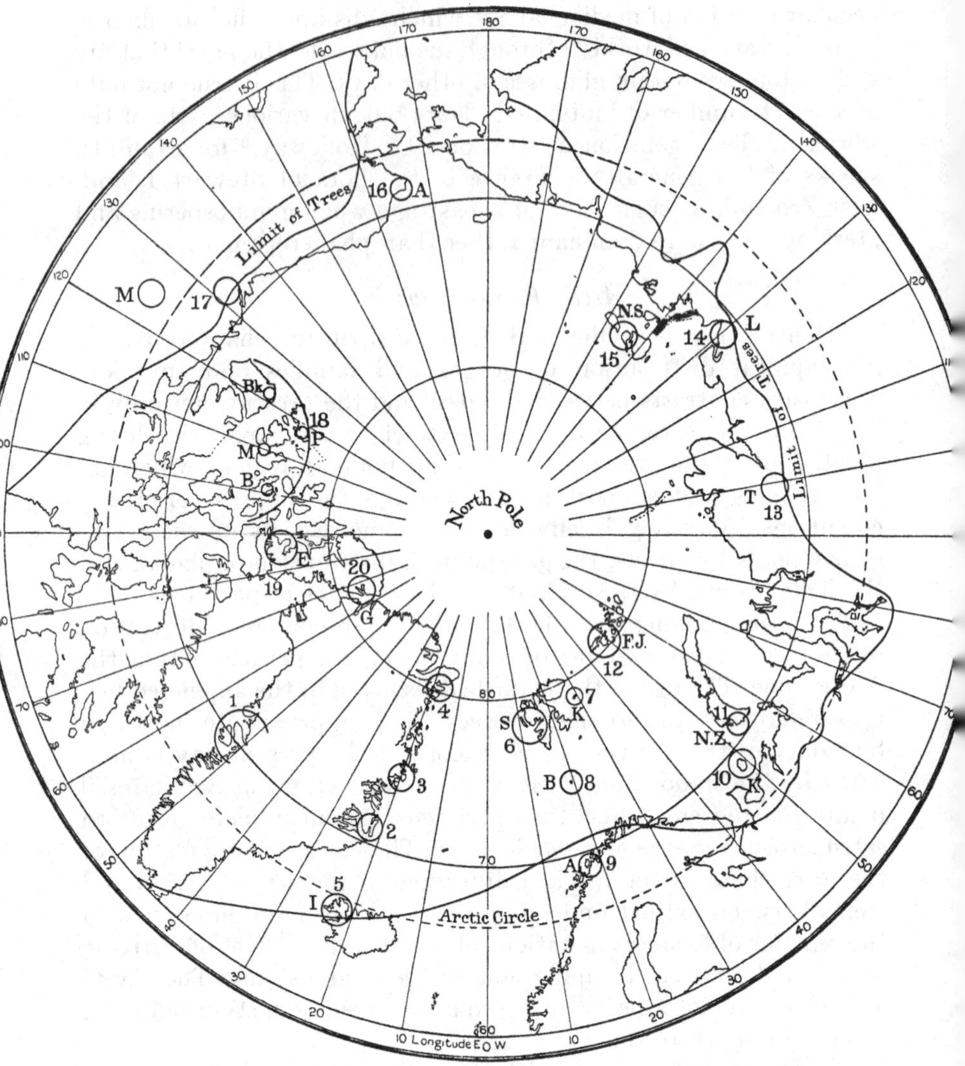

Fig. 137. Arctic floras. For explanation see the appended notes.

Numbers and Letters on Map			Geological age of the plant-beds	References
G.	1.	West Greenland (Disko Island, Upernivik Island, Hare Island, and the Nugsuak Peninsula)	Cretaceous and Tertiary	Seward (26), (29[2]) with references to earlier work; Walton (27[2])
	2, 3, 4.	East Greenland		
	2.	Scoresby Sound and the coast to the north	Rhaetic-Lias: Carboniferous	Harris (26) with references. Dr Harris found a few Carboniferous plants at a locality north of the Rhaetic beds: no description of the specimens has been published
	3.	Sabine Island	Tertiary	See Seward (26) for references
	4.	Ingolfs Fiord	Carboniferous	Nathorst (11[2])
I.	5.	Iceland	Tertiary	Heer (68); Gardner, J. S. (85)
S.	6.	Spitsbergen	Devonian (Upper); Carboniferous (Lower); Jurassic-Cretaceous; Tertiary. A few Triassic plants are recorded	For a general geological account of Spitsbergen, King Charles Land, and Bear Island see Nathorst (10). For descriptions of floras see Heer (68), (71), (77); Nathorst (94), (97), (14), (19), (20), (20[2]); Gothan (11)
K.	7.	King Charles Land	Jurassic (Coniferous wood)	Gothan (07), (11)
B.	8.	Bear Island	Devonian and Carboniferous	Nathorst (94), (02)
A.	9.	Lofoten Islands (Andö)	Jurassic	Johansson (20)
	10.	Kolguev Island	Mesozoic (?)	Feilden (96) with a note on fossil wood by Seward
N.Z.	11.	Novaya Zemlya	Carboniferous (a few fragmentary leaves of *Cordaites*)	Heer (78); Nathorst (94)
F.J.	12.	Franz Josef Land	Jurassic-Cretaceous	Newton and Teall (97), (98); Solms-Laubach (04); Nathorst (99); Gothan (11)
T.	13.	Siberia (Taimyrland)	Tertiary	Heer (78)
L.	14.	Siberia (Lena River and Ajakit River)	Jurassic	Heer (78)
N.S.	15.	New Siberian Islands	Triassic; Jurassic; Tertiary	Nathorst (07); Kryshtofovich (29)
A.	16.	Alaska	Jurassic	Knowlton (14)
M.	17.	Northern Canada: Mackenzie River	Tertiary—(localities along the Mackenzie River)	Heer (68), (82)
	18.	Arctic Archipelago Bk. Banksland P. Prince Patrick Island M. Melville Island B. Bathurst Land, etc.	Carboniferous (a few fragmentary specimens); Tertiary (many very well preserved, large dicotyledonous leaves and some conifers)	Heer (68)
E.	19.	Ellesmere Land	Devonian: Tertiary	Nathorst (04), (11), (15[2])
	20.	Grinnell Land	Tertiary	Heer (78), Feilden and Rance (78)

Climates of the Past

Through the whole course of geological history there must have been marked inequalities in the amount of heat received from the sun by different regions of the earth's surface: the equatorial zone must always have been warmer than the polar zone: on the one hand a vertical sun from which heat passes to the earth through a relatively narrow belt of atmosphere, on the other heat reaching the earth from a sun low in the heavens and through a greater thickness of atmosphere. It is inconceivable that the world can ever have had a climate that was uniform.[1] There must have been in the past as there is now a climatic zonation: our knowledge of the present correlation of zones of vegetation and zones of climate compels us to believe in a corresponding relationship in the past. The temperature at any place on the earth's surface is determined by the balance between radiation received from the sun and terrestrial radiation into space.

Fossil plants collected from rocks at localities in the Arctic or the temperate zone are found to resemble most closely plants that are now characteristic of sub-tropical or tropical countries: the inference is that the temperature was formerly much higher than it is now. Let us briefly consider whether or not an inference such as this, which seems reasonable and indeed obvious, is justifiable.

The present distribution of plants over the earth's surface cannot be regarded as a simple and direct expression of a correlation between plants and climate: there are many factors involved and our knowledge of some of them is still very incomplete. Vegetation in the Arctic regions differs widely from that of temperate regions, and a tropical rain-forest differs no less widely from forests in places where the mean annual temperature is lower. Similarly as we follow the belts of vegetation from the base to the summit of a high mountain in the Tropics we pass in a vertical direction through tropical to arctic conditions and even to a barrenness more barren than the most northerly land in the world. The lower slopes of Mount Kenya, which rise gradually from the plateau at about 6000 ft. above sea-level, are almost completely girdled by forest. On the north and west sides where the rainfall is from 35 to 40 inches a year the most abundant tree is a species of juniper (*Juniperus procera*);

[1] See discussion on geological climates by Dr G. C. Simpson and others (30).

on the south and east sides with a higher rainfall the dominant tree is the so-called camphor, a species of *Ocotea* (the laurel family) growing in company with tree-ferns, its stems and branches festooned with lianes. At an altitude of about 8000 ft. bamboos begin to replace the broad-leaved trees and they eventually form a belt reaching to 11,000 ft., a "mysterious and uncanny" forest.[1] Higher still the giant lobelias and senecios become more conspicuous; grassland and moorland pass upwards into bleak stony ground invaded by glaciers from the perpetual snow and ice of the summit. Climatic zones and zones of vegetation are equally well marked whether we travel from one latitude to another or climb the slopes of a tropical mountain. Many genera and species appear to have a wide range of tolerance to external factors, but vegetation in bulk reflects very definitely the conditions under which it grows.

There are many instances among flowering plants of ability to flourish over large areas in which the mean temperature as well as the extreme temperatures vary within wide limits. Area, it has been maintained, is an index of age: the wider the range in space the older a group of species. A study of the records of the rocks leads to preference for a contrary view: plants now restricted in range are frequently moribund survivals of families or genera which were formerly almost cosmopolitan. The conclusion clearly indicated by a comparative investigation of ancient floras is that in the heyday of their youth plants endowed with efficient means of dispersal rapidly spread from their original home and colonized wide stretches of country: as time passed, other competitors were evolved and the older types became enfeebled; the area of distribution became discontinuous through partial extinction or, at a later stage, the once dominant race passed into oblivion or left a few relics in regions such as Malaya, New Caledonia, China, the Altai Mountains and others which from their richness in peculiar and ancient types may be regarded as refuges exceptionally rich in links with the past.

Attention has been directed to the view that plants are subject to change not only, as is universally acknowledged, in the form and structure of their several parts but in their internal organization, in their response to external factors. The influence of the environment is indirect and acts through the living protoplasm, though the mechanism is unknown. Prof. Nathorst several years ago, in a very interesting paper on "Fossil floras of the Arctic regions as evidence

[1] Dutton (29).

of geological climates",[1] wrote: "Although to-day the cycads only occur in warm regions, it would be an error to conclude that the Cycadophyta of the past had always flourished under similar conditions". The differences between the extinct members of this class and the genera which still exist are so great that the conditions necessary for the requirements of the living species cannot be taken as guides to the requirements of their remotely related ancestors. Nathorst's remark implies belief in the possibility of change in the constitution of members of the same race. The American botanist, Dr Fernald, to whose recent work reference is made in Chapter XVII, definitely attributes the present restricted range of certain preglacial flowering plants on the Gaspé Peninsula to loss of vigour through senility. My own view is that we cannot regard plants as unalterable in their power of adjustment to circumstances: it seems reasonable to assume changes in constitution in the course of their long history. This assumption leads to the conclusion that estimates of the mean temperature of former periods based on comparisons of fossils and existing plants may be untrustworthy. It is legitimate to employ associations of fossil plants as tests of climate in a general sense: it is undesirable to base conclusions on comparisons of more or less dissimilar species. Fluctuations in climate must be accepted as a well-established fact based on the evidence of fossil plants and animals. Plants are less trustworthy as instruments for measuring changes in climate than is generally supposed, even though the extinct species may be closely related to living plants. Inferences have been deduced from the facts of geographical distribution without sufficient allowance being made either for the effect of competition of one kind of plant with another, or for the probability of decrease in power of adjustment to the factors conditioning existence as vigour becomes impaired.

Much ingenuity has been shown by authors of books and papers on climates of the past: many avenues have been explored in the hope of finding causes which may have produced variations in the balance between solar radiation received by the earth and the radiation from the earth's surface into space.[2] Neglecting many of the suggested explanations—some of which it is admitted are wholly inadequate or based on misconception, while others are based on extra-

[1] Nathorst (11). See also Berry (30[2]); Kubart (28).

[2] For a full discussion of the value of rings of growth as indices of climate see Antevs (17), (25), (25[2]); for a more general summary see Seward (92); also the references given on p. 515.

terrestrial phenomena of which we have insufficient knowledge—there remain two possible explanations: (i) alterations in the distribution of land and water and (ii) the theory of continental drift.

We know from the evidence of the rocks that the distribution of land and sea over the world has changed within wide limits in the course of geological periods: is it possible by altering the present arrangement of continents and oceans within limits permitted by geologists to provide such conditions as seem to be demanded by the facts supplied by fossil plants? We are told that no redistribution of land and water would substantially alter the mean annual temperature of any zone: in spite of the great difference in the proportion of land and water in the two hemispheres at the present day the mean annual temperature is practically the same in both, 15·2° C. in the northern, and 13·3° C. in the southern hemisphere. There are, however, considerable differences in the temperature of places on the same parallel of latitude: the tree-limit in the northern hemisphere is represented by a sinuous line (fig. 137) cutting across Cape Farewell at the extreme south of Greenland, rising in Alaska to beyond lat. 69° N. The white spruce (*Picea alba*) reaches a height of 50 ft. in Alaska: on the same parallel in Greenland the tallest willow does not exceed 3 ft. in height.[1] It has been pointed out that our common juniper (*Juniperus communis*), which occurs in Europe as far north as the North Cape, flourishes in the eastern hemisphere 20 to 25 degrees of latitude farther north, a limit well beyond that of the family to which it belongs. Granting that no rearrangement of land and water with the consequential shifting of ocean currents could convert a temperate into a sub-tropical climate throughout a complete zone, it is none the less clear that the temperature of regions within a zone could be substantially raised.

Meteorologists are not in complete agreement on the possible adequacy of redistribution of land and water as a cause of climatic differences large enough to satisfy some of the demands of geologists: Mr Brooks in his book on the evolution of climate[2] is prepared to create conditions in the Arctic regions fit for a temperate flora by increasing the area and depth of the northern sea, by increasing the inflow of warm surface-water and by diminishing the inflow of cold water. It cannot be doubted that some of the climatic changes

[1] Seward (25[2]); Huntington (14).

[2] Brooks, C. E. P. (22). See also Huntington and Visher (22); Douglass (19).

revealed by the floras of the past may be accounted for by rearrangement of currents, changes in area and in height of land-surfaces and alterations in the conditions governing rainfall. On the other hand, it is hard to believe that the explanation of all the anomalies revealed by palaeobotanical records can be found by even the most favourable readjustment of geographical features. But even if we take as our guide such meteorological opinion as is most favourable for our purpose, we cannot obtain from any redistribution of land and sea an amelioration of climate great enough to serve as an explanation of the occurrence of a luxuriant Rhaetic flora in East Greenland, the presence of a rich Upper Devonian flora in Ellesmere Land, of a rich and varied Jurassic flora in Graham Land, or of the presence of glaciers and ice-sheets in certain parts of Gondwanaland during the later stages of the Palaeozoic era. The evidence of substantial changes in climate furnished by these examples causes the palaeobotanist to adopt a favourable and hopeful attitude towards the Wegener hypothesis. The theory is attractive; it has many supporters, among them Dr Simpson whose papers should be consulted; it has also many opponents: as Lord Acton said, in speaking of human history, "the worst use of theory is to make men insensible to fact": the facts of palaeobotany though they cannot be regarded, in any sense, as proof of the correctness of Wegener's views, may be admitted as evidence in a case which for the present must remain *sub-judice*. The problem of climatic change like many other problems can be solved only by the co-operation of workers in different fields. Students of ancient floras should be grateful to the present head of the Meteorological Office for the stimulus and guidance he has given to his colleagues in other branches of science by his recent contributions to an exceptionally difficult and interesting subject.

We have seen that from the Pre-Cambrian epoch to the present climatic conditions in any one part of the world were far from uniform: evidence in support of recurrent ice-ages or of the occurrence of local glaciers is overwhelming. It has recently been stated by the Director of the Meteorological Department that widespread glaciation in the region that is now tropical India is "a physical impossibility": yet geological evidence demonstrates the spread of glaciers and ice-sheets over wide areas which are now within the Tropics. How then are we to reconcile the conclusions based on sound geological testimony with the authoritative statement that such things could not have happened with the continents where they are now?

The employment of plants as tests of climate is a wide and intricate subject and cannot be discussed in a section of a single chapter; but reference may be made to a few general considerations which should be borne in mind when fossils are used as "thermometers of the ages". There is the danger of ignoring the well-established fact that closely related species are able to live under very different climatic conditions. It does not in the least follow that because all living species of a genus are now confined to areas with a certain range of temperature, therefore extinct species, almost identical with the living species, were equally susceptible to limiting factors. It is a common practice to think of palms, tree-ferns, and many other plants as tropical or at least sub-tropical: the fact that the dwarf palm flourishes in southern Europe and the palmetto in Florida lends support to the suggestion that extinct palms may have been able to exist in still cooler climates. An exceptionally striking illustration of the association of sub-tropical vegetation with glacial conditions is shown in the frontispiece: the photograph was taken for this book by Dr Teichelmann, at the request of Prof. Cockayne, to both of whom I am very greatly indebted. It is a winter scene in Westland in the south island of New Zealand, almost a mile from the terminal face of the Franz Josef Glacier: in the foreground are plants of the fern *Hemitelia Smithii* (Cyatheaceae), the crown of the middle one pointing to the ice-filled valley, with *Coprosma* and other flowering plants of the sub-tropical flora. Prof. Cockayne tells me that he saw a young *Hemitelia* less than 100 yards from the end of the glacier. He writes: "The wealth of tree ferns along the track to this glacier is astonishing. In the valley of the Franz Josef there is a good deal of open river-bed near the terminal face, but in that of the Fox Glacier there is a dense shrubbery almost up to the ice". Another of Dr Teichelmann's photographs, reproduced in fig. 138, was taken in the forest on the north-east slope of the Baird Mountains looking across the Franz Josef Glacier, about one and a half miles above the terminal face: the tree fern is *Hemitelia Smithii.* These New Zealand landscapes enable us better to understand some at least of the apparent anomalies furnished by the records of the rocks.

The group of plants covering the ground sloping to the beach of Englishman's Harbour in Disko Island (Greenland; lat. 69°–70° N.) (fig. 139) affords another, though less striking, example of a

FIG. 138. Looking across the Franz Josef Glacier, Westland, New Zealand, from the north-east slope of the Baird Mountains, one and a half miles above the terminal face of the ice. A tree-fern (*Hemitelia Smithii*) in the foreground. (Photograph by Dr E. Teichelmann, obtained for the author by Prof. Cockayne.)

FIG. 139. Orchids and other flowering plants at Englishman's Harbour, Disko Island, Greenland. (Photograph by Mr R. E. Holttum.)

comparatively rich herbaceous vegetation growing on the edge of an ice-covered land. In this exceptionally favoured locality on Disko Island there are several flowering plants which reach their northern limit on the mainland considerably south of the latitude of Disko. The willows are unusually well grown; the sea-mertensia (*Mertensia maritima*) is scattered among the pebbles on the beach, and on the higher ground stems of *Archangelica*, reaching a yard in height, clasped by the large and handsome leaves and bearing candelabra-like umbels of flowers. In the left-hand upper edge of the photograph is a flowering plant of *Alchemilla glomerulans*, and scattered through the herbage the tall flowering spikes of an orchid, *Habenaria albida*, and the slightly smaller spikes of the circumpolar *Polygonum viviparum* are conspicuous features. The larger blooms are the flower heads of a *Taraxacum*, and a few fern fronds are visible at the lower edge of the photograph.

The interrelations of climate and vegetation are extremely complex: we cannot confine our attention to a single plant or to a group of plants; we must regard the flora as a whole, composed of individuals unequally endowed with qualities which make for success in the struggle for existence. The greater efficiency of a recently evolved type may often have been the cause responsible for the migration of a former member of the flora to another station where the conditions of life were easier. Climate is not the sole controlling agent, there is the relation of plant with plant in their rival claims for dominance or survival. When we compare the more uniform nature of the plant-world of certain geological periods with the more diversified covering of vegetation at the present day, it is important to remember that the contrasts which, after making allowance for our imperfect knowledge of ancient floras, are undoubtedly considerable, are not necessarily dependent on differences in climate. Had it not been for the fatal barrier of the Alps and Mediterranean which cut off the retreat to safety of plants driven south by the spread of the Quaternary ice, Europe would have remained the median province of a vast territory, stretching across North America to China and Tibet, differing comparatively little in its vegetation from the regions to the west and east of it. The marked contrasts presented by the present tropical, sub-tropical and temperate floras are the expression of a zoning of vegetation in rough correspondence with latitude, the sinuous boundaries of the zones being caused by the varying humidity and other conditioning factors. We are apt to forget that not a few genera and even species are equally at home in tropical, sub-tropical and temperate countries. We think of the bracken fern (*Pteridium aquilinum*) as characteristically British and temperate, though we meet with it at Zanzibar on the tropical shores of the Indian Ocean, on the lower slopes of Mount Kenya in company with tree-ferns and forest trees, in Cape Colony, South America, Australia and indeed almost all over the world. The common *Polypodium vulgare*[1] is widely distributed through the north temperate zone; it reaches latitude 70° N. in Norway: from northern Japan and Manchuria it follows the Altai Mountains and extends to the Pamir, Persia, Asia Minor, the coastal regions of the Mediterranean, the Canary Islands and the Azores: a form hardly distinguishable from the European fern grows on Kerguelen Island; and a variety ranges from Alaska to California. These two ferns, members of the same family, are only two among many examples of far travelled plants tolerant of wide ranges of temperature. Palaeobotanical data support the contention that very often the older plants have a more restricted geographical distribution than those of more recent origin. *Osmunda regalis* is one of the exceptions. The genus *Lycopodium* represented in the circumpolar flora by *Lycopodium Selago*, a plant widely spread in the temperate zone, has many more species in the Tropics than in any other part of the world: this is a type which seems to have acquired the attribute of perpetual youth; tracing its ancestors far back into the Palaeozoic era, it shows no signs of decadence. Geographical range is dependent on the existence of open roads—mountain-chains and other unimpeded highways—the presence, for example, of several plants in the southern provinces of South America specifically identical, or nearly so, with forms which range far across the northern hemisphere is no doubt correlated with the existence of the Andean chain.

Among plants which are now confined within narrow boundaries or have a discontinuous range and are members of ancient families or genera are *Thyrsopteris elegans* of Juan Fernandez, *Matonia*, *Dipteris*, *Cheiropleuria* (figs. 94–97), *Archangiopteris* and several other ferns; *Sequoia*, *Callitris*, *Sciadopitys*, *Araucaria*, *Libocedrus* and other conifers. All these are old types, some with a much longer history than others; but it does not follow that it is age alone and loss of vitality that have been the cause of interrupted or restricted

[1] Christensen (28).

range; there are other factors such as the nature of the soil and the effect of competition with other groups of plants.

As the science of ecology progresses we shall be better able to separate one influence from another and to formulate laws expressing in more definite terms the response of plants to environment. Experiments in acclimatization show that plants are often more tolerant than has been supposed.[1] If man is able by patient cultivation to establish plants in localities where the climate differs considerably from that of their normal surroundings, it is easy to believe that, were it not for the competition of better equipped neighbours, or because of other causes operative under natural conditions but not in the more peaceful circumstances of cultivation, plants may in the past have been able to grow in regions that are now beyond their range.

* * *

The wonder and admiration awakened by contemplation of the plant-world are the expression of our aesthetic sense; they have their source in a power of response to the stimulating influence of the beauty of nature's handicraft. No knowledge of botany or no backward glance into the past is needed to produce the response. In one of his books W. H. Hudson, writing of an apple tree in blossom, says "it is like nothing on earth, unless we say that, indescribable in its loveliness, it is like all other sights in nature which wake in us a sense of the supernatural". The present, it has been said, is the key to the past; but it may be urged with at least equal truth that the past is the key to the present. A knowledge of the past however imperfect adds to the attractiveness of the present: in the present we see the drama of nature in progress; as we decipher the lamentably incomplete though wonderful records of former ages, our enthusiasm and imagination are quickened. We long to know what was and how the present has been evolved from the past. The more we know the more conscious we become of the little we really know: the passion for the search grows as we read the story of creation. Increase in knowledge clears our vision only in part. To some minds is revealed evidence of nature's unchanging harmony, an ordered sequence of development; other seekers after truth are tempted to think,

> "The will has woven with an absent heed
> Since life first was; and ever will so weave."

[1] Seliber (29); White, O. E. (26).

Fig. 140. The Summit of Mount Kenya.

BIBLIOGRAPHY

The numbers in brackets after the names of authors indicate the year of publication, e.g. (29) = 1929; (61) = 1861

ABSALOM, R. G. (29). The Lower Carboniferous coal-ball flora of Haltwhistle, Northumberland. *Proc. Univ. Durham Phil. Soc.* **8**, ii.

AMALITZKY, V. (01). Sur la découverte, dans les dépôts permiens supérieurs du Nord de la Russie, d'une flore glossoptérienne, etc. *Comp. Rend.* **131**, p. 591.

ANTEVS, E. (13). Results of Dr E. Mjöberg's Swedish Scientific Expeditions to Australia, 1910–13. V. Mesozoic plants. *K. Svensk. Vetenskapsakad. Hand.* **52**, v.

—— (14). Die Gattungen *Thinnfeldia* und *Dicroidium*. *Ibid.* **51**, vi.

—— (17). Die Jahresringe der Holzgewächse und die Bedeutung derselben als klimatischer Indikator. *Progr. rei Bot.* **5**.

—— (19). Die liassische Flora des Hörsandsteins. *K. Svensk. Vetenskapsakad. Hand.* **59**, viii.

—— (25). Quaternary climates. *Carnegie Inst. Washington, Publ.* **352**.

—— (25^{2}). The Climatologic Significance of Annual Rings in Fossil Woods. *Amer. Journ. Sci.* **9**.

—— (28). The last glaciation, etc. *Amer. Geogr. Soc. N. York.*

ARBER, AGNES (28). The tree-habit in Angiosperms: its origin and meaning. *New Phyt.* **27**.

ARBER, E. A. N. (02). The Clarke Collection of fossil plants from New South Wales. *Quart. Journ. Geol. Soc.* **58**.

—— (04). *Cupressinoxylon Hookeri*. *Geol. Mag.* [5], **1**.

—— (04^{2}). The fossil flora of the Culm Measures of north-west Devon. *Phil. Trans. R. Soc.* **197**.

—— (05). *Catalogue of the fossil plants of the Glossopteris flora in the Department of Geology, British Museum, London.*

—— (06). On the past history of the Ferns. *Ann. Bot.* **20**.

—— (06^{2}). Bibliography of literature on Palaeozoic fossil plants, 1870–1905. *Progr. rei Bot.* **1**.

—— (09). On the fossil plants of the Waldershare and Fredville Series of the Kent Coalfield. *Quart. Journ. Geol. Soc.* **65**.

—— (09^{2}). On the affinities of the Triassic plant *Yuccites vogesiacus*. *Geol. Mag.* [5], **6**.

—— (10). Note on a collection of fossil plants from the neighbourhood of Lake Nyasa. *Quart. Journ. Geol. Soc.* **66**.

—— (10^{2}). Some fossil plants from Western Australia. *Bull. Geol. Surv. W. Australia*, **36**.

—— (11). *The Natural History of Coal.* Cambridge.

—— (12). Contributions to our knowledge of the floras of the Irish Carboniferous rocks. *Sci. Proc. R. Dublin Soc.* **13** [N.S.], xii.

—— (12^{2}). On *Psygmophyllum majus* sp. nov. from the Lower Carboniferous rocks of Newfoundland. *Trans. Linn. Soc. London*, **7**, xviii.

—— (12^{3}). Fossil plants from the Kent Coalfield. *Geol. Mag.* [5], **9**.

—— (12^{4}). On the fossil flora of the Forest of Dean Coalfield (Gloucestershire), etc. *Phil. Trans. R. Soc.* **202**.

—— (14). Fossil flora of the Kent Coalfield. *Quart. Journ. Geol. Soc.* **70**.

—— (14^{2}). On the fossil flora of the Wyre Forest, etc. *Phil. Trans. R. Soc.* **204**.

ARBER, E. A. N. (17). The earlier Mesozoic floras of New Zealand. *N.Z. Geol. Surv. Palaeont. Bull.* no. 6.

—— (21). *Devonian Floras.* Cambridge.

ARBER, E. A. N. and R. H. GOODE (15). On some fossil plants from the Devonian rocks of North Devon. *Proc. Camb. Phil. Soc.* **18**, iii.

ARLDT, T. (07). *Die Entwicklung der Kontinente und ihrer Lebewelt.* Leipzig.

ARNOLD, C. A. (29). On the radial pitting in *Callixylon. Amer. Journ. Bot.* **16**.

—— (30). The genus *Callixylon* from the Upper Devonian of Central and Western New York. *Papers of the Michigan Acad. Sci. etc.* **11**.

AUGUSTA, J. (29). Note Phytopaléontologique sur le niveau à *Palaeoniscus* du Permien de la Fosse de Boskovice. *Věstnik Stát Geol. ustavu Čsl. Rep.* **5**, ii–iii.

BAILEY, E. B. (26). Plant-migration across the Millstone Grit. *Geol. Mag.* **63**.

—— (28). The Palaeozoic mountain-systems of Europe and America. *Pres. Address, (Sect. C) Brit. Assoc.* Glasgow.

BAILEY, I. W. (22). Notes on neotropical ant-plants. I. *Cecropia angulata* sp. nov. *Bot. Gaz.* **74**.

BAILY, W. H. (61). *Explanations to accompany sheets* 147 *and* 157 *of the Map of the Geology of the South of Ireland.* Dublin.

—— (69). Notice of plant remains from beds interstratified with the basalt in the county of Antrim. *Quart. Journ. Geol. Soc. London,* **25**.

—— (80). Second Report of the Committee appointed for the purpose of collecting and reporting on the Tertiary (Miocene) flora of the basalt of the North of Ireland. *Brit. Assoc. Rep.*

BAIN, G. W. (25). Is the Huronic Gowganda Conglomerate of glacial origin? *Pan-Amer. Geol.* **44**.

BAKER, H. A. (22). Final Report on geological investigations in the Falkland Islands. (*Government Geologists' Report.*)

BALL, O. M. (30). A Partial Revision of Fossil Forms of *Artocarpus. Bot. Gaz.* **90**.

BANDULSKA, H. (23). A preliminary paper on the cuticular structure of certain dicotyledonous and coniferous leaves from the Middle Eocene flora of Bournemouth. *Journ. Linn. Soc.* **46**.

—— (24). On the cuticles of some recent and fossil Fagaceae. *Ibid.* **46**.

—— (26). On the cuticles of some fossil and recent Lauraceae. *Ibid.* **47**.

—— (28). A *Cinnamomum* from the Bournemouth Eocene. *Ibid.* **48**.

BARBER, C. A. (89). The structure of *Pachytheca. Ann. Bot.* **3**.

—— (90). The structure of *Pachytheca. Ibid.* **5**.

BARBOUR, G. B. (24). Cretaceous beds in North China. *Nature,* **113**.

—— (29). The geology of the Kalgan area. *Mem. Geol. Surv. China,* Ser. A, no. 56.

BARCLAY, G. W. W. (86). On some algoid lake-balls found in South Uist. *Proc. R. Soc. Edinb.* **13**.

BARNARD, C. (28). A note on the structure of a Lepidodendron stem from the Lower Carboniferous of New South Wales. *Ann. Bot.* **42**.

BARNES, B. and H. DUERDEN (30). On the preparation of celluloid transfers from rocks containing fossil plants. *New Phyt.* **29**.

BARRELL, J. (08). Relations between climate and terrestrial deposits. *Journ. Geol.* **16**.

—— (13). The Upper Devonian Delta of the Appalachian Geosyncline. *Amer. Journ. Sci.* **36**.

—— (17). Rhythms and the measurement of geologic time. *Bull. Geol. Soc. Amer.* **28**.

BARRELL, J., C. SCHUCHERT and others (18). *The Evolution of the Earth and its inhabitants.* New Haven.

BARTHOLIN, C. T. (92). Nogle i den bornholmske Juraformation forekommende Planteforsteninger. I. *Bot. Tidsskrift, Kjövenhavn,* **18**.

—— (94). *Ibid.* II. *Ibid.* **19**.

—— (10). Planteforsteninger fra Holsterhus paa Bornholm. *Danmarks geol. Unders.* R. 2, no. 24.

BASSLER, H. (16). A cycadophyte from the North American Coal Measures. *Amer. Journ. Sci.* **42**.
BATESON, W. (15). *Pres. Address, Brit. Assoc.* Melbourne.
BATHER, F. A. (07). Nathorst's use of collodion imprints in the study of fossil plants. *Geol. Mag.* **4**.
—— (08). Nathorst's methods of studying cutinized portions of fossil plants. *Ibid.* **5**.
—— (09). Visit to the Florissant Exhibition in the British Museum (Natural History). *Proc. Geol. Assoc.* **21**.
BAUMANN, E. (12). Vegetation des Untersees (Bodensee). Karsten und Schenck—*Vegetationsbilder*, R. 9, Taf. 13. Jena.
BECK, R. (20). Über *Protothamnopteris Baldaufi* n.sp. *Abh. K. Sächs. Akad. Wiss.* **36**, v.
BENSON, W. N. (27). Materials for the study of the Devonian palaeontology of Australia. *Rec. Geol. Surv. N.S. Wales*, **10**, ii.
BENSON, W. N., W. S. DUN and W. B. BROWN (21). The geology and petrology of the Great Serpentine Belt of New South Wales. *Proc. Linn. Soc. N.S. Wales*, **45**.
BERKEY, C. P. and F. K. MORRIS (27). *The Geology of Mongolia*, **4**. New York.
BERRY, E. W. (11). Lower Cretaceous. *Maryland Geol. Surv.*
—— (11²). A Lower Cretaceous species of Schizaeaceae from eastern North America. *Ann. Bot.* **25**.
—— (12). American Triassic *Neocalamites*. *Bot. Gaz.* **53**.
—— (14). The Upper Cretaceous and Eocene floras of South Carolina and Georgia. *U.S. Geol. Surv., Profl. Paper* 84.
—— (15). The age of the Cretaceous flora of southern New York and New England. *Journ. Geol.* **23**, no. 7.
—— (16). Upper Cretaceous. *Maryland Geol. Surv.*
—— (16²). The Lower Eocene floras of south-eastern North America. *U.S. Geol. Surv., Profl. Paper* 91.
—— (16³). The physical conditions and age indicated by the flora of the Alum Bluff formation. *Ibid., Profl. Paper* 98—E.
—— (16⁴). The physical conditions indicated by the flora of the Calvert formation. *Ibid., Profl. Paper* 98—F.
—— (17). Geologic history indicated by the fossiliferous deposits of the Wilcox Group (Eocene) at Meridian, Mississippi. *Ibid., Profl. Paper* 108—E.
—— (17²). Fossil plants from Bolivia and their bearing upon the age of uplift of the Eastern Andes. *Proc. U.S. Nat. Mus.* **54**.
—— (18). The fossil higher plants from the Canal Zone. *U.S. Nat. Mus. Bull.* **103**.
—— (18²). A Restoration of *Neocalamites*. *Amer. Journ. Sci.* **45**.
—— (19). Upper Cretaceous floras of the Eastern Gulf Region in Tennessee, etc. *U.S. Geol. Surv., Profl. Paper* 112.
—— (19²). The age of the Brandon lignite and flora. *Amer. Journ. Sci.* [4], **47**.
—— (19³). Miocene Fossil plants from northern Peru. *Proc. U.S. Nat. Mus.* **55**.
—— (20). Palaeobotany: a sketch of the origin and evolution of floras. *Smithson. Rep. for* 1918. Washington.
—— (21). A palm nut from the Miocene of the Canal Zone. *Proc. U.S. Nat. Mus.* **59**.
—— (21²). Tertiary fossil plants from Costa Rica. *Ibid.* **59**.
—— (21³). Tertiary fossil plants from Venezuela. *Ibid.* **59**.
—— (22). The flora of the Woodbine Sand at Arthur's Bluff, Texas. *U.S. Geol. Surv., Profl. Paper* 129—G.
—— (22²). The flora of the Cheyenne Sandstone of Kansas. *Ibid., Profl. Paper* 129—I.
—— (22³). Carboniferous plants from Peru. *Amer. Journ. Sci.* **3**.
—— (23). Miocene plants from southern Mexico. *Proc. U.S. Nat. Mus.* **62**, art. 19.
—— (23²). *Tree Ancestors*. Baltimore.
—— (24). The Middle and Upper Eocene floras of south-eastern North America. *U.S. Geol. Surv., Profl. Paper* 92.
—— (24²). Fossil plants from the Eastern Andes of Colombia. *Bull. Torr. Bot. Club*, **51**.

BERRY, E. W. (24³). A fossil flower from the Miocene of Trinidad. *Amer. Journ. Sci.* 7.
—— (24⁴). Mesozoic plants from Patagonia. *Ibid.* 7.
—— (24⁵). Mesozoic *Gleichenia* from Argentina. *Pan-Amer. Geol.* 41.
—— (25). The flora of the Ripley formation. *U.S. Geol. Surv., Profl. Paper* 136.
—— (25²). Flora and ecology of the so-called Bridger beds of Wind River basin, Wyoming. *Pan-Amer. Geol.* 44.
—— (26). *Terminalia* in the Lower Eocene of south-eastern North America. *Bull. Torr. Bot. Club,* 53.
—— (26²). The fossil seeds from the *Titanotherium* beds of Nebraska, etc. *Amer. Mus. Novitates,* no. 221.
—— (26³). A fossil palm fruit from the Middle Eocene of north-western Peru. *Proc. U.S. Nat. Mus.* 70.
—— (26⁴). Tertiary floras from British Columbia. *Canad. Dept. of Mines, Geol. Surv. Bull.* 42.
—— (27). Links with Asia before the mountains brought aridity to the western United States. *Sci. Monthly,* 25.
—— (27²). The flora of the Esmeralda formation in western Nevada. *Proc. U.S. Nat. Mus.* 72.
—— (28). An *Alethopteris* from the Carboniferous of Peru. *Journ. Washington Acad. Sci.* 18.
—— (28²). Tertiary fossil plants from the Argentine Republic. *Proc. U.S. Nat. Mus.* 73.
—— (29). The flora of the Frontier formation. *U.S. Geol. Surv., Profl. Paper* 158—H.
—— (29²). Eocene plants of the Restin Formation of Peru. *Pan-Amer. Geol.* 51.
—— (29³). A revision of the flora of the Latah Formation. *U.S. Geol. Surv., Profl. Paper* 154—H.
—— (29⁴). Tertiary fossil plants from Colombia, S. America. *Proc. U.S. Nat. Mus.* 75.
—— (30). A flora of Green River age in the Wind River basin of Wyoming. *Ibid. Profl. Paper* 165—B.
—— (30²). The past climate of the North Polar regions. *Smiths. Misc. Coll.* 82.
—— (30³). Revision of the Lower Eocene Wilcox Flora of the south-eastern States. *U.S. Geol. Surv., Profl. Paper* 156.
BERTRAND, C. E. (93). Les bogheads à Algues. *Bull. Soc. Belge Géol. etc.* 7.
—— (96). Nouvelles remarques sur le kérosene shale. *Ibid.* 9.
BERTRAND, C. E. and B. RENAULT (92). *Pila bibractensis* et le boghead d'Autun. *Bull. Soc. d'hist. nat. d'Autun,* 5.
—— (94). *Reinschia australis* et premières remarques sur le kérosene shale. *Ibid.* 6.
BERTRAND, P. (09). Les phénomènes glaciaires de l'époque Permo-Carbonifère. *Ann. Soc. géol. Nord,* 38.
—— (11). Structure des stipes d'*Asterochlaena laxa. Mém. Soc. géol. Nord,* 7, i.
—— (13). Note préliminaire sur les Psilophytons des grès de Matringhem. *Ann. Soc. géol. Nord,* 42.
—— (14). Étude du stipe de l'*Asteropteris noveboracensis. Congr. géol. internat. (Comp. Rend.).*
—— (26). Conférences de paléobotanique. *École nat. Mines St Étienne.* Lille.
—— (26²). Les gisements à *Mixoneura* dans la région de St Gervais-Chamonix. *Bull. Soc. géol. France* [4], 26.
—— (28). Valeur des flores pour la caractérisation des différentes assises du terrain houiller et pour les synchronisations de bassin à bassin. *Congr. stratig. Carbon.* Heerlen.
—— (28²). Stratigraphie du Westphalien et du Stéphanien dans les différents bassins houillers français. *Ibid.*
—— (28³). L'échelle stratigraphique du terrain houiller de la Sarre et de la Lorrain. *Ibid.*
BEWS, J. W. (25). *Plant Forms and their Evolution in South Africa.* London.

BIGOT, A. (25). Sur les calcaires cambriens de la région de Carteret et leur faune. *Bull. Soc. Linn. Normand.* [7], **8**.

BLACK, M. (29). Drifted plant-beds of the Upper Estuarine Series of Yorkshire. *Quart. Journ. Geol. Soc.* **85**.

—— (30). Exploring the Great Bahama Bank. *Discovery*, **11**.

BLACKWALDER, E. (13). Origin of the Bighorn Dolomite of Wyoming. *Bull. Geol. Soc. Amer.* **24**.

BLANCKENHORN, M. (86). Die fossile Flora des Buntsandsteins und des Muschelkalks der Umgegend von Commern. *Palaeont.* **32**.

—— (21). Aegypten. *Handbuch Regionaler Geologie*, **7**.

BODE, H. (27). *Palaeobotanisch-stratigraphische Studien im Ibbenbürener Carbon.* Dissertat. Berlin.

—— (29). Zur Kenntniss der Gattung *Porodendron* Nath. (non Zalessky). *Palaeont.* **72**.

BOMMER, C. (10). Contribution à l'étude du genre *Weichselia*. *Bull. Soc. R. bot. Bruxelles*, **47**.

BONČ-OSMOLOVSKIJ, G. (29). Le Paléolithique de Crimée. *Bull. Comm. pour l'étude Quatern.* **1**.

BOSWORTH, T. O. (12). *The Keuper Marls around Charnwood.* Leicester.

BOULE, M. (08). Sur l'existence d'une faune et flore permiennes à Madagascar. *Comp. Rend.* **146**.

BOWER, F. O. (08). *The Origin of a Land Flora.* London.

—— (23). The relation of size to the elaboration of form and structure of the vascular tracts in primitive plants. *Proc. R. Soc. Edinb.* **43**.

—— (23–28). *The Ferns (Filicales)*, **1–3**. Cambridge. **1**, 1923; **2**, 1926; **3**, 1928.

—— (30). *Size and Form in Plants with special reference to the Primary Conducting Tracts.* London.

BRADLEY, W. H. (29). Freshwater Algae from the Green River formation of Colorado. *Bull. Torr. Bot. Club*, **56**.

BREHMER, W. VON (14). Über eine Glossopteris Flora am Ulugurugebirge. *Bot. Jahrb.* **51**, 3–4.

BROCKMANN-JEROSCH, H. (14). Zwei Grundfragen der Paläophytogeographie. *Ibid.* Suppl. **50**.

BROILI, F. (28). Ein ? Pflanzenrest aus den Hunsrück-Schiefer. *Sitzungsber. Bay. Akad. Wiss.*

BROOKS, C. E. P. (22). *The Evolution of Climate.* London.

BROOKS, W. K. (94). The origin of the oldest fossils and the discovery of the bottom of the ocean. *Journ. Geol.* **2**.

BROWN, A. (94). On the structure and affinities of the genus *Solenopora*. *Geol. Mag.* **1**.

BROWN, T. C. (14). Origin of oolites and the oolitic texture in rocks. *Bull. Geol. Soc. Amer.* **25**.

BUBNOFF, S. VON (26). *Deutschlands Steinkohlenfelder.* Stuttgart.

BUCHER, W. H. (18). On oolites and spherulites. *Journ. Geol.* **26**.

BUREAU, E. (11). Sur la flore dévonienne du bassin de la Basse Loire. *Bull. Soc. Sci. nat. de l'ouest de la France* [3], **1**.

—— (14). Bassin de la Basse Loire. *Études Gîtes Min. de la France.* 2 vols. Paris.

BURLING, L. D. (23). Cambro-Ordovician section near Mount Robson, Brit. Columbia. *Bull. Geol. Surv. Amer.* **34**.

CAMPBELL, D. H. (29). The phylogeny of the Angiosperms. *Bull. Torr. Bot. Club*, **55**.

CAPELLINI, G. and GRAF ZU SOLMS-LAUBACH (91). I tronchi di Bennettitee dei Musei italiani. *Mem. R. Accad. Sci. inst. Bologna* [5], **2**.

CARPENTIER, A. (13). Contribution à l'étude du Carbonifère du nord de la France. *Mém. Soc. géol. Nord*, **7**.

—— (20). Notes d'excursions paléobotaniques à Chalonnes et Montjean. *Bull. Soc. géol. France*, **19**.

—— (27). La flore wealdienne de Féron-Glageon (Nord). *Mém. Soc. géol. Nord*, **10**.

CARPENTIER, A. (29). Sur les fructifications du *Rhodea Gutbieri*. *Comp. Rend.* **189**.

—— (29²). Empreintes de fructifications trouvées en 1929 dans le Westphalien du nord de la France. *Rev. Gén. Bot.* **41**.

—— (29³). Recherches sur les végétaux fossiles des Argiles Éocrétaciques du Pays de Bray. *Bull. Soc. géol. France* [4], **29**.

—— (30). Flore viséenne de la région de Kasba ben Ahmed. *Service des mines et de la carte géol. du Maroc, Notes et Mém.*

—— (30²). La flore permienne du Bon Achouch (Maroc central). *Ibid.*

—— (30³). Notes Paléophytologiques. *Ann. Soc. géol. Nord,* **54**.

CASE, E. C. (19). The environment of vertebrate life in the late Palaeozoic in North America: a palaeogeographic study. *Carnegie Inst. Washington, Publ.* 283.

CAYEUX, L. (09). *Les minérales de fer oolithique de France.* Paris.

CHAMBERLAIN, C. J. (19). *The Living Cycads.* Chicago.

CHAMBERLIN, T. C. (16). *The Origin of the Earth.* Univ. Chicago Press.

CHAMBERLIN, T. C. and A. D. SALISBURY (04–06). *Geology,* **1–3**. New York.

CHANDLER, M. E. J. (21). The Arctic flora of the Cam valley. *Quart. Journ. Geol. Soc.* **77**.

—— (22). *Sequoia Couttsiae* Heer at Hordle, Hants. A study of the characters which serve to distinguish *Sequoia* from *Athrotaxis*. *Ann. Bot.* **36**.

—— (23). Geological history of the genus *Stratiotes*. *Quart. Journ. Geol. Soc.* **79**.

—— (25). The Upper Eocene flora of Hordle, Hants. *Palaeontograph. Soc. London.*

CHANEY, R. W. (22). Notes on the flora of the Payette Formation. *Amer. Journ. Sci.* **4**.

—— (24). Quantitative studies of the Bridge Creek flora. *Ibid.* **8**.

—— (25). A comparative study of the Bridge Creek flora and the modern Redwood forest. *Carnegie Inst. Washington, Publ.* 349.

—— (25²). The Mascall flora—its distribution and climatic relation. *Ibid.* 349.

—— (27). Geology and paleontology of the Crooked River basin, etc. *Ibid.* 346.

CHAPMAN, F. (07). Newer Silurian fossils of eastern Victoria. I. *Rec. Geol. Surv. Vict.* **2**, i.

—— (09). Jurassic plant-remains from Gippsland. *Ibid.* **3**, i.

—— (18). On the age of the Bairnsdale Gravels; with a note on the included fossil wood. *Proc. R. Soc. Vict.* **31**.

CHAPMAN, F. and J. C. COOKSON (26). A revision of the Sweet Collection of Triassic plant-remains from Leigh's Creek, Southern Australia. *Trans. R. Soc. S. Australia,* **1**.

CHAUVET, G. and J. WELSCH (16). Les plantes miocènes de Péruzet, près de Larochefoucauld (Charente). *C.R.S. Soc. géol. France,* no. 13.

CHOW, T. C. (24). The Lower Liassic flora of Sofiero and Dompäng in Scania. *Arkiv Bot. Stockholm,* **19**, iv.

CHRIST, H. (10). *Die Geographie der Farne.* Jena.

CHRISTENSEN, C. (28). On the systematic position of *Polypodium vulgare*. *Dansk Bot. Arkiv,* **5**.

CHRISTY, M. (24). The hornbeam (*Carpinus Betulus* L.). *Journ. Ecol.* **12**.

CHURCH, A. H. (95). The structure of the thallus of *Neomeris dumetosa* Lemour. *Ann. Bot.* **9**.

—— (19). *Thallassiophyta and the subaerial transmigration.* Oxford.

—— (26). Reproductive mechanism in land flora. *Journ. Bot.* **64**.

CLARKE, F. W. and H. S. WASHINGTON (24). The composition of the earth's crust. *U.S. Geol. Surv., Profl. Paper* 127.

CLARKE, J. M. (00). The Water Biscuit of Squaw Island, Canandaigua Lake, N.Y. *N.Y. State Mus. Bull.* **8**, no. 37.

—— (21). The oldest of the forests. *Sci. Monthly* (January).

COCKERELL, T. D. A. (06). The fossil fauna and flora of the Florissant (Col.) shales. *Univ. Col. Studies,* **3**, iii.

—— (08). Florissant: a Miocene Pompeii. *Pop. Sci. Monthly,* **74**.

COCKERELL, T. D. A. (08[2]). The fossil flora of Florissant, Col. *Amer. Mus. Nat. Hist.* **24**.
—— (08[3]). Some results of the Florissant expedition 1908. *Amer. Naturalist*, **42**.
—— (16). A Lower Cretaceous flora in Colorado. *Journ. Washington Acad. Sci.* **6**, v.
—— (25). Plant and insect fossils from the Green River Eocene of Colorado. *Proc. U.S. Nat. Mus.* **66**.
COLANI, M. (19). Sur quelques végétaux paléozoiques. *Bull. Serv. géol. Indo-Chine*, **6**, Fasc. i.
COLEMAN, A. P. (26). *Ice Ages Recent and Ancient.* New York.
COLEMAN, A. P. and W. A. PARKS (22). *Elementary Geology with special reference to Canada.* London and Toronto.
COLLET, L. W. (26). The Alps and Wegener's theory. *Geogr. Journ.* (April).
—— (27). *The Structure of the Alps.* London.
COMPTON, R. H. and others (22). A systematic account of the plants collected in New Caledonia and the Isle of Pines by Mr R. H. Compton in 1914. Pt. II. *Journ. Linn. Soc. London*, **45**.
CONWENTZ, H. (86). *Die Flora des Bernsteins*, **2**. Danzig.
COOKSON, I. C. (26). On the occurrence of the Devonian genus *Arthrostigma* in Victoria. *Proc. R. Soc. Vict.* **38**.
COPELAND, E. B. (08). New genus and species of Bornean ferns. *Philipp. Journ. Sci., C, Botany*, **3**.
CORSIN, P. (27). Sur la position systématique du *Zeilleria avoldensis*. *Ann. Soc. géol. Nord*, **52**.
COTTER, G. DE P. (29). The late Palaeozoic glaciation. *Nature*, **124**, p. 723.
COTTON, A. D. (15). A biological survey of Clare Island in the County of Mayo, Ireland. *Proc. R. Irish Acad.* **31**.
CRAMPTON, C. B. and R. G. CARRUTHERS (14). The geology of Caithness. *Mem. Geol. Surv.*
CRÉPIN, F. (74). Description de quelques plantes fossiles de l'étage des Psammites du Condroz. *Bull. Acad. R. Belg.* [2], **38**, viii.
—— (75). Observations sur quelques plantes fossiles des dépôts dévoniens. *Bull. Soc. Roy. Bot. Belg.* **14**.
CRIÉ, L. (89). Beiträge zur Kenntniss der fossilen Flora einiger Inseln des südpacifischen und indischen Oceans. *Palaeont. Abhand.* (Dames and Kayser.)
CROOKALL, R. (25). On the fossil flora of the Bristol and Somerset Coalfield. *Geol. Mag.* **62**.
—— (29). *Coal Measure Plants.* London.
—— (30). *Crossotheca* and *Lyginopteris oldhamia*. *Ann. Bot.* **44**.
DACHNOWSKI, A. (11). The problem of xeromorphy in the vegetation of the Carboniferous period. *Amer. Journ. Sci.* [4], **32**.
DALY, R. A. (07). Limitless ocean of Pre-Cambrian time. *Ibid.* **23**.
—— (12). Geology of the North American Cordillera at the 49th parallel. *Canad. Dept. of Mines, Geol. Surv. Mem.* no. 38.
—— (26). *Our Mobile Earth.* New York and London.
DANDY, J. E. and R. D'O. GOOD (29). Magnoliaceae. *Die Pflanzenareale*, R. 2, v.
DAS-GUPTA, H. C. (29). Batrachian and reptilian remains found in the Panchet beds at Deoli, Bengal. *Journ. Proc. Asiat. Soc. Bengal*, **24**, iv.
DAVID, SIR T. W. EDGEWORTH (96). Evidence of glacial action in Australia in Permo-Carboniferous time. *Quart. Journ. Geol. Soc.* **52**.
—— (24). Discovery of glacial erratics and tillites by T. Blatchford and H. W. B. Talbot in the Kimberley area of Western Australia. *Austral. Assoc. Adv. Sci.* **17**.
DAVID, SIR T. W. E. and E. F. PITTMAN (93). On the occurrence of *Lepidodendron australe* in the Devonian rocks of New South Wales. *Rec. Geol. Surv. N.S. Wales*, **3**, iv.
DAVIES, D. (21). The ecology of the Westphalian and the lower part of the Staffordian series of Clydach Vale and Gilfach Goch (East Glamorgan). *Quart. Journ. Geol. Soc.* **77**.

DAVIES, D. (29). Correlations and Palaeontology of the Coal Measures in eastern Glamorganshire. *Phil. Trans. R. Soc.* **217**.
DAVIS, W. M. (06). Observations in South Africa. *Bull. Geol. Soc. Amer.* **17**.
DAWSON, SIR J. W. (59). On fossil plants from the Devonian rocks of Canada. *Quart. Journ. Geol. Soc.* **15**.
—— (71). The fossil plants of the Devonian and Upper Silurian formations of Canada. *Geol. Surv. Can.* Montreal.
—— (81). Notes on New Erian (Devonian) Plants. *Quart. Journ. Geol. Soc.* **37**.
—— (82). The fossil plants of the Erian and Upper Silurian formations of Canada. Pt. II. *Geol. Surv. Can.* Montreal.
—— (88). *The Geological History of Plants.* London.
—— (95). On collections of Tertiary plants from the vicinity of the city of Vancouver, B.C. *Trans. R. Soc. Can.* Sect. iv.
DEANE, H. (96). President's Address. *Proc. Linn. Soc. N.S. Wales*, **10**.
—— (00). Observations on the Tertiary flora of Australia, etc. *Ibid.* **25**.
DEBEY, M. H. and C. VON ETTINGSHAUSEN (59). Die urweltlichen Acrobryen des Kreidegebirges von Aachen und Maestricht. *Denks. K. Akad. Wiss. Wien*, **17**.
DEHAY, C. and G. DEPAPE (29). Découverte de nouveaux gisements de plantes landéniennes aux environs d'Arras. *Ann. Soc. géol. Nord*, **54**.
DEPAPE, G. (22). *Recherches sur la flore pliocène de la vallée du Rhône.* Paris.
—— (24). Végétaux fossiles des Argiles à Poissons de la Chaussairie et de Lormandière à Chartres. *Bull. Géol. et Min. Bretagne*, **5**, i.
—— (25). La flore des grès landéniens du Nord de la France. *Ann. Soc. géol. Nord*, **50**.
—— (28). Le monde des plantes à l'apparition de l'homme en Europe occidentale. *Ann. Soc. Sci. Bruxelles*, **48**.
DEPAPE, G. and P. FALLOT (28). Les gisements de Burdigalien à plantes de Majorque. *Ann. Soc. géol. Nord*, **53**.
DERBY, O. A. (13). Observations on the stem structure of *Psaronius brasiliensis*. *Amer. Journ. Sci.* **36**.
—— (14). Observations on the crown structure of *Psaronius brasiliensis*. *Ibid.* **38**.
—— (15). Illustrations of the stem structure of *Tietea singularis*. *Ibid.* **39**.
DIX, E. (28). Seeds associated with *Linopteris Muensteri*. *Ann. Bot.* **42**.
DIXEY, F. (26). The Sumbu Coal Measures, Lower Shire, Nyasaland. *Mining Mag.* (March).
DOKTUROWSKY, W. S. (29). Die interglaziale Flora in Russland. *Geol. Fören. Förhand.* **51**.
DON, A. W. R. and G. HICKLING (17). On *Parka decipiens*. *Quart. Journ. Geol. Soc.* **71**.
DONNAN, F. G. (28). The mystery of life. *Nature*, **122**.
DORF, E. (30). Pliocene floras of California. *Carnegie Publication*, **412**.
DOUGLASS, A. E. (19). Climatic cycles and tree-growth. *Ibid.* **289**.
DOUVILLÉ, H. and R. ZEILLER (08). Sur le terrain houiller du Sud-Oranais. *Comp. Rend.* **146**.
DREW, G. H. (14). On the precipitation of calcium carbonate in the sea by marine bacteria, and on the action of denitrifying bacteria in tropical and temperate seas. *Papers from the Tostugas Laboratory, Carnegie Instit. Washington*, **5**.
DRYGALSKI, E. VON (97). *Grönland-Expedition der Gesellschaft für Erdkunde zu Berlin.*
DUCKWORTH, W. L. H. (11). Notes on the Cromer Forest bed. *Camb. Antiq. Soc.* **15**.
DUN, W. S. (97). On the occurrence of Devonian plant-bearing beds on the Genoa River, county of Auckland. *Rec. Geol. Surv. N.S. Wales*, **5**, iii.
DUSÉN, P. (99). Über die Tertiäre Flora der Magellansländer. *Wiss. Ergeb. Schwed. Exped. nach den Magellansländern*, **1**, iv.
DUTTON, E. A. T. (29). *Kenya Mountain.* London.
EDDINGTON, A. S. (23). The borderland of astronomy and geology. *Nature*, **111**.

EDWARDS, T. N. (21). Notes on *Parka decipiens*. *Ann. Mag. Nat. Hist.* **7**.
—— (21²). On a small Bennettitalean flower from the Wealden of Sussex. *Ibid.* **7**.
—— (21³). Fossil coniferous wood from Kerguelen Island. *Ann. Bot.* **35**.
—— (22). An Eocene microthyriaceous fungus from Mull, Scotland. *Trans. Brit. Mycol. Soc.* **8**.
—— (23). On some Tertiary plants from south-east Burma. *Geol. Mag.* **60**.
—— (24). On the cuticular structure of the Devonian plant *Psilophyton*. *Journ. Linn. Soc. London*, **46**.
—— (26). Cretaceous plants from Kaipara, New Zealand. *Trans. N.Z. Instit.* **56**.
—— (26²). Carboniferous plants from the Malay States. *Malayan Branch R. Asiat. Journ.* **4**, ii.
—— (26³). On the occurrence of the Jurassic fern *Laccopteris* in North Africa. *Ann. Mag. Nat. Hist.* [9], **17**.
—— (26⁴). Fossil plants from the Nubian Sandstone of eastern Dafur. *Quart. Journ. Geol. Soc.* **82**.
—— (28). The occurrence of *Glossopteris* in the Beacon Sandstone of Ferrar Glacier, South Victoria Land. *Geol. Mag.* **65**.
—— (29). Lower Cretaceous plants from Syria and Transjordania. *Ann. Mag. Nat. Hist.* [10], **4**.
—— (29²). The Jurassic flora of Sardinia. *Ibid.* **4**.
ELKINS, M. G. and G. R. WIELAND (14). Cordaitean wood from the Indiana black shale. *Amer. Journ. Sci.* **38**.
ELLIS, D. (19). *Iron Bacteria*. London.
ENDÔ, S. (25). Nilssonia bed of Hokkaido and its flora. *Sci. Repts. Tôhoku Imp. Univ. Geol.* **7**, iii.
—— (28). A new Palaeogene species of *Sequoia*. *Jap. Journ. Geol. Geogr.* **6**, i–ii.
ENGELHARDT, H. (85). Die Tertiärflora des Jesuitengrabens bei Kundratitz in Nordböhmen. *N. Act. K. Leop.-Carol. Deutsch. Akad. Naturforsch.* **48**, iii.
—— (85²). Die Crednerien im unteren Quader Sachsens. *Festschrift der Isis, Dresden.*
—— (91). Über Tertiärpflanzen von Chile. *Abhand. Senckenb. naturforsch. Ges.*
—— (94). Über neue fossile Pflanzenreste vom Cerro de Potosi. *Abhand. naturwiss. Ges. Isis*, Abh. **1**.
—— (95). Über neue Tertiärpflanzen Süd-Amerikas. *Abhand. Senckenb. naturforsch. Ges.* **19**.
—— (11). Über Tertiäre Pflanzenreste von Florsheim am Main. *Ibid.* **29**.
—— (12). Neue Beiträge zur Kenntniss der fossilen Tertiärflora Bosniens. *Wiss Mitteil. Bosnia und Herzegovina*, **12**.
ENGELHARDT, H. and F. KINKELIN (08). I. Oberpliocäne Flora und Fauna des Untermaintales. II. Unterdiluviale Flora von Hainstadt am Main. *Abhand. Senckenb. naturforsch. Ges.* **29**, iii.
ENGLER, A. (04). Plants of the northern temperate zone in their transition to the high mountains of tropical Africa. *Ann. Bot.* **18**.
ERDTMAN, O. G. E. (24). Studies in micropalaeontology. *Geol. Fören. Stockholm Förhand.* **46**.
—— (24²). Studies in the micropalaeontology of post-glacial deposits in northern Scotland and the Scotch Isles, etc. *Journ. Linn. Soc. London*, **46**.
—— (27). Literature on pollen-statistics published before 1927. *Geol. Fören. Stockholm Förhand.* **49**.
—— (28). Studies in the post-Arctic history of the forests of north-western Europe. *Ibid.* **50**.
—— (30). Literature on pollen-statistics published during the years 1927–29. *Ibid.* **52**.
ETTINGSHAUSEN, C. VON (53). Die Tertiäre Flora von Häring in Tirol. *Abhand. Geol. Reichsanst. Wien*, **2**, ii.
—— (54). Die Eocene Flora des Monte Promina in Dalmatien. *Denks. Ak. Wiss. Wien*, **8**. [The date should be (55).]

ETTINGSHAUSEN, C. VON (58). Beiträge zur Kenntniss der fossilen Flora von Sotzka in Unter-Steiermark. *Sitzungsber. Ak. Wiss. Wien*, **28**.

—— (69). Die fossile Flora des Tertiär-Beckens von Bilin. *Denks. Ak. Wiss. Wien*, **26**.

—— (77). Die fossile Flora von Sagor in Krain. *Ibid.* **37**.

—— (79). Report on the phyto-palaeontological investigations of the fossil flora of Sheppey. *Proc. R. Soc.* **29**.

—— (87). Beiträge zur Kenntniss der fossilen Flora Neuseelands. *Denks. Ak. Wiss. Wien*, **53**.

—— (88). Contributions to the Tertiary flora of Australia. *Mem. Geol. Surv. N.S. Wales, Palaeont.* no. 2.

EVANS, J. W. (26). La corrélation des roches dévoniennes britanniques. *Soc. géol. Belg. Livre Jubilaire.*

—— (26²). Regions of compression. Presidential Address, Geol. Soc. London. *Quart. Journ.* **82**.

EVANS, J. W. and C. J. STUBBLEFIELD (29). *Handbook of the Geology of Great Britain.* A compilative work. London.

FAIRCHILD, H. L. (20). Pleistocene clays as a chronometer. *Science*, **52**.

FARROW, E. P. (25). *Plant-Life on East Anglian Heaths.* Cambridge.

FEILDEN, COL. H. W. (96). Notes on the glacial geology of Arctic Europe. *Quart. Journ. Geol. Soc.* **52**.

FEILDEN, COL. H. W. and C. E. RANCE (78). Geology of the coasts of Arctic lands visited by the late British Expedition under Capt. Sir G. Nares. *Ibid.* **34**.

FEISTMANTEL, O. (76). Jurassic (Oolitic) flora of Kach. *Mem. Geol. Surv. India: Foss. Flor. Gondwana Syst.* **2**, i.

—— (77). Jurassic (Liassic flora) of the Rajmahal group in the Rajmahal Hills. *Ibid.* **1**, ii.

—— (77²). Jurassic (Liassic) flora of the Rajmahal group from Golapili near Ellore, South Godaveri. *Ibid.* **1**, iii.

—— (77³). Flora of the Jabalpur group in the Son-Narbada region. *Ibid.* **2**, ii.

—— (79). Upper Gondwana flora of the outliers on the Madras coast. *Ibid.* **1**, iv.

—— (79²). The flora of the Talchir-Karharbari beds. *Ibid.* **3**, i.

—— (80). The flora of the Damuda-Panchet divisions. *Ibid.* **3**, ii.

—— (81). *Ibid.* (Supplement), **3**, i.

—— (81²). The flora of the Damuda-Panchet divisions. *Ibid.* **3**, iii.

—— (82). The fossil flora of the South Rewah Gondwana Basin. *Ibid.* **4**.

—— (86). The fossil flora of some of the coalfields in western Bengal. *Ibid.* **4**, ii.

—— (90). Geological and palaeontological relations of the coal and plant-bearing beds of Palaeozoic and Mesozoic age in eastern Australia and Tasmania. *Mem. Geol. Surv. N.S. Wales, Palaeont.* no. 3.

FELIX, J. (83). Die fossilen Hölzer Westindiens. *Sammlg. Paläont. Abhand.* Ser. 1, Heft 1.

FERMOR, L. L. (23). General Report, Geol. Surv. India (1921). *Rec. Geol. Surv.* **54**.

FERNALD, M. L. (25). Persistence of plants in unglaciated areas of boreal America. *Mem. Gray Herbarium, Harvard Univ.* II, *Mem. Acad. Arts Sci.* **15**, no. 3.

—— (28). Unverified geographic ranges. *Science*, **68**.

—— (29). Some relationships of the floras of the northern hemisphere. *Proc. Internat. Congr. Plant Sci.* **2**. Ithaca, N.Y.

—— (30). Unglaciated western Newfoundland. *Harvard Alumni Bull.* (January).

FLAMAND, G. B. M. (07). Observations nouvelles sur les terrains carbonifériens de l'extrême Sud-Oranais. *Comp. Rend.* **145**.

FLICHE, P. (99). Note sur quelques fossiles végétaux de l'Oligocène dans les Alpes françaises. *Bull. Soc. géol. France* [3], **27**.

—— (05). Note sur des bois fossiles de Madagascar. *Ibid.* [4], **5**.

FLICHE, P. (completed by M. R. Zeiller) (10). Flore fossile du Trias. *Bull. Soc. sci. Nancy.*

Florav, N. (27). Über Lössprofile in den Steppen am Schwarzen Meer. *Zeitsch. Gletscherkunde*, **15**.

Florin, R. (19). Zur Kenntnis der *Weichselia reticulata*. *Svensk Bot. Tidsk.* **13**, H. 3–4.

—— (19²). Eine Übersicht der fossilen *Salvinia*-Arten, etc. *Bull. Geol. Instit. Upsala*, **16**.

—— (20). Zur Kenntnis der jungtertiären Pflanzenwelt Japans. *K. Svensk. Vetenskapsakad. Hand.* **61**, i.

—— (20²). Einige chinesische Tertiärpflanzen. *Svensk. Bot. Tidsk.* **14**, ii–iii.

—— (22). On the geological history of the Sciadopitineae. *Ibid.* **16**.

—— (22²). Zur alttertiären Flora der südlichen Mandschurei. *Geol. Surv. China, Palaeont. Sinica*, Ser. A, **1**, i.

—— (25). Zur Kenntniss der paläozoischen Pflanzengattung *Dolerophyllum*. *Svensk Bot. Tidsk.* **19**, ii.

—— (26). Waren Eupodocarpeen (Konif.) in der alttertiären Flora Europas vertreten oder nicht? *Senckenbergiana*, **8**, ii.

—— (27). Preliminary descriptions of some Palaeozoic genera of Coniferae. *Arkiv Bot.* **21** A, 13.

—— (29). Palaeozoic conifers. *Proc. Internat. Congr. Plant Sci.* **1**. Ithaca, N.Y.

—— (29²). Über einige Algen und Koniferen aus dem mittleren und oberen Zechstein. *Senckenbergiana*, **11**.

—— (30). Die Koniferen-Gattung *Libocedrus* in Ostasien. *Svensk Bot. Tidsk.* **24**, i.

Fontaine, W. M. (83). Contributions to the knowledge of the older Mesozoic flora of Virginia. *U.S. Geol. Surv. Mon.* **6**.

—— (89). The Potomac or younger Mesozoic flora. *Ibid.* **15**.

Foslie, M. (29). Contributions to a monograph of the *Lithothamnia* (edited by Dr Printz). *Kong. Hort. Videnskab. Selskab. Mus. Trondhjem.*

Fox-Strangways, C. (92). The Jurassic rocks of Britain. I. *Mem. Geol. Surv.*

Fox-Strangways, C. and G. Barrow (15). The geology of the country between Whitby and Scarborough. *Ibid.*

Fraipont, C. (21). Contribution à la paléophytologie du Wealdien. *Ann. Soc. géol. Belg.* **44** (Mém.).

Frauenfelder, K. O. H. (24). Der Grafit in Finnland, etc. *Geol. Kommiss. Finland*, no. 38.

Frech, F. (95). Das Profil des grossen Colorado-Cañons. *Neues Jahrb. Min.* **2**.

—— (97–02). *Lethaea geognostica*. Stuttgart.

Frentzen, K. (15). *Die Flora des Buntsandsteins Badens*. Inaug. Diss. Heidelberg.

—— (20). Die Flora des Buntsandsteins Badens. *Mitt. badisch. geol. Landesanst.* **8**.

—— (22). Die Keuperflora Badens. *Verhand. naturwiss. Ver. Karlsruhe*, **28**.

—— (26). *Bernouillia franconica* n.sp. aus der Lettenkohle Frankens. *Centralbl. Min. Geol. etc.*

Frič, A. and E. Bayer (01). Studien im Gebiete der böhmischen Kreideformation. *Archiv naturwiss. Landesdurchforsch. Böhmen*, **2**, ii.

Fritel, P. H. (10). Étude sur les végétaux fossiles de l'étage Sparnacien du Bassin de Paris. *Mém. Soc. géol. France*, **16**, iv.

—— (25). Végétaux paléozoiques et organismes problématiques de l'Ouadai. *Bull. Soc. géol. France* [4], **25**.

Fritsch, F. E. (21). Thalassiophyta and the algal ancestry of the higher plants. *New Phyt.* **20**.

Fritsch, K. (29). Die systematische Gruppierung der Pteridophyten. *Ber. deutsch. bot. Ges.* **47**, Heft 10.

Fucini, A. (28). Perchè il verrucano della Verruca è Wealdiano? *Boll. soc. geol. Ital.* **47**.

Fujii, K. (95). On the nature and origin of the so-called Chichi (nipples) of *Ginkgo biloba*. *Bot. Mag. (Tokyo)*.

Gams, H. (27). Trapa. *Die Pflanzenareale*, R. 1, Heft 3.

GAMS, H. (27²). Die Ergebnisse der pollenanalytischen Forschung, etc. *Zeitsch. Gletscherkunde*, **15**.

GARDNER, J. S. (85). The Tertiary basaltic formation in Iceland. *Quart. Journ. Geol. Soc.* **41**.

—— (85²). Eocene ferns from the basalts of Ireland and Scotland. *Journ. Linn. Soc. London*, **21**.

—— (86). A monograph of the British Eocene flora. Vol. I. *Palaeont. Soc. London.*

GARDNER, J. S. and C. VON ETTINGSHAUSEN (82). Monograph of the British Eocene flora. *Ibid.*

GARWOOD, E. J. (13). On the important part played by Calcareous Algae at certain geological horizons, etc. *Geol. Mag.* [5], **10**.

—— (14). Rock-building organisms from the Lower Carboniferous beds of Westmorland. *Ibid.* [6], **1**.

GARWOOD, E. J. and E. GOODYEAR (19). On the geology of the old Radnor district, with special reference to an algal development in the Woolhope Limestone. *Quart. Journ. Geol. Soc.* **74**.

GATES, R. R. (28). Notes on the tundra of Russian Lapland. *Journ. Ecol.* **16**.

GEER, G. DE (12). A geochronology of the last 12,000 years. *Comp. Rend. XIe Congr. Géol. Internat. Stockholm*, **1**.

GEIKIE, A. (79). On the Old Red Sandstone of Western Europe. *Trans. R. Soc. Edinb.* **28**, ii.

—— (97). *The Ancient Volcanoes of Great Britain.* London.

—— (03). *Text-book of Geology.* London.

GEPP, A. and E. S. (11). *The Codiaceae of the Siboga Expedition.* Monograph 62. Leiden.

GERASIMOV, D. A. (30). On the age of the Russian peat-bogs. *Geol. Fören. Stockholm Förhand.* **52**.

GIBSON, W. (20). *Coal in Great Britain.* London.

GIGNOUX, M. (26). *Géologie stratigraphique.* Paris.

GILKINET, A. (75). Sur quelques plantes fossiles de l'étage des Psammites du Condroz. *Bull. Acad. Roy. Belg.* [2], **39**, iv.

—— (75²). Sur quelques plantes fossiles de l'étage du Poudingue de Burnot. *Ibid.* **40**, viii.

—— (09). Quelques plantes fossiles des terres Magellaniques. *Résult. voyage S.Y. "Belgica"*, 1897–99. Anvers.

—— (22). Flore fossile des Psammites du Condroz. *Ann. Soc. géol. Belg. Mém.* **2**.

—— (22²). Plantes fossiles de l'argile plastique d'Ardenne. *Ibid.* **11**.

—— (25). Flore fossile du Landénien de Huppaye (Éocène Inf.). *Mém. Soc. géol. Belg.*

GILLIGAN, A. (20). The petrography of the Millstone Grit of Yorkshire. *Quart. Journ. Geol. Soc.* **75**.

GLÜCK, H. (12). Eine neue gesteinsbildende Siphonee aus dem marinen Tertiär von Süddeutschland. *Mitt. Grossh. Badischen Geol. Landesanst.* **7**, i.

GOEPPERT, H. R. and A. MENGE (83). *Die Flora des Bernsteins.* Danzig.

GOLDRING, WINIFRED (24). The Upper Devonian forest of seed ferns in eastern New York. *N.Y. State Mus. Bull.* **251**. Albany.

—— (26). New Upper Devonian plant-material. *Ibid.* **267**.

—— (27). The oldest known petrified forest. *Sci. Monthly*, **24**.

GOODCHILD, T. G. (00). Desert conditions in Britain. *Trans. Geol. Soc. Glasgow*, **11**.

GORDON, M. M. OGILVIE (28). *Geologisches Wanderbuch der westlichen Dolomiten.* Vienna.

GORDON, W. T. (11). On the structure and affinities of *Metaclepsydropsis duplex. Trans. R. Soc. Edinb.* **48**, i.

—— (12). On *Rhetinangium Arberi*, a new genus of Cycadofilices from the Calciferous Sandstone series. *Ibid.* **48**, iv.

—— (20). Scottish national Antarctic expedition 1902–4: Cambrian organic remains from a dredging in the Weddell Sea. *Ibid.* **52**, iv.

Gothan, W. (07). Die fossilen Hölzer von König Karls Land. *K. Svensk. Vetenskapsakad. Hand.* **42**.

—— (07[2]). Pflanzengeographisches aus der paläozoischen Flora. *Naturwiss. Wochensch.* **32**.

—— (08). Die fossilen Hölzer von der Seymour und Snow Hill Insel. *Wiss. Ergeb. Schwed. Südpolar-Expedit.* 1901–3, **3**, viii.

—— (10). Die fossilen Holzreste von Spitzbergen. *K. Svensk. Vetenskapsakad. Hand.* **45**, viii.

—— (11). Das geologische Alte der Holzreste von König Karls Land. *Zeitsch. deutsch. geol. Ges.* **63**.

—— (13). Die oberschlesische Steinkohlenflora. *Jahrb. preuss. geol. Landesanst.* Heft 75.

—— (14). Die unter-liassische (rhätische) Flora der Umgegend von Nürnberg. *Abhand. naturhist. Ges. Nürnberg*, **19**, iv.

—— (14[2]). Die fossile Flora des Tete-Beckens am Sambesi. *Branca Festschrift.*

—— (15). Pflanzengeographisches aus der paläozoischen Flora mit Ausblicken auf der mesozoischen Folgefloren. *Bot. Jahrb.* **52**, iii.

—— (23). *Leitfossilien* (G. Gürich). Lief. 3, Karb. und Perm.

—— (24). Neue Ansichten über die Bildung von Braunkohlen-Flötzen. *Ber. deutsch. bot. Ges.* **42**, ii.

—— (25). Sobre Restos de Plantas fosiles procedentes de la Patagonia. *Bol. Acad. Nacl. Ciencias Cordoba*, **28**.

—— (25[2]). Gemeinsame Züge und Verschiedenheiten in den Profilen des Karbons der paralischen und limnischen (Binnen-) Kohlenbecken. *Zeitsch. deutsch. geol. Ges.* **77**, iii.

—— (27). Gondwanapflanzen aus der Sierra de Los Llanos. *Abhand. Senckenb. naturforsch. Ges.* **39**, iii.

—— (27[2]). Über einige Kulmpflanzen vom Kossberg bei Plauen. *Abhand. Sächs. geol. Landesamts*, Heft 5.

—— (27[3]). Die Tanner Grauwacke des Unterharzes. *Jahrb. preuss. geol. Landesanst.* **48**.

—— (27[4]). I. Ein araucarioider Coniferenzapfen aus den Tendaguru-Schichten. II. Fossile Pflanzen aus den Karru-Schichten der Umgebung des Uluguru-gebirges in Deutsch-Ostafrika. *Palaeontograph.* Supplt. **7**.

—— (27[5]). Strukturzeigende Pflanzen aus dem Oberdevon von Wildenfels. *Abhand. Sächs. geol. Landesamts*, Heft 3.

—— (28). Bemerkungen zur Alt-Carbonflora von Peru. *Neues Jahrb. Min.* Beilageband **59**.

—— (28[2]). Über einige pflanzenführende Geschiebe Norddeutschlands. *Zeitsch. Geschiebeforschung*, **4**, i.

—— (28[3]). Bemerkungen zu *Gomphostrobus* und *Crossotheca*. *Ber. deutsch. bot. Ges.* **46**, vii.

Gothan, W. and K. Nagalhard (22). Kupferschieferpflanzen aus dem niederrheinischen Zechstein. *Jahrb. preuss. geol. Landesanst.* **42**, i.

Gothan, W. and E. Zimmermann (19). *Pflanzliche und tierische Fossilien der deutschen Braunkohlenlager.* Halle.

Grabau, A. W. (23–28). *Stratigraphy of China.* Pt. I, 1923–4. Pt. II, 1928. Peking.

Grand'Eury, F. C. (12). *Recherches géobotaniques sur les Forêts et Sols fossiles et sur la végétation et la Flore houillères.* Pt. I. Paris.

—— (13). *Ibid.* Pt. II.

Grandori, L. (13). *La Flora dei Calcari Grigi del Veneto.* Pt. I. Padova.

Gray, Asa (78). Forest geography and archaeology. *Amer. Journ. Sci.* **16**.

—— (89). *Scientific Papers by Asa Gray*, **2**. London.

Greenly, E. (94). A Triassic land-surface. *Trans. Edinb. Geol. Soc.* **7**.

Gregory, J. W. (96). *The Great Rift Valley.* London.

—— (20). The African Rift Valley. *Nature*, **104**, p. 518.

GREGORY, J. W. (25). The geology and physical geography of Chinese Tibet, etc. *Phil. Trans. R. Soc.* **213**.

—— (26). The age of the Duruma Sandstone, East Africa. *Geol. Mag.* **63**.

GREGORY, J. W. and B. H. BARRETT (27). The major terms of Pre-Palaeozoic. *Journ. Geol.* **35**.

GRIFFITHS, B. M. (27). Modern pools and Carboniferous analogies. *Geol. Mag.* **64**.

GROSS, H. (30). Das Problem der nacheiszeitlichen Klima- und Florenentwicklung in Nord- und Mittel-Europa. *Beiheft. Bot. Cent.* **47**, i.

GROUT, F. F. and T. M. BRODERICK (19). Organic structure in the Bwalik iron-bearing formation of the Huronian in Minnesota. *Amer. Journ. Sci.* **198**.

GROVES, J. and G. R. BULLOCK-WEBSTER (24). *The British Charophyta*, **2**. London.

GRUNER, J. W. (25). Discovery of life in the Archaean. *Journ. Geol.* **33**.

GÜRICH, G. (06). Les Spongiostromides du Viséen de la Province de Namur. *Mém. Mus. roy. d'hist. nat. Belg.* **3**.

HALKET, A. C. (30). The Rootlets of *Amyelon radicans* Will.; their anatomy, their apices and their endophytic fungus. *Ann. Bot.* **44**.

HALL, J. (47). *Palaeontology of New York*, **1**. Albany.

—— (83). Description of Plate VI in the 32*nd Ann. Rep. N.Y. State Mus. Nat. Hist.* Albany.

HALLE, T. G. (07). Einige krautartige Lycopodiaceen paläozoischen und mesozoischen Alters. *Arkiv Bot. Stockholm*, **7**, v.

—— (08). Zur Kenntnis der mesozoischen Equisetales Schwedens. *K. Svensk. Vetenskapsakad. Hand.* **43**, i.

—— (10). A Gymnosperm with Cordaitean-like leaves from the Rhaetic beds of Scania. *Arkiv Bot. Stockholm*, **9**, xiv.

—— (11). On the geological structure and history of the Falkland Islands. *Bull. geol. Inst. Univ. Upsala*, **11**.

—— (11^2). On the fructifications of Jurassic fern leaves of the *Cladophlebis denticulata* type. *Arkiv Bot. Stockholm*, **10**.

—— (11^3). *Cloughtonia*, a problematic fossil plant from the Yorkshire Oolite. *Ibid.* **10**.

—— (13). Some Mesozoic plant-bearing deposits in Patagonia and Tierra del Fuego, and their floras. *K. Svensk. Vetenskapsakad. Hand.* **51**, iii.

—— (13^2). The Mesozoic flora of Graham Land. *Wiss. Ergeb. Schwed. Südpolar-Expedit.* 1901–3, **3**, xiv.

—— (15). Some xerophytic leaf-structures in Mesozoic plants. *Geol. Fören. Stockholm Förhand.* **37**.

—— (16). Lower Devonian plants from Röragen in Norway. *K. Svensk. Vetenskapsakad. Hand.* **57**, i.

—— (20). *Psilophyton* (?) *Hedii* n.sp. probably a land-plant from the Silurian of Gothland. *Svensk Bot. Tidsk.* **14**.

—— (21). On the sporangia of some Mesozoic ferns. *Arkiv Bot. Stockholm*, **17**.

—— (25). *Tingia*, a new genus of fossil plants from the Permian of China. *Geol. Surv. China*, **7**.

—— (27). Palaeozoic plants from central Shansi. *Geol. Surv. China, Palaeont. Sinica*, **2**, i.

—— (27^2). Fossil plants from south-western China. *Ibid.* fasc. ii.

—— (28). On leaf-mosaic and anisophylly in Palaeozoic Equisetales. *Svensk Bot. Tidsk.* **22**, i–ii.

—— (29). On the habit of *Gigantopteris*. *Geol. Fören. Stockholm Förhand.* **51**.

—— (29^2). Some seed-bearing pteridosperms from the Permian of China. *K. Svensk. Vetenskapsakad. Hand.* **6** [3], no. 8.

HARDER, E. C. (19). Iron-depositing bacteria and their geologic relations. *U.S. Geol. Surv., Profl. Paper* 113.

HARKER, A. (09). *The Natural History of Igneous Rocks*. London.

—— (14). Some remarks on geology in relation to the exact sciences, with an excursus on geological time. *Proc. Yorks. Geol. Soc.* **19**, i.

HARRIS, T. M. (26). The Rhaetic flora of Scoresby Sound. *Meddel. Grønland*, **68**.
—— (29). *Schizopodium Davidi* gen. et sp. nov., a new type of stem from the Devonian rocks of Australia. *Phil. Trans. R. Soc.* **217**.
HARTZ, N. (96). Planteforsteninger fra Kap Stewart Østgrønland. *Meddel. Grønland*, **19**.
HATSCHEK, E. (19). *An Introduction to the Physics and Chemistry of Colloids*. London.
—— (25). *Laboratory Manual of Elementary Colloid Chemistry*. London.
HAUG, E. (20). *Traité de Géologie*, **2**. Paris.
HAUGHTON, S. (60). On *Cyclostigma*, a new genus of fossil plants from the O.R.S. of Kiltorcan, Kilkenny, etc. *Ann. Mag. Nat. Hist.* [3], **5**.
HAVILAND, M. D. (26). *Forest, Steppe, and Tundra*. Cambridge.
HAYDEN, H. H. (04). The geology of Spiti. *Mem. Geol. Surv. India*, **36**, i.
HEARD, A. (27). Old Red Sandstone plants from Brecon. *Quart. Journ. Geol. Soc.* **83**.
HEDGES, E. S. and J. E. MYERS (26). *The Problem of Physico-Chemical Periodicity*. London.
HEER, O. (55–59). *Flora Teriaria Helvetiae*. 3 vols. Winterthur.
—— (62). On the fossil flora of Bovey Tracey. *Phil. Trans. R. Soc.* **152**.
—— (68). *Die fossile Flora der Polarländer: Flora Fossilis Arctica*, **1**.
—— (71). *Flora Fossilis Arctica*, **2**.
—— (76). *Flora Fossilis Helvetiae*. Zürich.
—— (77). *Flora Fossilis Arctica*, **4**.
—— (78). *Ibid.* **5**.
—— (82). *Ibid.* **6**.
HENRY, A. and M. G. FLOOD (20). The history of the London plane, *Platanus acerifolia*, with notes on the genus *Platanus*. *Proc. R. Irish Acad.* **35**.
HERITSCH, F. (29). *The Nappe theory in the Alps*. (Translated by Prof. Boswell.) London.
HICKLING, E. (08). The O.R.S. of Forfarshire. *Geol. Mag.* **5**.
—— (12). On the geology and palaeontology of Forfarshire. *Proc. Geol. Assoc.* **23**.
HINDE, G. J. (13). *Solenopora Garwoodi* sp. nov. from the Lower Carboniferous in the north-west of England. *Geol. Mag.* **10**.
HIRMER, M. (25). Ergebnisse der Forschungsreisen Prof. E. Stromers in den Wüstern Ägyptens. *Abhand. Bayer. Akad. Wiss.* **30**, iii.
—— (27). *Handbuch der Paläobotanik*, **1**. München and Berlin.
—— (28). Über Vorkommen und Verbreitung der Dolomitknollen und deren Flora. *Congr. stratig. Carbon. Heerlen.*
—— (30). Psilophyten-Reste aus deutschem Unterdevon. *Sitzungsber. Bay. Akad. Wiss.* Heft 1.
—— (30^2). Über ein zweites in den Hunsrück-Schiefern gefundenes Stück von *Maucheria gemündensis* Broili. *Ibid.*
HOBBS, W. H. (21). *Earth Evolution and its Facial Expression*. New York.
HØEG, O. (27). *Dimorphosiphon rectangulare:* preliminary note on a new Codiacea from the Ordovician of Norway. *Avhand. Norsk. Videnskapsakad. Oslo.*
—— (29). Studies in Stromatolites. I. A post-glacial marine Stromatolite from south-eastern Norway. *K. Norsk. Videnskap. Selskaps. Skrift.* no. 1.
HØEG, O. and J. KIAER (26). A new plant-bearing horizon in the marine Ludlow of Ringerike. *Ibid.*
HÖGBOM, A. G. (10). Pre-Cambrian geology of Sweden. *Bull. geol. Instit. Univ. Upsala*, **10**.
HÖRICH, O. (10). *Knorripteris mariana*. *Abbild. Beschreib. Foss. Pflanz. Potonié.* Lief. 8.
—— (12). *Knorripteris Jutieri*. *Palaeobot. Zeitsch.* **1**.
—— (15). Einige strukturbietende Pflanzenreste aus deutschem Culm und Devon. *Jahrb. preuss. geol. Landesanst.* **36**, Teil 1, Heft 3.
HOLDEN, H. S. (30). On the structure and affinities of *Ankyropteris corrugata*. *Phil. Trans. R. Soc.* **218**.

HOLDEN, R. (13). Some fossil plants from eastern Canada. *Ann. Bot.* **27**.
—— (16). A fossil wood from Burma. *Rec. Geol. Surv. India*, **47**.
—— (17). On the anatomy of two Palaeozoic stems from India. *Ann. Bot.* **31**.
HOLLICK, A. (06). The Cretaceous flora of southern New York and New England. *U.S. Geol. Surv. Mon.* 50.
—— (10). A new fossil fucoid. *Bull. Torr. Bot. Club*, **37**.
—— (27). The flora of the Saint Eugene Silts, Kootenay Valley, British Columbia. *Mem. N.Y. Bot. Gard.* **7**.
—— (29). New species of fossil plants from the Tertiary shales near De Begue, Colorado. *Contrib. N.Y. Bot. Gard.* no. 306.
HOLLICK, A. and E. C. JEFFREY (09). Studies of Cretaceous coniferous remains from Kreischerville, New York. *Mem. N.Y. Bot. Gard.* **3**.
HOLLICK, A. and G. C. MARTIN (30). The Upper Cretaceous Floras of Alaska. *U.S. Geol. Surv., Profl. Paper* 159.
HOLLOWAY, J. G. (18). The Prothallus and young plant of *Tmesipteris*. *Trans. N.Z. Instit.* **50**.
HOLM, T. (22). Contributions to the morphology, synonymy and geographical distribution of Arctic plants. *Rep. Canadian Arct. Exped.* 1913–18, **5**. Ottawa.
HOLMES, A. (15). Radioactivity and the measurement of geological time. *Proc. Geol. Assoc.* **26**.
—— (20). The measurement of geological time. *Discovery* (no. 4, April).
—— (27). *The Age of the Earth*. London.
HOLMES, A. and H. F. HARWOOD (28). The age and composition of the Whin Sill and the related dykes of the north of England. *Mining Mag.* **21**.
HOLMES, A. and R. W. LAWSON (27). Features involved in the calculation of the ages of radioactive minerals. *Amer. Journ. Sci.* **13**.
HOLMES, T. RICE (07). *Ancient Britain and the Invasions of Julius Caesar*. Oxford.
HOLTEDAHL, O. (19). On the Palaeozoic formations of Finnmarken in northern Norway. *Amer. Journ. Sci.* **47**.
—— (21). On the occurrence of structures like Walcott's Algonkian Algae in the Permian of England. *Ibid.* [5], **1**.
HOOKER, J. D. (53). Notice of the discovery of fossil plants in the Shetland Islands, etc. *Quart. Journ. Geol. Soc.* **9**.
—— (62). Outlines of the distribution of Arctic plants. *Trans. Linn. Soc. London*, **23**.
—— (91). *Himalayan Journals*. London. (Originally published in 1854.)
HORN, G. and A. K. ORVIN (28). Geology of Bear Island. *Skript. Svalbard Ishavet.* no. 15. Oslo.
HOSKINS, F. H. (30). Celloidin transfer-method for thin rock sections. *Bot. Gaz.* **89**.
HOWCHIN, W. (08). Glacial beds of Cambrian age in South Australia. *Quart. Journ. Geol. Soc.* **64**.
—— (10). Description of a new and extensive area of Permo-Carboniferous glacial deposits in South Australia. *Trans. Proc. and Rep. R. Soc. S. Australia*, **34**.
—— (12). Australian glaciations. *Journ. Geol.* **20**.
—— (24). Further discoveries of Permo-Carboniferous glacial features near Hallett's Cove. *Trans. R. Soc. S. Australia*, **48**.
—— (29). *The Geology of South Australia*. Adelaide.
HOWE, M. A. (12). The building of Coral Reefs. *Science* [N.S.], **35**.
—— (18). On some fossil and recent Lithothamnieae of the Panama Canal Zone. *Bull. U.S. Nat. Mus.* no. 103.
—— (19). Tertiary Calcareous Algae from the Islands of St Bartholomew, Antigua, and Anguilla. *Contribut. Geol. Palaeont. W. Indies, Carnegie Instit.* no. 291.
—— (22). Two new Lithothamnieae, Calcareous Algae, from the Lower Miocene of Trinidad. *Proc. U.S. Nat. Mus.* **62**.
HUNTINGTON, E. (14). The climatic factor. *Carnegie Publ.* no. 192.
HUNTINGTON, E. and S. S. VISHER (22). *Climatic Changes*. New Haven.

Hutchinson, J. (26). *The Families of Flowering Plants*, **1**. London.

Huxley, T. H. (96). Geological contemporaneity and persistent types of life. *Discourses, biological and geological.* London.

Hylander, C. J. (22). A Mid-Devonian *Callixylon*. *Amer. Journ. Sci.* **4**.

Ihering, H. V. (28). Die phytogeographischen Grundgesetze. *Engler's Bot. Jahrb.* **62**.

Ingvarson, F. (03). Om Drifveden i norra ishafvet. *K. Svensk. Vetenskapsakad. Hand.* **37**, i.

Irmscher, E. (22). Pflanzenverbreitung und Entwicklung der Kontinente. *Mittlg. Inst. Allg. Bot. Hamburg*, **5**.

Jack, R. L. and R. Etheridge (92). *The Geology and Palaeontology of Queensland and New Guinea.* Brisbane.

Jaennicke, F. (99). Studien über die Gattung *Platanus*. *N. Acta K. Leop.-Carol. Deutsch. Akad. Naturforsch.* **77**.

Jeans, Sir J. (29). *The Universe Around Us.* Cambridge.

Jeffrey, E. C. (09). On the nature of the so-called algal or boghead coals. *Rhodora*, **2**.

—— (10). The nature of some supposed algal coals. *Proc. Amer. Acad. Arts Sci.* **46**, no. 12.

—— (24). The origin and organization of coal. *Mem. Amer. Acad. Arts Sci.* **15**, i.

—— (25). *Coal and Civilization.* New York.

Jeffrey, E. C. and M. A. Chrysler (06). The Lignites of Brandon. *Fifth Report, Vermont State Geologist.*

Jeffreys, H. (24). *The Earth, its Origin, History and Physical Constitution.* Cambridge.

—— (26). On Prof. Joly's Theory of Earth History. *Phil. Mag.* **1**.

—— (26²). The Earth's Thermal History, etc. *Geol. Mag.* **63**.

Jessen, K. and V. Milthers (28). Stratigraphical and palaeontological studies of interglacial freshwater deposits in Jutland and north-west Germany. *Danmarks Geol. Unders.* Raek. 2, no. 48.

Johansson, N. (20). Neue mesozoische Pflanzen aus Andö in Norwegen. *Svensk Bot. Tidsk.* **14**.

—— (22). Die rhätische Flora der Kohlengruben bei Stabbarp und Skromberga in Schonen. *K. Svensk. Vetenskapsakad. Hand.* **63**, no. 5.

Johnson, T. (11). The occurrence of *Archaeopteris Tschermaki* and other species of *Archaeopteris* in Ireland. *Sci. Proc. R. Dublin Soc.* **13**.

—— (11²). Is *Archaeopteris* a pteridosperm? *Ibid.* **13**.

—— (12). *Forbesia cancellata*, gen. et sp. nov. (*Sphenopteris* sp. Baily). *Ibid.* **13**.

—— (14). *Bothrodendron kiltorkense*, its *Stigmaria* and cone. *Ibid.* **14**.

—— (14²). *Ginkgophyllum kiltorkense* sp. nov. *Ibid.* **14**.

Johnson, T. and J. G. Gilmore (21). The occurrence of *Dewalquea* in the Coal-bore at Washing Bay. *Ibid.* **16**.

—— (21²). The occurrence of a *Sequoia* at Washing Bay. *Ibid.* **16**.

—— (22). The lignite of Washing Bay, Co. Tyrone. *Ibid.* **17**.

Joly, J. (25). *The Surface History of the Earth.* Oxford.

Jongmans, W. J. (09). *The Flora of the Coalfields of the Netherlands.* The Hague.

—— (11). Anleitung zur Bestimmung der Karbonpflanzen West-Europas. *Meded. Rijksopspor Delfshoffen.* Freiberg.

—— (17). *Flora of the Carboniferous of the Netherlands, etc.* Vol. I. *A Monograph of the Calamites of Western Europe.* Kidston and Jongmans. The Hague.

—— (28). Congrès pour l'étude de la stratigraphie carbonifère. Heerlen (1927). *Comp. Rend.* Liége.

—— (28²). Stratigraphische Untersuchungen im Karbon von Limburg. *Ibid.*

—— (28³). Stratigraphie van het Karbon in het algemeen en van Limburg in het bijzonder. *Meded.* 6, *Geol. Bur. Nederland. Mijngebied te Heerlen.*

Jongmans, W. J. and W. Gothan (25). Beiträge zur Kenntnis der Flora des Oberkarbons von Sumatra. *Meded.* 2, *Ibid.*

KALKOWSKY, E. (08). Oolith und Stromatolith im norddeutschen Buntsandstein. *Zeitsch. deutsch. geol. Ges.*

KARPINSKY, A. (06). Die Trochilisken. *Mém. com. géol. St Pétersbourg*, **27**.

KAWASAKI, S. (25). Some older Mesozoic plants in Korea. *Bull. Geol. Soc. Chosen* (*Korea*), **4**, i.

—— (26). Additions to the older Mesozoic plants in Korea. *Ibid.* **4**, ii.

—— (27). The flora of the Heian System. Pt. I. *Ibid.* **6**.

KAYSER, E. (23–24). *Lehrbuch der allgemeinen Geologie* (4 vols.). Stuttgart.

KEIDEL, J. (22). Sobre la distribución de los depositos glaciares del Permico conocidos en la Argentine. *Bol. Acad. nac. cienc. Cordoba*, **25**.

KELLER, P. (28). Pollenanalytische Untersuchungen an Schweizer-Mooren und ihre Floren-geschichtliche Deutung. *Geobot. Instit. Rübel, Zürich*, Heft 5.

KENDALL, P. F. and H. E. WROOT (24). *Geology of Yorkshire*, **1**, **2**.

KERNER, F. VON (95). Kreidepflanzen von Lesina. *Jahrb. k. geol. Reichsanst.* **45**, i.

KERSHAW, E. M. (09). The structure and development of the ovule of *Myrica Gale*. *Ann. Bot.* **23**.

KIDSTON, R. (82). Report on fossil plants collected by the Geological Survey of Scotland in Eskdale and Liddersdale. *Trans. R. Soc. Edinb.* **30**.

—— (84). On a new species of *Lycopodites* from the Calciferous Sandstone Series of Scotland. *Ann. Mag. Nat. Hist.* **14**.

—— (86). On the fossil flora of the Radstock Series of the Somerset and Bristol Coalfield. *Trans. R. Soc. Edinb.* **33**, ii.

—— (88). On the fossil flora of the Staffordshire Coalfield. I. *Ibid.* **35**, i.

—— (89). On some fossil plants from Teilia Quarry, Gwaenysgor, near Prestatyn, Flintshire. *Ibid.* **35**, ii.

—— (91). On the fossil flora of the Staffordshire Coalfield. II. *Ibid.* **36**, i.

—— (93). On *Lepidophloios*, etc. *Ibid.* **37**, iii.

—— (94). On the various divisions of the British Carboniferous rocks as determined by the fossil flora. *Proc. R. Phys. Soc. Edinb.* **12**.

—— (96). On the fossil flora of the Yorkshire Coalfield. *Trans. R. Soc. Edinb.* **38**, ii.

—— (97). *Ibid.* **39**, i.

—— (01). The flora of the Carboniferous Period. *Proc. Yorks. Geol. Polytechnic Soc.* **14**, ii.

—— (02). *Ibid.* **14**, iii.

—— (03). The fossil plants of the Carboniferous rocks of Canonbie, etc. *Trans. R. Soc. Edinb.* **40**, iv.

—— (08). On a new species of *Dineuron* and of *Botryopteris* from Pettycur, Fife. *Ibid.* **46**, ii.

—— (11). Les végétaux houillers recueillis dans le Hainaut Belge. *Mém. Mus. roy. d'hist. nat. Belg.* **4**.

—— (14). On the fossil flora of the Staffordshire Coalfield. III. *Trans. R. Soc. Edinb.* **50**, i.

—— (17). The Forest of Wyre and the Tillerstone Clee Hill Coalfields. *Ibid.* **51**, iv.

—— (23–25). Fossil plants of the Carboniferous rocks of Great Britain. *Mem. Geol. Surv.* (*Palaeont.*) **2**.

—— (23). *Ibid.* **2**, Pt. 2.

—— (23^2). *Ibid.* **2**, Pt. 3.

—— (24). *Ibid.* **2**, Pt. 5.

KIDSTON, R. and D. T. GWYNNE-VAUGHAN (07). On the fossil Osmundaceae. Pt. I. *Trans. R. Soc. Edinb.* **45**, iii.

—— (08). *Ibid.* Pt. II. *Ibid.* **46**, ii.

—— (09). *Ibid.* Pt. III. *Ibid.* **46**, iii.

—— (11). On a new species of *Tempskya* from Russia. *Verhand. russ. min. Ges.* **48**.

—— (12). On the Carboniferous flora of Berwickshire. *Trans. R. Soc. Edinb.* **48**, ii.

KIDSTON, R. and W. J. JONGMANS (15–17). *Flora of the Carboniferous of the Netherlands and adjacent regions. I. A Monograph of the Calamites of Western Europe.* The Hague.

KIDSTON, R. and W. H. LANG (17). On Old Red Sandstone plants showing structure, from the Rhynie chert bed, Aberdeenshire. Pt. I. *Rhynia. Trans. R. Soc. Edinb.* **51**, iii.

—— (20). *Ibid.* Pts. II and III. *Rhynia, Hornea, Asteroxylon. Ibid.* **52**, iii.

—— (21). *Ibid.* Pt. IV. Restorations. *Ibid.* **52**, iv, p. 831.

—— (21^2). *Ibid.* Pt. V. Thallophyta. *Ibid.* **52**, iv, p. 855.

—— (23). On *Palaeopitys Milleri. Ibid.* **53**, ii.

—— (23^2). Notes on fossil plants from the O.R.S. of Scotland. I. *Hicklingia. Ibid.* **53**, ii.

—— (24). On the presence of tetrads of resistant spores in the tissue of *Sporocarpon furcatum* from the Upper Devonian of America. *Ibid.* **53**, iii.

—— (24^2). Notes on fossil plants from the O.R.S. of Scotland. II. *Nematophyton.* III. *Pachytheca. Ibid.* **53**, iii.

KINDLE, E. M. (21). Mackenzie River driftwood. *Geogr. Rev.*

KIRCHHEIMER, F. (30). Die fossilen Vertreten der Gattung *Salvinia. Planta: Archiv wiss. Bot.* **9**.

—— (30^2). Die fossilen Vertreten der Gattung *Salvinia. Ibid.* **11**, i.

KJELLMAN, F. R. (83). The Algae of the Arctic Sea. *K. Svensk. Vetenskapsakad. Hand.* **20**, no. 5.

KNOWLTON, F. H. (89). Description of a problematic organism from the Devonian at the Falls of Ohio. *Amer. Journ. Sci.* **37**.

—— (93). Fossil flora of Alaska. *Bull. Geol. Soc. Amer.* **5**.

—— (94). A revision of the fossil flora of Alaska, etc. *Proc. U.S. Nat. Mus.* **17**.

—— (98). Report on the fossil plants of the Payette formation. *U.S. Geol. Surv.* 18*th Ann. Rep.*

—— (99). Fossil flora of the Yellowstone National Park. *U.S. Geol. Surv. Mon.* 32.

—— (02). Fossil flora of the John Day Basin. *U.S. Geol. Surv. Bull.* 204.

—— (04). Fossil plants from Kukak Bay. *Harriman Alaska Expedit.* **4**.

—— (14). The Jurassic flora of Cape Lisburne. *U.S. Geol. Surv., Profl. Paper* 85—D.

—— (16). A review of the fossil plants in the United States National Museum from the Florissant Lake beds at Florissant. *Proc. U.S. Nat. Mus.* **51**.

—— (16^2). A Lower Jurassic flora from the Upper Matunuska valley, Alaska. *Ibid.* **51**.

—— (19). A catalogue of the Mesozoic and Cainozoic plants of North America. *U.S. Geol. Surv. Bull.* 696.

—— (22). The Laramie flora of the Denver basin. *U.S. Geol. Surv., Profl. Paper* 130.

—— (23). Revision of the flora of the Green River Formation, etc. *Ibid.* 131—F.

—— (24). Upper Cretaceous and Tertiary Formations of the western part of the San Juan Basin, Colorado and New Mexico. *Ibid.* 134.

—— (26). Flora of the Latah formation of Spokane, Washington, and Cœur d'Alene, Idaho. *Ibid.* 140—A.

—— (27). *Plants of the Past.* Princeton.

—— (30). The Flora of the Denver and Associated Formations of Colorado. Edited by E. W. Berry. *U.S. Geol. Surv., Profl. Paper* 155.

KON'NO, E. (29). On the genera *Tingia* and *Tingiostachya* from the Lower Permian and the Permo-Triassic beds in northern Korea. *Jap. Journ. Geol. Geogr.* **6**, iii–iv.

KOOPMANS, R. G. (28). Researches on the flora of the coal-balls from the Finefrau-Nebenbank horizon in the Prov. Limburg. *Geol. Bur. Nederland. Mijngebied.*

KÖPPEN, W. and A. WEGENER (24). *Die Klimate der geologischen Vorzeit.* Berlin.

KRÄUSEL, R. (19). Die Pflanzen des schlesischen Tertiärs. *Jahrb. preuss. geol. Landesanst.* **38**, Tl. II, Heft. i–ii.

—— (20). Nachträge zur Tertiärflora Schlesiens. I. *Ibid.* **39**, Tl. I, Heft iii.

Kräusel, R. (20²). *Ibid.* II. *Ibid.* **39**, Tl. I, Heft iii.
—— (20³). *Ibid.* III. *Ibid.* **40**, Tl. I, Heft iii.
—— (21), (21²). Über einige Pflanzen aus dem Keuper von Lunz. *Jahrb. preuss. geol. Landesanst.* **41**, Teil I, i.
—— (22). Fossile Hölzer aus dem Tertiär von Süd-Sumatra. *Verhand. Geol.-Mijnbouwkundig Genootschap voor Nederlanden Koloniën* (*Geol. Ser.*), **5**.
—— (22²). Beiträge zur Kenntnis der Kreideflora. *Meded. 's Rijks Geol. Dienst*, Ser. A, no. 2. Leiden.
—— (23). *Nipadites borneensis* n. sp., eine fossile Palmenfrucht aus Borneo. *Senckenberg.* **5**, iii–iv.
—— (24). *Archaeoxylon Krasseri*, ein Pflanzenrest aus den böhmischen Präkambrium. *Lotos*, **72**.
—— (24²). Beiträge zur Kenntnis der fossilen Flora Südamerikas. I. Fossile Hölzer aus Patagonien, etc. *Arkiv Bot. Stockholm*, **19**, ix.
—— (25). Der Stand unserer Kenntnisse von der Tertiärflora Niederländisch-Indiens. *Verhand. Geol.-Mijnbouwkundig Genootschap Nederland. Koloniën* (*Geol. Ser.*), **8**.
—— (25²). *Raumeria Reichenbachiana* und ihre verwandten ausgestorbenen Cycadophyten des Mesozoikums. *Senckenberg. naturforsch. Ges. Bericht* 55.
—— (26). Über einige fossile Hölzer aus Java. *Leidsche geol. Meded.* **11**, i.
—— (28). Paläobot. Notizen, X–XI. *Senckenberg.* **10**.
—— (28²). Paläobot. Braunkohlenstudien. *Abh. naturforsch. Ges. Görlitz*, **2**, ii.
—— (29). *Die paläobotanischen Untersuchungsmethoden, ein Zeitfaden für die Untersuchung fossiler Pflanzen sowie der aus ihnen aufgebauten Gesteine.* Jena.
—— (29²). Fossile Pflanzen aus den Tertiär von Süd-Sumatra. *Verhand. Geol.-Mijnbouwkundig Genootschap Nederland. Koloniën* (*Geol. Ser.*), **9**.
Kräusel, R. and W. J. Jongmans (23). Über pflanzenführende Kreideschichten aus der Umgebung von Heerlen und die Verbreitung des Aachenen Sandes in den südlichen Niederlanden. *Senckenberg.* **5**, v–vi.
Kräusel, R. and P. Range (28). Beiträge zur Kenntnis der Karruformation Deutsch-Südwest-Afrikas. *Beit. geol. Erforsch. deutsch. Schutzgebiete*, Heft 20.
Kräusel, R. and G. Schönfeld (24). Fossile Hölzer aus der Braunkohle von Süd-Limburg. *Abhand. Senck. naturforsch. Ges.* **38**, iii.
Kräusel, R. and E. Stromer (24). Die fossilen Floren Ägyptens. *Abhand. Bayer. Akad. Wiss.* **30**, ii.
Kräusel, R. and H. Weyland (26). Beiträge zur Kenntnis der Devonflora. *Abhand. Senck. naturforsch. Ges.* **40**, ii.
—— (29). *Ibid.* III. *Ibid*, **41**.
—— (30). Über Pflanzenreste aus dem Devon Deutschlands. *Senckenbergiana*, **12**.
Krasser, F. (91). Über die fossile Flora der rhätischen Schichten Persiens. *Sitzungsber. Ak. Wiss. Wien*, **100**.
—— (96). Beiträge zur Kenntniss der fossilen Kreideflora von Kunstadt in Mähren. *Beit. Paläont. geol. Österreich-Ungarns und des Orients, Wien*, **10**, iii.
—— (00). Die von W. A. Obrutschew in China und Centralasien 1893–94 gesammelten fossilen Pflanzen. *Denks. Ak. Wiss. Wien*, **70**.
—— (05). Fossile Pflanzen aus Transbaikalien, der Mongolei und Mandschurei. *Ibid.* **78**.
—— (06). Über die fossile Kreideflora von Grimbach in Niederösterreich. *Sitzungsber. Ak. Wiss. Wien, Akad. Anzeig.* **3**.
—— (08). Kritische Bemerkungen und Übersicht über die bisher zutage geförderte fossile Flora des unteren Lias der österreichischen Voralpen. *Wiesner-Festschrift.*
—— (09). Zur Kenntniss der fossilen Flora der Lunzer Schichten. *Jahrb. k. k. geol. Reichs.* **59**, i.
—— (09²). Die Diagnosen der von D. Stur in der obertriadischen Flora der Lunzerschichten als Marattiaceenarten unterschiedenen Farne. *Sitzungsber. Ak. Wiss. Wien*, **118**, i.

KRASSER, F. (18). Die fossile Flora der Williamsonien bergenden Juraschichten von Sardinien. *Anzeig. Ak. Wiss. Wien*, **4**.

—— (19). Studien über die fertile Region der Cycadophyten aus den Lunzerschichten. *Deutsch. Ak. Wiss. Wien*, **97**.

—— (20). Die Doggerflora von Sardinien. *Sitzungsber. Ak. Wiss. Wien*, **129**, I, Heft. i–ii.

KRIGE, L. J. (29). Magmatic cycles, continental drift and Ice Ages. *Proc. Geol. Soc. S. Africa.*

KRYSHTOFOVICH, A. (10). Jurassic plants from Ussuriland. *Mém. com. géol. St Pétersb.* [N.S.] Livr. 56.

—— (11). Über problematische Algenreste *Taonurus-Spirophyton* aus Juraablagerungen des Ussuri-golfes. *Bull. com. géol. Russie*, **30**.

—— (12). Mesozoic plant-remains from the eastern Urals. *Ibid.* **31**.

—— (13). Die Pflanzenreste aus den Juraablagerungen der Krym. *Soc. nat. Amis de la Nature en Crimée, Bull.* 11.

—— (14). The latest finds of remains of a Sarmatian flora in southern Russia. (Russian.) *Bull. Acad. Imp. Sci. St Pétersb.*

—— (14^2). A discovery of remains of an angiospermous flora in the Cretaceous beds of the Ural Prov. (Russian.) *Ibid.*

—— (15). The butter nut (*Juglans cinerea*) from freshwater deposits of the Province of Yakoutsk. *Mém. com. géol.* Livr. 24.

—— (15^2). Jurassic plants from Amurland. (Russian.) *Trav. Mus. géol. Pierre le Grand près l'Acad. Imp. Sci.* **8**.

—— (15^3). Plant remains from Jurassic lake-deposits of Transbaikalia. *Mém. Soc. imp. Russ. Min.* **51**.

—— (18). On the Cretaceous flora of Russian Sakhalin. *Journ. Coll. Sci. Imp. Univ. Tokyo*, **40**.

—— (18^2). Occurrence of the palm *Sabal nipponica* n. sp. in the Tertiary rocks of Hokkaido and Kyûshû. *Journ. Geol. Soc. Tokyo*, **25**, no. 303.

—— (18^3). On the Cretaceous age of the "Miocene" flora of Sakhalin. *Amer. Journ. Sci.* **46**.

—— (23). *Pleuromeia* and *Hausmannia* in eastern Siberia. *Ibid.* **5**.

—— (26). Geology. *The Pacific Russian Scientific Investigations.* Leningrad.

—— (26^2). Some fossil plants from the Jurassic strata of the northern Caucasus. *Bull. com. géol. Leningrad*, **45**.

—— (27). *Nipadites Burtinii* from the Eocene of south-western Ukraina. *Ibid* **45**.

—— (27^2). Some plant impressions of the Tertiary Sandstone near Adjamka, western Ukraina. *Ibid.* **46**.

—— (27^3). Some traces of the old Devonian flora in Urals, Turkestan and Siberia. *Ibid.* **46**.

—— (29). The oldest Angiosperms in the Cretaceous of Asia, etc. *Amer. Journ. Sci.* **18**.

—— (29^2). Evolution of the Tertiary flora in Asia. *New Phyt.* **28**.

—— (29^3). Discovery of the oldest dicotyledons of Asia in the equivalents of the Potomac group in Suchan, Ussuriland, Siberia. *Bull. com. géol. Russie*, **48**.

KRYSHTOFOVICH, A. and J. PALIBIN (15). Tertiary plants from the Province Tourgai (Russian). *Bull. Acad. Imp. Sci.*

KUBART, B. (14). Über die Cycadofilicineen *Heterangium* und *Lyginodendron* aus dem Ostrauer Kohlenbecken. *Österreich. bot. Zeitsch.*

—— (24). Beiträge zur Tertiärflora der Steiermark. *Arb. Phytopal. Laboratoriums Univ. Graz*, **1**.

—— (28). Das Problem der tertiären Nordpolarfloren. *Ber. deutsch. bot. Ges.* **46**, vi.

KURTZ, F. (94). Contribuciones a la palaeophytologia Argentina. *Rev. Mus. de la Plata*, **6**.

—— (99). *Ibid.* **10**.

KURTZ, F. (01). Sur l'existence d'une flore Rajmahalienne dans le Gouvernement du Neuquen. *Ibid.* **10**.
—— (21). Atlas de plantas fosiles de la Republica Argentina. *Act. Acad. nac. cienc. Cordoba*, **7**.
LAKE, P. (22). Wegener's Displacement Theory. *Geol. Mag.* **59**.
—— (23). Wegener's hypothesis of Continental Drift. *Nature*, **111**.
LAKE, P. and R. H. RASTALL (13). *A text-book of Geology.* London.
LANE, A. C. (08). Van Hise on the division of the Pre-Cambrian. *Geol. Mag.* [5], **5**.
LANG, W. H. (24). Notes on fossil plants from the O.R.S. of Scotland. *Trans. R. Soc. Edinb.* **53**, iii.
—— (25). Contributions to the study of the O.R.S. flora of Scotland. I, II. *Ibid.* **54**, ii.
—— (26). *Ibid.* III, IV, V (*Hostimella Thomsoni*, *Protolepidodendron*, etc.). *Ibid.* **54**, iii.
—— (26²). A cellulose-film transfer method in the study of fossil plants. *Ann. Bot.* **40**.
—— (27). Contributions to the study of the O.R.S. flora of Scotland. VI–VII (*Zosterophyllum*, *Pseudosprochnus*). *Trans. R. Soc. Edinb.* **55**, ii.
—— (29). On fossil wood (*Dadoxylon Hendriksi*, n. sp.) and other plant-remains from the clay-slate of southern Cornwall. *Ann. Bot.* **43**.
LANG, W. H. and I. COOKSON (27). On some early Palaeozoic plants from Victoria, Australia. *Mem. Proc. Manchester Lit. Phil. Soc.* **71**.
LANGERON, M. (99). Contributions à l'étude de la flore fossile de Sézanne. I. *Bull. Soc. d'hist. nat. d'Autun*, **12**.
—— (00). *Ibid.* II. *Ibid.* **13**.
—— (02). *Ibid.* III. *Ibid.* **15**.
—— (02²). Note sur une empreinte remarquable provenant des cinérites du Cantal *Paliurites Martyi* (Langeron). *Ibid.* **15**.
—— (07). Note préliminaire sur la résine fossile de Leval. *Mém. Mus. roy. d'hist. nat. Belg.* **5**.
LAURENT, L. (99). Flore des calcaires de Célas. *Ann. Mus. d'hist. nat. Marseille*, **1**, ii.
—— (07). Les progrès de la Paléobotanique angiospermique dans la dernière décade. *Progr. rei Bot.* **1**.
—— (12). Flore fossile des schistes de Menat (Puy-de-Dôme). *Ann. Mus. d'hist. nat. Marseille (Géol.)*, **14**.
—— (19). Les Liquidambars. *Ibid.* **17**.
—— (19²). Contributions à l'étude des flores fossiles du centre de la France. *Ibid.* **17**.
LAURENT, L. and P. MARTY (23). Flore foliaire Pliocène des Argiles de Reuver, etc. *Meded. 's Rijks Geol. Dienst*, Ser. B, no. 1.
LECLERCQ, S. (28). Les végétaux à structure conservée du houiller Belge. II. Sur un *Stigmaria* à bois primaire centripète. *Ann. Soc. géol. Belg.* **51**.
—— (28²). *Psygmophyllum Gilkineti* sp. n. du Dévonien moyen à facies Old Red Sandstone de Malonne. *Journ. Linn. Soc. London*, **48**.
LEE, W. T. and F. H. KNOWLTON (17). Geology and paleontology of the Raton Mesa and other regions in Colorado and New Mexico. *U.S. Geol. Surv., Profl. Paper* **101**.
LEMOINE, P. (11). Le rôle des Algues dans la formation des dépôts calcaires. *Rev. Gén. Sci.* Ann. **22**, no. 16.
—— (17). Contribution à l'étude des Corallinacées fossiles. *Bull. Soc. géol. France* [6], **17**.
—— (29). Les Corallinacées de l'archipel des Galapagos et du Golfe de Panama. *Arch. Mus. d'hist. nat. Paris* [6], **4**.
LESQUEREUX, L. (79). Atlas to the coal flora of Pennsylvania, etc. *Second Geol. Surv. Pa.*
—— (91). The flora of the Dakota group (a posthumous work ed. F. H. Knowlton). *U.S. Geol. Surv. Mon.* 17.

LEUTHARDT, F. (03). Die Keuperflora von Neuewelt bei Basel. I. *Abhand. Schweiz. paläont. Ges.* **30**.
—— (04). *Ibid.* II. *Ibid.* **31**.
LEWIS, F. J. (11). The plant-remains in the Scottish peat mosses. IV. *Trans. R. Soc. Edinb.* **47**, iv.
LIESEGANG, R. E. (13). *Geologische Diffusionen.* Dresden and Leipzig.
—— (15). *Die Achate.* Dresden and Leipzig.
LIGNIER, O. (95). Végétaux fossiles de Normandie. *Mém. Soc. Linn. Normand.* **18**
—— (07). *Ibid.* **22**.
—— (07²). Nouvelles recherches sur le *Propalmophyllum liasinum. Mém. Soc. Linn. Normand.* **23**.
—— (13). *Ibid.* **24**.
LILIENSTEIN, A. R. VON (28). "*Dioonites pennaeformis* Schenk" eine Cycadee aus der Lettenkohle. *Palaeont. Zeitsch.* **10**, i.
LINDENBEIN, H. A. R. (21). Une flore marine sapropélitique de l'Ordovicien moyen de la Baltique. *Comp. Rend. Soc. physique d'hist. nat. Genève*, **38**, ii, p. 60.
—— (21²). La Kuckersite. Étude d'un dépôt marin phytogène du Silurien inférieur de la Baltique, etc. *Ibid.* **38**, ii, p. 71.
—— (21³). Les Protophycées (*Gloeocapsomorpha prisca* Zalessky), une flore marine du Silurien inférieur de la Baltique. *Bull. Soc. bot. Genève.*
LIPPS, T. (23). Über die Unter-Kreideflora Nordwest-Deutschlands. *Bot. Archiv*, **4**, v.
LLOYD, F. E. and V. MORAVEK (28). Studies in periodic precipitation. *Plant Physiol.* **3**.
LORENZ, T. (04). Ascosomaceae, eine neue Familie der Siphoneen aus dem Cambrium von Schantung. *Centralbl. Min. Geol. Palaeont.*
LOUBIÈRE, A. (29). Étude anatomique et comparée du *Leptotesta Grand'Euryi* n. gen. et sp. (graine silicifiée du *Pecopteris Pluckeneti*). *Rev. gén. Bot.* **41**.
LÜCK, H. (13). Beiträge zur Kenntnis des älteren Salzgebirges im Berlepsch-Bergwerk bei Stassfurt, etc. *Inaugur. Dissert.* Leipzig.
LUNDQUIST, G. (19). Fossile Pflanzen der Glossopteris-Flora aus Brasilien. *K. Svensk. Vetenskapsakad. Hand.* **60**, iii.
MCCOY, F. (74). Prodromus of the Palaeontology of Victoria. Dec. I. *Geol. Surv. Victoria.*
MACGREGOR, A. M. (27). The problem of the Pre-Cambrian atmosphere. *S. Afr. Journ. Sci.* **24**.
MACVICAR, S. M. (26). *The Student's Handbook of British Hepatics.* Eastbourne.
MÄGDEFRAU, K. (30). Beiträge zur Kenntniss des thüringischen Buntsandsteins. *Beit. geol. Thüringen*, **2**.
MARR, J. E. (16). *The Geology of the Lake District, etc.* Cambridge.
—— (20). The Pleistocene deposits around Cambridge. *Quart. Journ. Geol. Soc.* **75**.
MARR, J. E. and E. W. GARDNER (16). An Arctic flora in the Pleistocene beds of Barnwell, Cambridge. *Geol. Mag.* [6], **3**.
MARTY, P. (07). Études sur les végétaux fossiles du Trieu de Leval (Hainaut). *Mém. Mus. roy. d'hist. nat. Belg.* **5**.
MASON, E. (28). Note on the presence of Mycorrhiza in the roots of Salt Marsh plants. *New Phyt.* **27**.
MATHIEU, F. F. (21). Flore fossile du Bassin houiller de Kaiping (Chine). *Ann. Soc. géol. Belg.* **44**.
—— (22). L'âge géologique des Charbons de la Chine. *Ibid.* **45**.
MATLEY, C. A. (21). On the stratigraphy, fossils and geological relationships of the Lameta beds of Jubbulpore. *Rec. Geol. Surv. India*, **53**.
MATSON, G. C. and E. W. BERRY (16). The Pliocene Citronelle formation of the Gulf Coastal plain and its flora. *U.S. Geol. Surv., Profl. Paper* 98—L.
—— (16²). The Catahoula Sandstone and its flora. *Ibid.* 98—M.
MATTFELD, J. (26). Abies. *Die Pflanzenareale*, **1**, Heft 2.
MATTHEWS, J. R. (23). The distribution of certain portions of the British flora. *Ann. Bot.* **37**.

MAUFE, H. B. (22). The Dwyka tillite near Palapye, Bechuanaland Protectorate. *Trans. Geol. Soc. S. Afr.*

—— (24). An outline of the geology of Southern Rhodesia. *S. Rhodes. Geol. Surv.* **17**.

MAWSON, SIR D. (29). Some South Australian algal limestones in process of formation. *Quart. Journ. Geol. Soc.* **85**.

MELLOR, E. T. (05). A contribution to the study of the glacial (Dwyka) conglomerate in the Transvaal. *Ibid.* **61**.

MENZEL, P. (21). *See* Potonié (21).

MESCHINELLI, A. and X. SQUINABOL (92). *Flora Tertiaria Italica.* Padua.

METZGER, A. A. T. (24). Die jatulischen Bildungen von Suojärvi in Ostfinnland. *Bull. com. géol. Finlande,* no. 64.

MILLER, H. (49). *Footprints of the Creator.* London.

—— (57). *The Testimony of the Rocks.* Edinburgh.

MÖLLER, H. (02). Bidrag till Bornholms fossila Flora. *Lunds Univ. Årsskrift,* **38**, Afd. 2, v.

—— (03). *Ibid. K. Svensk. Vetenskapsakad. Hand.* **36**, vi.

MÖLLER, H. and T. G. HALLE (13). The fossil flora of the coal-bearing deposits of south-eastern Scania. *Arkiv Bot.* **13**, vii.

MOLISCH, H. (27). *Im Lande der aufgehenden Sonne.* Wien.

MOORE, E. S. (18). The Iron-formation on Belcher Islands, Hudson Bay, etc. *Journ. Geol.* **26**.

—— (25). Sources of Carbon in Pre-Cambrian formations. *Trans. R. Soc. Can.* **19**, iv.

MORELLET, L. and J. (13). Les Dasycladacées du Tertiaire Parisien. *Mém. Soc. géol. France, Paléont.* **21**, i.

—— (22). Nouvelle contribution à l'étude des Dasycladacées Tertiaires. *Ibid.* **25**, ii.

MYRES, J. L. (24). Primitive man in geological time. *Cambridge Ancient Hist.* **1**.

NATHORST, A. G. (78). *Beiträge zur fossilen Flora Schwedens.* Stuttgart.

—— (83). Contribution à la flore fossile du Japon. *K. Svensk. Vetenskapsakad. Hand.* **20**, ii.

—— (78–86). Floran vid Bjuf. *Sver. geol. Unders. Afh. och uppsatser,* Ser. C, no. 27.

—— (90). Über die Reste eines Brotfruchtbaums, *Artocarpus Dicksoni,* n. sp., aus den Cenomanen Kreideablagerungen Grönlands. *K. Svensk. Vetenskapsakad. Hand.* **24**, i.

—— (90[2]). Beiträge zur mesozoischen Flora Japans. *Denks. Ak. Wiss. Wien,* **57**.

—— (92). Fresh evidence concerning the distribution of Arctic plants during the glacial epoch. *Nature,* **45**.

—— (93). *Beiträge zur Geologie und Palaeontologie der Republik Mexico.* (Felix and Lark.) **2**, i. Leipzig.

—— (94). Zur paläozoischen Flora der Polarländer. *K. Svensk. Vetenskapsakad. Hand.* **26**, iv.

—— (97). Zur mesozoischen Flora Spitzbergens. *Ibid.* **30**, i.

—— (97[2]). Nachträgliche Bemerkungen über die mesozoische Flora Spitzbergens. *K. Vetenskapsakad. Förh.* no. 8.

—— (99). Über die oberdevonische Flora der Bären-Insel. *Bull. geol. Inst. Univ. Upsala,* no. 8, **4**, ii.

—— (99[2]). Fossil plants from Franz Josef Land. *The Norwegian North Polar Exped.* 1893–96. *Sci. Results* (ed. F. Nansen).

—— (02). Zur fossilen Flora der Polarländer. *K. Svensk. Vetenskapsakad. Hand.* **36**, iii.

—— (04). Die oberdevonische Flora des Ellesmere-Landes. *Rep. Second Norwegian Arctic Exped.* ("Fram"), 1898–1902, no. 1.

—— (06). Über *Dictyophyllum* und *Camptopteris spiralis. K. Svensk. Vetenskapsakad.* **41**, v.

—— (06[2]). Bemerkungen über *Clathropteris meniscoides,* etc. *Ibid.* **41**, ii.

NATHORST, A. G. (07). Über Trias- und Jurapflanzen von der Insel Kotelny. *Résult. Sci. Expéd. Pol. Russe* (1900–03). *Sect. C. Géol. Paléont.* Livr. 2. *Mém. Acad. Imp. Sci. St Pétersb.* [8], **21**, ii.

—— (07²). Paläobotanische Mitteilungen. I, II. *K. Svensk. Vetenskapsakad. Hand.* **42**, v.

—— (08). *Ibid.* III. *Ibid.* **43**, iii.

—— (08²). *Ibid.* VII. *Ibid.* **43**, viii.

—— (09). Über die Gattung *Nilssonia* Brongn. *Ibid.* **43**, xii.

—— (10). Beiträge zur Geologie der Bären-Insel, Spitzbergens und des König-Karl-Landes. *Bull. geol. Instit. Univ. Upsala*, **10**.

—— (11). Fossil floras of the Arctic regions as evidence of geological climates. *Geol. Mag.* [5], **8**.

—— (11²). Contributions to the Carboniferous flora of north-eastern Greenland. *Meddel. Grønland*, **43**.

—— (11³). Paläobot. Mittn. IX. *K. Svensk. Vetenskapsakad. Hand.* **46**, iv.

—— (11⁴). *Ibid.* X. *Ibid.* **46**, viii.

—— (13). Die pflanzenführenden Horizonte innerhalb der Grenzschichten des Jura und der Kreide Spitzbergens. *Geol. Fören. Stockholm Förhand.* **35**.

—— (14). *Nachträge zur paläozoischen Flora Spitzbergens.* Stockholm.

—— (15). Zur Devonflora des westlichen Norwegens. *Bergens Mus. Aarbok*, no. 9.

—— (15²). Tertiäre Pflanzenreste aus Ellesmere-Land. *Rep. Second Norwegian Arct. Exped.* ("Fram"), 1898–1902, no. 35. Kristiania.

—— (19). *Ginkgo adiantoides* im Tertiär Spitzbergens nebst einer kurzen Übersicht der übrigen fossilen Ginkgophyten desselben Landes. *Geol. Fören. Stockholm Förhand.* **41**, Heft 3.

—— (20). *Zur fossilen Flora der Polarländer. Zur Kulm-Flora Spitzbergens.* Stockholm.

—— (20²). Einige Psygmophyllum-Blätter aus dem Devon Spitzbergens. *Bull. geol. Instit. Univ. Upsala*, **18**.

NATHORST, A. G. and V. M. GOLDSCHMIDT (13). Das Devongebiet am Röragen bei Röros. *Videnskapsselskapets Skrifter*, no. 9. Kristiania.

NATHORST, A. G. and C. F. KOLDERUP (15). Zur Devonflora des westlichen Norwegens. *Bergens Mus. Aarbok*, no. 9.

NĚMEJC, F. (29). On some discoveries of fossil plant remains in the Carboniferous districts of Central Bohemia. *Bull. Int. Acad. Sc. Bohême.* (See Abstract in *Bot. Centralblatt*, **159**, p. 310.)

NEUMANN, R. (07). Beiträge zur Kenntnis der Kreideformation in Mittel-Peru. *Neues Jahrb. Min., etc.* Beilag. **24**.

NEWBERRY, J. S. (88). Rhaetic plants from Honduras. *Amer. Journ. Sci.* **36**.

—— (90). Devonian plants from Ohio. *Journ. Cincinnati Nat. Hist. Soc.* **12**.

—— (95). The flora of the Amboy Clays. A posthumous work ed. A. Hollick. *U.S. Geol. Surv., Mon.* 26.

—— (98). The later extinct floras of North America. Ed. A. Hollick. *Ibid. Mon.* 35.

NEWTON, E. T. and J. J. H. TEALL (97). Notes on a collection of rocks and fossils from Franz Josef Land. *Quart. Journ. Geol. Soc.* **53**.

—— (98). Additional notes. *Ibid.* **54**.

NICHOLSON, H. A. (69). On the occurrence of plants in the Skiddaw Slates. *Geol. Mag.* **6**.

NOBLE, L. F. (22). A section of the Palaeozoic formations of the Grand Canyon at the Bass Trail. *U.S. Geol. Surv., Profl. Paper* 131—B.

NOÉ, A. C. (23). A Palaeozoic Angiosperm. *Journ. Geol.* **31**.

NORTH, F. J. (26). *Coal and the coalfields in Wales.* Nat. Mus. Wales, Cardiff.

NOVOPOKROVSKIJ, I. (12). Beiträge zur Kenntnis der Jura-flora des Tyrma-Tal. *Explor. géol. Min. du Chemin de fer, Sibérie*, Livr. 32.

NOWAK, M. J. (07). Kopalna flora Senońska z Potylicza. *Bull. Acad. Sci. Cracovie.*

OBERSTE-BRINK, K. (14). *Beiträge zur Kenntniss der Farne und farnähnlichen Gewächse des Culms von Europa.* Inaug. Dissert. Münster-i.-W.

Obrutschew, W. A. (26). Geologie von Sibirien. *Fortschrit. geol. Palaeont.* Heft 15.

Ogura, Y. (27). On the structure and affinities of some fossil tree-ferns from Japan. *Journ. Coll. Sci. Imp. Univ. Tokyo*, Sect. 3, *Bot.* **1**.

Ôishi, S. (30). Notes on some fossil plants from the Upper Triassic beds of Nariwa Province Bitchû, Japan. *Jap. Journ. Geol. Geogr.* **7**, ii.

Oltmanns, F. (04–05). *Morphologie und Biologie der Algen.* 2 vols. Jena.

Osborn, H. F. and C. A. Reeds (22). Old and New Standards of Pleistocene divisions in relation to the pre-history of man in Europe. *Bull. Geol. Soc. Amer.* **33**.

Osborne, T. G. B. (09). The lateral roots of *Amyelon radicans* and their Mycorrhiza. *Ann. Bot.* **23**.

Ostenfeld, C. H. (25). Flowering plants and ferns from north-western Greenland. *Meddel. Grønland*, **68**.

—— (26). The flora of Greenland and its origin. *K. Dansk. Videnskab. Selskab. Biol. Meddel.* **6**, iii.

Overbeck, F. (28). Studien zur postglazialen Waldgeschichte der Rhön. *Zeitsch. Bot.* Jahrg. 20.

Palibin, J. (04). Notice sur la flore Tertiaire dans la Steppe Kirghize. *Bull. com. géol.* **23**.

—— (04^2). Pflanzenreste von Sichota-Alin Gebirge. *Verhand. russ. min. Ges.* **41**, i.

—— (06). Über die Flora der Sarmatischen Ablagerungen der Krym und Kaukasus. *Ibid.* **43**, i.

—— (06^2). Die fossilen Pflanzenreste der Küsten des Aralsees. (Russian.) *Bull. Turkestan Sect. Russian Geogr. Soc.* **4**.

—— (06^3). Fossile Pflanzen aus den Kohlenlagern von Fuschun in der südlichen Mandshurei. *Verhand. russ. min. Ges.* **44**, i.

—— (22). On the Pliocene flora of Transcaucasia. *Geol. Mag.* **59**.

Parkin, J. (27). The evolution and classification of flowering plants. *Rep. Bot. Exchange Club for* 1926.

Parsons, E. (28). The origin of the Rift Valley as evidenced by the geology of Coastal Kenya. *Trans. Geol. Soc. S. Afr.* **31**.

Pax, F. (26–27). Acer. I, II. *Die Pflanzenareale*, **1**, Hefte i und iv.

Peach, B. N. and J. Horne (07). The geological structure of the north-west Highlands of Scotland. Ed. Sir A. Geikie. *Mem. Geol. Surv. Gt Brit.*

Pearson, H. H. W. (29). *Gnetales.* Cambridge.

Pelourde, F. (13). Sur quelques végétaux fossiles du Tonkin. *Bull. Serv. géol. Indochine*, **1**, i.

Penhallow, D. P. (89). Notes on Devonian plants. *Trans. R. Soc. Canad.* Sect. iv.

—— (93). Additional notes on Devonian plants from Scotland. *Ibid.* **5**.

—— (96). *Nematophyton Ortoni. Ann. Bot.* **10**.

—— (97). *Nematophyton crassum. Canad. Rec. Sci.* **7**.

—— (00). Notes on the North American species of *Dadoxylon. Trans. R. Soc. Can.* Sect. iv, **6**.

—— (02). Notes on Cretaceous and Tertiary plants of Canada. *Ibid.* **8**.

—— (03). Notes on Tertiary plants. *Ibid.* **9**.

—— (08). Report on Tertiary plants from British Columbia. *Canad. Dept. of Mines Geol. Surv.* no. 1013.

Pfender, J. (26). Les Mélobésiées dans les Calcaires Crétacés de la Basse-Provence. *Mém. Soc. géol. France*, **6**.

Pia, J. (20). *Die Siphoneae verticillatae vom Karbon bis zur Kreide.* Vienna.

—— (24). Geologisches Alter und geographische Verbreitung der wichtigsten Algengruppen. *Oest. bot. Zeitsch.* nos. 7–9.

—— (25). Die Diploporen der Trias von Süddalmatien. *Sitzungsber. Ak. Wiss. Wien*, **133**, vii–viii.

—— (25^2). Die Gliederung der alpinen Mitteltrias auf Grund der Diploporen. *Ibid.* Anzeig. no. 23.

—— (26). *Pflanzen als Gesteinsbildner.* Berlin.

PIA, J. (26²). Die Diploporen der deutschen Trias und die Frage der Gleichsetzung der deutschen und alpinen Triasstufen. *Zeitsch. deutsch. geol. Ges.* (*Monatsberichte*).
—— (27). *See* Hirmer, M.
—— (28). Die vorzeitlichen Spaltpilze und ihre Lebensspuren. *Palaeobiologica*, **1**. Wien.
—— (30). Neue Arbeiten über fossile Solenoporaceae und Corallinaceae. *Neues. Jahrb. Minerologie, etc.*
PIRSSON, L. V. and C. SCHUCHERT (20). *A Text-book of Geology*. New York.
POSTHUMUS, O. (27). Some remarks concerning the Palaeozoic flora of Djambi, Sumatra. *K. Akad. Wetenschappen, Amsterdam*, Proc. **30**, vi.
—— (29). On Palaeobotanical investigations in the Dutch East Indies and adjacent regions. *Bull. jard. bot. Buitenzorg* [3], **9**, iii.
POTONIÉ, H. (93). Die Flora des Rothliegenden von Thüringen. *Jahrb. preuss. geol. Landesanst.* Heft 9.
—— (96). Die floristische Gliederung des deutsche Carbon und Perm. *Ibid.* Heft 21.
—— (00). Fossile Pflanzen aus Deutsch- und Portugiesisch-Ostafrika. *Deutsch-Ostafrika*, **7**.
—— (01). Die Silur- und die Culm-Flora des Harzes und des Magdeburgischen. *Jahrb. preuss. geol. Landesanst.* Heft 36.
—— (02). Fossile Hölzer aus der oberen Kreide Deutsch-Ostafrikas. *Mitt. Deutsch-Schutzgebiet*, **15**, iv.
—— (04). Abbildungen und Beschreibung fossiler Pflanzen-Reste. Lief. ii. *kön. preuss. geol. Landesanst.*
—— (08). Zur Genesis der Braunkohlenlagern der südlichen Provinz Sachsen. *Ibid.* **29**, Tl. 1, iii.
—— (09). Die Tropen-Sumpflachmoor-Natur der Moore des Produktiven Carbon. *Ibid.* **30**, Tl. 1, Heft iii.
—— (21). *Lehrbuch der Paläobotanik* (Ed. 2). Revised by W. Gothan. Berlin.
POTONIÉ, H. and C. BERNARD (04). *Flore Dévonienne de l'étage II. de Barrande.* Leipzig.
PRINADA, U. (28). Sur des restes de plantes des dépôts mésozoïques de la Samarskaya Louka. *Bull. com. géol.* **46**, viii.
PRINGLE, J. and R. CROOKALL (30). Palaeobotany of the Kent Coalfield. *Proc. Geol. Assoc.* **41**.
PROSSER, C. S. (94). The Devonian system of eastern Pennsylvania and New York. *Bull. U.S. Geol. Surv.* **19**.
RACIBORSKI, M. (92). Przyczynek do Flory Retyckiej Polski. *Akad. Umiejet. Krakow*, **22**.
—— (94). *Flora Kopalna. I. Pamietnik Wydz. mat. przyr.* Krakow.
—— (09). Über eine fossile *Pangium*-Art aus dem Miozän Javas. *Bull. Acad. Sci. Cracovie.*
RADLEY, E. G. (29). The preservation of pyritised and other fossils. *The Naturalist.*
RASTALL, R. H. (29). On continental drift and cognate subjects. *Geol. Mag.* **66**.
REED, F. R. C. (28). A Permo-Carboniferous marine fauna from the Umaria coalfield. *Rec. Geol. Surv. India*, **60**.
REED, F. R. C., DE P. COTTER, and H. M. LAHIRI (30). The Permo-Carboniferous succession in the Warcha Valley, Western Salt Range, Punjab. *Rec. Ind. Geol. Surv.* **62**, iv, p. 412.
REID, C. (99). *The Origin of the British Flora.* London.
—— (01). Notes on plant-remains of Roman Silchester. *Archaeologia*, **57**.
—— (02). *Ibid.* **58**.
—— (13). *Submerged Forests.* Cambridge.
REID, C. and E. M. (08). On the Pre-Glacial Flora of Britain. *Journ. Linn. Soc.* **38**.
—— (10). The lignite of Bovey Tracey. *Phil Trans. R. Soc.* **201**.
—— (15). The Pliocene floras of the Dutch-Prussian Border. *Meded. Rijksopsporing van Delfstoffen*, no. 6. The Hague.

Reid, E. M. (07–09). On a method of disintegrating peat and other deposits containing fossil seeds. *Journ. Linn. Soc. London*, **38**.

—— (20). A comparative revision of Pliocene floras based on the study of fossil seeds. *Quart. Journ. Geol. Soc.* **76**.

—— (20[2]). On two Pre-glacial floras from Castle Eden. *Ibid.*

—— (23). Nouvelles recherches sur les graines du Pliocène inférieur du Pont-de-Gail (Cantal). *Bull. Soc. géol. France* [4], **32**.

—— (30). Tertiary fruits and seeds from Saint Tudy (Finistère). *Bull. Soc. géol. Min. Bretagne*, **8**.

Reid, E. M. and M. E. J. Chandler (26). The Bembridge flora. *Catalogue of Cainoz. Plants*, **1**. British Mus. (Nat. Hist.).

—— (29). Palaeobotany: Tertiary plants. *Encyc. Brit.* ed. 14.

Reid, E. M. and P. Marty (20). Recherches sur quelques graines pliocènes du Pont-de-Gail (Cantal). *Bull. Soc. géol. France* [4], **20**.

—— (23). Nouvelles recherches sur les graines du pliocène inférieur du Pont-de-Gail. *Ibid.* **23**.

Renault, B. (96). Bassin houiller et permien d'Autun et d'Épinac. *Études Gîtes Min. de la France.* Paris. Vol. of Plates. *Ibid.* 1893.

Renault, B. and R. Zeiller (88–90). *Études sur le Terrain houiller de Commentry.* St Étienne.

Rendle, A. B. (04) and (25). *The Classification of Flowering Plants*, **1** (1904); **2** (1925).

Renier, A. (10). L'origine raméale des cicatrices ulodendroides. *Ann. Soc. géol. Belg.* **2**.

—— (10[2]). Note sur quelques végétaux fossiles du Dinantien moyen de Belgique. *Ibid.*

—— (10[3]). *Documents pour l'étude de la Paléontologie du terrain houiller.* Liége.

—— (26). Étude stratigraphique du Westphalien de la Belgique. *Comp. Rend. Congr. géol. Internat.* (1922). Liége.

—— (26[2]). La morphologie générale des *Ulodendron*. *Comp. Rend. Acad. Sci. Paris*, **182**.

Reynolds, S. H. (21). The lithological succession of the Carboniferous Limestone of the Avon section at Clifton. *Quart. Journ. Geol. Soc.* **77**.

Richter, P. B. (06). *Beiträge zur Flora der Unterkreide Quedlinburg*, **1**. Leipzig.

—— (09). *Ibid.* **2**.

Rickmers, W. R. (13). *The Duab of Turkestan.* Cambridge.

Roddy, H. J. (15). Concretions in streams formed by the agency of Blue-Green Algae and related plants. *Proc. Amer. Phil. Soc.* **54**.

Rogers, I. (26). On the discovery of fossil fishes and plants in the Devonian rocks of North Devon. *Trans. Dev. Assoc. Adv. Sci. Lit. and Art*, **58**.

Romanes, F. M. (16). Note on an algal limestone from Angola. *Trans. R. Soc. Edinb.* **51**, iii.

Rothpletz, A. (08). Über Algen und Hydrozoen im Silur von Gotland und Oesel. *K. Svensk. Vetenskapsakad. Hand.* **43**, v.

—— (13). Über die Kalkalgen, Spongiostromen und einige andere Fossilien aus dem Obersilur Gotlands. *Sverig. geol. Unders. Afh. och uppsatser*, Ser. C.

—— (16). Über die systematische Deutung und die stratigraphische Stellung der ältesten Versteinerungen Europas und Nordamerikas mit besonderer Berücksichtigung der Cryptozoon und Oolithe. Teil. II. Über *Cryptozoon*, *Eozoon* und *Artikokania*. *Abhand. Bayer. Akad. Wiss.* **28**, iv.

Rothpletz, A. and K. Giesenhagen (22). *Ibid.* Teil III. Über Oolithe. *Ibid.* **29**, v.

Rudolph, K. (30). Grundzüge der nacheiszeitlichen Waldgeschichte Mitteleuropas. *Beiheft. Bot. Cent.* **47**, i.

Ruedemann, R. (04) and (08). Graptolites of New York. Pts. I and II. *N.Y. State Mus. Mem.* 7 and 11.

—— (09). Some marine Algae from the Trenton Limestone of New York. *N.Y. State Mus. Bull.* 133.

RUTTEN, L. (20). On the occurrence of *Halimeda* in Old-Miocene coast-reefs of east Borneo. *K. Akad. Wetenschap. Amsterdam*, **23**, iv.
RYDZEWSKI, B. (13). Sur l'âge des couches houillères du bassin Carbonifère de Cracovie. *Bull. Acad. Sci. Cracovie* (July).
—— (19). Flora Węglowa Polski. I. *Paleont. ziem Polskich*, no. 2.
SAHNI, B. (22). Presidential Address, Indian Sci. Congress. *Proc. Asiat. Soc. Bengal*, **17**.
—— (23). On the structure of the cuticle in *Glossopteris angustifolia*. *Rec. Geol. Surv. Ind.* **54**, iii.
—— (26). Presidential Address, Ind. Sci. Congress. The southern fossil floras. *Proc. 13th Ind. Sci. Congr.*
—— (28). On *Clepsydropsis australis*. *Phil. Trans. R. Soc.* **217**.
—— (28²). Revision of Indian fossil plants. Pt. I. Coniferales. *Mem. Geol. Surv. Ind. Pal. Ind.* **11**.
—— (30). On *Asterochlaenopsis*, a new genus of Zygopterid Tree-ferns from western Siberia. *Phil. Trans. R. Soc.* **218**.
SAHNI, B. and T. C. N. SINGH (26). On some specimens of *Dadoxylon Arberi* from New South Wales and Queensland. *Journ. Ind. Bot. Soc.* **5**, iii.
SAKISAKA, M. (29). On the seed-bearing leaves of *Ginkgo*. *Jap. Journ. Bot.* **4**.
SALFELD, H. (07). Fossile Land-Pflanzen der Rät- und Juraformation Südwestdeutschlands. *Palaeontograph.* **54**.
—— (09). Beiträge zur Kenntnis jurassischer Pflanzenreste aus Norddeutschland. *Ibid.* **56**.
—— (09²). Versteinerungen aus dem Devon von Bolivien, dem Jura und der Kreide von Peru. *Wiss. Veröffentl. Ges. Erdk.* **7**. Leipzig.
SALISBURY, R. D. (10). *Outlines of geologic history with especial reference to North America. A Symposium.* Chicago.
SALTER, J. W. (57). On some remains of Terrestrial plants in the O.R.S. of Caithness. *Quart. Journ. Geol. Soc.* **14**.
SANDFORD, K. S. (29). On the erratic rocks and the age of the southern limit of glaciation in the Oxford district. *Ibid.* **85**.
SAPORTA, G. DE (62–73). Études sur les végétaux du Sud-Est de la France à l'époque tertiare. *Ann. Sci. Nat.* (*Bot.*).
—— (68). Prodrome d'une flore fossile des Travertins anciens de Sézanne. *Mém. Soc. géol. France* [3], **8**.
—— (73–91). Plantes Jurassiques. *Paléont. Française* **1**, 1873; **2**, 1875; **3**, 1884; **4**, 1891.
—— (94). Flore fossile du Portugal. *Direct. Trav. géol. Port.*
SAPORTA, G. DE and A. F. MARION (76). Recherches sur les végétaux fossiles de Meximeux. *Arch. Mus. d'hist. nat. Lyon.*
—— (78). Revision de la Flore Heersienne de Gelinden. *Mém. cour. et mém. savants étrangers.* Bruxelles.
SARLE, C. J. (06). I. *Arthrophycus* and *Dædalus* of burrow origin. II. Preliminary note on the nature of *Taonurus*. *Proc. Rochester Acad. Sci.* **4**.
SAURAMO, M. (23). Studies on the Quaternary Varve sediments in South Finland. *Bull. com. géol. Finlande*, no. **60**.
—— (29). The Quaternary geology of Finland. *Ibid.*
SAYLES, R. W. (14). The Squantum Tillite. *Bull. Mus. Comp. Zool. Harvard Coll.* **56**, ii.
—— (19). Seasonal deposition in aqueo-glacial sediments. *Ibid.* **47**, i.
SCHENK, A. (67). *Die fossile Flora der Grenzschichten des Keuper und Lias Frankens.* Wiesbaden.
—— (83). Pflanzen aus der Steinkohlenformation. *Richthofen's China*, **4**.
—— (87). Fossile Pflanzen aus der Albourskette. *Bibl. Bot.* Heft 6, vi.
SCHIMPER, W. P. and A. MOUGEOT (44). *Monographie des Plantes fossiles du grès bigarré de la Chaîne des Vosges.* Leipzig.
SCHLAGINTWEIT, O. (19). *Weichselia Mantelli* in nordöstlichem Venezuela. *Centralbl. Min. Geol. Paläont.*

SCHLÜTER, H. and H. SCHMIDT (27). *Voltzia, Yuccites* und andere neue Funde aus dem südhannoverschen Buntsandstein. *Neues Jahrb. Min.* Beilag. 57.

SCHMALHAUSEN, J. (84). Beiträge zur Tertiär-Flora Süd-West-Russlands. *Palaeont. Abhand.* **1**, iv.

—— (87). Über tertiäre Pflanzen aus dem Thale des Flusses Buchtorma am Fusse des Altaigebirges. *Palaeont.* **33**.

—— (94). Über Devonische Pflanzen aus dem Donetz-Becken. *Mém. com. géol.* **8**, iii.

SCHMALHAUSEN, J. and BARON E. v. TOLL (90). Tertiäre Pflanzen der Insel Neusibirien. *Mém. Acad. Imp. Sci. St Pétersb.* [7], **37**.

SCHMIDT, W. L. (28). A voyage to the island home of Robinson Crusoe. *Natl. Geogr. Mag.* **54**, no. 3.

SCHOUTE, J. C. (25). La nature morphologique du bourgeon féminin des *Cordaites*. *Rec. Trav. Bot. Néerland.* **22**.

—— (25^2). Rectification de mon article sur le bourgeon féminin des *Cordaites*. *Ibid.* **22**.

SCHROETER, C. (80). *Untersuchung über fossile Hölzer aus der arctischen Zone*. Zürich.

SCHUCHERT, C. (23). Presidential Address. *Bull. Geol. Soc. Amer.* **34**.

—— (28). Review of the late Palaeozoic formations and faunas, with special reference to the Ice Age of Mid-Permian time. *Ibid.* **39**.

SCHUSTER, J. (11). Über Goeppert's *Raumeria* in Zwinger zu Dresden. *Sitzungsber. Bayer. Akad. Wiss.* Heft 3.

—— (11^2). *Osmundites* von Sierra Villa Rica in Paraguay. *Ber. deutsch. bot. Ges.* **29**, viii.

—— (11^3). *Xylopsaronius*, der erste Farn mit sekundärem Holz. *Ibid.*

—— (11^4). Monographie der fossilen Flora der *Pithecanthropus* Schichten. *Abhand. Bayer. Akad. Wiss.* **25**, vi.

SCHWARZ, E. H. L. (06). The three Palaeozoic Ice Ages of South Africa. *Journ. Geol.* **14**.

—— (06^2). South African Palaeozoic fossils. *Rec. Albany Mus.* **1**, vi.

SCOTT, D. H. (97). On the structure and affinities of fossil plants from the Palaeozoic rocks. On *Cheirostrobus*. *Phil. Trans. R. Soc.* **189**.

—— (01). On the structure and affinities of fossil plants from the Palaeozoic rocks. IV. The seed-like fructification of *Lepidocarpon*. *Ibid.* **194**.

—— (02). On the primary structure of certain Palaeozoic stems with the *Dadoxylon* type of wood. *Trans. R. Soc. Edinb.* **40**, ii.

—— (02^2). The old wood and the new. *New Phyt.* **1**.

—— (12). On *Botrychioxylon paradoxum* sp. nov., a Palaeozoic fern with secondary wood. *Trans. Linn. Soc. London* [2], **7**.

—— (17). The Heterangiums of the British Coal Measures. *Journ. Linn. Soc. London*, **44**.

—— (20) and (23). *Studies in Fossil Botany*, **1**, 1920; **2**, 1923. London.

—— (22). The early history of the land flora. *Nature*, **110**.

—— (24). *Extinct Plants and Problems of Evolution*. London.

—— (26). New discoveries in the Middle Devonian flora of Germany. *New Phyt.* **25**.

—— (29). Aspects of fossil botany. Ferns and Seed-ferns. *Nature*, **123**.

SCOTT, D. H. and E. C. JEFFREY (14). On fossil plants, showing structure, from the base of the Waverley shale of Kentucky. *Phil. Trans. R. Soc.* **205**.

SEDERHOLM, J. (99). Über eine archäische Sedimentformation im südwestlichen Finland. *Bull. com. géol. Finlande*, no. 6.

—— (10). Sur les vestiges de la vie dans les formations progonozoïques. *Comp. Rend. Congr. Géol. Internat.* (Stockholm).

—— (25). Nochmals das *Corycium*. *Centralbl. Min. Geol. Palaeont.* Abt. B, no. 11.

—— (26). Archäikum. (W. Salomon, *Grundzüge der Geologie*, **2**, p. 21.) Stuttgart.

SELIBER, G. (29). Le milieu extérieur et le développement des plantes (analyse des Trav. G. Klebs). *Rev. gén. Bot.* **41**.

SELLARDS, E. H. (00). A new genus of ferns from the Permian of Kansas. *Cont. Palaeont. Laboratory No.* 57. *Kansas Univ. Quart.* **9**.
SEWARD, A. C. (92). *Fossil Plants as Tests of Climate.* Cambridge.
—— (94). The Wealden flora, **1**. *Brit. Mus. Catalogue.*
—— (95). *Ibid.* **2**.
—— (97). On *Cycadeoidea gigantea*, a new Cycadean stem from the Purbeck beds of Portland. *Quart. Journ. Geol. Soc.* **53**.
—— (97²). On the association of *Sigillaria* and *Glossopteris* in South Africa. *Ibid.*
—— (98). *Fossil Plants*, **1**. Cambridge. (This volume is out of print.)
—— (99). Notes on the Binney collection of Coal-Measure plants. (*Megaloxylon.*) *Proc. Camb. Phil. Soc.* **10**.
—— (00). The Jurassic flora, **1**. *Brit. Mus. Catalogue.*
—— (00²). La flore Wealdienne de Bernissart. *Mém. Mus. roy. d'hist. nat. Belg.* **1**.
—— (03). Fossil floras of Cape Colony. *Ann. S. African Mus.* **4**.
—— (03²). On the occurrence of *Dictyozamites* in England, etc. *Quart. Journ. Geol. Soc.* **59**.
—— (04). The Jurassic flora, **2**. *Brit. Mus. Catalogue.*
—— (04²). On a collection of Jurassic plants from Victoria. *Rec. Geol. Surv. Vict.* **50**, iii.
—— (07). Fossil plants from South Africa. *Geol. Mag.* [4], **4**.
—— (07²). Permo-Carboniferous plants from Kashmir. *Rec. Geol. Surv. Ind.* **56**, ii.
—— (07³). Jurassic plants from Caucasia and Turkestan. *Mém. com. géol.* Livr. 38.
—— (07⁴). Fossil plants from Egypt. *Geol. Mag.* [5], **4**.
—— (09). Fossil plants from the Witteberg series of Cape Colony. *Ibid.* **5**.
—— (10). *Fossil Plants*, **2**.
—— (11). The Jurassic flora of Sutherland. *Trans. R. Soc. Edinb.* **47**, iv.
—— (11²). Jurassic plants from Chinese Dzungaria. *Mém. com. géol.* Livr. 75.
—— (11³). *Links with the Past in the Plant World.* Cambridge.
—— (12). A petrified *Williamsonia* from Scotland. *Phil. Trans. R. Soc.* **203**.
—— (12²). Mesozoic plants from Afghanistan, etc. *Mem. Geol. Surv. Ind.* **4**, iv.
—— (12³). Jurassic plants from Amurland. *Mém. com. géol.* Livr. 81.
—— (12⁴). Dicotyledonous leaves from the Coal Measures of Assam. *Rec. Geol. Surv. Ind.* **42**.
—— (13). A British fossil *Selaginella*. *New Phyt.* **12**.
—— (13²). Contribution to our knowledge of Wealden floras. *Quart. Journ. Geol. Soc.* **69**.
—— (14). Antarctic fossil plants. *Brit. Ant. Exped. Nat. Hist. Rep. Geology*, **1**, i.
—— (17). *Fossil Plants*, **3**.
—— (19). *Ibid.* **4**.
—— (22). A study in contrasts: the present and past distribution of certain ferns. *Journ. Linn. Soc. London*, **46**.
—— (22²). *A Summer in Greenland.* Cambridge.
—— (22³). On a small collection of fossil plants from the Tanganyika Territory. *Geol. Mag.* **59**.
—— (22⁴). Carboniferous plants from Peru. *Quart. Journ. Geol. Soc.* **78**.
—— (23). The earlier records of plant life. *Presid. Add. Geol. Soc. London.*
—— (23²). A supposed Palaeozoic Angiosperm. *Bot. Gaz.* **76**.
—— (24). On a new species of *Tempskya* from Montana. *Ann. Bot.* **38**.
—— (24²). A collection of fossil plants from south-eastern Nigeria. *Geol. Surv. Nigeria, Bull.* 6.
—— (25). Notes sur la Flore Crétacique du Greenland. *Livre Jubil. Soc. géol. Belg.* Liège.
—— (25²). Arctic vegetation: past and present. *Journ. Hort. Soc.* **50**, i.
—— (26). The Cretaceous plant-bearing rocks of western Greenland. *Phil. Trans. R. Soc.* **215**.
—— (29). Botanical records of the rocks. *Presid. Add. Sect. K, Brit. Assoc. S. Africa.*

SEWARD, A. C. (29[2]). Greenland: as it is and as it was. *Nature*, **213**.

—— (29[3]). Palaeobotany: Mesozoic. *Encyc. Brit.* ed. 14.

SEWARD, A. C. and E. A. N. ARBER (03). Les Nipadites des couches Éocènes de la Belgique. *Mém. Mus. d'hist. nat. Belg.* **11**.

SEWARD, A. C. and N. BANCROFT (13). Jurassic plants from Cromarty and Sutherland, Scotland. *Trans. R. Soc. Edinb.* **48**, iv.

SEWARD, A. C. and S. O. FORD (06). The Araucarineae, recent and extinct. *Phil. Trans. R. Soc.* **198**.

SEWARD, A. C. and A. W. HILL (00). On the structure and affinities of a Lepidodendroid stem from the Calciferous Sandstone of Dalmeny, Scotland. *Trans. R. Soc. Edinb.* **39**, iv.

SEWARD, A. C. and R. E. HOLTTUM (21). On a collection of fossil plants from south Rhodesia. *S. Rhodes. Geol. Surv. Bull.* 8.

—— (22). Jurassic plants from Ceylon. *Quart. Journ. Geol. Soc.* **78**.

—— (24). Tertiary plants from Mull. *Mem. Geol. Surv. Scotland.*

SEWARD, A. C. and T. N. LESLIE (08). Permo-Carboniferous plants from Vereeniging. *Quart. Journ. Geol. Soc.* **44**.

SEWARD, A. C. and B. SAHNI (20). Indian Gondwana plants: a revision. *Mem. Geol. Surv. India, Pal. Ind.* **7**, i.

SEWARD, A. C. and H. H. THOMAS (11). Jurassic plants from the Balagansk district, Govt. Irkutsk. *Mém. com. géol.* Livr. 73.

SEWARD, A. C. and J. WALTON (23). On fossil plants from the Falkland Islands. *Quart. Journ. Geol. Soc.* **79**.

SEWARD, A. C. and A. SMITH WOODWARD (05). Permo-Carboniferous plants and vertebrates from Kashmir. *Mem. Geol. Surv. India, Pal. Ind.* **2**, ii.

SEYLER, C. A. and W. J. EDWARDS (29). The microscopical examination of coal. *Dept. Sci. Indust. Research (Fuel Research), London.*

SHREVE, F. (16). The weight of physical facts in the study of plant distribution. *The Plant World*, **19**.

SIMPSON, E. C. (29). Past climates. *Manchester Lit. Phil. Soc.* **74**. Also *Nature* (Dec. 28).

—— (30). The Climate during the Pleistocene Period. *Proc. R. Soc. Edinburgh*, **50**, iii.

SIMPSON, E. C. and others (30). Discussion on geological climates. *Proc. R. Soc.* **106**.

SINNOTT, G. W. (14). Some Jurassic Osmundaceae from New Zealand. *Ann. Bot.* **28**.

SKOTTSBERG, C. (12). Einige Bemerkungen über die Vegetationsverhältnisse des Graham Landes. *Wiss. Ergb. Schwed. Südpolar-Exped.* 1901–3, **4**, xiii.

SLATER, G. (29). *See* Evans and Stubblefield.

SMITH, G. O. and D. WHITE (05). The geology of the Perry Basin in south-eastern Maine. *U.S. Geol. Surv., Profl. Paper* 35.

SMUTS, GENERAL J. C. (26). *Holism and Evolution.* London.

SOLLAS, W. J. (05). *The Age of the Earth and other Geological Studies.* London.

—— (09). Anniversary Address of the President. *Proc. Geol. Soc. London*, **65**.

SOLMS-LAUBACH, H. GRAF ZU (84). Die Coniferenformen des deutschen Kupferschiefers und Zechsteins. *Palaeont. Abhand.* **2**, ii.

—— (93). Über die in den Kalksteinen des Kulm von Glätzisch-Falkenberg in Schlesien enthaltenen structurbietenden Pflanzenreste. *Bot. Zeit.* **51**.

—— (95). Über devonische Pflanzenreste aus den Lenneschiefern der Gegend von Gräfrath am Niederrhein. *Jahrb. preuss. geol. Landesanst.*

—— (96). Über die seinerzeit von Unger beschriebenen strukturbietenden Pflanzenreste des Unterculm von Saalfeld in Thüringen. *Ibid.* Heft 23.

—— (99). Das Auftreten und die Flora der rhätischen Kohlenschichten von La Ternera (Chile). *Neues Jahrb. Min.* Beilageband. 12.

—— (04). Die strukturbietenden Pflanzengesteine von Franz Josef Land. *K. Svensk. Vetenskapsakad. Hand.* **37**, vii.

SOLMS-LAUBACH, H. GRAF ZU (04^2). Über die Schicksale der als *Psaronius brasiliensis* beschriebenen Fossilreste unserer Museen. *Festschrift P. Ascherson.* Berlin.

—— (13). *Tietea singularis* ein neuer fossiler Pteridinenstamm aus Brasilien. *Deutsch. bot. Jahrg.* **5**, ix.

STAPF, O. (14). The southern element in the British flora. *Bot. Jahrb.* **50** (Supplt. Bd.).

STAUB, M. (87). Die aquitanische Flora des Zsilthales im Comitate Hunyad. *Mitt. Jahrb. Ungarn Geol. Anst.* **7**, vi.

STEINMANN, G. (99). Ueber fossile Dasycladaceen vom Cerro Escamela, Mexico. *Bot. Zeit.* Heft 8.

—— (21). Rhätische Floren und Landverbindungen auf der Südhalbkugel. *Geol. Rundschau,* **11**, vii–viii.

STEINMANN, G. and W. ELBERSKIRCH (29). Neue bemerkenswerte Funde im ältesten Unterdevon des Wahnbachtales bei Siegburg. *Sitzungsber. niederrheinisch. geol. Ver.* (1927–1928).

STENZEL, G. (04). Fossile Palmenhölzer. *Beit. Paläont. Geol.* **16**, iii–iv.

STEPHENSON, L. W. and E. W. BERRY (29). Marine shells in association with land plants in the Upper Cretaceous of Guatemala. *Journ. Paleont.* **3**, ii.

STERZEL, J. T. (07). Die Karbon und Rotliegendfloren im Grossherzogtum Baden. *Mitt. Grossh. Bad. geol. Landesanst.* **5**, ii.

STEVENSON, J. J. (11–13). Formation of coal beds. *Proc. Amer. Phil. Soc.* **50–52**.

STOJANOFF, N. and E. STEFANOFF (29). Beitrag zur Kenntnis der Pliozänflora der Ebene von Sofia. *Bot. Centralbl.* **157**, p. 187.

STOMPS, T. J. (23). A contribution to our knowledge of the origin of the British flora. *Recueil Trav. bot. néerland.* **20**.

STOPES, M. C. (07). The flora of the Inferior Oolite of Brora (Sutherland). *Quart. Journ. Geol. Soc.* **63**.

—— (12). Petrifications of the earliest European Angiosperms. *Phil. Trans. R. Soc.* **203**.

—— (13). The Cretaceous flora. Pt. I. *Brit. Mus. Catalogue.*

—— (14). The "Fern ledges" Carboniferous flora of St John, New Brunswick. *Canad. Dept. of Mines, Geol. Surv. Mem.* **41**.

—— (14^2). A new *Araucarioxylon* from New Zealand. *Ann. Bot.* **28**.

—— (15). The Cretaceous flora. Pt. II. *Brit. Mus. Catalogue.*

—— (16). An early type of the Abietineae (?) from the Cretaceous of New Zealand. *Ann. Bot.* **30**.

—— (18). New Bennettitean cones from the British Cretaceous. *Phil. Trans. R. Soc.* **208**.

STOPES, M. C. and K. FUJII (10). Studies on the structure and affinities of Cretaceous plants. *Phil. Trans. R. Soc.* **201**.

STOPES, M. C. and D. M. S. WATSON (08). On the present distribution and origin of the calcareous concretions in coal seams, known as "coal balls". *Ibid.* **200**.

STRAHAN, SIR A. (10). Guide to the geological model of Ingleborough and district. *Mem. Geol. Surv. Eng. and Wales.*

STUART, M. (25). The Eocene lignites and amber deposits of Burmah, etc. *Journ. Inst. Petroleum Technologists,* **11**, no. 52.

STUR, D. (75–77). Beiträge zur Kenntniss der Flora der Vorwelt, **1**. *Abhand. geol. Reichsanst.* **8**.

—— (85). Die Carbon-flora Schatzlarer Schichten. *Ibid.* **11**.

SUESS, E. (04–09). *The Face of the Earth.* Oxford.

SÜSSMILCH, C. A. (22). *An Introduction to the Geology of New South Wales.* Sydney.

SÜSSMILCH, C. A. and SIR T. W. E. DAVID (20). Sequence, glaciation and correlation of the Carboniferous rocks of the Hunter River district, New South Wales. *Journ. Proc. R. Soc. N.S.W.* **53**.

SZAJNOCHA, L. (88). Über fossile Pflanzenreste aus Cacheuta in der Argentinische Republik. *Sitzungsber. Ak. Wiss. Wien,* **97**, i.

SZAJNOCHA, L. (91). Über einige carbone Pflanzenreste aus der Argentinischen Republik. *Ibid.* **100**, i.

THIESSEN, R. (25). Origin of the boghead coals. *U.S. Geol. Surv., Profl. Paper* 132—I.

THOMAS, D. (29). The late Palaeozoic glaciation. *Nature*, **124**.

THOMAS, H. H. (11). The Jurassic flora of Kamenka in the district of Isium. *Mém. com. géol. St Pétersb.* Livr. 71.

—— (12). *Stachypteris Hallei*, a new Jurassic fern. *Proc. Camb. Phil. Soc.* **16**, vii.

—— (12^2). On the leaves of Calamites. *Phil. Trans.* **202**.

—— (13). On some new and rare Jurassic plants from Yorkshire: *Eretmophyllum*, a new type of Ginkgoalean leaf. *Proc. Camb. Phil. Soc.* **17**, iii.

—— (13^2). The fossil flora of the Cleveland district of Yorkshire. I. The flora of the Marske Quarry. *Quart. Journ. Geol. Soc.* **69**.

—— (15). On *Williamsoniella*, a new type of Bennettitalean flower. *Phil. Trans. R. Soc.* **207**.

—— (15^2). The Thinnfeldia leaf-bed of Roseberry Topping. *Yorks. Naturalist.*

—— (21). An *Ottokaria*-like plant from South Africa. *Quart. Journ. Geol. Soc.* **77**.

—— (22). On some new and rare Jurassic plants from Yorkshire. V. Fertile specimens of *Dictyophyllum rugosum*. *Proc. Camb. Phil. Soc.* **21**, ii.

—— (25). The Caytoniales, a new group of angiospermous plants from the Jurassic rocks of Yorkshire. *Phil. Trans. R. Soc.* **213**.

—— (30). Further observations on the cuticle structure of Mesozoic Cycadean fronds. *Journ. Linn. Soc. London*, **48**.

THOMAS, H. H. and N. BANCROFT (13). On the cuticles of some recent and fossil Cycadean fronds. *Trans. Linn. Soc. London*, **8**, v.

TOIT, A. L. DU (21). Land connections between the other continents and South Africa in the past. *S. Afr. Journ. Sci.* **18**.

—— (26). *The Geology of South Africa.* London.

—— (27). A geological comparison of South America with South Africa. *Carnegie Inst. Publ.* 381.

—— (27^2). The fossil flora of the Upper Karroo beds. *Ann. S. Afr. Mus.* **22**, ii.

TOMLINSON, C. W. (16). The origin of red beds, etc. *Journ. Geol.* **24**.

TURNTANOVA-KETOVA, A. I. (29). Jurassic flora of the Chain Kara-Tan (Tian-Shan). *Trav. Mus. géol. Acad. Sci. de l'Urss.* **6**. (Russian.)

TUYL, F. M. VAN (16). A contribution to the oolite problem. *Journ. Geol.* **24**.

TUZSON, J. (14). Beiträge zur fossilen Flora Ungarns. *Mitt. Jahrb. K. Ungar. geol. Reichs.* **21**, viii.

TWENHOFEL, W. H. and others (26). *Treatise on Sedimentation.* London.

UMBGROVE, J. H. F. (27). Over Lithothamnia in het Maastrichtsche Tufrijt. *Leidsch. geol. Meded.* **2**, ii.

UNGER, F. (51). Die fossile Flora von Sotzka. *Denks. Ak. Wiss. Wien*, **2**.

—— (67). Die fossile Flora von Kumi. *Ibid.* **27**.

—— (69). Die fossile Flora von Radoboj. *Ibid.* **29**.

VELENOVSKÝ, J. (81). Die Flora aus den ausgebrannten Tertiären Letten von Vršovic bei Laun. *Abhand. böhm. Ges. Wiss.* [6], **11**.

VELENOVSKÝ, J. and L. VINIKLÁŘ (26–27). *Flora Cretacea Bohemiae.* Pts. I and II. Prague.

VERNON, R. O. (10). On the occurrence of *Schizoneura paradoxa* in the Bunter of Nottingham. *Proc. Camb. Phil. Soc.* **15**, v.

WADIA, D. N. (19). *Geology of India.* London.

WALCOTT, C. D. (99). Pre-Cambrian fossiliferous formations. *Bull. Geol. Soc. Amer.* **10**.

—— (12). Cambrian geology and paleontology. No. 6. Mid-Cambrian Branchiopoda, etc. *Smiths. Misc. Coll.* **57**, vi.

—— (13). *Ibid.* No. 12. Cambrian formations of the Robson Peak district. *Ibid.* **57**, xii.

—— (14). Pre-Cambrian Algonkian algal flora. *Ibid.* **64**, ii.

WALCOTT, C. D. (15). Discovery of Algonkian bacteria. *Proc. Nat. Acad. Sci. U.S.A.* **1**.
—— (19). Cambrian geology and paleontology. No. 5. Middle Cambrian Algae. *Smiths. Misc. Coll.* **67**, v.
WALKOM, A. B. (15). Mesozoic floras of Queensland. I. Ipswich and Walloon series. *Queensland Geol. Surv. Publ.* 252.
—— (16). Notes on a specimen of *Annularia* from near Dunedoo, New South Wales. *Mem. Queensland Mus.* **5**.
—— (17). Mesozoic floras of Queensland. *Queensland Geol. Surv. Publ.* 257.
—— (17^2). *Ibid. Publ.* 259.
—— (18). *Ibid. Publ.* 262.
—— (18^2). The geology of the Lower Mesozoic rocks of Queensland. *Proc. Linn. Soc. N.S.W.* **43**, i.
—— (19). Queensland fossil floras. *Proc. R. Soc. Queensland*, **31**, i.
—— (19^2). On a collection of Jurassic plants from Bexhill, near Lismore, New South Wales. *Proc. Linn. Soc. N.S.W.* **44**, i.
—— (19^3). Mesozoic floras of Queensland. III, IV. *Queensland Geol. Surv. Publ.* 263.
—— (21). *Nummulospermum bowenense* gen. et sp. nov. *Quart. Journ. Geol. Soc.* **77**.
—— (21^2). On the occurrence of *Otozamites* in Australia. *Proc. Linn. Soc. N.S.W.* **46**, i.
—— (21^3). Mesozoic floras of New South Wales. I. *Mem. Geol. Surv. N.S.W. Palaeont.* **12**.
—— (22). Palaeozoic floras of Queensland. *Queensland Geol. Surv. Publ.* 270.
—— (24). On fossil plants from Bellevue, near Esk. *Mem. Queensland Mus.* **8**, i.
—— (28). Lepidodendroid remains from Yalwal, New South Wales. *Proc. Linn. Soc. N.S.W.* **53**, iii.
—— (28^2). Notes on some additions to the Glossopteris flora in New South Wales. *Ibid.* **53**, v.
—— (28^3). Fossil plants from the Upper Palaeozoic rocks of New South Wales. *Ibid.* **53**, iii.
—— (28^4). Fossil plants from the Esk district, Queensland. *Ibid.* **53**, iv.
—— (28^5). Fossil plants from Plutoville, Cape York Peninsula. *Ibid.* **53**, ii.
—— (29). Note on a fossil wood from Central Australia. *Ibid.* **54**, iii.
WALTHER, J. (09). Über algonkische Sedimente. *Zeit. deutsch. geol. Ges.* **61**.
—— (12). *Lehrbuch der Geologie Deutschlands*. Leipzig.
—— (19–22). *Allgemeine Palaeontologie*. Berlin.
WALTON, J. (23). On the structure of a Middle Cambrian Alga from British Columbia. *Proc. Camb. Phil. Soc. (Biological Sciences)*, **1**, i.
—— (23^2). On *Rhexoxylon*, a Triassic genus of plants exhibiting a liane-type of vascular organisation. *Phil. Trans. R. Soc.* **212**.
—— (25). Carboniferous Bryophyta. *Ann. Bot.* **39**.
—— (25^2). On a Calcareous Alga belonging to the Triploporelleae from the Tertiary of India. *Rec. Geol. Surv. India*, **56**, iii.
—— (26). On some Australian fossil plants referable to the genus *Leptophloeum* Daws. A note on the structure of the plant cuticles in the paper-coal from Toula in Central Russia. *Mem. Proc. Manchester Lit. Phil. Soc.* **70**.
—— (26^2). Additions to our knowledge of the fossil flora of the Somabula beds, southern Rhodesia. *Trans. Geol. Soc. S. Afr.* **29**.
—— (27). Contributions to the knowledge of Lower Carboniferous plants. *Phil. Trans. R. Soc.* **215**.
—— (27^2). On some fossil woods of Mesozoic and Tertiary age from the Arctic zone. *Ann. Bot.* **41**.
—— (28). Recent developments in Palaeobotanical technique. *Congr. stratig. Carbon. Heerlen.* Liége (1927).
—— (28^2). On the structure of a Palaeozoic cone-scale and the evidence it furnishes of the primitive nature of the double cone-scale in the conifers. *Mem. Proc. Manchester Lit. Phil. Soc.* **73**.

WALTON, J. (28[3]). A method of preparing sections of fossil plants contained in coal balls, etc. *Nature*, **122**.

—— (28[4]). Carboniferous Bryophyta. *Ann. Bot.* **42**.

—— (28[5]). A preliminary account of the Lower Carboniferous flora of North Wales and its relation to the floras of some other parts of Europe. *Congr. stratig. Carbon. Heerlen.* Liége (1927).

—— (29). Palaeobotanical evidence for the age of the late Palaeozoic glaciation in South Africa. *Nature*, **124**.

—— (29[2]). The fossil flora of the Karroo system in the Wankie district, southern Rhodesia. *Geol. Surv. S. Rhod. Bull.* 15.

—— (29[3]). Palaeobotany: Palaeozoic. *Encyc. Brit.* ed. 14.

—— (30). Improvements in the peel-method of preparing sections of fossil plants. *Nature*, **125**.

WARD, L. F. (99). The Cretaceous formation of the Black Hills as indicated by fossil plants. 19*th Ann. Rep. U.S. Geol. Surv.*

WARD, L. F. and others (00). Status of the Mesozoic floras of the United States. I. The older Mesozoic. 20*th Ann. Rep. U.S. Geol. Surv.*

—— (05). *Ibid. U.S. Geol. Surv. Mon.* **48**.

WASHINGTON, H. S. (22). Deccan Traps and other plateau basalts. *Bull. Geol. Soc. Amer.* **33**.

WATELET, A. (66). *Description des plantes fossiles du Bassin de Paris.* Paris.

WATTS, W. W. (03). Charnwood Forest: a boreal Triassic landscape. *Geogr. Journ.*

—— (22). Carboniferous Nomenclature. *Geol. Mag.* **59**.

WEGENER, A. (24). *The Origin of Continents and Oceans.* (Trans. J. G. A. Skerl.) London.

WEIGELT, J. (28). Die Pflanzenreste des mitteldeutschen Kupferschiefers und ihre Einschaltung ins Sediment. *Fortschrit. Geol. Palaeont.* **6**, xix.

WEISS, E. (85). Zur Flora der ältesten Schichten des Harzes. *Jahrb. preuss. geol. Landesanst.*

WEISS, F. E. (04). A probable parasite of Stigmarian rootlets. *New Phyt.* **3**.

WELSCH, J. (15). Les Vallées pliocènes avec lignite de Bidart, etc. *Bull. Soc. géol. France* [4], **15**.

WERTH, E. (29). Zur Klimatologie, Pflanzengeographie und Geschichte des Europäischen Ackerbaues. *Ber. deut. bot. Ges.* **47**, i.

WEST, G. S. and F. E. FRITSCH (27). *A Treatise on the British Freshwater Algae.* Cambridge.

WETTSTEIN, R. VON (92). Die fossile Flora der Höttinger Breccie. *Denks. Ak. Wiss. Wien*, **59**.

—— (23–24). *Handbuch der systematischen Botanik.* Leipzig and Wien.

WHITE, D. (95). The Pottsville series along New River, W. Virginia. *Bull. Geol. Soc. Amer.* **6**.

—— (99). Fossil flora of the Lower Coal Measures of Missouri. *U.S. Geol. Surv. Mon.* **37**.

—— (01). Two new species of algae of the genus *Buthotrephis* from the Upper Silurian of Indiana. *Proc. U.S. Nat. Mus.* **24**.

—— (02). Description of a fossil alga from the Chemung group of New York. *N.Y. State Palaeontologist.* Albany.

—— (02[2]). A new name for *Buthotrephis divaricata. Proc. Biol. Soc. Washington*, **15**.

—— (04). Permian elements in the Dunkard flora. *Bull. Geol. Soc. Amer.* **14**.

—— (07). A remarkable fossil tree trunk from the Mid-Devonic of New York. *N.Y. State Mus. Bull.* **107**.

—— (08). Report on the fossil flora of the Coal Measures of Brazil. Pt. III of the *Final Rep. of Dr I. C. White.* Rio de Janeiro.

—— (09). The Upper Palaeozoic floras, their succession and range. *Journ. Geol.* **17**.

—— (12). The characters of the fossil plant *Gigantopteris* Schenk, and its occurrence in North America. *Proc. U.S. Nat. Mus.* **41**.

White, D. (13). Excursion in eastern Quebec and the Maritime Provinces. *Guidebook, Internatl. Congr.* **1**, Pt. I. Ottawa.
—— (29). Flora of the Hermit Shale, Grand Canyon, Arizona. *Carnegie Inst. Publ.* 405.
White, D. and C. Schuchert (98). Cretaceous series of the western coast of Greenland. *Bull. Geol. Soc. Amer.* **9**.
White, D. and T. Stadnichenko (23). Some mother plants of petroleum in the Devonian black shales. *Econom. Geol.* **18**.
White, O. E. (26). Geographical distribution of the cold resisting characters of certain herbaceous perennials and woody plant groups. *Brooklyn Bot. Gard. Rec.* **15**, i.
Wieland, G. R. (06). American fossil Cycads. I. *Carnegie Inst. Publ.* 34.
—— (14). Further notes on Ozarkian seaweeds and oolites. *Bull. Amer. Mus. Nat. Hist.* **33**.
—— (14^2). La Flora liasica de la Mixteca Alta. *Bolet. Inst. Geol. Mexico*, **31**.
—— (16). American fossil Cycads. II. *Carnegie Inst. Publ.* 34, vol. ii.
—— (19). Classification of the Cycadophyta. *Amer. Journ. Sci.* **47**.
—— (21). Monocarpy and Pseudomonocarpy in the Cycadeoids. *Amer. Journ. Bot.* **8**.
—— (26). The El Consuelo Cycadeoids. *Bot. Gaz.* **81**.
—— (29). The world's greatest petrified forests. *Science*, **69**.
—— (29^2). Antiquity of the Angiosperms. *Proc. Internat. Congr. Plant Sci.* (*Ithaca*), **1**.
—— (29^3). Views of higher seed-plant descent since 1879. *Science*, **70**.
—— (30). A reef-forming Phormidioid Alga. *Amer. Journ. Sci.* **19**.
Williamson, W. C. (80). On the organization of the fossil plants of the coal measures. Pt. X. *Phil. Trans. R. Soc.* **171**.
Willis, B. (22). Index to the stratigraphy of North America. *U.S. Geol. Surv., Profl. Paper* 71.
Willis, B. and others (07). *Research in China.* 3 vols. Washington.
—— (28). *Theory of Continental Drift. A Symposium.* Published by the Amer. Assoc. Petroleum Geologists, Tulsa, Oklahoma.
Willis, J. C. (22). *Age and Area.* Cambridge.
Wills, L. J. (10). The fossiliferous Lower Keuper rocks of Worcestershire. *Proc. Geol. Assoc.* **21**.
Wilson, E. H. (13). *A Naturalist in Western China.* London.
Witham, H. (31). A description of a fossil tree discovered in the quarry at Craigleith near Edinburgh in Nov. 1830. *Trans. Nat. Hist. Soc. Northumberland, etc.* Newcastle.
Woodhead, T. W. (28). The forests of Europe and their development in post-glacial times. *Empire Forestry Journ.* **7**, ii.
—— (29). History of the vegetation of the southern Pennines. *Journ. Ecol.* **17**.
Woods, H. (22). Note on *Pygocephalus* from the Upper Dwyka shales of Kimberley. *Trans. Geol. Soc. S. Afr.*
Woodworth, J. B. (12). Geological expedition to Brazil and Chile, 1908–9. *Bull. Mus. Comp. Zool. Harvard Coll.* **56**, i.
Woolnough, W. G. and Sir T. W. E. David (26). Cretaceous glaciation in central Australia. *Quart. Journ. Geol. Soc.* **82**.
Wright, W. B. (14). *The Quaternary Ice Age.* London.
Yabe, H. (12). Über einige gesteinbildende Kalkalgen von Japan und China. *Sci. Rep. Tôhoku Imp. Univ., Second Ser.* (*Geol.*), **1**, i.
—— (13). Mesozoische Pflanzen von Omoto. *Ibid.* **1**, iv.
—— (17). Geological and geographical distribution of *Gigantopteris*. *Ibid.* **4**, ii.
—— (22). Notes on some Mesozoic plants from Japan, Korea and China. *Ibid.* **7**, i.
—— (27). Cretaceous stratigraphy of the Japanese Islands. *Ibid.* **11**, i.
Yabe, H. and S. Endo (21). Discovery of stems of a *Calamites* from the Palaeozoic of Japan. *Sci. Rep. Tôhoku Imp. Univ., Second Ser.* (*Geol.*). **5**, iii.

YABE, H. and S. ENDO (27). *Salvinia* from the Honkeiko group of the Honkeiko coalfield, south Manchuria. *Jap. Journ. Geol. Geogr.* **5**, iii.

YABE, H. and S. OISHI (28). Jurassic plants from the Fang-tzu coalfield of Shantung. A new species of *Protoblechnum* from the Hei-shan coalfield in Shantung. A new species of *Sphenophyllum*. *Ibid.* **6**.

YABE, H. and S. TOYAMA (28). On some rock-forming algae from the younger Mesozoic of Japan. *Sci. Rep. Tôhoku Imp. Univ.* [2] (*Geol.*), **12**, i.

YAPP, R. H. (23). The history of the present flora and vegetation of the British Isles. *Pure Sci. Mag. Univ. Birmingham*, **3**.

YOKOYAMA, M. (89). Jurassic plants from Kaga, Hida and Echizen. *Journ. Coll. Sci. Imp. Univ. Japan*, **3**, i.

—— (94). Mesozoic plants from Kōzuke, Kii, Awa and Tosa. *Ibid.* **7**, iii.

—— (05). Mesozoic plants from Nagato and Bitchû. *Ibid.* **20**.

—— (06). Mesozoic plants from China. *Ibid.* **21**.

—— (08). Palaeozoic plants from China. *Ibid.* **23**.

YOUNGHUSBAND, SIR F. (26). *The Epic of Mount Everest.* London.

ZABLOCKI, J. (28). Tertiäre Flora des Salzlagers von Wieliczka. *Act. Soc. Bot. Poloniae*, **5**, ii.

ZALESSKY, M. (07). Contributions à la flore fossile du terrain houiller du Donetz. *Bull. com. géol. St Pétersb.* **26**.

—— (09). Note sur les débris végétaux du terrain carbonifère de la Chaîne de Mugodžary. *Ibid.* **28**.

—— (11). Étude sur l'anatomie du *Dadoxylon Tchikatcheffi*. *Mém. com. géol. St Pétersb.* Livr. 68.

—— (18). Flore paléozoïque de la série d'Angara. *Ibid.* Livr. 174.

—— (18^2). Sur le sapropélite marin de l'âge Silurien formé par une alge cyanophycée. *Soc. Paléont. Russie* (November).

—— (20). Über einen durch eine Zyanalge gebildeten marinen Sapropel silurischen Alters (Kuckersit.). *Centralbl. Min.*

—— (24). On new species of Permian Osmundaceae. *Journ. Linn. Soc. London*, **46**.

—— (26). Sur les nouvelles algues découvertes dans le sapropélogène du lac Beloe et sur une algue sapropélogène. *Rev. gén. Bot.* **38**.

—— (27). Flore permienne des limites ouraliennes de l'Angaride. Atlas. *Mém. com. géol.* Livr. 176.

—— (28). Sur la flore stéphanienne découverte dans la crête Naryn-taon au Turkestan. *Bull. géol. com.* no. 5.

—— (28^2). Sur l'extension du continent de l'Angaride et premières données sur la flore de ses limites oussouriennes. *Ann. Soc. géol. Nord*, **53**.

—— (28^3). Étude de la structure microscopique du charbon sapropélien de Cassianovka dans le bassin de Tchéremkhovo, en Sibérie. *Com. géol.* Livr. 92.

—— (28^4). Essai d'une division du terrain houiller au bassin du Donetz. *Congr. stratig. Carbon. Heerlen* (1927).

—— (29). Sur des débris de nouvelles plantes permiennes. *Bull. Acad. Sci. de l'Urss.*

—— (29^2). Observations sur de nouveaux spécimens du *Psygmophyllum expansum* et sur une nouvelle plante fossile, *Idelopteris elegans*. *Ibid.*

—— (29^3). Sur le *Syniopteris nesterenkoi* nov. gen. et sp. et le *S. demetriana* nov. gen. et sp., nouveaux végétaux permiens. *Ibid.*

ZALESSKY, M. and G. ZALESSKY (21). Structure du rameau du *Lepidodendron caracubense*. *Ann. Soc. Paléont. Russie*, **3**.

ZEILLER, R. (88). Bassin houiller de Valenciennes. *Études Gîtes Min. de la France.* Paris.

—— (90). Bassin houiller et permien d'Autun et d'Épinac. *Ibid.*

—— (92). Bassin houiller et permien de Brive. *Ibid.*

—— (95). Note sur la flore fossile des gisements houillers de Rio Grande do Sul. *Bull. Soc. géol. France* [3], **23**.

—— (96). Remarques sur la flore fossile de l'Altai. *Ibid.* **24**.

ZEILLER, R. (96[2]). Étude sur quelques plantes fossiles, en particulier *Vertebraria* et *Glossopteris* des environs de Johannesburg. *Ibid.*

—— (98). Contributions à l'étude de la flore ptéridologique des schistes permiens de Lodève. *Bull. Mus. Marseille*, **1**, fasc. 11.

—— (99). Étude sur la flore fossile du bassin houiller d'Héraclée. *Mém. Soc. géol. France*, no. 2.

—— (00). Sur quelques plantes fossiles de la Chine méridionale. *Comp. Rend.* **130**.

—— (00[2]). Sur les végétaux fossiles recueillis par M. Villiaume dans les gîtes charbonneux du nord-ouest de Madagascar. *Ibid.* (June 5).

—— (00[3]). *Éléments de Paléobotanique.* Paris.

—— (02). Observations sur quelques plantes fossiles des Lower Gondwana. *Mem. Geol. Surv. India*, **2**.

—— (02[2]). Sobre algunas impresiones vegetales del Kimeridgense de Santa María de Meyá. *Mem. R. Acad. Cienc. Barcelona*, **4**, xxvi.

—— (03). *Flore fossile des Gîtes de Charbon du Tonkin.* Paris. (Plates, 1902.)

—— (05). Sur quelques empreintes végétales de la formation charbonneuse supracrétacée des Balkans. *Ann. Mines.*

—— (06). Bassin houiller et permien de Blanzy et du Creusot. *Études Gîtes Min. de la France.*

—— (11). Sur une flore triasique découverte à Madagascar par M. Perrier de la Bâthie. *Comp. Rend.* **153**.

—— (11[2]). Études sur le *Lepidostrobus Brownii.* Paris.

—— (11[3]). Note sur quelques végétaux infraliasiques des environs de Niort. *Bull. Soc. géol. France* [4], **11**.

—— (14). Sur quelques plantes wealdiennes recueillies au Pérou. *Rev. gén. Bot.* **25** *bis.*

ZIGNO, A. DE (56–85). *Flora fossilis formationis Oolithicae*, **1** and **2**. Padua.

ZIMMERMANN, W. (26). Die Spaltöffnungen der Psilophyta und Psilotales. *Zeitsch. Bot.* **19**.

—— (30). *Die Phylogenie der Pflanzen.* Jena.

ADDENDA

The following papers are selected from several which were published or came under my notice after the chapters of this book had been printed; with one exception (Lang and Cookson), the full titles are given in the Bibliography but their contents are not referred to in the text.

Chapter v, p. 46. Hoskins (30) described a method of examining fossil plants.

" IX, p. 123. *Zosterophyllum* etc. Lang, W. H. and I. C. Cookson (30). Some Fossil Plants of early Devonian type from the Walhalla series, Victoria, Australia. *Phil. Trans. R. Soc.* **219**.

" IX, p. 154. GERMANY: Broili (28); Hirmer (30), (30^2); Kräusel and Weyland (30).

" IX, p. 154. AUSTRALIA: Lang and Cookson (30).

" XII, p. 208. Fungus in the rootlets of *Amyelon* (root of *Cordaites*): Halket (30).

" XII, p. 226. *Dicranophyllum* and Bohemian Coal Measures (p. 288): Němejc (29).

" XII, p. 269. *Lyginopteris*: Crookall (30).

" XII, p. 288. Kent Coalfield: Pringle and Crookall (30).

" XIII, p. 303. *Pleuromeia*: Mägdefrau (30).

" XV, p. 411. Cretaceous floras of N. America: Berry, E. W. (30^3); Hollick and Martin (30); Knowlton (30).

" XV, p. 391. *Artocarpus*: Ball (30).

" XVII, p. 486. Climate during the Quaternary Period: Simpson (30).

" XVIII, p. 527. Relation between the size and anatomical structure of plants: Bower (30).

" XXII, p. 501. The Post-Glacial Stage. Several important additions have been made to our knowledge of Post-Glacial changes in climate and the development of forests since this chapter was written. Attention is called to two papers of more general interest: Gross (30); Rudolph (30).

Plant distribution and the development of continents. A general treatise by Irmscher (22) including an account of recent Cretaceous and Cainozoic plants with distribution maps, also a reference to the Wegener hypothesis.

Algae, Chapter X, p. 183, Chapter XVI, p. 424. Recent work on the Solenoporaceae and Corallinaceae: Pia (30).

INDEX

CAMBRIDGE: PRINTED BY W. LEWIS, M.A., AT THE UNIVERSITY PRESS

Printed in the United States
By Bookmasters